Manual of

AIRBORNE TOPOGRAPHIC LIDAR

Edited by Michael S. Renslow

Michael S. Renslow
Manual of AIRBORNE TOPOGRAPHIC LIDAR

ISBN 979-8-89691-435-8

Spines

TABLE OF CONTENTS

FOREWORD

Michael S. Renslow

In the fall of 1996, I was asked if I was interested in learning about and working with data using laser technology for mapping. The opportunity to experience new technology has always been of interest to me, so saying yes was easy, and I began a pursuit to understand everything about Light Detection and Ranging (lidar), the instrumentation, and the data characteristics. Six months later I was fully involved with a prototype lidar system full of technical challenges, which lasted about a year.

My first lidar epiphany occurred in the summer of 1997 when I draped color-coded, multiple-return lidar points onto a digital stereo pair of steep terrain, forested-covered terrain in Oregon on a softcopy workstation. I could "see" that the lidar had penetrated into the forest and "mapped" the structure of the trees and also the ground that I could not see on the stereo pair. I realized that this lidar technology would rapidly change the way the geospatial community approaches mapping the surface(s) on the Earth. That fall, I sat in on the first ASPRS Lidar Workshop.

The following year the instrumentation was stabilizing and the development of reliable direct georeferencing technology added confidence to the data collection process. The practitioners' focus shifted to software development, handling/processing "huge" data sets, and trying to quantify lidar map products in the context of existing map standards. Just about the time the software and computer processing time was acceptable, the vendors would introduce new lidar systems with increased hertz rate and even larger files. Even with these issues, lidar was creating quite a lot of interest in the mapping community, and resulted in (in fact, still is) exciting times. At the 1998 ASPRS Fall Conference, I taught my first Lidar Workshop.

Fast forward to 2012, lidar has become an accepted, often preferred, technology for mapping solutions. Enacting the technology, the mapping community throughout the world acquires, processes, validates, fuses with other sensor data, experiments, and archives the data (in a format standard), and visualizes the Earth and its features like never before. Lidar has its own vocabulary, definitions, and acronyms. Furthermore, there are new types of lidars being developed, and finally we have standards and metadata templates

The need for this Manual was recognized in 2010 during the Annual Spring Conference by ASPRS Executive Director, Jim Plasker, and ASPRS Communications Director/Associate Executive Director, Kim Tilley. The amount of interest in lidar technology and applications at the ASPRS conferences and the workshop attendance indicated that a manual was required. Within the ASPRS organization the Lidar Committee had been very active in organizing technical sessions and developing the ASPRS LAS Standard. In 2011, the ASPRS Lidar Division was created to establish the presence of lidar with photogrammetry, remote sensing, and GIS technologies. ASPRS is the leading professional association supporting lidar science and technology.

Because of my experience with lidar as a practitioner, teacher, and consultant, I was asked to become the editor for the ASPRS *Manual of Airborne Topographic Lidar*. For the content, I called upon technical experts to contribute to the individual chapters. From the onset of this project, I recognized that a reference book was needed for general use to document how this technology developed especially within the past 15 years, and introduce new sensors and applications. Additionally, I felt the need to emphasize "best practices" and real-world solutions for the user community. Also, since quality control and quality assurance are so critical to a successful data collection, processing, and validation, the reader will find several subchapters to support QA/QC throughout the Manual. The arrangement of the content was designed to first review the basics, describe the approach for flight planning, completion of the collection, data processing, and product development. The authors responded well with figures, diagrams, and tables for visualization and ease of understanding.

Operationally, airborne lidar is fundamentally different from photogrammetry or surveying: all of the data collection, control, ground support, postprocessing, and QA/QC results in a "point cloud" which is "classified" and processed into product(s). The driving value of the point cloud is that it is natively 3D, georeferenced, and in the correct coordinate space. When necessary, other technologies (photogrammetry, breaklines, image fusion, lidargrammetry, static lidar, mobile mapping data) are introduced to generate specialized products. If required, lidar data can be processed very quickly.

The following summarizes the contents for each chapter:

- Chapter 1 (Introduction) is an overview of lidar technology, historical events leading to current technologies, and a list of Commonly Used Lidar Terms.

- Chapter 2 (An Overview of ALS Technology) begins with a detailed description of Operating Principals starting with the Electromagnetic Spectrum describing how lidar works with light followed by a summary of Laser Emissions, Absorption and Reflectance. The Features of a Laser Beam details the characteristics of a laser pulse, pulse frequency, beam divergence, and target interaction.

  The Key Elements of ALS Technology is a basic overview of laser and lidar history, technology, various scan patterns, and how lidar systems work.

  In Chapter 2 several types of lidar systems are described:
  · Fiber Optic Lidar Scanners and Applications
  · Full Waveform Lidar including key applications and processing
  · Flash Imaging Lidar – a detailed description, operational procedures, and applications
  · Geiger Mode Lidar – technology description, operations, and applications

- Chapter 3 (Enabling Technologies) is a thorough description of direct georeferencing (DG) as it relates to lidar: overview of GNSS positioning, sources of error, flight planning considerations, processing of DG data and quality control/quality assurance. The Ground Segment section addresses base stations and causes of error.

- Chapter 4 (Airborne Installation and Integration of ALS Systems) begins with a in-depth review of Rotary Wing Platforms based on many years of work to achieve engineering grade lidar data products and imagery; every aspect of low-altitude collection and processing is described in detail.

  The Fixed Wing Platform section covers typical installations of current systems and calibration procedures.

- Chapter 5 (Guidelines for Lidar Data Collection) is a comprehensive summary of "Best Practices" starting with mission planning, in-flight calibration, mapping various types of terrain types, and guidelines for pilots.

- Chapter 6 (Data Processing) begins with a suggested work flow (step-by-step) once the lidar data is collected in the field to delivery of the final product. Preprocessing and Post-processing and handling specific feature types are also discussed.

- Chapter 7 (Status of Lidar Industry Support Systems) starts with a sampling of the latest lidar system technologies available as of June 2012.

  The Lidar Data Management section is a review of the general management of lidar data acquisition through data delivery. Several figures illustrate the data flow and suggested organizational approach as well as best practices. This section ends with an extensive, detailed description of the LAS 1.4 Standard.

- Chapter 8 (Lidar Data and Complimentary Technologies) covers Lidargrammetry with excellent graphics illustrating workflow. Combined Photogrammetric and Lidar Mapping Processes is a comparison of strengths and weaknesses of each technology. Lidar/Image Combinations explains the process to achieve desired co-registration. The Quality Control section focuses on contours, 3D modeling, and urban applications.

- Chapter 9 (Accuracy Standards and Guidelines) address all of the factors that can affect lidar data quality.

  The section on Metadata Requirements is from the USGS-NGP Lidar Base Specification and is a suggested template for a Lidar-Specific Metadata Tags document.

  The Existing Standards and Guidelines section is the core section of the USGS Lidar Base Specification Version 1.0. which contains Conversion Factors, Acronym List, Glossary, and all operational aspects of the Base Specification. A specific section on QA/QC Reporting finishes this chapter.

- Chapter 10 (Applications) is a series of practical lidar applications:

  <u>Digital Elevation Models</u> describes six commonly derived products and definitions of surface interpolation methods, hydro-flattening, use of breaklines, hydro-enforcement, and lidar intensity data. This section contains many color figures illustrating the lidar data products and various terrain treatments.

  <u>Forestry</u> discusses the interaction of lidar pulses with forest vegetation cover, mapping the bare earth, measurement principals, and point density.

  <u>Corridor Mapping</u> is an overview of two examples: Transmission Line lidar mapping and software for vegetation management, and hazard identification; and Transportation Engineering which addresses DOT applications with airborne, static, and mobile mapping lidar data.

  <u>Flood-Prone Area Mapping</u> presents a case study for using lidar for FEMA Floodplain Assessment and Surface Water Modeling.

  <u>Building Extraction and Reconstruction</u> from Lidar Point Clouds describes the current state of reliably classifying and extracting buildings and roof vertices.

  <u>Airport Surveying</u> with lidar is described for a variety of applications in the context of the most current FAA Advisory Circulars with several specific examples.

Coastal Applications are described for coastal management, wetlands and habitat, storm damage, shoreline extraction, with an explanation of the USGS Experimental Advanced Airborne Research Lidar (EAARL) program.

Hydrologic Applications is a literature review for water resources and hydrological issues.

Natural Hazards addresses the constantly changing, dynamic environment of the Earth and recurring hazards with several applications for earth

- Appendix A (Geoid Height Models and Vertical Datums) focuses on geoid/datum applications specifically for lidar work and the models available, including USGG2009 and GEOID09.

- Appendix B (QC Process for Data Users) describes post-mission procedures for evaluating the quality of the final lidar data product. It was taken from and upcoming ASPRS Guideline being developed in conjunction with the USGS.

- Appendix C (Lidar Metadata) is a metadata example document extracted from the USGS Lidar Base Specification Version 1.0

- Appendix D (Acronyms) is a compilation of the acronyms used in the Manual.

I sincerely thank all of the contributing authors for their professional input, technical knowledge, experiences, and time spent writing and reviewing: all of which was volunteered. Several authors made significant contributions which are listed at the end of each chapter; the reader will recognize the author frequency of Jamie Young, Chris Parrish, Qassim Abdullah, and Karl Heidemann. The primary lidar system providers, Leica Geosystems, Optech, Riegl, and Trimble, generously provided information and graphics.

I appreciate the continued support from ASPRS, especially Kim Tilley to gently push me towards the "finish line" and Matthew Austin for his excellent graphic skills and layout for the entire book. I also recognize the love, support, and patience from my wife and life partner, Linda, who has been so understanding during this project and all the other projects during the last 36 years.

Finally, I would like to express my gratitude to all of my ASPRS colleagues for the past 40 years of professional involvement, technology applications, friendship, and collective progress towards our mutual interests. Often we have been "many in body with lots of ideas, but always of one mind" for the advancement and betterment of our profession. I also appreciate my students; I am the benefactor of their efforts and learn so much from them.

Lidar is basically about light, time, and knowing something about where you are. Be there.

Michael S. Renslow, CP, RPP, Editor
August 2012

ABOUT THE EDITOR

Michael S. Renslow is a photogrammetrist with 45 years of geospatial experience. His career began in 1967 as a civilian land surveyor for the U.S. Department of Defense. In 1971, he went to work for the U.S. Forest Service in California as a photogrammetric cartographer and often field surveyor; he was instrumental in the development and implementation for the FIRESCOPE Project in Southern California and the Incident Command System for disaster response to wildland fires. In 1983, Mike left government service and entered the private sector as the general manager of Pacific Aerial Surveys and vice-president of Hammon, Jensen, and Wallen, Inc. His next position was to move to Oregon in 1989 and work for WAC Corporation in market development and operations management of Digital Geographic Systems writing digital aerial imagery on CD-ROM. Mike started his first mapping firm in 1995, Renslow Image Mapping, which specialized in digital imagery and orthophotography. After one year, he sold the company to Spencer B. Gross Engineering and became the firm's vice president managing the Eugene office; and became interested in lidar. Also, in 1997, Mike took a position at the University of Oregon, Geography Department as adjunct faculty, teaching a class in Fundamentals of Remote Sensing for five years. In 2007, he established Renslow Mapping Services specializing in consulting for emerging technologies and QA/QC procedures. In 2009, he accepted a position with Penn State University, John A. Dutton e-Education Institute as a senior lecturer in the Geography Department teaching a new course on Lidar Technology and Applications.

He graduated with an AA in Civil Engineering in 1967 from Solano College, and a BS in Geography in 1971 from San Francisco State University.

Mike has been a member of ASPRS for 40 years, and served nationally on several committees and technical divisions, the Board of Directors, the Executive Committee, and as President in 1999-2000. He is currently the technical editor for *Photogrammetric Engineering & Remote Sensing*, Treasurer for the ASPRS Foundation, Chair of the Evaluation for Certification Committee, and the ASPRS Representative to the Council of Engineering and Scientific Specialty Boards.

He is active with ISPRS serving as a Working Group Chair 2000 to 2004, active member of Commission I, and elected to Council as the Treasurer for 2008-2012.

He is an ASPRS Certified Photogrammetrist and an Oregon Registered Professional Photogrammetrist.

1
INTRODUCTION

MICHAEL S. RENSLOW

The word "lidar" began as an acronym for LIght Detection And Ranging; today it is considered as a mature mapping technology in the same company as radar and sonar. Airborne lidar mapping systems provides 3D information for the surface of the Earth which includes terrain surface models, vegetation characteristics, and man-made features. Lidar is an active sensing technology, i.e., it generates its own pulses of light and detects the reflections from those pulses (very similar to radar and sonar). Depending on the specific application, the data acquisition and the type of lidar system employed are customized to capture specific elements leading to classification of the data and ultimately mapping products.

One of the key elements that has made lidar a "technology of choice" for modern mapping and integration into GIS is the fact that the data is natively collected in 3D and georeferenced; furthermore it can be processed to classify 3D elements within the "cloud of points" representing the Earth and the features above the Earth's surface. There are many direct uses of lidar such as feature mapping, vegetation mapping, transportation mapping, transmission corridor mapping, 3D building models, natural hazard detection, and natural disaster evaluation. From a mapping standpoint, lidar is also quite accurate (5 to 15 cm vertical accuracy) and the completeness of the data sets is impressively rich compared to traditional mapping technologies and methodologies. Also, when required, lidar data can be processed quickly to yield 3D information and lidar intensity "imagery" for spatial referencing.

Similar to many remote sensing technologies, lidar developed generally from the defense industry and was used for classified programs. The very first commercial application for lidar was to measure contaminates in smoke plums. These first systems where physically very large and operated very slow. The first commercial "lidar mapping systems" as discussed in this Manual began their development in the early 1990's and operated from 5,000 to 10,000 pulses per second (also known as repetition rate). Typically, the first systems provided last return or first return data, i.e., the last object or first object the pulse hit. The information would be returned to the system and recorded. Advances in electronics and laser technology in the late 1990's and early 2000's increased the repetition rates to 25,000 to 50,000 pulses per second, providing multiple returns from individual pulses that could be recorded from the system. Currently, the airborne systems can operate between 150,000 to 400,000 pulses per second, employ multiple lasers, and when necessary provide 10 to 12 points per square meter. The systems and components are also smaller, portable, and very reliable.

This Manual focuses on the most current lidar systems, data management, processing, best practices, quality control/quality assurance, standards, reporting, and applications. The operation of a lidar system and the collection processes are critical components resulting in useful data and the basis for analysis. The data processing follows traditional survey principals, and the reporting is expressed in traditional statistical terms. Although lidar mapping has progressed for 15 years without specific standards, the development of the ASPRS LAS data standard and the U.S. Geological Survey's Base Lidar Specification have now established formats for lidar data classification and digital elevation model production, respectively.

In addition to the traditional airborne lidar applications, two emerging lidar technologies are presented: FLASH Lidar and Gieger Mode Lidar. Again, emerging from the defense industry, these two lidar technologies offer the opportunity for increased airborne mapping efficiency and additional applications.

With the transition from traditional mapping techniques to airborne lidar technology, the geospatial community adopted a new vocabulary. The following is a list for some of these newer terms.

1.1 COMMONLY USED LIDAR TERMS

Airborne Laser Scanning (ALS)

ALS is also referred to as laser altimetry or lidar; an active remote sensing technology that measures topography of the Earth's surface and landscape feature heights and reflectance intensity. An ALS is comprised of a Direct Georeferencing System (often termed a Position Orientation System [POS]) to accurately determine the position and orientation of the airborne platform and the Laser Scanner System to emit laser pulses and receive pulse-returns using the laser scanner and lidar mechanisms.

Along Track Resolution

The spacing of the pulses from the lidar system in the flight direction of the platform.

Artifacts

Detectable surface remnants of buildings, trees, towers, telephone poles, or other elevated features in a bare earth elevation model.

Attitude (pitch, roll yaw)

Pitch: the vertical rotation of the aircraft (nose up, nose down)
Roll: rotation of the aircraft around the flight vector (wing up, wing down)
Yaw: horizontal rotation of the aircraft (nose left, nose right)

Avalanche Photodiode (APD)

At the time of pulse transmission and reception, the APD converts the light signal into an electronic voltage pulse.

Backscatter

Electromagnetic energy that is reflected back toward its source by terrain or particles in the atmosphere.

Base Station

A GPS or GNSS receiver set up at an accurately-known fixed location which is used to derive correction information for nearby GNSS receivers, i.e., onboard the ALS system. The base station receiver records the raw GNSS data which are applied during the postprocessing of the raw lidar data.

Bare Earth

Digital elevation data of the terrain, free from vegetation, buildings, and other man-made structures.

Bean Divergence

Divergence of a light beam is the increase in beam diameter with distance from the aperture. Typically measured in mrad (milliradians). Higher frequency beams generally have lower divergence.

Boresight
Calibration of a sensor system equipped with an IMU to determine the accurate orientation (roll, pitch, yaw) of active sensor pulses.

Canopy Height Model (CHM)
A canopy height model is the representation of the difference between the top of the canopy surface and the underlying ground topography. Derived by filtering and classifying lidar point clouds to separate ground from canopy hits.

Collimated
Light that does not disperse and therefore has low beam divergence is termed collimated. Lidar laser beams are highly collimated. It is often said that collimated light rays are parallel and focused at infinity.

Cross Track Resolution
The spacing of the pulses from the lidar system in the scanning direction or perpendicular to the direction of which the platform is moving in the case of airborne systems.

Direct Georeferencing
The direct measurement of exterior orientation parameters, i.e., position (x-y-z coordinates) and attitude (roll-pitch-yaw) of a sensor to determine position and orientation.

Echo
Each of the multiple returns from an emitted laser pulse in a multiple-pilse-return laser scanning system (e.g., first, intermediate....last).

GLONASS
Acronym for *Globalnaya navigatsionnaya sputnikovaya sistema* or Global Navigation Satellite System, a radio-based satellite navigation system operated for the Russian government by the Russian Space Forces.

GNSS
A satellite navigation system with global coverage is termed a global navigation satellite system.

Inertial Measurement Unit (IMU)
An IMU monitors angular accelerations (using accelerometers) and rotations (using gyroscopes) of the aircraft (and sensor) with respect to the three primary axis: x = long axis (roll); y = horizontally perpendicular to x (pitch); and z = vertical axis perpendicular to x (yaw or heading). Integrating these measurements with time allow for the precise orientation of the sensor platform for each increment of time. Orientation measurements are typically recorded at rates of 50 Hz to 200 Hz.

Lever Arm Offsets
The lever arm is the perpendicular distance from the axis of rotation to the line of action of the force. Lever arm offsets between system components are measured precisely when setting up the ALS to enable the integration of information from all subcomponents of the system during postprocessing.

Near Infra-red (NIR)

Near Infra-red wavelengths fall between 800 nm and 2500 nm of the electromagnetic spectrum. The majority of commercial ALS sensors use NIR wavelengths of 1000 nm, 1047 nm, 1064 nm, and 1550 nm. This choice is mainly due to the availability of stable and efficient lasing materials at these wavelengths, since natural surfaces are sufficiently reflective at these wavelengths, and since NIR has a low signal-to-noise ratio in sunlight and is more eye-safe than other visible wavelengths.

Nominal Point Spacing (NPS)

The NPS is important depending on the specific application because lidar sensors are random sampling systems. Lidar technology is based on a random distribution of mass points that result in a "point cloud" of data. The flight planning process can design a project to estimate the NPS and the density of the resulting data set.

Point Cloud

The cluster of points that comprise lidar data achieved from reflections of the laser beams off of various landscape features, especially complex vegetation, thereby representing feature height and the 3D relationship between features.

Point Dropout

Laser pulses for which no energy was returned to the sensor. These can occur if the aircraft (sensor) was at too high an altitude, the surface material is absorbing the radiation, or the ground level energy is reflected away from the sensor.

Point Spacing

The average ground distance between successive pulse returns. All considerations being equal, close (lower) point spacing can allow for higher resolution descriptions of the landscape.
Positioning Orientation System (POS)
POS accurately determines the position and orientation of the laser sensor platform in 3D space at the time of laser pulse transmission and reception. Comprised of a Differential GPS System (DGPS) to determine the geographic position of the platform to within 5 cm of the true trajectory, and the Inertial Measurement Unit (IMU) to measure the orientation of the platform through time based on changes in the platform's attitude.

Pulse Footprint

The area of ground intersected by the laser pulse; it is a function of range, angle of incidence, and beam divergence. As an example, a pulse from 2,500 m AGL from a typical ALS system will cover an area of ~60 cm diameter (Note that the x, y, z from that reflection can be located anywhere within that circular area.)

Pulse Footprint Smearing

Smearing or elongation of the pulse footprint caused when the laser beam is reflected from sloped terrain (especially away from the center of the scan) causing increased uncertainty of the horizontal and vertical position).

Pulse Repetition Frequency (PRF)

The frequency of transmitted laser pulses (i.e., points per second): High PRF enables dense point-spacing on the ground providing high-resolution representations of the landscape.

Pulse Return

Laser pulses reflected off surfaces encountered below the sensor and received by the lidar sensor. First pulse returns measure the range to the first surface encountered (i.e., vegetation, canopy, building roofs); last pulse returns measure the range to the last surface encountered (i.e., the ground).

Pulse Return Intensity

The reflective NIR intensity of pulse returns can be measured by most ALS systems which can all improved discrimination and classification of scanned features. Intensity data is a function of pulse range, pulse footprint size, angle of incidence at the point of return, and spectral characteristics of the encountered surface.

Range

The distance between the laser aperture and the detected object or surface.

Repetition Rate

The repetition rate is the pulses-per-second at which the laser is pulsing denoted in KHz. A 200 KHz system indicates that the lidar system will pulse at 200,000 times per second, which means that the laser transceiver generates out this number of pulse every second and the receiver can (but usually does not) receive information from these 200,000 pulses.

Scan Angle

Half the angle of the full sweep of a scanning mirror scanner. Large scan angles are generally not chosen at high altitudes due to high dropout rates, increased error, and obstruction shadowing at edges of the scan. Typically, scan angles do not exceed 30 degrees.

Scan Rate

The frequency of cross-track sweeps of a mirror scanner in Hz.

Swath Width

The width of the survey area covered by a complete sweep of the scanner which is related to flying height and scanner angle.

Time Interval Meter (TIM)

The method used to time-stamp the pulse transmission and reception points to determine the time difference between pulse transmission and reception, and resulting in the range distance and position of the reflective surface.

REFERENCES

Applied Geomatics Research Group, 2009. URL: http://agrg.cogs.nscc.ca/resources/lidar-glossary (last date accessed: 16 May 2012).

Maune, D.F.(editor), 2007. *Digital Elevation Model Technologies and Applications: The DEM Users Manual*, Second edition, American Society for Photogrammetry and Remote Sensing, 655 p.

RIEGL Laser Measurement Systems, 2012. *RIEGL Glossary*, 07 March, 11 p.

2
AN OVERVIEW OF ALS TECHNOLOGY

ROBERT T. PACK, VALERIE BROOKS, JAMIE YOUNG, NUNO VILAÇA, SVEIN VATSLID, PETER RINDLE, SVEN KURZ, CHRISTOPHER E. PARRISH, REX CRAIG, and PHILIP W. SMITH

2.1 OPERATING PRINCIPLES

2.1.1 Electromagnetic Spectrum

2.1.1.1 Light Spectrum

Lidar works with light in the electromagnetic spectrum from ultraviolet through infrared. The wavelength for a given lidar is chosen according to considerations of required range-performance, atmospheric or water absorption, eye-safety requirements, and the reflectivity of materials to be lased. There is a common consensus on how to subdivide the visible spectrum into specific labels such as blue, green, and red. However, in the infrared (IR) spectrum, several different conventions exist. Along with the visible spectrum, Table 2.1 shows an IR subdivision governed by a "detector response scheme" (Miller, 1994). The detector response governs the choice of a given laser wavelength used in lidar systems. For this reason an IR classification based on this is used for lidar sensors (as well as in IR imagers). The IR spectrum continues beyond the limits of this table up to a wavelength of 1 mm.

Table 2.1 Classification Of The Light Spectrum Based on Detector Response (Miller, 1994).

Color	Acronym	Wavelenght (nm)	Detector Material
Near Ultraviolet	NUV	200–400	Si
Violet	V	400–450	Si
Blue	B	450–475	Si
Cyan	C	475–495	Si
Green	G	495–570	Si
Yellow	Y	570–590	Si
Orange	O	590–620	Si
Red	R	620–750	Si
Near Infrared	NIR	750–1000	Si
Short-Wave Infrared	SWIR	1000–3000	InGaAs
Mid-Wave Infrared	MWIR	3000–5000	InSb, HgCdTe, PbSe
Long-Wave Infrared	LWIR	8000–12000	HgCdTe
Very-long Wave Infrared	VLWIR	12000–30000	doped Si

2.1.1.2 Laser Emission

Ordinary light consists of a mixture of many different wavelengths. For example, a prism is able to disperse white light into many component wavelengths (colors). By contrast, lasers emit an extremely narrow range of wavelengths within a given portion of the spectrum. In other words, it is nearly "monochromatic" or single colored. The reason for this is that a laser produces light through a process termed "stimulated emission." The term laser is actually an acronym for "Light Amplification by Stimulated Emission of Radiation."

Types of lasers include gas lasers, solid state lasers, liquid dye lasers, and semiconductor lasers. Solid state lasers and semiconductor lasers are the most common in terrestrial lidar applications. Also, there are lasers that emit light continuously (continuous wave or CW) and lasers that emit pulses of light. In all cases, the emitted wavelength is a function of the medium or material in which the light is stimulated. Therefore, it is possible to produce a table of laser materials and their associated wavelength as shown in Table 2.2 (Weber, 1999). Note that the materials most commonly used in terrestrial lidar transmitters are identified with an asterisk.

Table 2.2 List of common solid state and semiconductor materials and their associated emission wavelengths (Weber, 1999).

Laser Material	Type**	Wavelength (nm)	Color
Nd:YAG (tripled)	SS	354	Near UV
Nd:YAG (doubled)*	SS	523	Green
InGaAIP	SC	635 660	Red
AlGaInP	SC	635–900	Red-NIR
Ti:sapphire	SS	650–1100	Red-NIR
GaAlAs	SC	670–905	Red-NIR
AlGaAs	SC	720–850	Red-NIR
GaAs	SC	835–905	NIR
InGaAs*	SC	900–1100	NIR-SWIR
InGaAsP/InP	SC	1000–2100	SWIR
Yb:YAG	SS	1030	SWIR
Nd:YLF	SS	1047–1053	SWIR
Nd:Glass	SS	1062	SWIR
Nd:YAG*	SS	1064	SWIR
Nd:YAP	SS	1080	SWIR
Er-doped glass*	SS	1540	SWIR
Tm:TAG	SS	2000	SWIR
CrZnSe	SS	2200–2800	SWIR

* Most common in airborne mapping applications

** SS = Solid State; SC = Semiconductor

2.1.1.3 Absorption and Reflectance

When laser light strikes a surface, some of the light is transmitted, some of the light is absorbed, and the rest is reflected. The amount of each is a function of the type of material along with the wavelength of the light. For example, Figure 2.1 shows an example of the relationship between the reflectance, absorptance, and transmittance spectra of a plant leaf (Knipling, 1970). Once light enters the medium, the amount of absorption is a function of the distance travelled within the medium. Hence absorption is expressed as a ratio per unit distance, e.g., units of m^{-1}. In the case of Figure 2.1, the amount of absorption is a function of the typical thickness of a leaf. The fact not all of the light is absorbed is manifest in the observation that you can see light come through the leaf when holding it up against the sun.

Table 2.3 shows a graph of the absorption of light by water as a function of wavelength (Bukata et al., 1995; Bertie and Lan, 1996). The table shows that at a wavelength of approximately 460 nm (dark blue) the amount of light absorption is at a minimum. If one wanted to choose a lidar material capable of penetrating water and detecting a lake bottom, one could choose from Table 2.2 the frequency-doubled neodymium-doped yttrium aluminum garnet (Nd:YAG) that emits a wavelength of 523 nm. Between the NIR wavelengths of 1000 and 1500 nm, the absorption takes a big jump from 60 to 5000 m^{-1}.

Using a laser wavelength that is strongly absorbed by water has disadvantages and advantages. The disadvantage is that it is incapable of penetrating water and detecting the bottom of a stream or lake. An advantage is that the retinal fluid of the human eye absorbs most of the radiation and therefore helps protect the retina from radiation damage. Therefore, urban scenes with many spectators can be scanned by such a laser without worry. Because of the absorption properties of water, Nd:YAG lidars that emit 1064 nm are not considered eye-safe. On the other hand, Er-doped glass lasers (sometimes called fiber lasers) that emit 1540 nm are considered eye-safe. Some of the most popular lidar systems manufactured today use fiber lasers because of this eye-safe property.

Reflectance is another important material property that affects lidar performance. The amount of energy that arrives back at the lidar receiver is directly proportional to the percentage of the energy that reflects off the object, or in other words the object's reflectance. Wavelength-dependent reflectance is a material property that is unique to the material. Since lidar systems are monochromatic, the reflectance at that particular wavelength determines how detectable an object is given the laser power. Figure 2.2 shows the reflectance of a few common materials at different wavelengths (Clark et al., 2007).

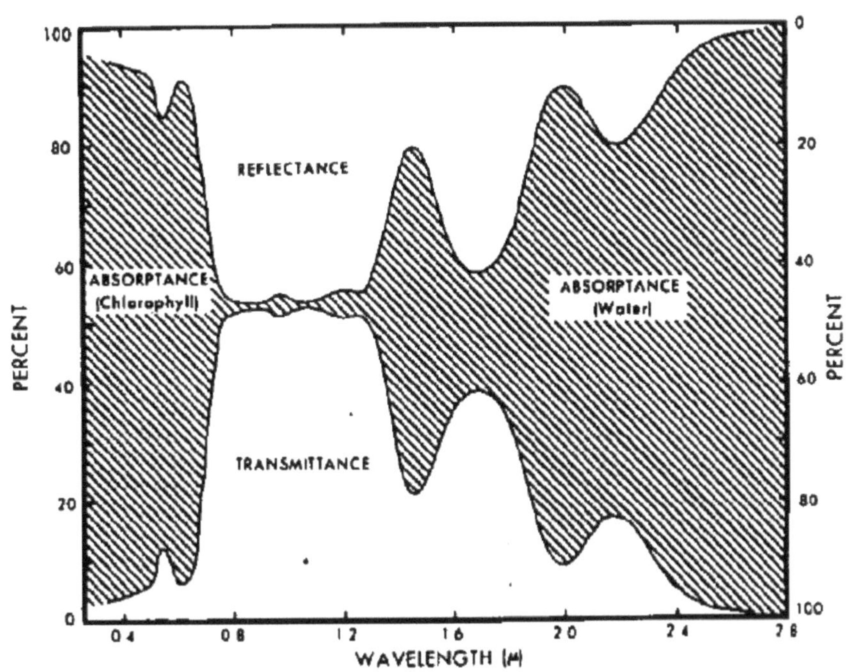

Figure 2.1 The relationship between absorptance (absorption), reflectance, and transmittance through a typical leaf as a function of lidar wavelength (Knipling, 1970).

Figure 2.3 shows some interesting properties. For example, in the visible spectrum, wet snow has relatively high reflectivity and therefore appears white to the human eye. However, at the 1540 nm wavelength used by Er-glass lidar systems, the reflectivity is very low, i.e., it appears to the lidar to be dark grey to black. On the other hand, lidars based on a 1064 nm Nd:YAG laser can easily see snow because it has a reflectance of 0.25 to 0.30 at that wavelength. Other materials such as dry grass have a relatively high reflectance at all VIS and NIR wavelengths.

Charts are sometimes developed by lidar manufactures that facilitate the estimation of lidar performance when scanning various materials. For example, Figure 2.3 shows a chart developed by RIEGL Laser Measurement Systems, Ltd. showing allowable operating altitudes at various pulse repetition frequencies for various materials for their LMS-Q560 lidar (RIEGL, 2007). This lidar operates at 1540 nm and therefore shows poor performance against snow. A chart like this is useful for planning missions. Obviously, the chart would look quite different if the lidar operated at a different wavelength.

Table 2.3 The absorption of light by water as a function of the wavelength of the light (Bukata et al., 1995; Bertie and Ian 1996).

Color	Acronym	Wavelength (nm)	Absorption (m^{-1})
Near Ultraviolet	NUV	250	0.190
Near Ultraviolet	NUV	300 – 350	0.040–0.012
Violet	V	400 – 440	0.006–0.004
Blue	B	460 – 500	0.002–0.006
Green	G	520 – 560	0.140–0.039
Yellow	Y	580	0.074
Orange	O	600 – 620	0.200–0.240
Red	R	640 – 680	0.270–0.380
Red	R	700	0.6
Near Infrared	NIR	740 – 800	2
Near Infrared	NIR	900	9
Short-Wave Infrared	SWIR	1000	60
Short-Wave Infrared	SWIR	1500	5000
Mid-Wave Infrared	MWIR	4000	20000
Long-Wave Infrared	LWIR	10000	80000
Very-Long Wave Infrared	VLWIR	21000	300000

REFERENCES

Bertie, J.E., and Z.D. Lan. 1996. The intensity of the OH stretching band of liquid water revisited, and the best current values of the optical constants of H20 at 25 degrees C between 15,000 and 1 cm^{-1}, *Applied Spectroscopy*, 50:1047–1057.

Bukata, R.P., J.H. Jerome, K.Y. Kondratyev, and D.V. Pozdnyakov, 1995. *Optical Properties and Remote Sensing of Inland and Coastal Waters*, New York, CRC Press.

Clark, R.N., G.A. Swayze, R. Wise, E. Livo, T. Hoefen, R. Kokaly, and S.J. Sutley, 2007. USGS digital spectral library splib06a, U.S. Geological Survey, *Digital Data Series* 231.

Knipling, E.B., 1970. Physical and physiological basis for the reflectance of visible and near-infrared radiation from vegetation, *Remote Sensing of Environment*, 1:155–159.

Miller, J.L., 1994. *Principals of Infrared Technology: a Practical Guide to the State of the Art*, New York, Van Nostrand Reinhold.

RIEGL Laser Measurement Systems, Ltd., 2007. *Airborne Laser Scanner LMS-Q560 User Instructions*, Horn, Austria.
Weber, M.J., 1999. *Handbook of Laser Wavelengths*, CRC Press.

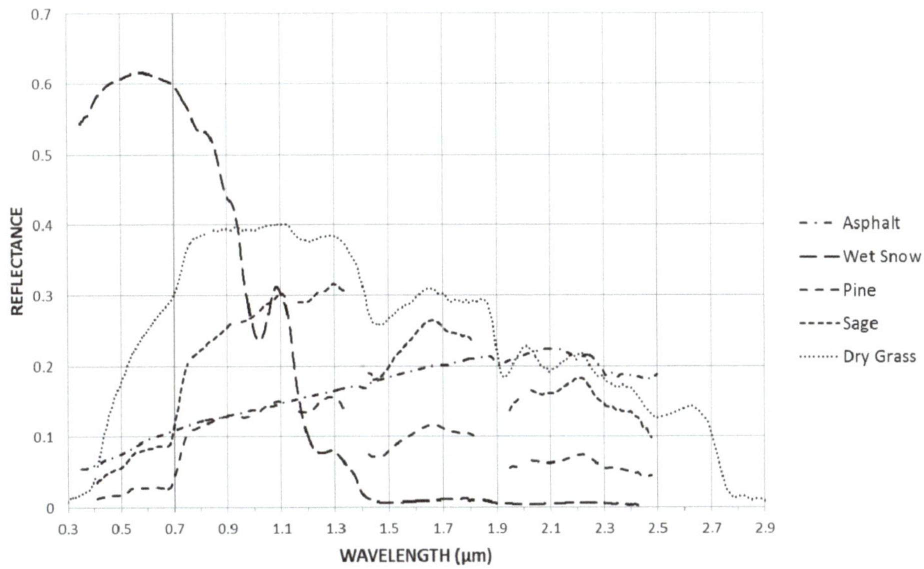

Figure 2.2 Spectral reflectance of various common landscape materials (Clark et al., 2007).

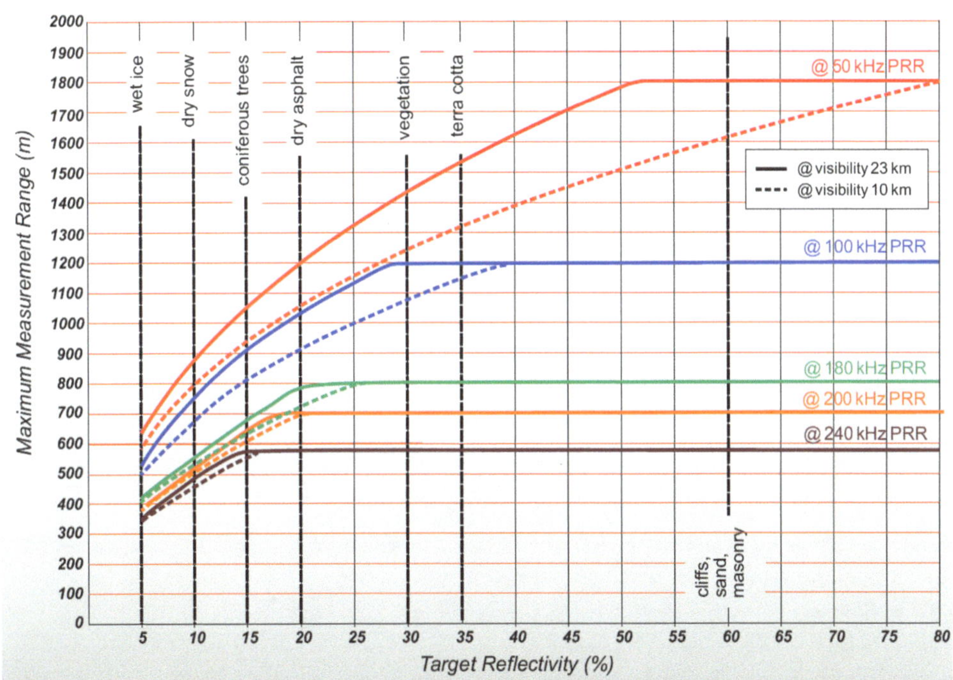

Figure 2.3 Maximum measurement range plotted against target reflectivity (reflectance) for the RIEGL LMS Q560 lidar system (RIEGL, 2007). This system operates at the 1540 nm wavelength.

2.1.2 Features of a Laser Beam

2.1.2.1 Laser ranging

Any laser ranging unit contains the laser transmitter emitting laser pulses and the optical receiver detecting the backscattered pulse. With a pulsed laser, the laser transmitter generates a sequence of narrow laser pulses. Many important performance parameters of the entire lidar system are determined by these laser pulses.

A simplified description of the principle of operation of a pulsed lidar is often based on a mathematical representation of a laser pulse as a rectangular signal (Figure 2.4) with zero rise/fall time and typical pulse width of 10–15 ns (Wehr and Lohr, 1999).

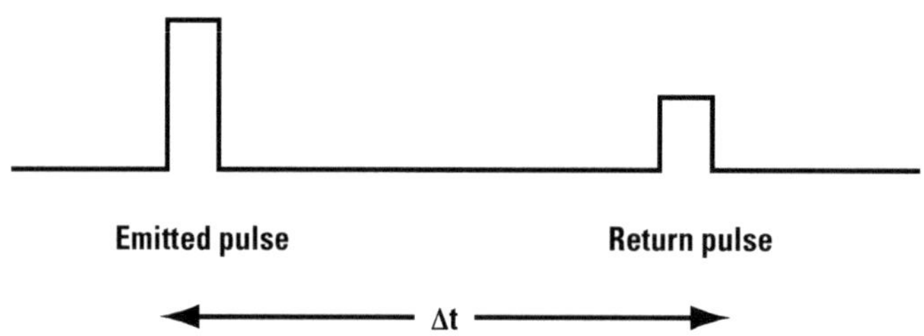

Figure 2.4 Simplified representation of the basic operational principle of a laser rangefinder.

The time difference between the laser pulse when it leaves the transmitter (emitted pulse) and when it is detected by the receiver (return pulse) is measured by the receiver electronics and converted to the measured range by data processing software:

$$\Delta t = 2R/c \tag{2.1}$$

Here Δt is the time-of-flight of the round trip of the laser beam to and from the target; R is the range (distance) between the laser transmitter and the target, and c is the speed of light in the atmosphere. Since advanced time-measurement electronics are capable of taking very precise time measurements, and the speed of light is very well determined, the range measurements taken by a laser rangefinder are considered to be highly accurate.

However, in reality the laser pulses are not rectangular; not in the time domain nor in the spatial domain, and the speed of light and the optical path of a laser beam might be significantly affected by atmospheric refraction and interactions with various atmospheric particles (clouds, aerosols). Moreover, due to physical interactions between the laser pulse and the target (ground), temporal and spatial properties of the backscattered laser pulse are usually distorted. All these factors are very important for considerations concerning not only lidar measurement accuracy but also selection of appropriate operational parameters during data acquisition and processing.

2.1.2.2 Laser Pulse Characteristics

For most typical types of lasers used in commercial airborne lidar systems, power distribution of the emitted laser pulse could be approximated by Gaussian distribution both in time and spatial domains.

Time characteristics of the laser pulse itself, rise time, fall time and pulse width, are of crucial importance for rangefinder applications. The smaller the pulse width, and the sharper the pulse front, the more precise the range measurements which could be taken by a laser rangefinder. Figure 2.5 represents a screen shot of an oscilloscope showing a typical laser pulse shape captured in time domain. One of the most frequently used parameters of a laser pulse, full width half maximum (FWHM), is the pulse duration at 50% of intensity of its peak value.

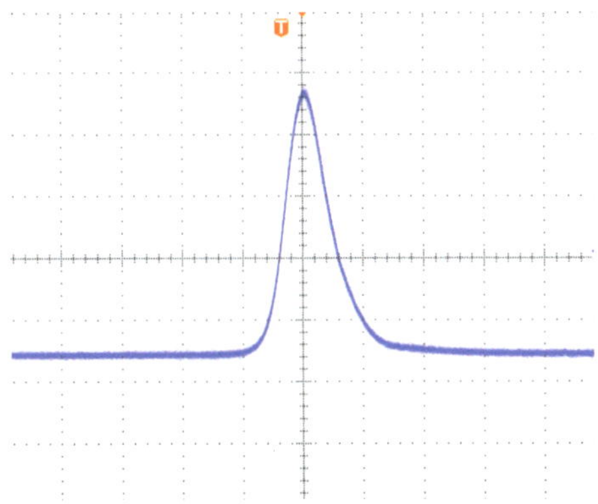

Figure 2.5 Power distribution within the emitted laser pulse in time domain captured as an oscilloscope screen shot.

Quantitative relations between pulse width, pulse energy and pulse peak power as well as relations between these parameters and range measurement precision and maximum measurement range are given elsewhere (Measures, 1984, Baltsavias, 1999). Some of the most important relations are represented below.

For a pulsed laser of P_{av} power operating at a pulse repetition frequency PRF and emitting pulses of t_p FWHM, pulse energy E_p and pulse peak power P_{peak} can be calculated as:

$$E_p = \frac{P_{av}}{PRF} \tag{2.2}$$

$$P_{peak} = \frac{E_p}{t_p} = \frac{P_{av}}{t_p PRF} \tag{2.3}$$

For example, for a 5W laser operating at 100 kHz and emitting 10-ns pulses, E_p = 50 μJ and P_{peak} = 5 kW Relations (2, 3) indicate that the higher the laser PRF and the wider the laser pulses, the lower the pulse energy and peak power, which are critical for both maximum operational range and measurement accuracy of a laser rangefinder

Maximum operational range, R_{max}, of a laser rangefinder is determined by several lidar system parameters and external factors including target size and reflective properties, atmospheric attenuation, etc. R_{max} could be calculated and/or modeled through the Lidar Equation (Measures, 1984, Baltsavias, 1999), which formulates the relationship between the detected lidar signal power P_{det} at the distance R from the target. Considering a full-area target (Jelalian, 1992) this equation could be represented in the following simplified form:

$$P_{det}(R) = \rho \frac{TQA_r}{R^2} P_{peak} \tag{2.4}$$

Here ρ is a characteristic of target reflective properties; it can be expressed as a fraction of the laser incident energy reflected by the target; T is a characteristic of atmospheric optical properties, which can be expressed as a fraction of transmitted and reflected laser energy attenuated by atmosphere (atmospheric transmission); A_r is the effective area of the receiver optical aperture, and Q is a characteristic of optical efficiency of the lidar system design. Equation (4) indicates that maximum operational range for a given lidar system is basically determined by the main system parameters: laser pulse peak power, effective area of the system receiver optics, and the quality of system design determining how efficiently optical power is collected and detected.

2.1.2.3 Laser Pulse Repetition Frequency

Most of pulsed lasers utilized in commercially available lidar systems are capable of generating narrow pulses with very high repetition rate, which is practically selected by the user within 20–250 kHz. Pulse repetition rate, or pulse repetition frequency (PRF) is often considered to be the ultimate parameter determining data collection efficiency of a lidar system, which is not exactly the case. In fact, while considering data collection efficiency, achievable ground point density and area coverage rate, other system characteristics such as the beam deflection mechanism, scan pattern, and point distribution on the ground should be taken into account (Ussyshkin, 2008). In some cases the manufacturer would specify laser PRF and data collection rate separately if they are distinctively different (Reigl, 2009), but the impact of scan pattern on the ground point density should be carefully examined by the system user.

2.1.2.4 Laser Beam Divergence

Even a highly coherent laser beam is diverging with the distance from the emitting source, where its diameter is minimal, and known as laser beam waist (CVI-Melles Griot, 2009). The parameter, which is called laser beam divergence, is a measure of how fast the beam expands with distance from its waist. The beam divergence and the diameter of the beam waist characterize the quality of a laser beam. For most laser rangefinder applications only far-field consideration of laser beam propagation is important, and in this case the half-angle beam divergence ϑ of an ideal Gaussian laser beam can be determined as:

$$\vartheta = \frac{w(R)}{R} + \frac{\lambda}{\pi w_0} \tag{2.5}$$

Here w(R) is the radius of a laser beam at distance R from the source; λ is the wavelength, and w_0 is the radius of the beam waist. Laser beam divergence ϑ could also be expressed in full-angle terms, if instead of the beam radius w_0, beam diameter is considered. A higher beam divergence for a given beam radius, is related to an inferior beam quality, which essentially means a lower potential for focusing the beam to a very small spot. For most lasers utilized for rangefinder applications, the beam quality could be characterized with a certain M^2 factor, and then the beam divergence could be calculated as:

$$\vartheta = M^2 \frac{\lambda}{\pi w_0} \tag{2.6}$$

The closer the M-factor is to 1, the better the quality of the laser beam. Lasers used in advanced commercially available lidar systems usually have an M factor of about 1.2–1.3.

2.1.2.5 Laser Pulse Energy Distribution

Laser pulse energy distribution in the spatial domain is represented in Figure 2.6; it is often approximated by Gaussian distribution and could be characterized by 1/e, 1/e^2 or FWHM (full width half maximum) beam size. The commonly adopted definition of laser beam size is the diameter at which the beam irradiance (intensity) has fallen to 1/e^2 (13.5%) of its peak value.

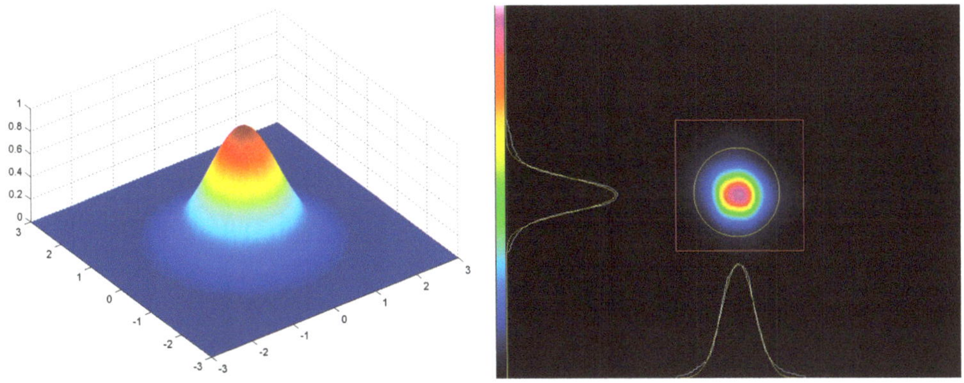

Figure 2.6 Power distribution within the emitted laser pulse could be approximated by a Gaussian distribution.

The equations below show the simplified relations between 1/e, $1/e^2$ and FWHM beam diameter values:

$$1/e \text{ beam diameter} = 1.2 \text{ FWHM beam diameter (Gaussian beam)} \tag{2.7}$$
$$1/e^2 \text{ beam diameter} = 1.7 \text{ FWHM beam diameter (Gaussian beam)} \tag{2.8}$$

2.1.2.6 Laser Beam – Target Interaction

The laser footprint size on the ground D_L, typically referred to in the literature as $1/e^2$ or 1/e diameter, could be estimated using a basic mathematical relationship connecting flying height above the ground, h and full-angle laser beam divergence, ϑ (Baltsavias,1999):

$$D_L = 2h \tan \vartheta/2\pi \tag{2.9}$$

The laser beam divergence, ϑ, could be expressed in $1/e^2$ or 1/e terms; therefore the diameters of the laser footprint on the ground could be characterized as $1/e^2$ or 1/e size. The mathematical relation (9) is valid only for normal incidence of the laser beam with respect to the target (ground). While collecting data, the angle of incidence (AOI) of the laser beam is continuously changing due to changing scan angle, α, and local terrain slope, β. The combined effect of these changes results in elongation of the laser footprint on the ground (Figure 2.7), which may contribute to the range measurement error (Schaer, 2007, Ussyshkin, 2009).

Figure 2.8 illustrates an airborne lidar measurement scenario in which one range measurement is affected by AOI error and the other is not. Here the blue beam represents a range measurement being made over a flat and level surface from directly below the airplane; this scenario is not affected by AOI error. The red beam is firing at the swath edge onto the same surface and is affected by AOI error. The two beams have the same divergence but, the blue beam's footprint is a circle whereas the red beam's footprint is an oval.

The simplified geometry presented in Figures 2.7 and 2.8 becomes more complicated if the actual laser pulse power distribution within the waveform is taken into account (Figure 2.9). Depending on the AOI, the power distribution in the return pulse would be distorted, and the magnitude of this distortion would be proportional to the AOI, and eventually translated into the range measurement error (Schaer et al., 2007, Ussyshkin et al., 2009).

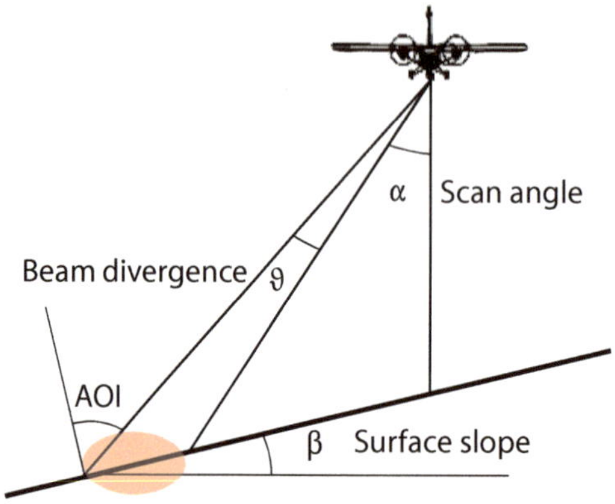

Figure 2.7 Simplified illustration of elongation of laser footprint on the ground for sloped terrain and large scan angles.

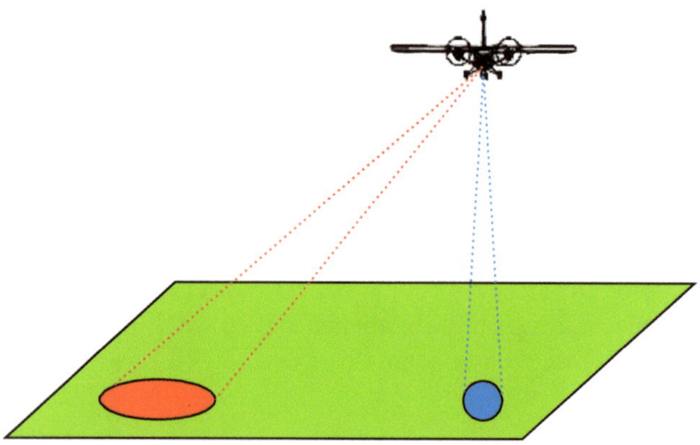

Figure 2.8 Laser mapping scenario plagued by Angle of Incidence (AOI) effects.

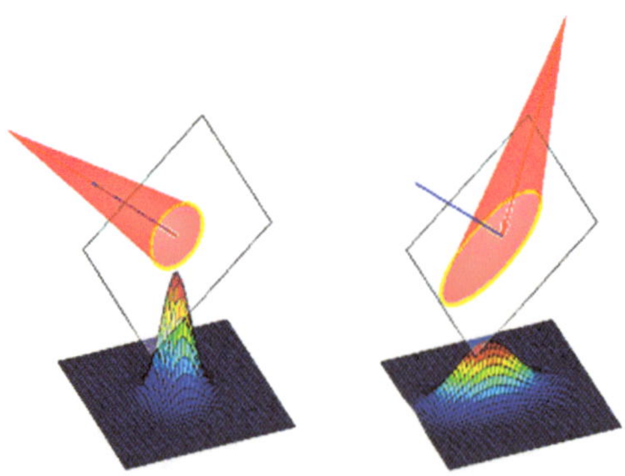

Figure 2.9 Impact of the AOI error on laser pulse power distribution. Influence of incident angle on relative power distribution (after Schaer, 2007).

This error may become a significant part of overall lidar error budget at large scan angles even in the case of flat uniform terrain (Glennie, 2007, Ussyshkin et al, 2009), and that is one of the reasons for poor lidar data accuracy at the swath edges, especially for wide scan FOV. (Ussyshkin et al, 2008). Consideration of the AOI error then is always beneficial while considering details of a lidar data collection mission, but it obviously becomes critical for highly non-uniform terrain in mountainous areas, where steep slopes cause large AOI resulting in increased horizontal and vertical errors. Further detailed discussion on the range measurement error linked to scanning geometry and laser beam – target interaction could be found in (Glennie, 2007, Schaer et al., 2007, Ussyshkin et al, 2009).

REFERENCES

Baltsavias, E.P., 1999. Airborne laser scanning: basic relations and formulas, *ISPRS Journal of Photogrammetry and Remote Sensing*, 54(2/3):199–214

CVI-Melles-Griot Optical Guide. 2009. *Gaussian Beam propagation; Beam waist and divergence*, URL: http://www.mellesgriot.com/products/optics/gb_2_1.htm, CVI Melles Griot, Albuquerque, New Mexico (last date accessed: 15 June 2009).

Glennie, C.L., 2007. Rigorous 3D Error Analysis of Kinematic Scanning LIDAR systems. *Journal of Applied Geodesy*, 1:147–157

Jelalian, A.V., 1992. *Laser Radar Systems*, Artech House, Boston, Massachusetts, 308 p.

Measures, R.M., 1984. *Laser Remote Sensing: Fundamentals and Applications*, John Wiley & Son, Hoboken, N.J., 524 p.

RIEGL Laser Measurement Systems GmbH. Product Datasheet, URL: http://www.riegl.com/uploads/tx_pxpriegldownloads/10_Datasheet_LMS-Q560_05-03-2009.pdf, RIEGL Laser Measurement Systems GmbH, Horn, Austria (last date accessed: 15 June 2009).

Schaer, P., J. Skaloud, S. Landtwing, K. Legat, 2007. Accuracy estimation for laser point cloud including scanning geometry, *Proceedings of the 5th International Symposium on Mobile Mapping Technology*, 29–31 May, Padua, Italy, unpaginated CD-ROM.

Ussyshkin, V., Boba, M., Sitar, M., 2008. Performance characterization of airborne lidar system; bridging system specifications and expected performance, *Proceedings of the XXIst ISPRS Congress, Commission I, Vol. XXXVII, part B1*, 2–4 July 2008, Beijing, China, unpaginated CD-ROM.

Ussyshkin, R.V., M. Ilnicki, R. Ravi, M. Pokorny, 2009. Maintaining airborne lidar data accuracy for sloped and non-uniform terrain, *ILMF Conference Proceedings*, New Orleans, Louisiana, 26–28 January 2009, unpaginated CD-ROM.

Wehr A. and Lohr U., 1999. Airborne laser scanning—an introduction and overview, *ISPRS Journal of Photogrammetry and Remote Sensing*, Vol. 54(2/3): 68–82

2.2 KEY ELEMENTS OF ALS TECHNOLOGY

2.2.1 Lasers, Lidar, and What makes it Work

Lasers have been used in commercial measuring equipment for more than 35 years. Today, laser scanning is a mature technology. This technology is a complex integration of the fields of optics, opto-mechanics, and electronics. It is applied in many different fields, such as barcode scanning, as-built documentation, non-contact inspection, 3D modeling, rapid prototyping, civil engineering, oil exploration, architecture, topography, mining, health sciences, reverse engineering, archaeology, quality assessment, and mechanical dimensional inspection. The list of applications is vast, but this chapter will focus on the use of lidar for topographic mapping and field surveying.

Lidar makes use of several of the laws of physics which will be reviewed briefly.

2.2.1.1 Gravity

Gravity is a phenomenon with which we are all familiar, although scientists still struggle with the precise nature of this energy of attraction. There is still considerable conjecture in the scientific world whether there is such a thing as a "graviton," the theoretical element of gravitational energy. If it does exist, the force is so weak it is practically un-measurable. Despite popular science fiction, it is thought there is no anti-graviton as gravity acts on the total energy or mass not on the charge states of a mass. Gravity is, of course, what keeps us on the ground, but it has the peculiar property of varying depending on where on the ground one happens to be. The density of the rocks you stand on can affect the "amount" of gravity you experience. While the human brain cannot discern small changes in gravity, scientifically it is measurable and it does make a difference to certain sensitive equipment. Because the density of the earth is not consistent, the center of gravity of the Earth is different from the center of its shape.

Furthermore, while for years scientists have known the center of the Earth is molten, until the advent of GPS satellites it was not definitively known that the Earth's center of gravity is not in a constant location and that, indeed, it moves around. In fact, some claim it can move up to one meter per day because of the molten center of the Earth, coupled with the distortion of the Earth's surface from the pull of the moon. GPS has also allowed us to prove that not only are the oceans subjected to tidal effects but the Earth's hard surfaces (ground) are also affected by diurnal tides.

The land in Washington D.C., for example, rises and falls approximately 20 to 30 centimeters (8 to 12 inches) twice a day. Because everything on the Earth in our immediate surroundings moves up and down by the same amount, this is not at all obvious and really makes no difference to anything: unless one is conducting accurate measurements using satellite technology.

See Appendix A for further information and references on gravity and geoid modeling.

2.2.1.2 Time

Time is another required element in the lidar package. However, time as an element is arbitrary. There is a lot of cosmology which posits that time and space are interwoven (Einstein) and may be manifestations of the same phenomena. This is not at all intuitive, and it is doubtful that many outside of professional cosmologists understand this! But for practical purposes here on Earth, time is a construction based on the periodicity of the duration of the orbit of the earth around the Sun, manifested and measured in base 6 mathematics, originating from the ancient Sumerian civilization. There is considerable argument that the base 6 math was derived from the approximate time it takes the earth to circumnavigate the sun (360 days), and also the Sumerians apparent dislike of fractions (360 is neatly divisible by 1,2,3,4,5,6,8,9,10, and multiples thereof). But while no one knows for sure why the Sumerians came up with a base 6 rather than base 5 or 8 or 10 math; their legacy is we have both time based on days broken into hours, minutes, and seconds, and circles that have 360 degrees, 60 minutes in a degree, and 60 seconds in a minute, which works very handily in geometry.

2.2.1.3 Light

Light is both a wave and a discrete particle of energy. The particle is known as a photon. Coherent light is the phenomenon generated by the laser part of the lidar package. Visible light is just a small portion of the electromagnetic spectrum which encompasses a wide range of what we often tend to consider different energies, from radio waves to gamma rays. However light is really no different from any of the other parts of the spectrum. Light is simply the portion of the electromagnetic spectrum which is visible to human eyes. It is a narrow part of the spectrum ranging in wavelength from 400 to 700 nanometers (one nanometer is 10^{-9} meters or 0.000000001 m). Our eyes have developed the ability to correlate this part of the spectrum and provide us with information about the physical world that enables us to function relatively efficiently.

Interestingly, the speed of light was first measured by a Scandinavian, Olaus Roemer, in 1676. He calculated it from observing aberrations in the predictions of the orbits of the moons of Jupiter.

He was not very accurate i.e., coming up with an answer of about 225,000 Km/second, but it was a remarkably innovative approach and result for that period of history. Yet, only 50 years later an Englishman, James Bradley, calculated the speed of light to be 301,000 Km/second which is amazingly very close. In fact, the speed of light was fixed in international agreement only in 1983 at 299,792.458 Km/second, based on the time it takes light to travel 1 meter through a vacuum. (1/299,792,458 second). As a final historical note, it should be said that Roemer did not have an accurate measure of the diameter of the Earth for use in his computations, but if he had, his calculations would have yielded a figure very close to the correct value.

All of the electromagnetic spectrum travels at this speed, and it is the frequency or length of the waveforms in parts of a meter which vary and provide us with different properties (for example colors). In terms of energy, the shorter the wavelength, the larger the energy. At one end of the spectrum, radio waves are long and relatively weak forms of energy, and at the other end of the spectrum gamma rays are very short but disastrously (for life) very strong forms of energy.

Infrared waves are longer in wavelength than visible light and thus slightly outside the visible spectrum. Most airborne and terrestrial lidars operate on or close to 1.064 microns (1064 nanometers) in the infrared portion of the spectrum, but there are quite a number of systems using lasers in the green wavelength at 532 nanometers.

Since the mid-1990s, Airborne Laser Scanning (referred to henceforth as ALS) has become fully accepted and proven technologies because of its ability to supply the surveying and mapping communities with precise and directly georeferenced spatial information about the shape of the Earth or objects. Using these laser-based remote sensing techniques, it is possible to conduct a 3D survey mission in a more cost-effective and speedier way, especially when data cannot be collected easily by any other means.

There are several features which make laser-based remote sensing techniques very attractive.

First, these are active systems, generating their own energy to survey the target. This means a survey mission can be carried out any time of day and night and, subject to some general weather conditions, during any season.

Second, typically all ALS systems are integrated with a positioning and orientation system, which can include an onboard GPS (Global Positioning System), and a type of Inertial Measurement Unit (IMU). Once a GPS receiver locks on to four or more GPS satellites, it can triangulate its location on the Earth from the known positions of the GPS satellites.

The IMU, which also includes accelerometers, measures the exterior orientation of the sensor in its own coordinate system related to gravity, as well as velocity. By combining the results of calculations from the GPS receiver and the IMU, the X, Y, Z coordinates of the location of a footprint of a laser beam in 3D space can be determined with a very high accuracy. In the case of ALS, the accuracy is typically better than 15 cm (6 inches) vertically, and 30 cm to 2 meters (1 to 7 feet), depending on the system, horizontally. See Chapter 3 for details on GPS and IMU technologies.

Third, these laser-based remote sensing techniques have provided mapping professionals with a powerful means for high-speed acquisition of data in forests, vegetated and rugged areas which were previously difficult or impossible to access. It is relatively easy to derive a bare earth DEM from a collection of laser points. Therefore, forest managers and hydrologists can obtain accurate and complete vegetation cover maps and surface models for further development and improvement of slope and hydrology models.

2.2.1.4 Background

In 1917, Albert Einstein first proposed the process that makes lasers possible, called stimulated emission. In 1954, the first papers about the maser (or MASER) were published as a result of investigations carried out simultaneously and independently by Townes and his co-workers at Columbia University in New York and by Basov and Prokhorov at the Lebedev Institute in Moscow. MASER stands for Microwave Amplification by Stimulated Emission of Radiation. A laser is a maser that works with higher frequency photons in the ultraviolet (UV), visible light (VIS) or infrared (IR) spectrum. For this work they were awarded the 1964 Nobel Prize in Physics. The optical maser or the laser dates from 1958, when the possibilities of applying the maser principle in the optical region were analyzed by Schawlow and Townes as well as in the Lebedev Institute.

Laser spectroscopy was developed by Schawlow and his associates at Stanford University. The first laser was operating in 1960. It was a ruby laser generating strong pulses of red light.

With respect to laser-based remote sensing, in 1964 NASA launched the Beacon B satellite followed by Lunar laser ranging with the tracking of a corner cube reflector on the moon. Since the 1960s, a global network of ground stations has been developed to measure ranges to satellite-borne reflectors, and satellite laser ranging is still an integral part of NASA's space geodesy program.

In 1975, a joint NASA, NAVOCEANO, and NOAA program sponsored the Airborne Oceanographic Lidar (AOL). This was conceived as an airborne test bed for new applications of laser remote sensing. The AOL was a laser fluorosensor that measured a variety of oceanographic properties such as chlorophyll and other biological and chemical substances in the world's oceans. The original design also provided some topographic information. In 1994, a separate airborne lidar system dedicated to topographic mapping was developed and called the Airborne Topographic Mapper (ATM). The AOL remained for collecting the laser-induced fluorescence (LIF) spectrum in a contiguous region covering the spectral band from 370 nm to 740 nm, for each laser pulse. In addition to the LIF, the AOL collects passive (solar-induced) ocean color radiance data and sea surface temperature (from an ancillary infrared radiometer). Further investigations and development of laser-based remote sensing systems were carried out at the Laser Altimeter Processing Facility (LAPF) at NASA's Goddard Space Flight Center. Through the results of those efforts, SLICER (Scanning Lidar Imager of Canopies by Echo Recovery), RASCAL (RAster SCanning Airborne Laser Altimeter), SLA-01 (Shuttle Laser Altimeter I), and SLA-02 (Shuttle Laser Altimeter II) appeared.

Traditionally, terrestrial field surveyors and engineers have used a group of instruments for measuring horizontal and slope distances, vertical distances, and horizontal and vertical angles.

Angular measurements were produced using theodolites. These, used with separate microwave instruments (Tellurometers) in the 1940s for measuring distance, allowed horizontal distances to be calculated. Later in the 1970's, Electronic Distance Measuring Equipment (EDME) devices penetrated the market. EDMEs were originally often mounted on top of a theodolite. Their operating principle is the same as a rangefinder which transmits a narrow laser beam to a reflecting surface, and measures the time it takes to go and return. Eventually the total station was developed which combined the features of a theodolite and rangefinder in a single package.

2.2.2 Technology Overview

In theory, any light source can be used to create a lidar instrument, but in practice all modern lidar instruments use a definable laser as the source. In a nutshell, light detection and ranging is the science of using a laser to measure distances to specific points. Lasers can technically use any wavelength in what is considered the "light portion" of the electromagnetic spectrum.

A laser can be thought of as a tube with a mirror at each end. One of the mirrors has a small hole in its center. If the outer shell of the tube includes a light source which is flashed on and off, the photons from the light source are emitted in all directions. Inevitably a number of the photons will start to bounce back and forth between the two mirrors. If two photons collide, they also have the ability to split and create two more photons. Also, inevitably, some photons will escape through the hole in the mirror, and they will create a virtually (but not completely) parallel light source at the specific frequency of the emitting light source. While this may sound as if the laser has managed to solve the impossible by creating more photons than were initially generated, in fact there is considerable atrophy and indeed lasers are notorious energy hogs. It is common for the emitted laser light to be between only 5 to 20 percent of the source power. It should be remembered, however, that all lasers are potentially damaging, even the laser pointers used by teachers and lecturers. Lidar manufacturers are very cognizant of these dangers, and lidar units are manufactured to be used within certain parameters and circumstances.

Laser scanning systems can be classified in different ways. They can be pulsed and continuous wavelength laser systems or non-contact optical profilometers depending on their transmitting and receiving components. Different beam steering devices can produce a scan of a laser beam across-track or scanning can be spot-wise vertical profiling. From the operational point of view, laser scanning systems can be installed on different carrying platforms. Depending on the chosen laser sources and the reflective measuring features of the objects, different laser-based remote sensing units operate at different wavelengths.

Laser-based remote sensing systems, which are used for topographic purposes, can be classified in the following ways, by:

- Operating Principles
- Mounting
- Scanning Techniques
- Operating Wavelengths.

2.2.2.1 Operating Principles

Figure 2.10 shows a classification path of different operating modes of the laser-based remote sensing systems from the point of view of operating/measuring principles.

The pulsed laser system works in the following way. It transmits laser pulses, then amplifies the light that is scattered back through an optical telescope receiver and the photomultiplier tube. The distance to the object is determined by recording the time the transmitted pulse travels to the target and back by using the speed of light.

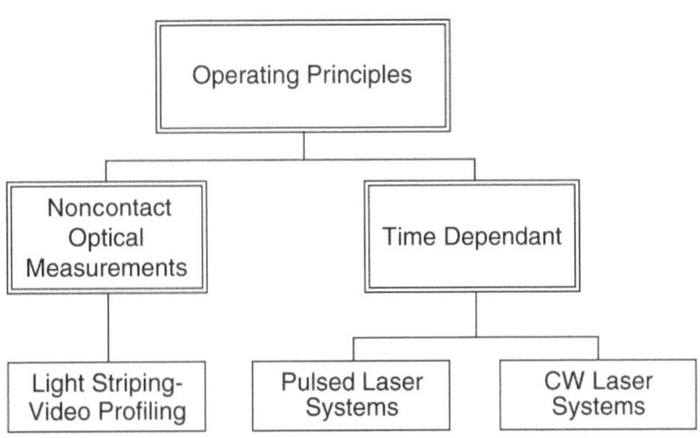

Figure 2.10 Operating principles of laser-based remote sensing.

The continuous wave (CW) laser system transmits a continuous signal, and ranging is carried out by modulating the light intensity of the laser light. Typically, it is a sinusoidal signal which is received with a time delay. The travel time is directly proportional to the phase difference between the received and transmitted signal.

There is also one type of laser measurement principle which is based on a combination of a laser light stripe generator and a video camera. It is a so-called, non-contact optical measurement. In this case the laser source is separate from the video camera, which can be digital. The laser light is visible and follows along the target surface on a continuing basis. This line is considered as a surface profile. Then, during the movement of a carrying platform, the profile maps are registered to the same 3D coordinate system by an iterative surface matching algorithm. The laser light is transmitted through a cylindrical lens. This lens produces the laser beam which is very narrow in one direction and expanded in the other direction. Typically, this form of the laser beam is so-called a line. The laser light sheet is usually perpendicular to the illuminated surface. Thus, if an observer looks down on the flat surface, they will see a straight line. If the surface is flat, then the profile is a straight line. If the surface is rough, then the profile is a curve. The digital image processing is based on a projective transformation between the image plane of the camera and the plane of the laser sheet, and also the direction of scanning with respect to the plane of the laser sheet. The refinement is obtained through weighted least squares matching of multiple profile maps acquired from different points of view, and registered previously using an approximate calibration.

Laser scanning systems can be classified in different ways. They can be pulsed and continuous wavelength laser systems or non-contact optical profilometers depending on their transmitting and receiving components. Different beam steering devices can produce a scan of a laser beam across-track or scanning can be spot-wise vertical profiling. From the operational point of view, laser scanning systems can be installed on different carrying platforms. Depending on the chosen laser sources and the reflective measuring features of the objects, different laser-based remote sensing units operate at different wavelengths.

In current lidar systems, the pulsed laser systems are most frequently used because they can produce a high power output at a very high pulse repetition rate. Therefore, the operating distance can be more than 1 km (0.62 mi).

2.2.2.2 Mounting

The laser sensor is used for collecting a dataset about a remotely located object. All of these sensors are differentiated by how they are mounted as noted in Figure 2.11 and are basically classified as airborne or ground based.

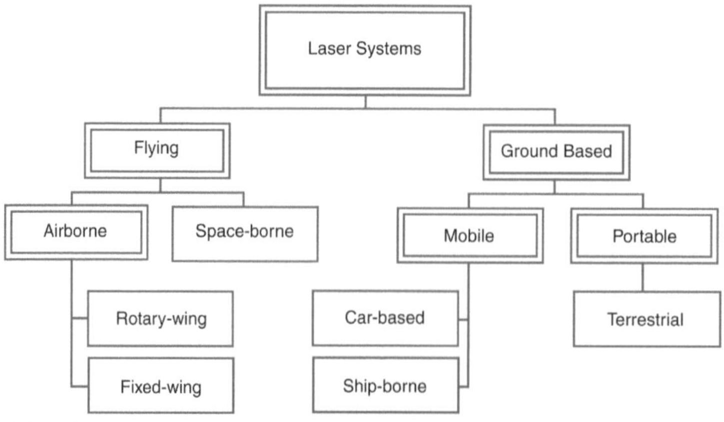

Figure 2.11 Mounting types of laser-based remote sensing systems.

The laser systems installed onboard airborne carrying platforms are typically known as airborne laser scanners. They can be mounted onboard a fixed-wing aircraft or a helicopter (rotary-wing).

Usually, airborne laser scanners are across-track or push-broom scanners. An airborne laser profiling system, which is essentially a laser altimeter, has very limited use. This type of airborne laser profiling system can be still applied in remote sensing of forest and forestry-oriented applications, although these systems are no longer economically efficient.

Regarding airborne applications, there are those that work over water called bathymetric laser scanning systems, and those that work over land called airborne terrain mapping systems or topographic systems. The bathymetric ALS systems operate mainly in the blue/green portion of the spectrum. Infrared radiation is mostly absorbed by water, so near-infrared lasers do not reflect many returns when used over water bodies. The blue-green lasers usually frequency-double the wavelength to create one pulse which is reflected off the water's surface, and one which penetrates to the sea floor. The differentiated distance is the bathymetric depth. Current systems provide a return for depths of up to three-four times what can be seen by the naked eye. So, simply stated, if you can see down through water to a depth of 3 meters, a laser will provide a return up to a depth of 10 to 45 feet (providing it hits a reflecting surface in that distance).

Space-borne topographic laser systems are typically laser altimeters which are applied for surface profiling of the Earth. This remote sensing technique is widely used for the study of geosciences such as measuring of forested ecosystems, global monitoring of ice fields, mapping of polar sea-ice freeboard and thickness, high-resolution mapping of ocean eddies, glacier topography, and lake and river levels. For forest land-cover applications, estimating forest canopy heights and canopy closure is based on the analyses of the waveform of the returned laser signal.

It should be noted that the footprint of space-borne lasers is usually large (in the tens of meters) and absolute accuracy is not appropriate for engineering.

2.2.2.3 Laser Scanning Techniques

Laser beam steering is one of the key technologies implemented in the laser-based remote sensing systems used for topographic mapping purposes. This is a well established technology that is thoroughly described in numerous literature sources. Figure 2.12 shows how laser-based remote sensing units can be grouped from the point of view of scanning techniques.

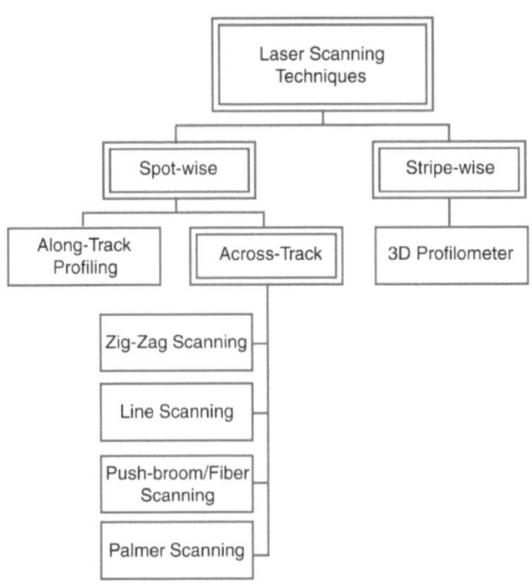

Figure 2.12 Laser Scanning Techniques.

In theory there are no special reasons why one scanning technique is preferable to another. Historically, spot-wise along-track profiling was the first laser use in remote sensing. Continuous research and development in the opto-mechanics and electro-optical industries led to improvements in beam steering technologies. Subsequently, laser beam steering optomechanical devices became robust and very well controlled. The most common scanner is a polygonal scanner, which consists of a polygonal mirror, a drive system, and a bearing system. Typically, because of the forward motion of the aircraft a polygonal scanner produces a zigzag pattern of scanning.

2.2.2.3.1 Oscillating Mirror (Zig-Zag Scanning)

In systems using an oscillating mirror, the mirror rotates back and forth between limited extents. Thus, a zigzag (i.e., a sinusoidal pattern) line is produced on the surface of the target area (Figure 2.13). The major advantage of this method is that the mirror is always pointing towards the ground so data collection can be continuous and theoretically all pulses of the laser are used. In most systems the user can generally control the mirror's field of view and scan rate (the angle through which the mirror swings and the number of times the mirror oscillates per second). Changing the field of view provides additional flexibility as it allows points to be collected over a shorter span (thus effectively closer together) or a wider span (further apart). But there are a number of disadvantages to the oscillating mirror principle. Some of these are due to the mirror having to start from one end of its travel, accelerate to full scan speed, and then stop at the other end of its travel. The changing velocity and acceleration of the mirror cause torsion between the mirror and the angular encoder.

The changing velocity also means that the measured points are not equally spaced at the beginning and ends of travel. The point density increases at the edge of the scan field where the mirror slows down, and decreases at nadir. Manufacturers have solved these problems by essentially ignoring the outlier points on a scan and modeling the distortions caused by changing speed using a computer algorithm.

2.2.2.3.2 Rotating Mirror (Line Scanning)

In this approach, the mirror is rotated continuously at a constant velocity in one direction by a motor. The measurements of the angles are carried out either directly from the motor rotation or from an angular encoder directly mounted to the mirror. The result is measured points are in parallel lines on the ground (Figure 2.14). The main advantage is that the constant velocity does not induce any acceleration type errors in the angle observation. However, there are also several disadvantages. The major disadvantage is that observations cannot be taken during a significant amount of time during each mirror rotation because the mirror is pointing away from the target. So while the laser may be firing at 200,000 times per second it could be that 140,000 of those pulses are not aimed at the target area. The other disadvantage is the field of view cannot be changed as the mirror must rotate 360 degrees.

2.2.2.3.3 Push Broom (Fiber Scanning)

Figures 2.15 and 2.16 show a push broom, fiber-optical or nutating scan pattern used by Toposys and Falcon. The 127 fibers are arranged at one end in a circle and at the other end in a fixed linear array.

The laser pulsed energy is transmitted into one of the fibers arranged in a circle. There it is formed into a narrow laser beam and transmitted to the ground. The reflected light is focused to the corresponding fiber in a linear array at the receiving side. There it is guided to the circular array and coupled into the center fiber leading to the detector. The direction of a transmitted laser beam is defined by each individual fiber. It is a stable system. Any misalignment within this arrangement will lead to a loss of energy but not to a false direction. The fiber scanner guarantees a precise knowledge about the beam direction with respect to the fiber scanner itself. The fiber scanner and IMU are mounted close together; thus, the orientation between them becomes invariable against any mechanical forces induced by the aircraft or forced by thermal variations. The bore-sight alignment is established by at minimum one specific calibration flight. This can be a flight over a large building where the coordinates of the roof corners are well known or can be measured conventionally to centimeter accuracy. In order to reduce the influence of errors caused by the weather conditions to the positioning, such a calibration flight is done during the most favorable conditions.

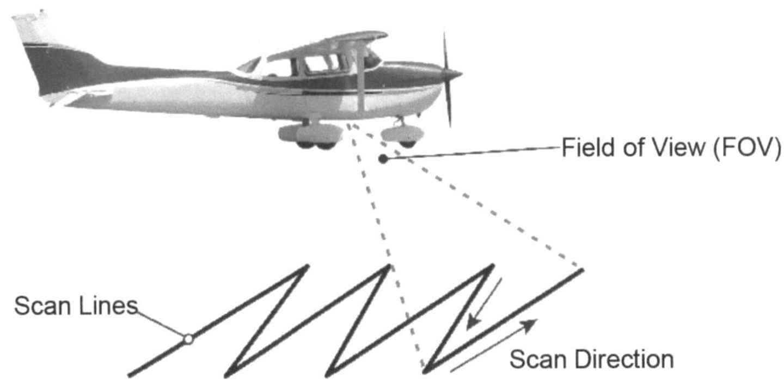

Figure 2.13 Zig-zag or sinusoidal pattern (ALS, ALTM, Topeye) Courtesy of Dewberry.

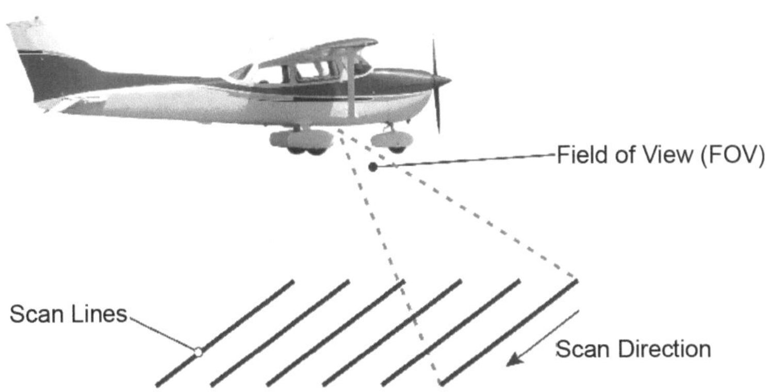

Figure 2.14 Rotating mirror line scan pattern – Courtesy of Dewberry.

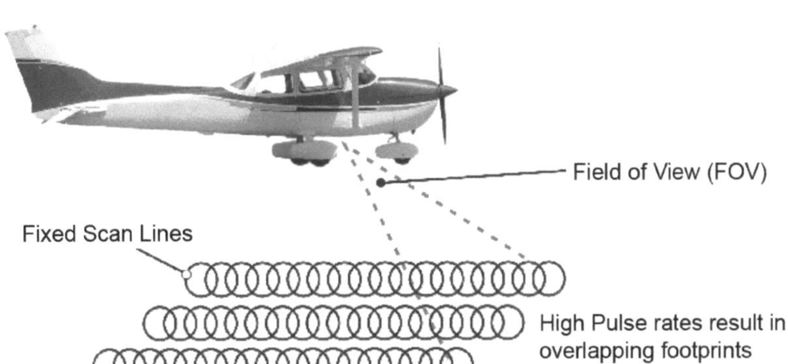

Figure 2.15 Pushbroom, fiber optical or nutating scan pattern (Toposys and Falcon). Courtesy of Dewberry

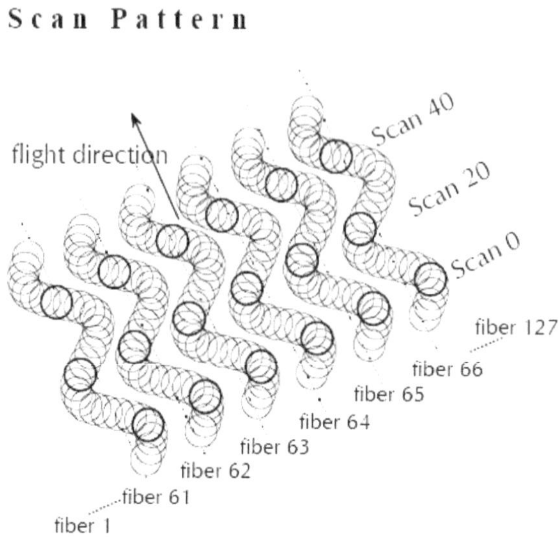

Figure 2.16 Toposys/Falcon fiber optical scan pattern. Courtesy of Toposys

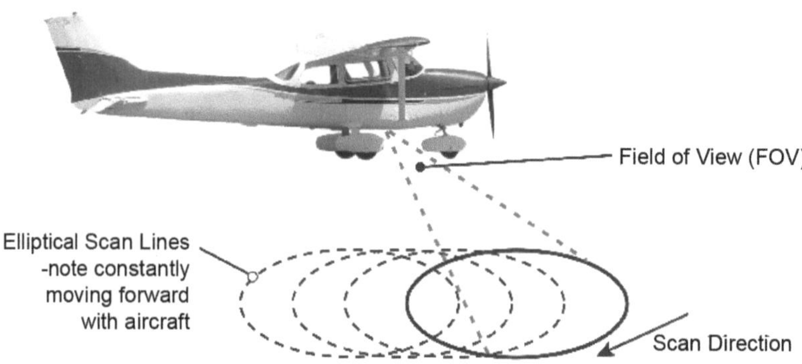

Figure 2.17 Palmer or elliptical scan pattern (ScaLARS). Courtesy of Dewberry.

The scan pattern of the Toposys/Falcon laser scanning system (Figure 2.16) looks like push-broom ranging from the right to the left side of a swath with a constant separation between adjacent fibers. Due to the high scan rate of 650 Hz and a slight swiveling with amplitude of about 1 mrad (0.06 degree), adjacent footprints are overlapped. The nominal scan lines are only about 6 cm. apart. In a flat area, adjacent measurements can not largely differ. If there are single measurements deviating from their neighbors, they can be considered as erroneous. Unlike widely separated spot-wise measurements this method provides a high reliability of the resulting DEM coming close to the reliability of EDME.

2.2.2.3.4 Palmer (Elliptical) Scanning

Figure 2.17 shows the Palmer scanning pattern, used by ScaLARS, which produces an elliptical pattern with redundant data that can be used for calibration.

2.2.2.3.5 Stripe-Wise Scanning

As shown in Figure 2.18, non-optical measurements are performed by means of a combination of a laser line generating unit and a video camera in a way similar to a structured light

3D scanning in close-range photogrammetry. The laser projector and the video camera are apart from each other.

The video camera looks at the profile, the deformation of the pattern on the subject, and uses a technique similar to triangulation to calculate the distance of every point on the line. Due to the triangulation angle and the fixed optical image formation the *X*, *Y* coordinates of one profile of the object can be calculated. The system is moved in defined steps along the moving direction, i.e., the flight direction. Thus, the 3D shape of the object is scanned slice by slice. In general, the scanned pattern may be 1D or 2D. An example of a 1D pattern is a single line. The 2D pattern is formed when single profiles are collected in a grid manner. The deformation of the pattern is usually registered on the CCD sensor. Then, a fairly complex algorithm is used to calculate the distance at each point in the pattern. In the simplest case, one could analyze an image and assume that the left-to-right sequence of stripes reflects the sequence of the lasers in the array, so that the leftmost image stripe is the first laser, the next one is the second laser, and so on. In non-trivial targets having holes, occlusions, and rapid depth changes, however, this sequencing breaks down as stripes, and are often hidden and may even appear to change order, resulting in laser stripe ambiguity. One of the big advantages of using such a 3D profilometer is speed. Stripe-wise scanning scans multiple points or the entire field of view at once. Therefore, the problem of distortion from motion can be significantly reduced or even eliminated. However, the commercial application of the 3D profilometer is limited because of the complex data processing and analyses regarding surface reconstruction.

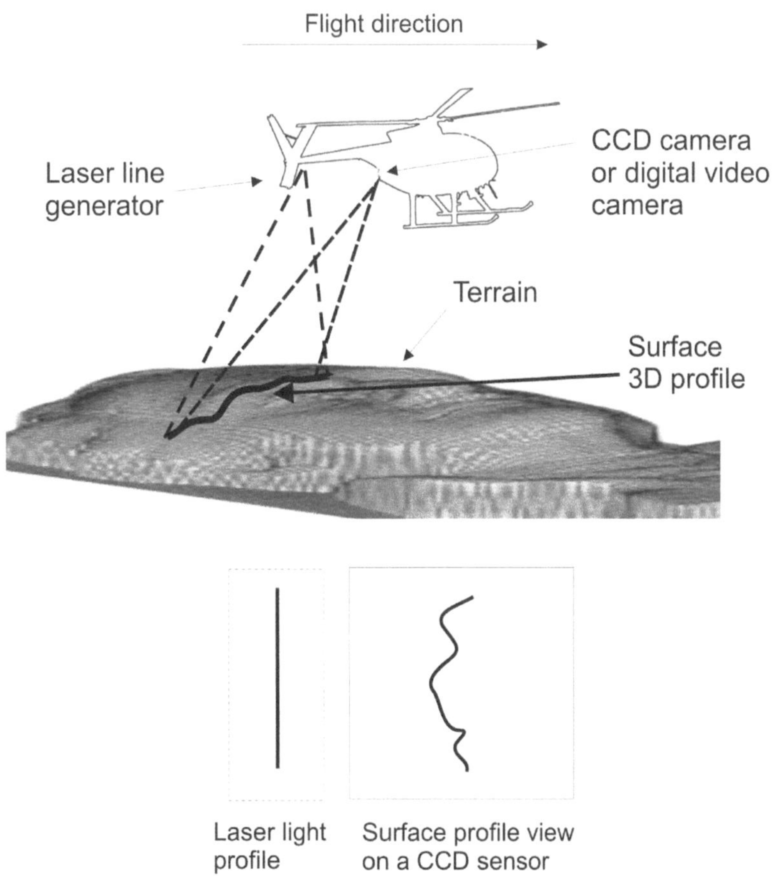

Figure 2.18 The principle of strip-wise scanning (3D profilometer).

2.2.3 Main Operating Principles

Laser rangefinder (LRF) technology is very suitable for applications where there is a demand for the speedy gathering of very accurate 3D information. The accuracy of distance measurement can be on the order of 20 to 30 mm accuracy for applications depending on the distance between the laser sensor and the study object. Having very high accuracy is possible because of low divergence of the laser light beam, which is typically on the order of 1 to 2 milliradians. From the physical point of view, there are no major differences between the operating principles of commercial airborne and terrestrial systems. The main measuring principles are the same as in a LRF. They can be divided into two groups: pulsed wave LRF and continuous wave (CW) LRF. The pulsed wave LRF can measure the Time-Of-Flight (TOF) of the transmitted pulse or determine the distance between the transceiver and the object surface using a Frequency-to-Distance Conversion (FDC). Continuous wave LRF can be CW-AM (Amplitude Modulation) and CW-FM (Frequency Modulation). In reality, the operating principle of the typical commercial laser scanning systems, either airborne or terrestrial, is pulsed-wave based.

The typical pulsed terrestrial LRF consists of a laser transmitter, the transmitting and receiving optics, and a time-to-digital converter. The transmitter emits a short optical pulse (2 to 20 nanoseconds) to the surface of the target, and back at the speed of light. A quartz-stabilized clock is used to measure the time difference between the transmitted pulse and received echo. The range is calculated by halving the two-way travel time and multiplying by the speed of light. An image of the object is created point by point by sweeping the transceiver through the specified zenith and azimuth angles.

In the FDC LRF, a time period is proportional to the distance to the target. The measurement is usually accomplished either by comparing two very short pulses, or using a continuous amplitude or frequency modulated wave, and measuring the phase difference or beat frequency.

Furthermore, in the measured signal the amplitude of the received waveform is proportional to the intensity of the received light. The phase difference between the waveform reflected from the surface of the target and the reference signal gives a direct measurement of the distance to the object point.

The use of a phase-sensitive detector can give a more accurate indication of TOF, and hence distance, than a simple pulsed system, in which a direct time measurement is made on an intermittent burst of laser energy. In this case, the accuracy can be on the order of 1 mm (close range measurements).

In CW-AM LRF, a sinusoidal signal modulates in amplitude the transmitted signal. The phase shift between the transmitted and received modulated signals is proportional to the distance to the target. The phase measurement is periodical in time. Thus, the measuring result is valid for range values which are less than one wavelength of the modulated signal determining a distance ambiguity interval. If there are no ambiguity resolving mechanisms (i.e., instructions from an operator, or from some heuristic rule), the depth of field is constrained to less than one wavelength.

If higher modulating frequencies are used, better depth resolution is obtained. If lower modulating frequencies are used, a larger depth of field is obtained.

In CW-FM LRF, a transmitting signal is modulated by a linearly time-varying frequency. The returned signal is mixed with a reference signal producing a beat frequency which is proportional to the distance. The advantage of this method is that the absence of phase measurement in this method excludes the existence of an ambiguity interval.

All laser scanning devices operate by directing a laser light beam to the object to be measured, then detecting and measuring the signal from the light reflected by the object surface. Unlike conventional photography, which records images reflected from ambient light, laser scanning systems record the strength of the echoes (reflectance). The intensity value of the returned echo can be useful for postprocessing purposes, but it should be remembered that laser light is monochromatic (see Chapter 10.1).

The operating distance of laser scanning systems depends on several factors such as the output energy of the laser source, the operating wavelength, and the eye safety limitations at a particular wavelength. Additionally, light mist, fog or clouds will cause a situation where the received signals are severely attenuated, as much as 100 db less than the transmitted signal. In heavy fog or clouds lidar systems are unusable. A rule of thumb is: if you cannot see the target because of haze or fog, then the laser will also not penetrate to the target.

Current advanced airborne systems can be operated above the ground surface, depending on the system, from 80 to 6,100 meters (260 to 20,000 feet). The flight altitude depends on the strength of the laser, weather conditions, and in terms of the final deliverables, the mission planning parameters. Terrestrial systems typically measure ranges from 3 to 1,000 meters (10 to 3,280 feet) with range uncertainties of the order of millimeters to centimeters. Ground based systems operate by measuring range distances sweeping through both azimuthal and zenith angles in small steps on the order of tens of microradians.

The main tradeoff between TOF scanners and phase-based scanners is between the speed of acquisition and dynamic range. TOF devices typically capture data at a rate between hundreds and thousands of points per second. By contrast, phase-based devices capture hundreds of thousands of points per second. Phase-based devices are faster, and in many applications this makes them the tool of choice. However, TOF devices have a much greater dynamic range of useful measurement than phase-based devices.

The tradeoff of acquisition speed versus dynamic range stems not only from the measurement principles but from other operational considerations. The pulse repetition rate of a TOF laser system governs in part how the device is classified in terms of eye safety because the mean output power is proportional to the pulse rate. Eye safety is a substantial issue affecting acceptance of laser scanners in some industries. Of course, the speed of light provides a physical upper bound. Each laser pulse must have time to return to the detector before the next one is sent out. If the signal is making a 2-kilometer round trip, the speed of light would limit the signal rate to 150,000 pulses per second, even before allowing for latency in the detector.

Phase-based scanners are also restricted in their output power for eye-safety reasons. Increasing the measurement range of phase-based devices requires higher output power. Another limitation on the measurement range of phase-based scanners is that the modulation waveform increases with range. Accuracy suffers over longer ranges because of the resolution of the measurement of the phase angles.

2.2.3.1 Sensor Position and Orientation

While it is fairly simple to measure distances using a laser, the positioning of the footprint of the laser beam in 3D space creates additional problems if the laser source itself is moving. If the laser is mounted in an aircraft, ship, or ground vehicle, then the relationship between the observer and the observed is not easy to solve. It would be even worse if the surveyor is on a moving vehicle and the object surface being surveyed is also moving. While the distance between the surveyor and object will indicate if it is getting closer or further away, there is no way of knowing whether the object is moving at the same speed or at an angle towards or away from the surveyor, or at a different speed, or possibly not moving at all.

The movement issue was long a problem with mounting a laser ranging system in an aircraft. The distance from the aircraft to a point on the ground could be accurately measured, but this distance was not relevant because there was no method of determining whether the laser was vertical or ranging off-vertical. In addition, the actual position and attitude of the aircraft in relation to a ground coordinate system was impossible to determine. The result was no one could say for certain exactly what spot on the ground was being measured.

Consequently, airborne laser technology requires two additional technologies to enable it to become a useful tool for terrain modeling and mapping. The first is inertial measurement technology and the second is global positioning systems. Both of these technologies are discussed in detail in Chapter 3.

Inertial technology uses two fairly old and well-known laws of physics. The first is that an object spinning very rapidly tends to keep its relative orientation in space; and the second is that on Earth, a rapidly spinning object typically aligns itself with gravity. This can be checked by making a toy top spin very fast on a board. The top will rotate around the true vertical, pulled by the force of gravity. If the board is tilted, the top will still stay upright. Modern, robust gyroscopes (the scientific spinning top) rotate a mass within a gimbal or cage. Recently, the mass in gyroscopes is kept in electromagnetic suspension to reduce the influence of friction. If the cage is attached and aligned to the laser head, then the angle of the laser, from vertical, can be measured.

Typical inertial navigation systems (INS) also contain accelerometers to measure the velocity of a moving object. An accelerometer may be visualized as a weight suspended on a string from the top center of a cardboard box. If the box is given a shove, the box itself moves and the weight takes a fraction of a second to catch up with the movement. Incorporating a high precision clock and a method of measuring angles, an accelerometer can measure both the speed and direction of movement. However, no accelerometers can be relied upon to remain accurate for long periods of time. There are a number of reasons for this, some of which are mechanical, and some of which relate to external influences such as gravity and rotation of the earth. Consequently, even if the precise starting point is known, once it has been moving for a while, the accelerometer tends to lose its precise location. In order to solve this problem, GPS measurement of position is used together with INS orientation of the lidar sensor.

The GPS satellites orbit the center of gravity of the Earth and transmit radio signals that are received by the GPS receivers. Using a minimum of four satellites whose precise orbits are known, it is theoretically possible to compute the position of the receiver. However, in practice it has been determined that for the very precise locations required for accurate positioning, lock on at least six GPS satellites is desirable. There are a number of other considerations that have to be addressed with the GPS technology. The GPS is relatively simple in theory (basic 3D geometry), but it is complicated by irregularities in the Earth's gravity, the shape of the Earth, the map projection to be used, as well as inconsistencies in atmospheric conditions and other phenomena such as sun spots (KP factor).

2.2.3.2 Airborne Lidar Systems

There are a variety of different systems in use around the world, some of which were developed for specific purposes and some developed for more generalized purposes. Earlier in the chapter it was mentioned that the commercial laser scanning systems utilize mostly pulsed wave lasers. Regardless of the manufacturer of laser scanning systems, all laser scanners are characterized by describing their features which are as follows: pulse energy measured in μJ, pulse duration measured in nanoseconds, pulse repetition rate (PRR) measured in kHz, field of view of scan measured in degrees, and scan rate measured in Hz. The Encyclopedia of Laser Physics and Technology defines those parameters as follows: Optical pulses are basically flashes of light, in most cases delivered in the form of laser beams. The pulse energy is simply the total energy content of a pulse. Pulse duration is most frequently defined as the Full Width at Half Maximum (FWHM) of the optical power versus time. The pulse repetition rate (or pulse repetition frequency) of a light source which periodically emits pulses is defined as the number of emitted pulses per second. For example, a 50 kHz system generates 50,000 pulses of laser energy in one second. The scan rate is the number of times per second a scanning device samples its field of view. Scan period is the period taken by a scanner to complete a scan pattern and return to its starting point. The field of view is the part of the observable world that can be sampled. Varying some of these parameters will change the effective results of the survey.

Some airborne systems have been developed specifically for very low level surveys (typically using helicopter platforms) for corridor applications, and some are for more generalized mapping. NASA has continued development of its own space-borne systems: their SLICER (Scanning Lidar Imager of Canopies by Echo Recovery) and LVIS (Laser Vegetation Imaging Sensor) are both tools for measuring the detailed canopy structure of vegetation. These units have an unusually broad footprint of about 9 meters (30 feet) for data analysis of forest cover. Commercial off-the-shelf (COTS) units, such as those offered by most lidar manufacturers, provide a middle-of-the-road "general" unit for accurate mapping of topography.

Compared to the earlier versions of laser scanning systems, which were capable of recording only one pulse at low pulse repetition rates, today's lidar systems are much more advanced. Manufacturers have developed sensors which can record simultaneously multiple echoes (currently up to 5 returns per pulse or first, second, and last of many) for each transmitted laser pulse, and reflected intensity for each echo, and recently full waveform digitization has been introduced.

A case of multiple returns from a single pulse can happen when a laser beam reaches the roof of a building. If the beam is broad enough, then it can partly hit the roof edge of a house and the rest of the laser beam travels to the ground. The return beam detector registers one reflection of the rooftop edge, and one reflection from the ground, subsequently providing two different elevations from the single pulse. The elevation computed for the rooftop is called the "first return," and the elevation computed for the ground is called the "last return." Thus, a digital surface model (DSM) and a digital terrain model (DTM) can be created. Figure 2.19 shows elevations from first-return lidar data of Baltimore, Maryland as a TIN, with variable elevations color-coded from dark blue (lowest elevations) to dark red (highest elevations).

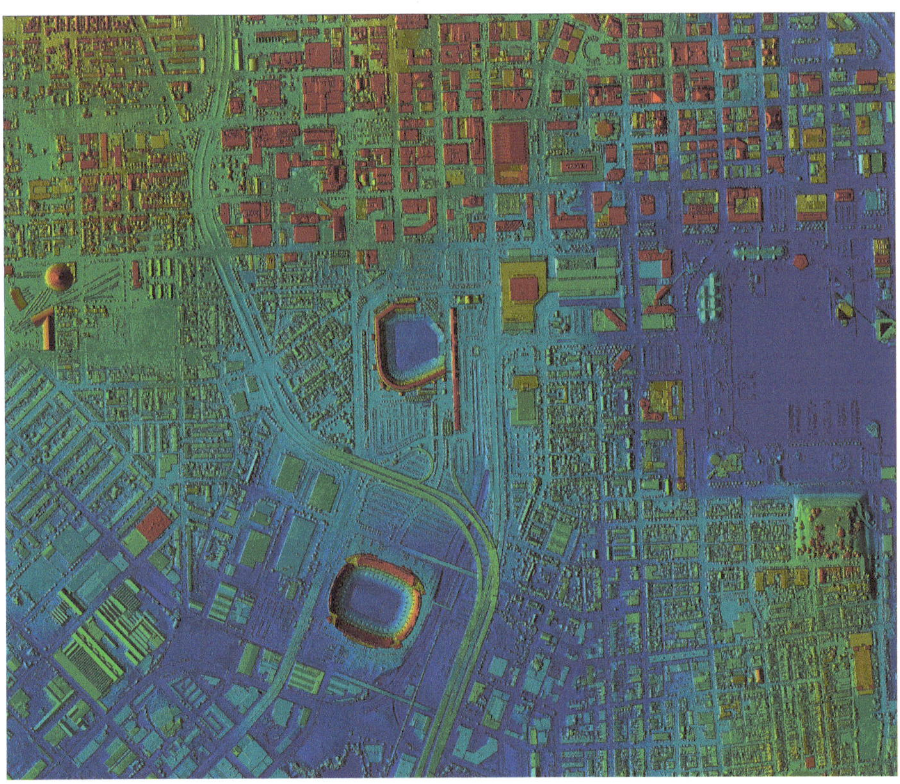

Figure 2.19 First return lidar data of Baltimore Maryland.

In a forest, the situation is more complicated. When the laser beam from a multi-return system hits a tree, then the first return is usually assumed to arrive from near the top of the canopy. The last return, perhaps the eighth, is assumed to come from the ground. The intermediate returns, i.e., perhaps 2nd, 3rd, or 4th, are expected to be caused by tree branches and understory vegetation between the top of the canopy and the ground. However, sometimes, the vegetation is so thick that the laser beam never penetrates through to the ground. Typically the rule of thumb is: if a person walking through a forest cannot see the sky above, then the lidar system probably cannot "see" the ground beneath the canopy (see Chapter 10.2).

Full waveform digitization offers even more choices in that it allows for many multiple returns with short separation (possibly at every 15 cm of pulse travel) to be collected from a single laser shot. The down side is the more returns collected the more data there are in storage, and more complex data processing is required (typically 32 times more processing). See Chapter 2.4 for an in-depth review of full waveform capture and applications.

In addition, most laser scanning systems are capable of storing the information about the intensity of each detected echo. Intensity capture of the return pulse, either through waveform digitization or return pulse peak capture, is becoming an increasingly common practice. This information is valuable for postprocessing and classification purposes, and considerable research and study is taking place to make more use of these data (see Chapter 8.1 on Lidargrammetry). Regardless, by reviewing the intensity returns it is possible to distinguish between different objects, general vegetation cover, and land use. As a general rule, objects with high reflectivity of visual light, such as a metal roof, show a higher return energy than objects such as a newly paved, black-tarred roadway. Continued development and improvement of software are needed in order to transfer the knowledge about the intensity into more useful information about different types of surfaces. Also, this information can be valuable for automated feature extraction. Figures 2.20 and 2.21 provide two examples of lidar intensity return images. However, it should be remembered as stated previously, lidar systems are monochromatic i.e., they record reflected energy from a single wavelength, compared to photography which collects reflective data from a wide variety of wavelengths.

Figure 2.20 Lidar intensity return image of Baltimore , Maryland.

For reference, the characteristics of "typical" commercial lidar instruments in operation today are included in Table 2.4. It should be noted that the new capabilities and functionality are being introduced on a regular basis so "typical" characteristics are changing over time. Also individual lidar systems may be developed for specific purposes, so not all lidar units have the range of characteristics noted.

(a)

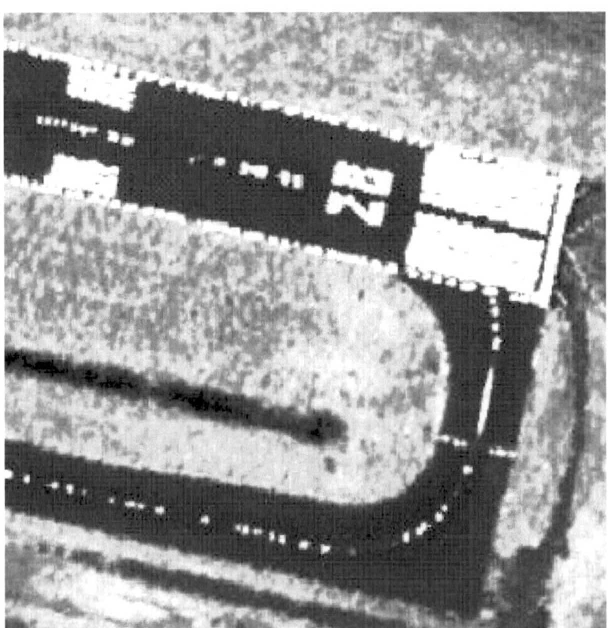

(b)

Figure 2.21 (a) example of intensity data over an airport, and (b) enlargement for a portion of the runway.

Table 2.4 Characteristics of typical commercial lidar systems.

Specification	Typical Value
Wavelength	1064 nm
Pulse Repetition Rate	100 to 500kHz
Pulse Energy	100 μJ
Pulse Width	10 ns
Beam Divergence	0.25 to 2 milliradian
Scan Angle	40° typical (75° max)
Scan Rate	100 to 400 Hz
Scan Pattern	Zig-zag, elliptical, sinusoidal, triangle, and rectangle
GPS Frequency	0.1 to 10 times per second
INS Frequency	50 to 200 times per second
Operating Altitude	80 m to 5,000 m
Footprint	0.25 to 2 m (from 1,000 m AGL)
Multiple Elevation Capture	1 to 7; or 1st, 2nd, last of many; or first and last
Grid Spacing	0.2 to 2 m
Vertical RMSEz	5+ cm
Horizontal RMSEr	10 to 100 cm
Postprocessing Software	Proprietary

2.2.3.3 State of the Art

The earliest versions of laser remote sensing systems were more scientific and prototype than commercial instruments useful for topographic purposes, but in the 1980s early commercial endeavors involved the use of laser profilers to generate a single line profile of the ground beneath an aircraft. However, it wasn't until the 1990s that the development of the enabling technologies reached a degree of sophistication for accurate airborne use. The 1990s also saw the development of laser scanners (as opposed to profilers) and the development of the current lidar instruments.

Two predominant companies were involved in the initial development of commercial systems: Azimuth Corporation of Westford, Massachusetts (now owned by Leica Geosystems), and Optech, Incorporated of Toronto, Canada. Also, RIEGL Laser Measurement Systems, Ltd. from Austria entered the commercial lidar systems market emphasizing portability and waveform capability. Today, in the number of current operating units, the market is dominated by these three firms.

Azimuth Corporation developed laser rangefinders as subsystems for profilers and military target tracking starting in the late-1970s and through the 1980s. Azimuth supplied rangefinder subsystems to several early entrants into the laser mapping field. They subsequently delivered a complete Aeroscan system in 1998, a 15 kHz system capable of flying up to 6,000 meters above ground. Leica has subsequently delivered progressively more advanced systems of increasing pulse rates. As of 2012, Leica's state-of-the-art lidar system is the ALS70-HP, shown at Figure 2.22, with a maximum pulse repetition rate of 500,000 pulses per second.

In the early 1980s, Optech, Inc. developed airborne profilometers based on a series of compact, rugged laser rangefinders for industrial applications. Optech, Inc. was also involved in the development of several airborne laser bathymeters: the LARSEN 500 in 1984 with Terra Surveys, Ltd. (now Terra Remote Sensing Inc.) and the Canadian Hydrographic Service, FLASH in1988, and ALARMS in 1988. They next built the prototype ALTM (Airborne Laser Terrain Mapper) in 1993, which was a 2 kHz instrument, and delivered the system to TopoScan GmbH in Germany. A re-packaged commercial design was offered for sale in 1995 as the 1020 model. Optech, Inc. delivered an enhanced model 1020 (5 kHz) in 1996 and upgraded the earlier 2 kHz models. A custom 1025 (25 kHz) system was delivered in Japan in 1997. Optech subsequently produced an ALTM 1210 system (10 kHz) system for commercial sales in 1998 and a new 1225 enhanced model (25 kHz) in 1999. In 2000, they delivered a 33 kHz lidar system followed by a 50 kHz system and most recently the ALTM Pegasus HD500, a 500 kHz system (see Figure 2.23). Optech. Inc. is also the only commercial source of continued development of bathymetric lidar systems.

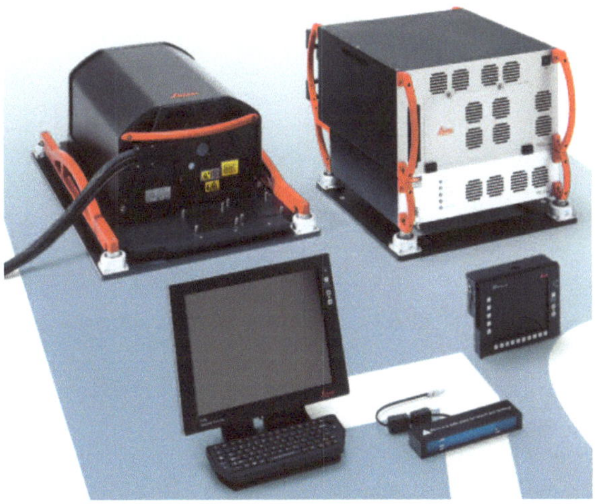

Figure 2.22 Leica ALSD70 lidar system.

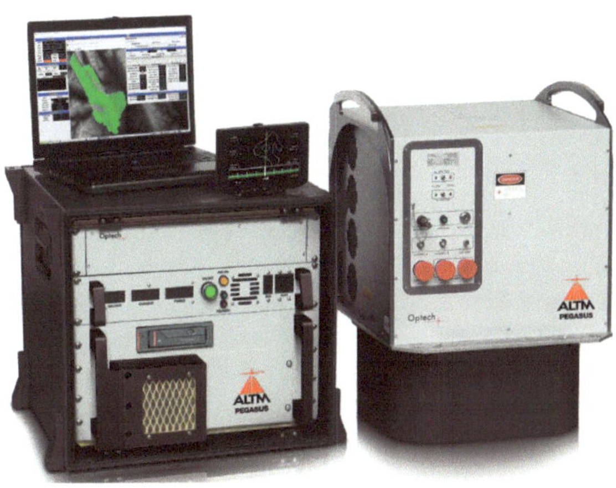

Figure 2.23 Optech, Inc. ALTM Pegasus HD500 lidar system.

RIEGL Laser Measurement Systems, GmbH (Austria) and RIEGL USA has been in the field of the development of laser radar systems for more than two decades. Currently, the company supplies a wide variety of laser scanners and supporting software for specific applications. RIEGL has also emphasized compact portability and the collection and processing of waveform data with their lidar systems. Their current latest release is the LMS-Q680i which offers 400 kHz performance and full waveform analysis see Figure 2.24.

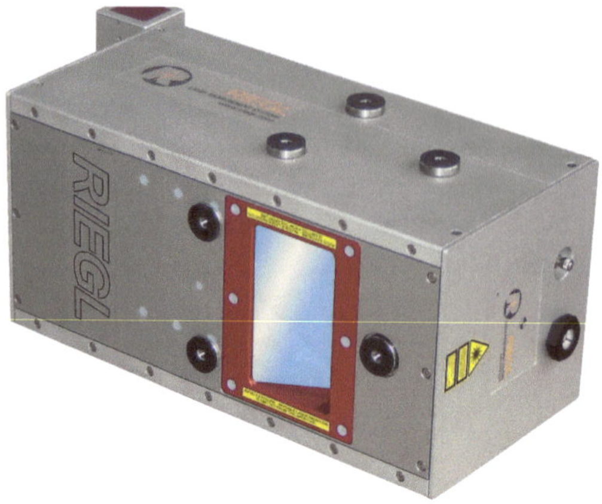

Figure 2.24 RIEGL LMS-Q680i Scanner.

SAAB Survey Systems AB in Jönköping (Sweden) delivered a commercial version of the TopEye system in 1996. This system was a result of a joint development of a hydrographic airborne laser scanning system, the hardware of which was originally designed by Saab Dynamics (Sweden) and Optech, Inc. Later, TopEye AB was spun off from Saab. It continues to offer a helicopter-based system, although a fixed-wing version of TopEye is also available and the company is now part of the Blom group.

About the same period of time, TopoSys GmbH (Germany) introduced the TopoSys-1 airborne laser scanning system. This type of ALS system is different from traditional ALS systems because it is based on a glass fiber technology. Today, it has been replaced by the latest development called the Falcon-III. The TopoSys system has also a polygon scanner, which employs the wide angle full wave form digitization of a laser scanner from RIEGL GmbH. Since the beginning of the 21st century, GeoLas Consulting (Germany) together with IGI GmbH (Germany) has been providing its own commercial airborne laser scanning system called LiteMapper.

There are a number of other companies and organizations which have developed and built customized laser scanning systems for supporting specific applications. One of the earliest was FliMap developed by John Chance Land Surveys, Inc. (USA). FliMap is a rotary-wing laser scanning system used for the corridor mapping purposes where a very high point density on the ground and a very high precision of a terrain model is required. Terrapoint, Terra Remote Sensing, Aeroscan and a number of others have all developed in-house systems. Another example is ScaLARS, which has been developed in the Institute for Navigation, University of Stuttgart (Germany). ScaLARS uses the Palmer scanning pattern (see Figure 2.17).

The advantage of in-house lidar development (compared to commercial off-the-shelf, manufactured systems) is a group can decide on its research program or market niche and develop a system that will meet that specifically by combining the type and quality of laser, inertial system and GPS to meet its specific needs. The disadvantage is a high degree of knowledge in optical, optical-mechanical, and electrical systems is required to engineer the disparate parts into a working whole, and then develop software to ensure all of the systems are working as a unit, and able to process the data efficiently and accurately. While considerable monetary savings can be achieved by buying all of the hardware (compared to a manufactured unit), development costs of making all the components work together can be extremely high.

While specific objectives are usually the reason for in-house development of airborne lidar systems, the major manufacturers are constantly expanding the capabilities of their systems and in reality almost all airborne lidar systems are, in effect, custom built.

There are also existing 3D FLASH laser imaging systems which are capable of producing 3D imagery of the scene (see Chapter 2.5), and GIEGER Mode lidar (see Chapter 2.6).

ACKNOWLEDGMENT

This subchapter is an updated version of a portion of Chapter 7 – Topographic and Terrestrial Lidar for the *Digital Elevation Model Technologies and Applications: The DEM Users Manual*, Second edition, edited by David F. Maune, 2007, pp. 119–218.

REFERENCES

Bachman, C.G., 1979. *Laser Radar Systems and Technologies*, Artech House, Inc.

Beiser, L., 1992. *Laser Scanning Notebook*, SPIE.

Farrell, J.A., and M. Barth, 1999. *The Global Positioning System and Inertial Navigation*, McGraw-Hill.

Jelalian, A.V., 1992. *Laser Radar Systems,* Artech House.

Jenkins, T.E., 1987. *Optical Sensing Techniques and Signal Processing*, Prentice Hall International.

Karim, A., 1990. *Electro-optical Devices and Systems*, PWS-KENT Publishing Company.

Leica Geosystems, 2012. URL: www.leica-geosystems.com (last date accessed: 17 May 2012).

Marshall, G.F., 1985. *Laser Beam Scanning* (G.F. Marshall, editor), Marcel Dekker, Inc.

Measures, R.M., 1984. *Laser Remote Sensing*, Krieger Publishing Company.

Merhav, S., 1996. *Aerospace Sensor Systems and Applications*, Springer-Verlag.

Optech, Inc., 2012. www.optech.com (last date accessed: 17 May 2012).

Renslow, M.S., Development of a bare ground DEM and canopy layer in NW forestlands using high performance lidar, URL: http://gis.esri.com/library/userconf/proc00/professional/papers/PAP808/p808.htm (last date accessed: 17 May 2012).

RIEGL Laser Measurement Systems, 2012. URL: www.riegl.com (last date accessed: 17 May 2012).

Samberg, A., 1997. The design of an airborne laser 3D imaging instrument, Proceedings of the ACSM/ASPRS Annual Convention and Exposition, 07–10 April, Seattle, Washington, ASPRS, ACSM and RTI, pp. 359–363.

Wehr, A., and U. Lohr, 1999. Airborne laser scanning - An introduction and overview, *ISPRS Journal of Photogrammetry and Remote Sensing*, 54:68–82.

2.3 FIBER OPTIC LIDAR SYSTEMS: TOPOSYS FALCON SERIES

TopoSys GmbH was founded in the spring of 1995 as a spin-off of Dornier GmbH, an air- and spacecraft manufacturer of worldwide reputation. When it was founded, TopoSys took over the technology of the Dornier's Airborne Fiber Optic Scanner and developed it further for civilian applications. This Dornier fiber optic scanner was used for military purposes, namely for obstacle avoidance and autonomous vehicle guidance.

TopoSys GmbH was acquired by Trimble Navigation in November 2008 and is now the Airborne Data Acquisition Solutions section, an integral part of Trimble's GeoSpatial Division, Engineering and Construction Sector.

One year after its founding, TopoSys GmbH presented its first airborne laser scanning system, the "Falcon I", also known as "TopoSys Sensor I". In 2000 TopoSys GmbH released the second member of the Falcon series family, the Falcon II, which was a significant contribution for the airborne laser scanning world.

2.3.1 Falcon I

The Falcon I ALS was based on fiber optic technology, becoming the first of its kind worldwide and including a complete software package for laser data pre- and post-processing called "TopPIT" (TopoSys Processing and Imaging Tools).

The Falcon I was solely designed to be used by TopoSys for laser data acquisition and not for commercialization, and therefore a Falcon I equipment was never sold.

2.3.1.1 Falcon I Technology and Specifications

In its time, the Falcon I was a revolutionary fiber optic system with a blasting pulse rate and a very fast scan frequency, especially when considering the year in which it was put into operation, i.e., 1996.

Although the Falcon I system could not simultaneously record first and last echo data, it was the first airborne laser scanner sensor to offer the recording of last echo data. Before a mission was flown, the type of echo recording had to be selected by the flight operator. Table 2.5 summarizes the main characteristics of the Falcon I sensor:

Table 2.5 Falcon I System Parameters (Source: TopoSys GmbH, 1998).

Parameters	Description
Sensor Type	Pulsed Fiber Scanner
Range (m)	<1,000 m
Wavelength (nm)	1540 nm
Pulse length (ns)	5 ns
Scan Frequency (Hz)	653 Hz
Pulse Repetition Rate (Hz)	83,000 Hz
Resolution of distance Measurements (m)	0.06 m
Field of View (°)	14° (±7°)
Swath (at maximum range) (m)	220 m
Number of Fibers	128 (one as reference)
Data Recording	First or Last Echo
Additional Sensors	SVHS Video camera or first generation of RGB line scanner
Laser Peak Power	2.5kW

The laser scanner fiber optic system was constituted by an array of 128 optical fibers with identical transmitting and receiving optics technology. These fibers were arranged in a linear pattern at the transmitter's end, and at a circular pattern at the receiver's end. The light of the transmitting fibers was sent to the homonymous receiving fiber by means of a small high-speed rotating mirror, as seen in Figure 2.25 (Artés, 2006; Schnadt and Katzenbeißer, 2004).

Each of the laser scanner fibers is scanned in an ordered fashion and automatically synchronized in order to obtain an extremely accurate laser beam deflection with an accompanying small viewing angle. The returning pulse will be directly read by its corresponding receiving fiber guaranteeing an absolute precision, which is in fact a feature derived from the stable orientation of the fiber array (Katzenbeißer, 2004).

A single fiber is directly fed from the transmitter to the receiver acting then as a dedicated reference. If any inconsistency arises within the measurement electronics, then these can be detected and immediately corrected, thus maintaining the data accuracy in a harmonized fashion (Katzenbeißer, 2004; Wiechert, 2004).

Although the Falcon I system rotating mirrors are very small components, no moving parts associated with the beam deflection exist. This reflected itself on the extreme scanning speeds of the Falcon I system, which could go as far as 653 Hz. Another advantage of the Falcon systems is the fact that the laser pulse rate is not dependent on the viewing angle and flight altitude, which allowed maintaining a constant pulse rate at all flight altitudes (Wiechert, 2004).

Schnadt and Katzenbeißer (2004) refer that the rigidity of the Falcon systems' fiber optics component is not prone to beam deflection errors due to mechanics, acceleration, or wear, and does not require constant calibration procedures. Figure 2.25 shows the schematics of the Falcon I fiber optics mechanism.

The Falcon I system did not have a roll angle stabilizer neither a working swing mode to compensate for the aircraft movement. This resulted on a linear scanning pattern on the ground, as seen in Figure 2.26.

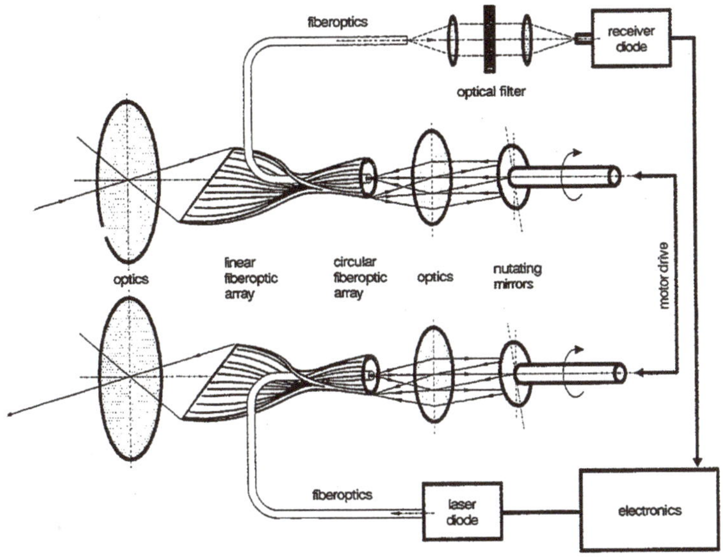

Figure 2.25 The full working fiber optics mechanism of the Falcon I system (Source: TopoSys GmbH, 1998).

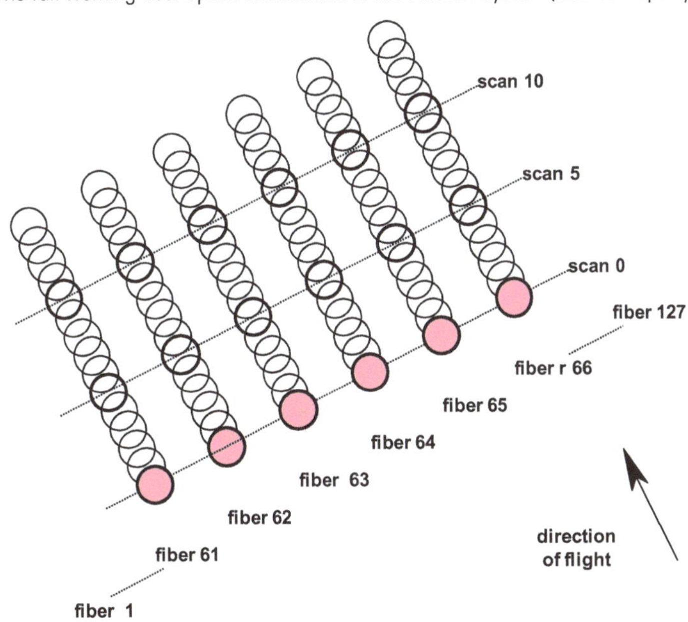

Figure 2.26 The Falcon I system linear scanning pattern (Source: TopoSys GmbH, 1998).

2.3.1.2 Falcon I Components

The Falcon I system was installed in a support plate along with a line scanner camera and a LINS unit (Laser Inertial Measurement system). Along with these instruments, a computer rack (PC-Rack) was used for the laser data recording as well as for the navigation data recording (GPS and LINS). A second rack (UPS/RGB-Rack), served as aircraft power input connector and distributed the power through the whole system. This rack also had an inbuilt Uninterruptible Power Supply (UPS), it incorporated a GPS receiver and controlled the imagery data recording. In order to control the various flight mission parameters, a double-screen flight operator console was as well available.

The first generation of the Falcon I equipment had a video camera for imagery data recording, which was later substituted by a line scanner camera in the second generation of the Falcon I system.

Figure 2.27 illustrates the main components of the Falcon I installed in a helicopter.

Figure 2.28 shows the mounting of the flight operator panel (C) on the PC-Rack (A).

The PC-Rack can be seen in Figure 2.29. This computer rack upper tray was used for the laser data recording, whilst the lower tray served for the navigation data recording.

Figure 2.30 shows the Falcon I UPS/RGB-Rack. Note the numerical screen which served as an indicator for imagery acquisition, the various fuses, cabling, and command buttons of the system.

The flight operator console can be seen in Figure 2.31. The upper screen was used to control the laser scanner, while the lower screen was used for navigation guidance.

Figure 2.32 shows the upper display of the flight operator console. Metadata such as number of recorded flight lines, free disk space, necessary time for flight line scanning, among others, were provided by this display.

Figure 2.33 shows the lower display from the flight operator console where the various parameters concerning navigation and other diagnostics are presented.

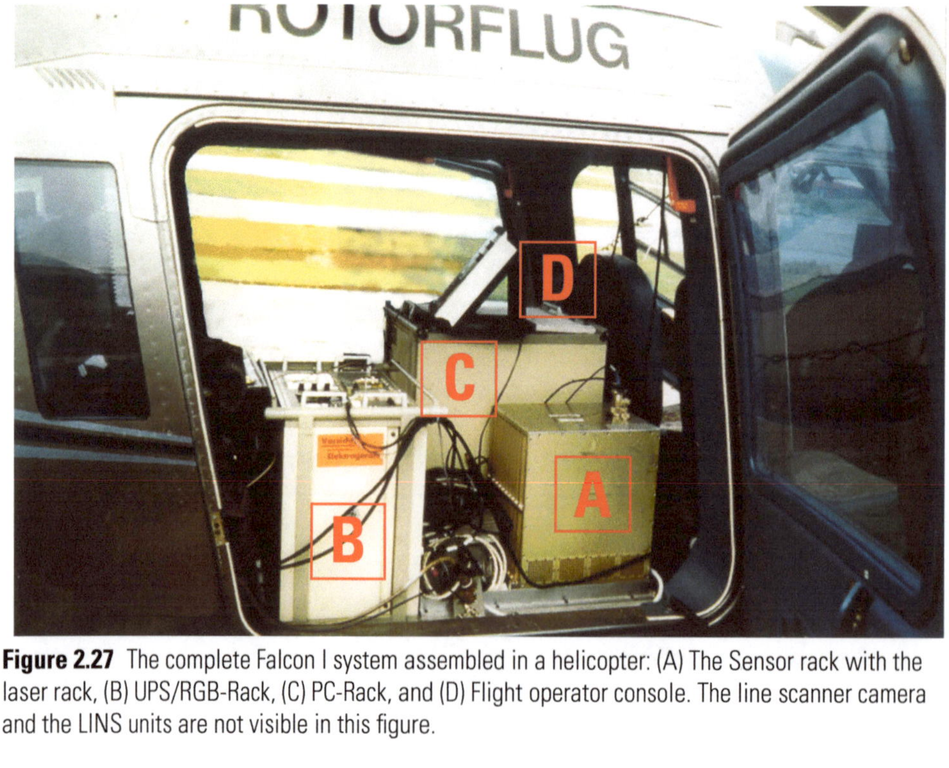

Figure 2.27 The complete Falcon I system assembled in a helicopter: (A) The Sensor rack with the laser rack, (B) UPS/RGB-Rack, (C) PC-Rack, and (D) Flight operator console. The line scanner camera and the LINS units are not visible in this figure.

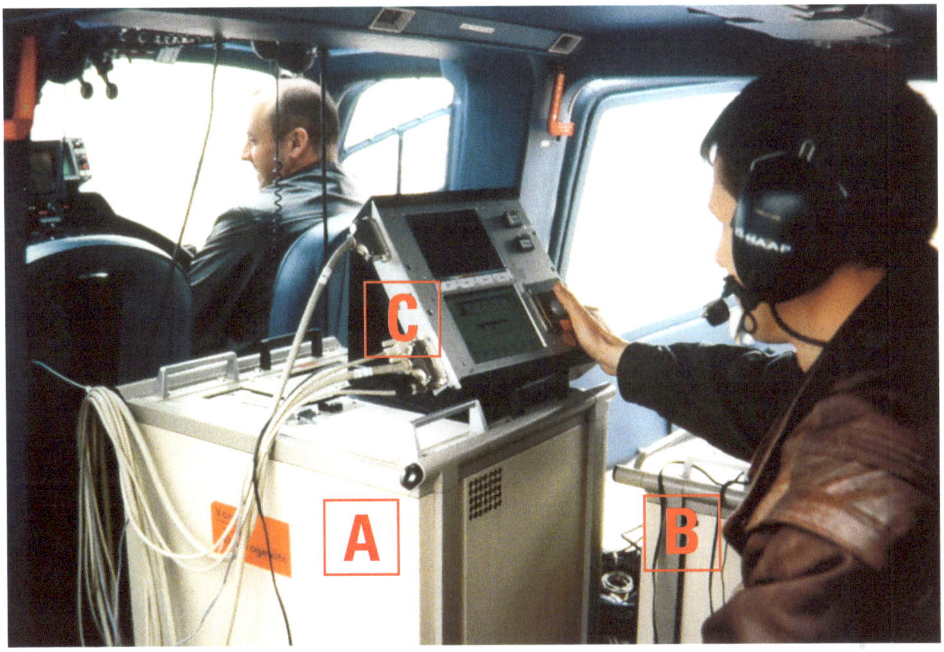

Figure 2.28 (A) The PC-Rack, (B) UPS/RGB-Rack, and (C) Flight operator console.

Figure 2.29 The Falcon I PC-Rack.

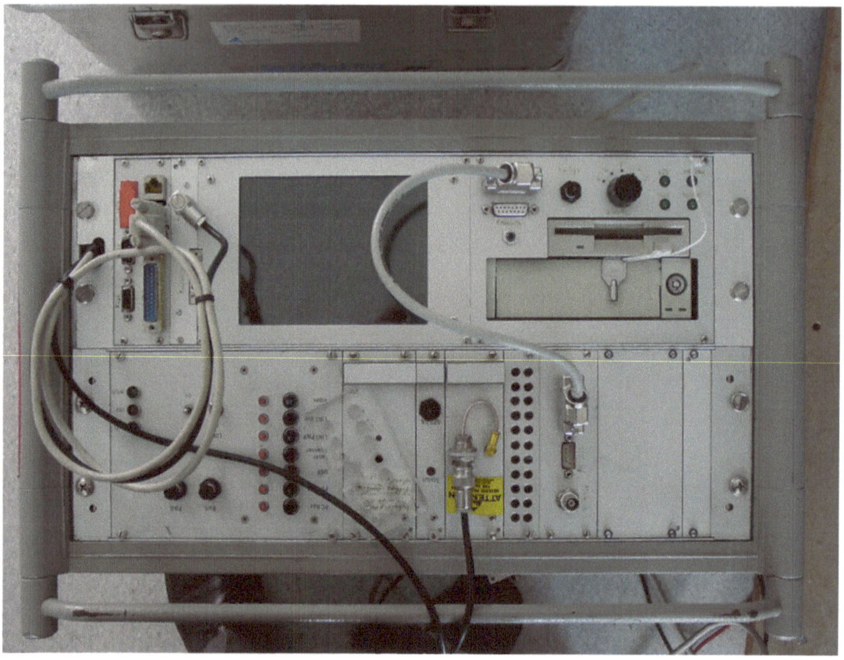

Figure 2.30 The Falcon I UPS/RGB Rack.

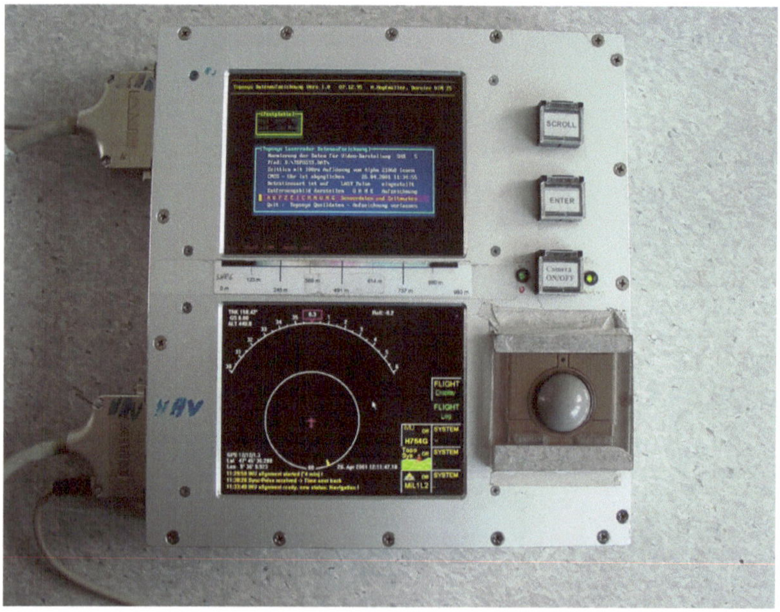

Figure 2.31 Falcon I system flight operator console.

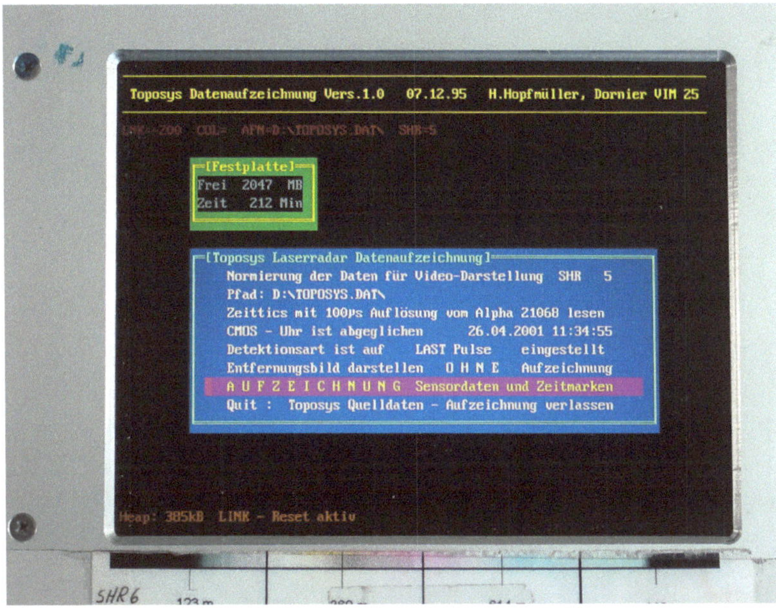

Figure 2.32 Upper display of the flight operator console where the laser scanner parameters were presented.

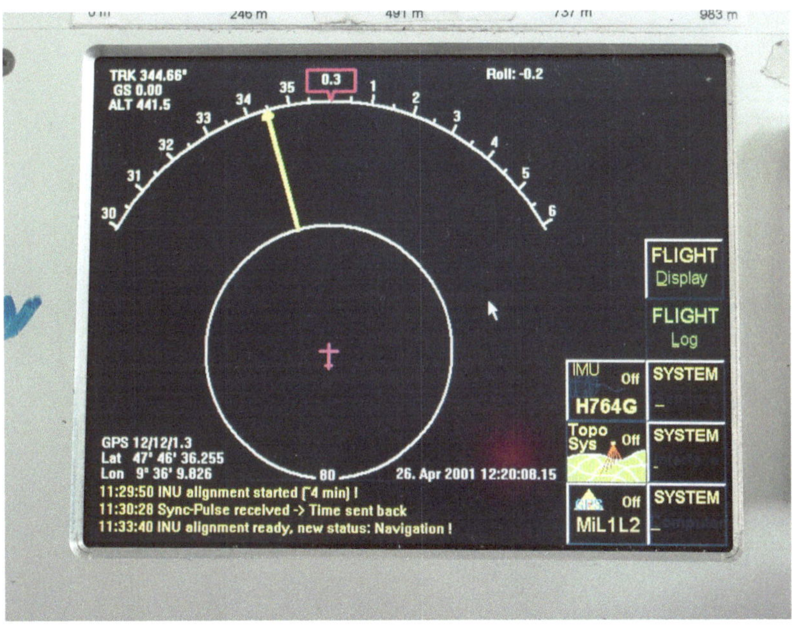

Figure 2.33 Lower display of the flight operator console where the navigation data parameters were presented.

2.3.2 Falcon II

With all the acquired knowledge and testing done with the Falcon I, in 2000 TopoSys developed the Falcon II sensor. This new sensor introduced many innovations, although the technology behind its laser scanner was still based on the Falcon I equipment design.

The Falcon II equipment brought a new set of improved features when compared with the Falcon I system. The Falcon II system is a fully integrated ALS, with only a scanner rack, a computer rack, and a flight operator console.

The Falcon II had three generations, with the first generation (G1) being launched in 2000, the second generation (G2) being launched in 2004 and the third generation (G3 or Falcon IId) in 2007. The improvements between the generations were mostly related with laser optics improvements and mechanics control.

2.3.2.1 Falcon II Technology and Specifications

The Falcon II equipment was capable of simultaneously recording first and last echo data and could operate at higher flight altitudes (Löffler, 2003); Table 2.6 shows its main characteristics.

As with the Falcon I equipment, the laser scanner fiber optic system is constituted by an array of 128 optical fibers with identical transmitting and receiving optics technology. These fibers are as well arranged in a linear pattern at the transmitter's end, and at a circular pattern at the receiver's end (Artés, 2006; Schnadt and Katzenbeißer 2004).

Upon its release, the Falcon II set again new quality and technology standards in the airborne laser scanning world. Its laser scanner was able to record up to nine return echoes, and the pulse repetition rates as well as the scan frequency speeds are still blasting values. The fact that it incorporates a line scanner camera in such a small space is as well a major advantage of this equipment.

Table 2.6 Falcon II system Parameters (Source: Lohr, 2003; Hyyppä et al., 2004 and 2005).

Parameters	Description
Sensor Type	Pulsed Fiber Scanner
Range (m)	<1,600 m
Wavelength (nm)	1560 nm
Pulse length (ns)	5 ns
Scan Frequency (Hz)	653 Hz
Pulse Repetition Rate (Hz)	83,000 Hz
Resolution of distance Measurements (m)	G1 and G2: 0.02 m G3: 0.007 m
Field of View (°)	14.5°
Swath (at maximum range) (m)	390 m
Number of Fibers	128 (one as reference)
Laser Peak Power	10 kW
Data Recording	First and Last Echo G3: Up to 9 Echoes
Additional Sensors	Integrated RGB/ CIR Line Scanner Camera
Additional features	G2. Swing Mode G2: Roll compensation G3: Enhanced data storage

Another striking benefit inherent to the Falcon systems is the fact that the pulse repetition rate is constant no matter the flight altitude (<1,600 m AGL), neither the required number of echoes (Katzenbeißer and Lohr, 2000). A scheme showing the Falcon II fiber scanner technology can be seen in Figure 2.34.

As previously explained, the Falcon series does not use a polygon or swiveling mirror to deflect the laser beam, but an array of tightly coupled glass fibers forming a linear array, as shown in Figure 2.35. Subsequently, laser pulses are transmitted through the neighboring fiber (TopoSys GbmH, 2004c).

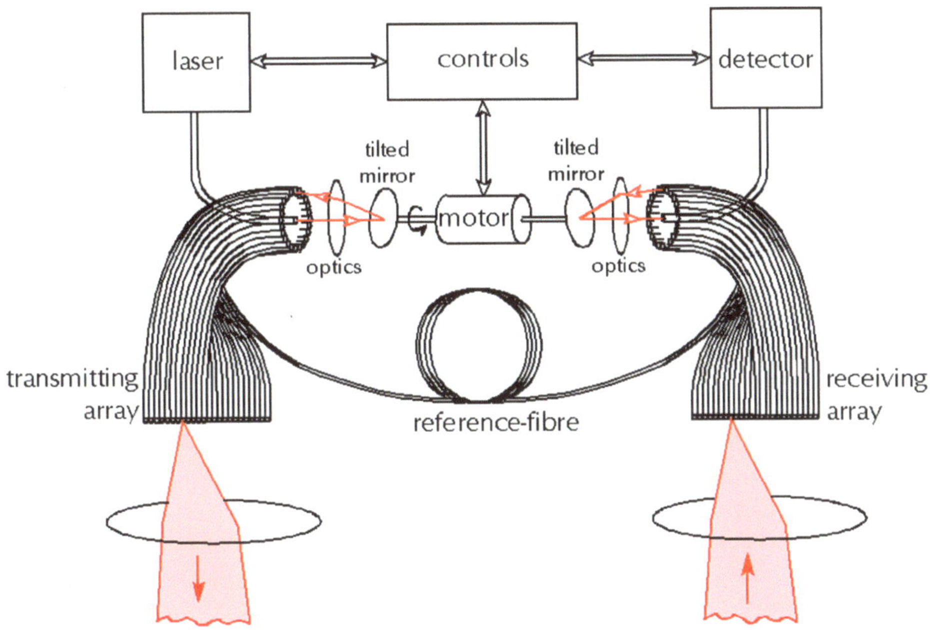

Figure 2.34 The full working fiber optics mechanism of the Falcon II system (Source: Schnadt and Katzenbeißer, 2004).

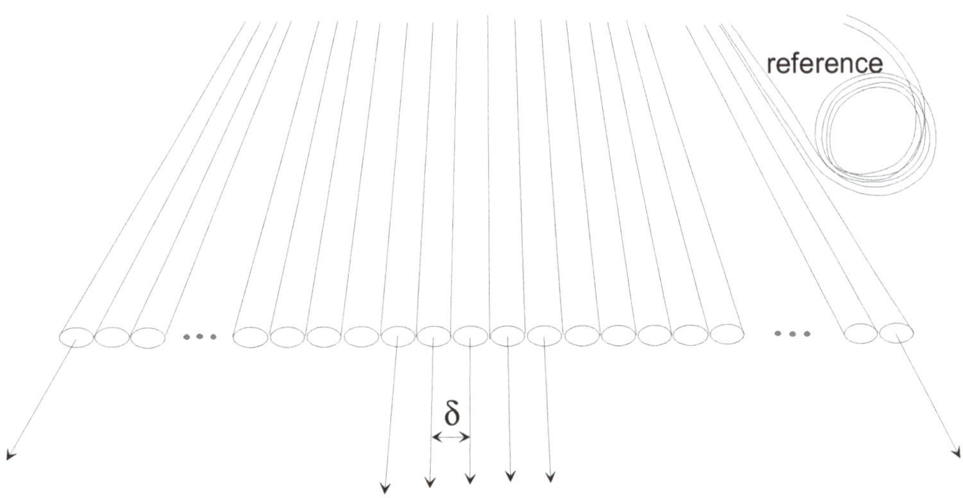

Figure 2.35 Linear array disposition of the Falcon series fiber laser. δ represents the angle between two fibers (Source: TopoSys GmbH, 2004c).

Using a total of 127 fibers (plus one as a reference), the result is a viewing angle of 14.5 degrees. As shown in Figure 2.35, the last laser pulse at the end of the array (a scan) is always coupled into a reference fiber (over 500 m long). The reference fiber allows a continuous control and correction ("calibration") of the distance measurements, and this fiber is directly fed from the circular transmitting side to the receiving side. This fixed arrangement of glass fibers is a stable configuration assuring a high repetition accuracy of the laser beam deflection, and assures that an immediate detection and posterior correction of variations in the measurement electronics for each scan line are automatically conducted (Schnadt and Katzenbeißer, 2004; TopoSys GmbH, 2004b and 2004c).

If at any time a misalignment of the tilted mirrors would happen, then this would only lead to a loss of energy, but no influence on the precision of the distance measurements would be registered.

The footprints of all beams of a scan give a straight line of illuminated circles (Figure 2.36). The distance between neighboring illuminated circles depends on the survey altitude above ground level. Adjacent scans form parallel lines and a non-complicated zigzag pattern with an unequal measurement density (TopoSys GmbH, 2004c).

Figure 2.37 explains the transmitting array mechanism from the optic fibers, passing through the optics' mirrors and the way the laser beam is diverted into the ground. The arrangement between the fibers is fixed and does not vary over time, thus providing a stable and repeatable pattern on the ground (TopoSys GmbH, 2004c). The resulting scan pattern is also shown in Figure 2.37.

It is obvious that for the fiber scanner the swath width is invariable as the fiber array is fixed and provides together with the focusing optics a constant field of view, T. For the Falcon II sensors' scan width is approximately one quarter of the flight altitude. The scan rate is 653 Hz providing a scan distance on ground of about 0.1 m along track.

The distribution of measurements is stable over the complete swath and does not vary over time. Figure 2.38 shows a small section out of the pattern indicating that the spots of subsequent scans overlap in flight direction. Measured distances in adjacent swaths should not vary widely (unless there is an edge). This over-sampling is used to prove the reliability of a measurement and to detect and discard erroneous measurements. Therefore, the distance measurements do not become more precise, but the overall results are much more reliable than fairly spaced single measurements as provided by other scan methods (TopoSys GmbH, 2004a and 2004c).

Roll angle compensation technology was integrated into the Falcon II G2 in October 2004. The main advantage is related with the fact that the whole system is not being rotated, but just the special stabilizing plate which holds the IMU, the RGB/NIR line scanner, and the fiber scanner, as seen in Figure 2.39. These three units are therefore collectively compensated. And as the internal orientation of these sub-units is invariant and remains constant even when the sensor system is integrated into another aircraft, only one single calibration is needed to determine the relationship of these three components to each other. This lead to real economical advantages as the Falcon series only required a yearly calibration to ensure the system well-being.

The following technical data is correspondent to the Falcon II roll angle compensation system:

- Compensation angle: ±6°
- Angular rate: 20° per second
- Angle error: <0.5°
- No degradation of viewing angle, pulse rate (83 kHz), echo count (up to nine) or echo separation (approximately 1.0 m)

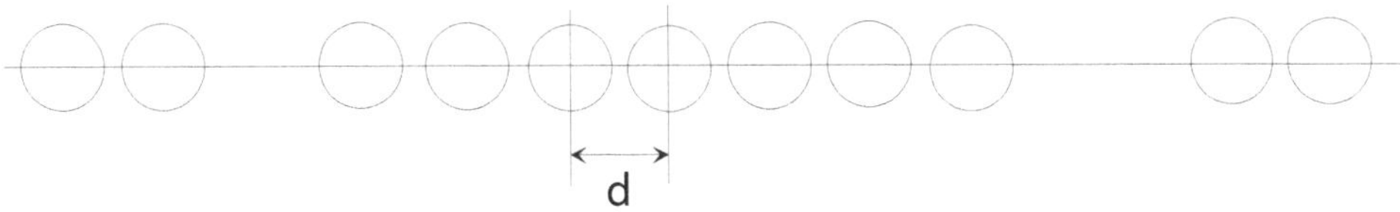

Figure 2.36 The laser beam footprint *d* is the distance between neighboring illuminated circles, and is dependent on the flight altitude. Source: TopoSys GmbH 2004c.

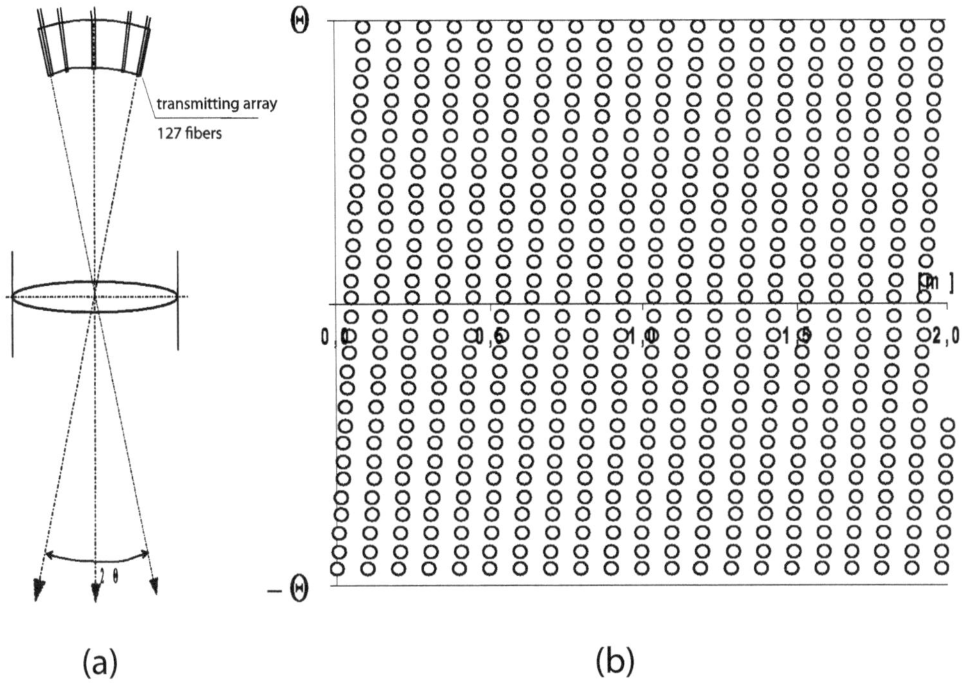

Figure 2.37 (a) Transmitting array mechanism of the Falcon II equipment, and (b) Resulting scan pattern (Source: TopoSys GmbH, 2004a and 2004b).

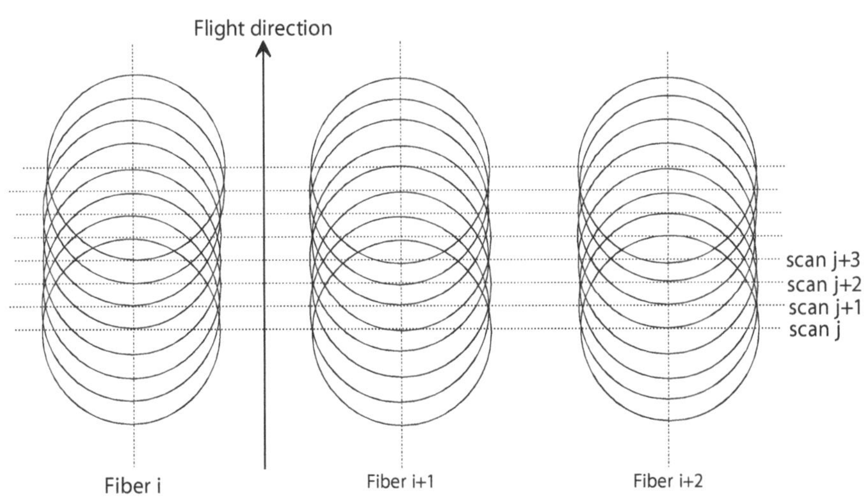

Figure 2.38 The Falcon II scanning pattern showing the subsequent scans overlap in flight direction (Source: TopoSys GmbH, 2004c).

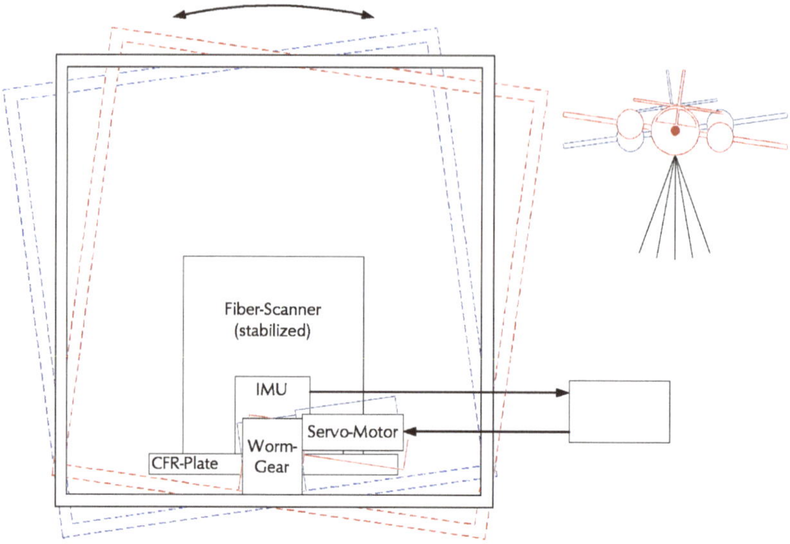

Figure 2.39 The roll angle compensation system of the Falcon II (Source: Source: TopoSys GmbH, 2004d).

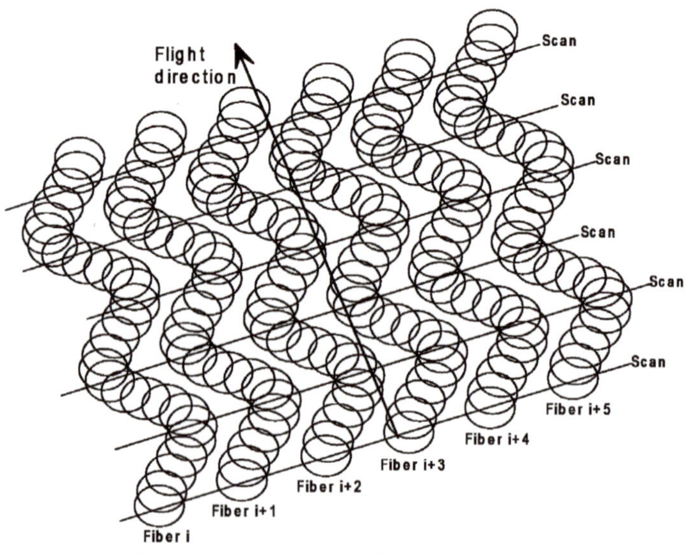

Figure 2.40 Improved scanning pattern obtained by using the Falcon II swing mode (Source: TopoSys GmbH, 2004c).

The Falcon II G2 and G3 system possess roll angle compensation as well as an incorporated swing mode function. With the combination of the swing mode and of the glass fiber technology, the distance of measurement spots on ground is regular in across and along-track direction. In the direction of flight, the scan lines slightly swing, which increases the overall distribution of measurement spots by generating regular wave patterns on the ground (Schnadt and Katzenbeißer, 2004; TopoSys GmbH, 2004c). With the addition of the swing mode, the gap between the fibers is closed by an additional small beam deflection, while the high reliability and accuracy is always kept. This results in a dense and homogeneous scanning pattern in the ground. This swing mode movement can be seen in Figure 2.40.

The effect of the Falcon II swing mode can be seen in Figure 2.41, where the edges of buildings and small objects oriented in flight direction are being precisely detected due to the new scan pattern. The gap between the fibers is closed while the high reliability is constantly kept (TopoSys, 2004a and 2004b).

The Falcon equipment always had the advantage of being built with a very small number of moving parts, which resulted in reduced need for calibrations and minimal risk of mechanical wear and failure.

The small viewing angle of the Falcon II leads to an almost absence of shadowing in the data, as well as it maintains a high quality standard at the edge of the flown stripes. This aspect is also productive-wise as less strips with higher overlap are required for a normal mission.

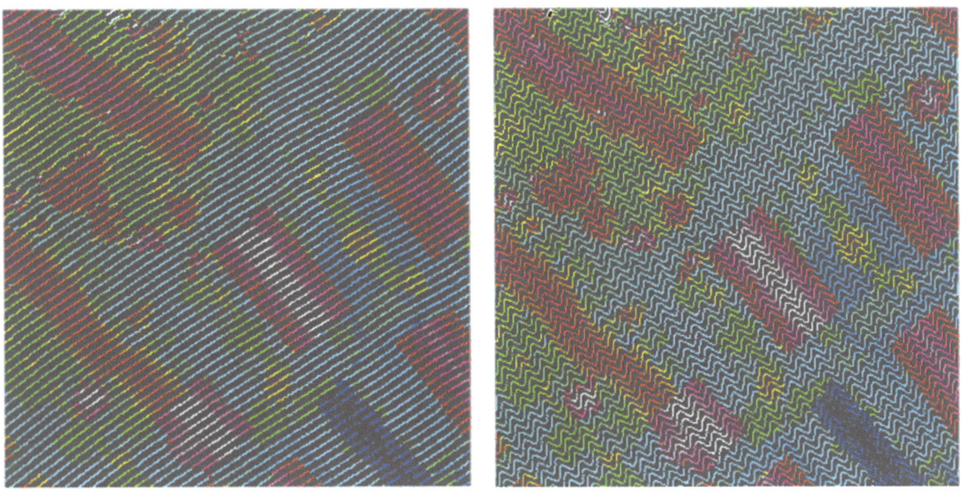

(a) **(b)**

Figure 2.41 (a) Absence of the swing mode in the Falcon II, and (b) Swing mode activated on Falcon II.

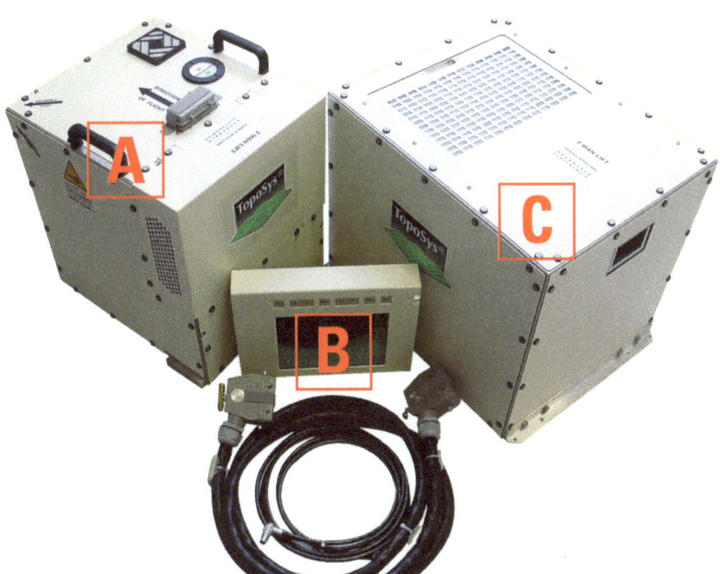

Figure 2.42 (A) Sensor Rack, (B) Flight operator console, (C) Computer Rack.

2.3.2.2 Falcon II Components

The Falcon II system is composed of the following equipment/parts, and the whole system can be seen in Figure 2.43:

- A sensor rack, where the laser scanner and a line scanner camera are installed together with the IMU.
- A computer rack for laser, imagery and navigation data recording, with a power supply unit, a Uninterruptible Power Supply (UPS) and with the necessary navigation instruments.
- A flight operator control panel.

The sensor rack possesses all the electronics concerning the laser scanner and the RGB/CIR line scanner. In this sensor rack, the laser powering unit and the fiber laser are installed along with an IMU and the line scanner camera. All Sensors are mounted inside the Sensor Rack on a CFRP (Carbon Fiber Ruggedized Plate) plate, which confers the necessary stability to the electronic components to handle any aircraft harder movements. A simple scheme showing the mounting of the Falcon II internal hardware devices is show in Figure 2.43.

The Falcon II computer rack incorporates the navigation system, plus the central mainframe computer and the necessary structure for the laser and imagery data recording. A power supply system and a UPS unit are as well a part of the computer rack.

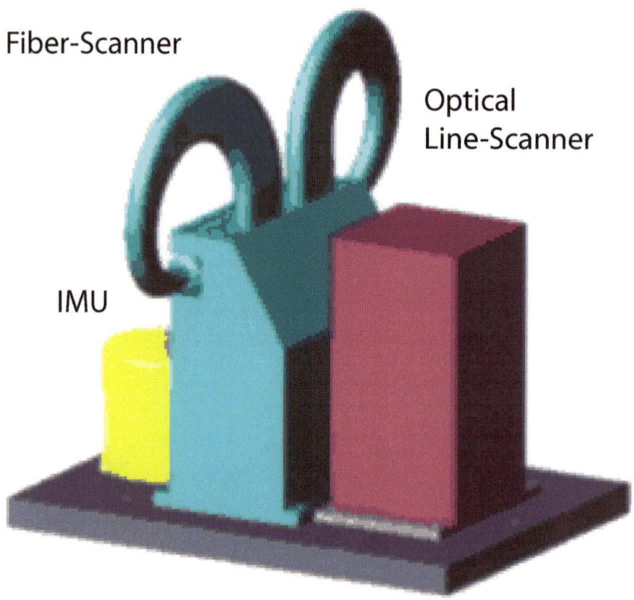

Figure 2.43 Scheme showing the internal mounting of the IMU, the fiber scanner laser, and the line scanner laser of the Falcon II equipment (Source: Katzenbeißer, 2001).

The Falcon II flight operator panel allows the flight operator to control all the mission parameters, such as project selection, camera parameters adjustment (e.g. aperture or ISO values), real time visualization of first and last echo laser data, as well as real-time visualization of acquired RGB and CIR imagery. Other information such as satellite signal quality, roll angle, heading, etc. are displayed as well in this console. The flight operator panel has a laser data "waterfall" visualization mode, in order to better display the height differences of the terrain, as well as for detection of water bodies, or possible data failures.

The Falcon systems are flexible and can be installed in both fixed- and rotary-wing platforms. The versatility and dimensions of the scanner rack allows its installation on both interior cabin and outside platforms of helicopters. A good example is the mounting of the Falcon II inside the storage compartment of a Bell Jet Ranger helicopter, as seen in Figure 2.44.

The Falcon II can be as well easily installed aboard any aircraft since the computer rack and sensor rack only share one single cable connection, and the Falcon II dimensions are quite modest. Figure 2.45 shows the Falcon II equipment installed aboard a Piper Seneca II.

Figure 2.44 The Falcon II sensor rack installed in the storage compartment of a Bell Jet Ranger helicopter.

Figure 2.45 The Falcon II lidar system installed aboard a Piper Seneca II.

The Falcon II is suitable for wide area mapping, corridor mapping, urban city mapping, but especially forestry applications. In terms of forestry applications, the Falcon II is well suited with its field of view and laser beam canopy penetration.

Compared with other scanning methods' the Falcon II provides an excellent probability of penetrating the forest canopy. Assuming a regular spacing of 1 m and an illuminated circle of 0.3 m diameter (beam divergence of 1 mrad at 1,000 m AGL), it results in about 14 percent of the subject area being scanned area, as seen in Figure 2.46 (TopoSys GmbH, 2004b).

If the gaps between the foliage are 20 percent, then the probability to penetrate the foliage becomes about 3 percent (0.14 × 0.2) indicated by the green spots (Figure 2.47a). The Falcon II fiber laser illuminates about 75 percent of the scanned area and thus its laser beam might penetrate through all existing gaps as shown by the green circles, as seen in Figure 2.47b.

(a) (b)

Figure 2.46 (a) Typical 1 m spacing grid, and (b) Graphic of a forest canopy (Source: TopoSys GmbH, 2004b).

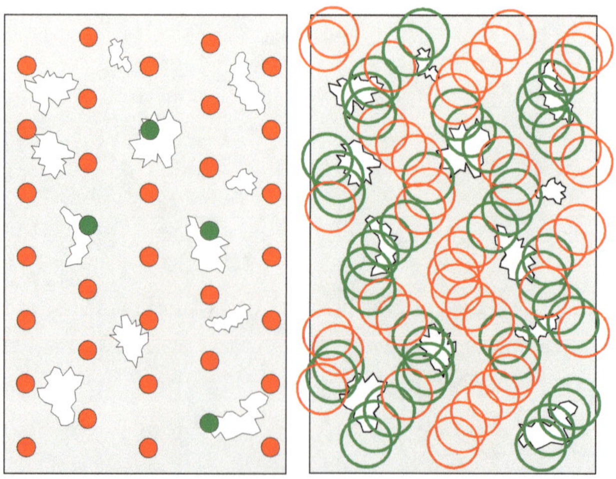

(a) (b)

Figure 2.47 (a) Forest penetration with a normal laser beam, and (b) with the Falcon II wide laser beam (Source: Schnadt and Katzenbeißer, 2004).

The narrow viewing angle of the Falcon II equipment is as well another major source driving its success in forestry applications. A nearly vertical viewing angle is necessary for penetration of a fir or a spruce stand, and the farther the beam direction diverges from vertical, the higher the probability that the laser beam is "stuck" at the lower branches (Figure 2.48). The advantageous combination of a wide beam, a high penetration ratio, and a fine separation is very well suited for forestry applications.

Figure 2.48 Laser beam divergence on a forest canopy (Source: Schnadt and Katzenbeißer, 2004).

REFERENCES

Artés, F. 2006. *Fiber-Based Laser Technology on the Cutting-Edge of Today's LiDAR Systems.* GEOInformatics, Oct/November.

Katzenbeißer, R., and U. Lohr. 2000. *Data Acquisition and Processing using the TopoSys Laser Scanner - Internal Documentation*, TopoSys GmbH.

Katzenbeißer, R. and TopoSys. 2004. *LIDAR Scanning Devices, Scan Patterns and Area Coverage - Internal Documentation,* TopoSys GmbH.

Löffler, G., 2003. Aspects of Raster DEM data derived from laser measurements, *Proceedings of the ISPRS Working Group III/3 Workshop, 3-D Reconstruction From Airborne Laser-scanner and InSAR Data*, Dresden, Germany, 08–10 October, International Archives of Photogrammetry, Remote Sensing and Spatial Information Sciences.

Lohr, U., 2003. Precision lidar data and true-ortho images, *Proceedings of the Map Asia Conference 2003.*

Schnadt, K., and R. Katzenbeißer, 2004. Unique airborne fibre scanner technique for application-oriented LIDAR products, *Proceedings of the ISPRS Working Group VIII/2,Laser-Scanners for Forest and Landscape Assessment*, Freiburg, Germany, 03–06 October 2004, International Archives of Photogrammetry, Remote Sensing and Spatial Information Sciences.

TopoSys GmbH, 1998. *Falcon I Architecture - Internal Documentation*, TopoSys GmbH.

TopoSys GmbH, 2004a. *Area Coverage and Point Density - Internal Documentation*, TopoSys GmbH.

———, 2004b. *Echo Detectability and Range - Internal Documentation,* TopoSys GmbH.

———, 2004c. *Edge Detection - Internal Documentation*, TopoSys GmbH.

———, 2004d. *Falcon II Airborne Lidar Sensor System – System Description,* TopoSys GmbH.

Wiechert, A. 2004. Production of ALS and RGB/CIR true ortho images, GIS@development, 8(5):30–32.

Yu, X., J. Hyyppä, H. Hyyppä and M. Maltamo. 2004. Effects of flight altitude on tree height estimation using airborne laser scanning, *Proceedings of the ISPRS Working Group VIII/2, Laser-Scanners for Forest and Landscape Assessment*, Freiburg, Germany, 03–06 October. International Archives of Photogrammetry, Remote Sensing and Spatial Information Sciences.

Yu, X., J. Hyyppä, H. Kaartinen, H. Hyyppä, M. Maltamo, and P. Rönnholm. 2005. Measuring the growth of individual trees using multi-temporal airborne laser scanning point clouds, *Proceedings of the ISPRS Workshop Laser Scanning 2005*, Enschede, The Netherlands, 12–14 September, International Archives of Photogrammetry, Remote Sensing and Spatial Information Sciences.

2.4 FULL-WAVEFORM LIDAR

The first decade of commercial, topographic mapping lidar (roughly, 1993-2003) was primarily limited to so-called "discrete-return" lidar systems, which record only a few—typically one to five—individual range measurements per transmitted pulse. In contrast, currently-available, full-waveform (FW) systems digitize the entire backscattered laser pulse at very high sampling rates (e.g., 1 GHz), providing a complete record of each received signal for subsequent processing. Figure 2.49 shows a digitized lidar waveform (amplitude vs. time) from an Optech ALTM topographic lidar system generated by sampling the received signal at one-nanosecond intervals. These digitized waveforms can be utilized to extract information about target characteristics and, in particular, about vertical structure. A related benefit is that FW systems enable improved target resolution, defined as the minimum separation of objects in the range direction, such that each can be individually detected. However, this additional information comes at the expense of increased data storage and post-processing requirements, so FW is not necessarily advantageous for all applications. In this section, we provide a brief history of FW lidar, an examination of why and how FW lidar can provide enhanced information, and a look at some of the applications for which FW is particularly beneficial. While a strong background in digital signal processing (DSP) is needed for a rigorous discussion of waveform post-processing algorithms, we include a high-level overview of some of the post-processing algorithms that can be applied to FW lidar.

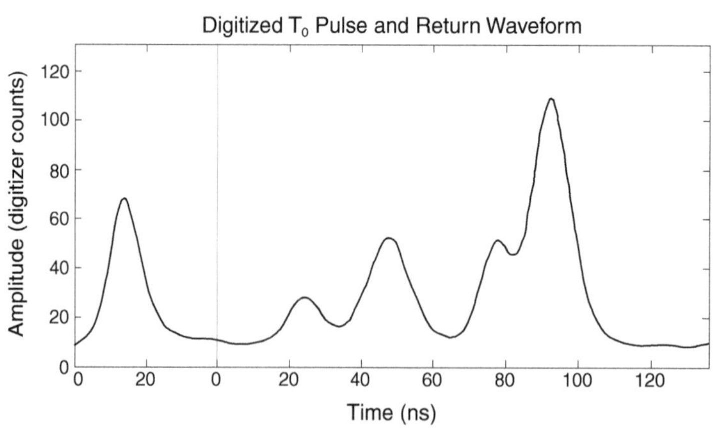

Figure 2.49 Example of an Optech ALTM lidar waveform (amplitude vs. time) digitized with 1-ns sampling interval using an 8-bit analog-to-digital (A/D) converter. The first 40 samples, to the left of the grey vertical line, are the transmitted, or T_0, pulse. (Note: if shown at the same scale, the T_0 pulse would, of course, be of significantly greater amplitude than the return. To avoid saturation, the T_0 pulse is digitized separately with a different receiver.)

2.4.1 Brief History and Background

The earliest non-military FW systems were generally:

1. Bathymetric lidar systems, designed for measuring water depths, or
2. Large-footprint, experimental, research systems, primarily intended for canopy mapping.

An example of a system in the first category is the National Aeronautics and Space Administration (NASA) Airborne Oceanographic Lidar (AOL), built in the 1970s (Hoge et al., 1980). In fact, nearly all bathymetric lidar systems perform ranging to the seafloor through post-processing of recorded waveforms (see, e.g., summary and development history in Guenther, 2007). Systems in the second category include NASA's Scanning Lidar Imagery of Canopies by Echo Recovery (SLICER) (Means et al., 1999), and Laser Vegetation Imaging Sensor (LVIS) (Blair et al., 1999). These systems employ large (5–30 m) footprints, and have been demonstrated to be very effective for a variety of forestry applications (e.g., Lefsky et al., 1999; Drake et al., 2002; Parker et al., 2001).

In contrast, the first commercial, topographic mapping systems were so-called discrete-return systems, as mentioned above. Because these systems employed much smaller (typically, < 0.5 m) footprints and were designed and used primarily for mapping terrain, rather than canopy or water depths, recording full waveforms was not deemed critical. By instead measuring and storing just a few, individual ranges, manufacturers could greatly reduce data storage and processing requirements. The basic idea behind these discrete-return systems is that each recorded range measurement corresponds, at least in theory, to an individual laser "echo"—that is, the received reflection from an individual object. Using other stored measurements, such as scan angles and post-processed GPS/IMU data, each recorded range can be mapped to a point in the output lidar point cloud.

In discrete-return systems, ranging is performed by measuring the time interval between the T_0 (transmitted) pulse and one or more distinct peaks in the return signal, typically by passing the analog signal output from the photodetector through a constant-fraction discriminator (CFD), followed by a time-interval counter (Harding, 2008; Baltsavias, 1999). The earliest discrete-return systems recorded only a single range measurement—corresponding to one individual backscattered echo (or "return" to use the more widely-adopted, but somewhat misleading, term)—per transmitted pulse. However, systems recording two to five returns per transmitted pulse soon followed (Harding, 2008). Additionally, many of these later discrete-return systems added the ability to store the amplitude of each recorded return, which could then be included as "intensity" values in the output point clouds.

While the addition of multiple returns and associated intensity values greatly improved the capabilities of discrete-return systems, there are still a number of reasons why FW systems are potentially advantageous. First, FW systems assist in overcoming the "minimum pulse separation limitation" of discrete-return systems. Briefly, due to the inherent limitations of the electronics, most discrete-return systems have a sizable "dead zone" of up to 2–3 m following each recorded return (Nayegandhi et al., 2006; Optech, 2007). In FW systems, the target resolution (minimum separation of targets in the range direction, such that they can be individually resolved) is greatly improved (Parrish et al., 2011). This, in turn, assists in resolving vertical structure of tree canopy and manmade vertical objects, such as radio transmission towers, comprising many different components at different elevations. FW systems also enable a variety of ranging techniques based on different pulse-detection methods, with the potential to improve ranging accuracy and/or customize the ranging technique to the specific application (Wagner et al., 2004; Wehr, 2007). Additionally, metrics derived from the digitized waveforms can be used to estimate parameters related to terrain slope and roughness, and to improve segmentation and classification (e.g., Mallet and Bretar, 2009 and references therein). Extraction of radiometric information from the digitized waveforms is also possible.

More broadly speaking, FW lidar opens the door to a multitude of new processing and analysis strategies by replacing fixed hardware subsystems with customizable DSP-based solutions. This entails a fundamental paradigm shift, wherein focus is transferred from sophisticated hardware to sophisticated software. Associated with this shift is a transfer of control and responsibility from the sensor (hardware) manufacturer to the user, at least for those users who are savvy enough to develop and implement their own waveform post-processing strategies.

Due to increasing recognition of these advantages, small-footprint, commercial, topographic FW lidar systems began appearing on the market around 2004, starting with the Riegl LMS-Q560 laser scanner and turnkey systems based on the Riegl laser scanners by other manufacturers, such as TopoSys and IGI (e.g., Hug et al., 2004). Shortly thereafter, other manufacturers, such as Optech and Leica, also began offering waveform digitizers with their commercial systems (Gutierrez et al., 2005; Toth et al., 2011). To date, at least six different models of FW topographic lidar systems have been introduced in the commercial market (Lemmens, 2007; 2009), and recent trends suggest that these numbers are likely to increase. Additionally, significant advancements are being made through newer experimental full-waveform systems, such as the NASA-developed (now USGS-owned), Experimental Advanced Airborne Research Lidar (EAARL) (Nayegandhi et al., 2006). Additional information related to the history and advancements in FW lidar can be found in Mallet and Bretar (2009).

2.4.2 Applications of FW Lidar

Table 2.7 lists a few applications for which FW lidar is particularly beneficial, along with a brief summary of the key benefits for these applications, and a small sampling of the related scientific literature. Forestry applications—listed first in the table—have been the most widely reported to date, but certainly do not comprise the only application area for FW technology. In fact, the list of applications and the corresponding references given in the table are far from comprehensive; new applications are being discovered at a high rate, and the volume of scientific literature related to FW lidar is already quite large and growing rapidly.

Table 2.7 A few applications for which FW lidar is particularly beneficial.

FW Lidar Application Area	Summary of Key Benefits	Small Sampling of Related Literature
Forestry/vegetation mapping	Enhanced ability to resolve vegetation layers and estimate forest biophysical parameters; improved vegetation/ground segmentation; potential for tree species classification; ability to map dense, tall coastal marsh grass and underlying terrain	Lefsky et al., 1999; Anderson et al., 2006; Reitberger et al., 2006; Wagner et al., 2008; Reitberger et al., 2008
Airport obstruction surveying	More information about vertical objects, such as poles and antennas, on and around airfields; improved obstacle detection and recognition	Parrish and Nowak, 2009; Parrish, 2007
Target detection (military/recon)	Improved target resolution; detection of objects (e.g., vehicles, machinery) below canopy	Tolt and Larsson, 2007
Habitat mapping/analysis	Enhanced information regarding habitat heterogeneity/habitat quality; improved estimation of biodiversity metrics	Goetz et al., 2007; Hyde et al., 2005

One common feature of the applications of FW lidar is that they are primarily concerned with analysis of objects extending above the ground surface, such as vegetation or human-made objects, such as towers or military equipment. Applications requiring bare-earth terrain models of relatively hard, flat areas tend not to benefit significantly from FW data, as the returns from these surfaces typically consist of a single peak, and the ranging can be done sufficiently well in hardware. In fact, FW can actually be somewhat disadvantageous in these applications, because, as noted earlier, the extra data storage and processing requirements may not be justified.

2.4.3 Waveform Processing

A frequent—though by no means universal—goal in waveform processing is to exploit the additional information in the waveforms to produce dense, detailed point clouds in which vertical structure is well characterized. A requisite first step in this process is to identify the positions of individual backscattered echoes in the waveform data. The motivation for this step is portrayed graphically in Figure 2.50. The cartoon in the center portion of this figure depicts an individual laser shot in a forested area. Portions of the incident laser energy are backscattered by various reflecting surfaces, including branches, twigs, and leaves within the tree canopy, and, finally, the ground below. The ideal return signal (impulse response) is shown on the left hand side of Figure 2.50. This ideal signal consists of a set of spikes (impulses) at different heights, each corresponding to a different laser reflection. For example, the first spike might correspond to a reflection from the highest branch or leaf intercepted by the laser beam, whereas the next might correspond to the next lowest branch, and so on, down to the ground. The location of each spike on the time axis is directly proportional to the range to that particular reflecting surface, and these ranges can be used in the laser geolocation equation, along with the post-processed GPS/IMU data, scanner angles, and sensor calibration, to produce the dense, detailed point cloud desired.

In practice, however, digitized lidar waveforms do not look like that shown on the left in Figure 2.50. An actual digitized T_0 (transmitted) pulse and return waveform are shown on the right-hand side of the figure. Even from this simplified diagram, the problem is evident: how do we estimate the locations of individual laser reflections (and, possibly, additional parameters of interest) from this observed waveform?

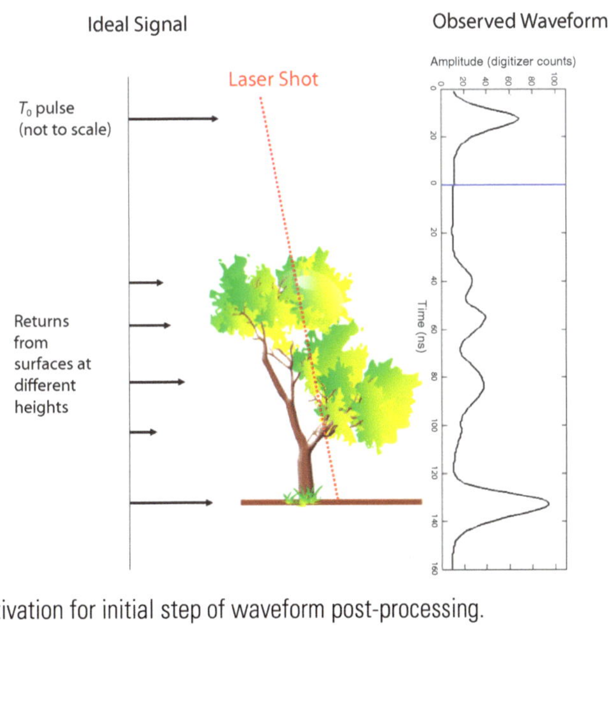

Figure 2.50 Motivation for initial step of waveform post-processing.

One well-documented waveform post-processing strategy is Gaussian decomposition (Hofton et al., 2000; Persson et al., 2005; Reitberger et al., 2006; Wagner et al., 2006), in which the received signal (right hand side of Figure 2) is modeled as a linear combination of Gaussians (i.e., an *N*-component Gaussian mixture):

$$\hat{y}(t) = \sum_{i=1}^{N} \alpha_i \exp\left\{-\frac{1}{2\sigma_i^2}(t-\mu_i)^2\right\} \tag{2-1}$$

In Equation 1, N is the number of Gaussian components, α_i is the amplitude of the ith Gaussian component, μ_i is the location on the time axis of its mean, and σ_i is its "width" (standard deviation). The objective is to obtain estimates of the unknown parameters $\{\alpha_i, \mu_i, \sigma_i\}$ that satisfy some pre-defined statistical optimality criterion using N Gaussian components, where N can be adjusted until the fit of the model to the data is within a prescribed tolerance. The means of the Gaussian components (i.e., the positions on the time axis) can be directly converted to ranges, which, in turn, can be mapped, via the laser geolocation equation, to points in the output point cloud. Furthermore, other estimated parameters, namely, the amplitudes and widths of the Gaussian components, may provide additional information about the scattering properties of targets (Wagner et al., 2006). Figure 2.51 shows an example of performing Gaussian decomposition on a lidar waveform using the approach described in Hofton et al., 2000. An extension of this approach using refined peak detection and log-normal or generalized Gaussian functions is described in Chauve et al, 2007.

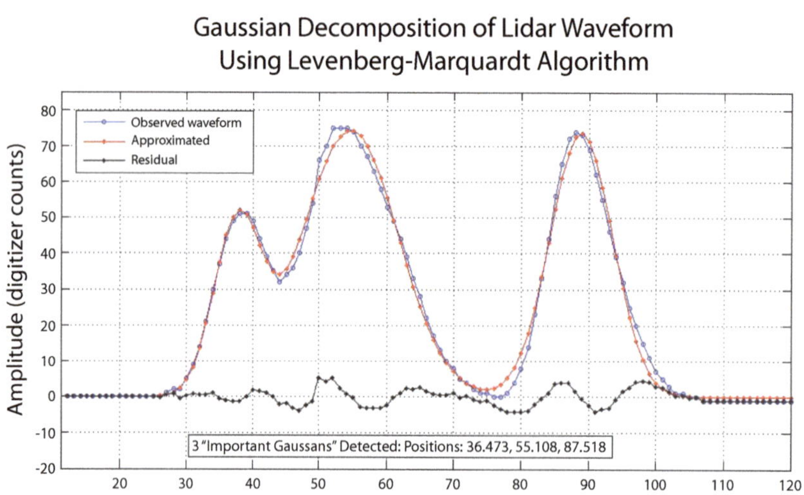

Figure 2.51 Example of Gaussian decomposition of a lidar waveform using the approach outlined in Hofton et al., 2000. In this case, the observed waveform is modeled as a combination of three Gaussian components.

An alternative approach to waveform post-processing is deconvolution. In this approach, the observed signal depicted on the right-hand side of Figure 2.50 is considered a blurry, noisy version of the ideal signal shown on the left-hand side of the figure. Mathematically (and in the common signal-processing notation for discrete-time signals), the relationship between the two is expressed as:

$$y[n] = g[n] * x[n] + \eta[n] \tag{2-2}$$

where $x[n]$ is the desired waveform, $y[n]$ is the observed waveform, $g[n]$ is the blurring function, $\eta[n]$ is additive noise, and denotes convolution. The objective is to recover the ideal signal from the received waveform and then to perform the georeferencing, as described above. Based on other starting assumptions—which are beyond the scope of this manual—various deconvolution algorithms can be applied to the observed waveform to "restore" the ideal signal.

Another strategy for waveform post-processing is a combination approach, comprising a deconvolution step to estimate the "surface response function," followed by fitting of Gaussian(s) to the output (Jutzi and Stilla, 2006). And, new and enhanced strategies continue to be reported, such as the B-splines algorithm presented in Roncat et al. (2010). Comparison of the different general approaches to waveform post-processing and of specific algorithms within each category is well beyond the scope of this manual. However, all of the approaches mentioned here can be used to attain densification of point clouds over what is achievable with discrete-return systems. Figure 2.52 shows an example of the achievable densification of point data on vertical objects achieved using full-waveform data and the deconvolution approach presented in Parrish (2007) and Parrish and Nowak (2009) (right-hand column), as compared with discrete-returns (middle column) from the same system.

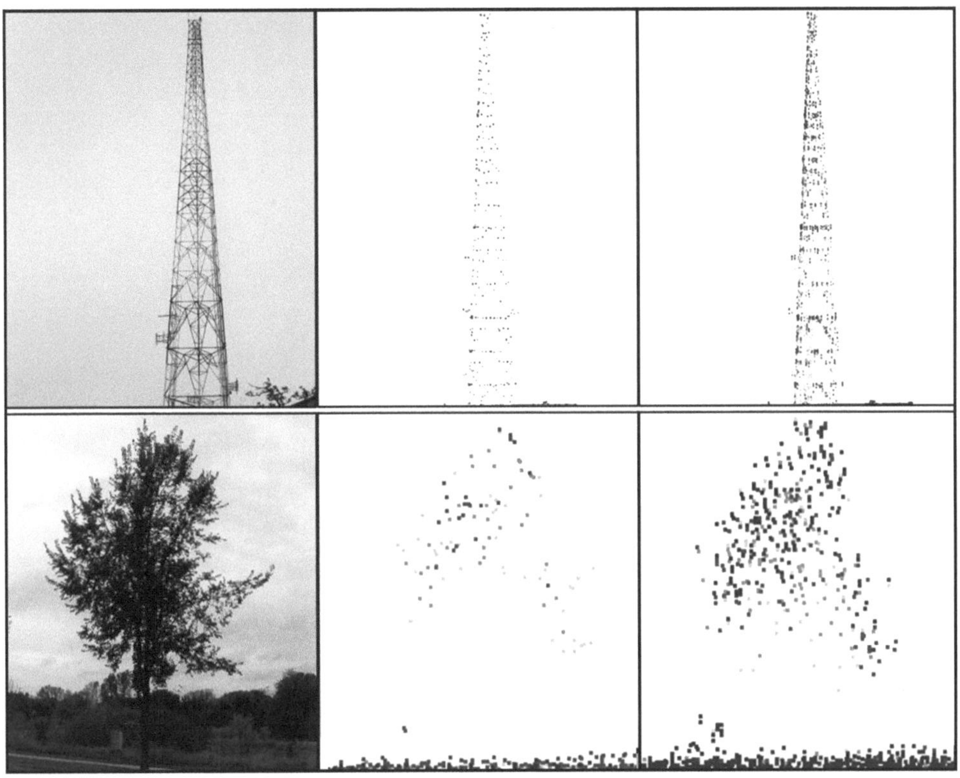

Figure 2.52 Example of point densification on vertical objects using full-waveform data. The left column shows photos of two objects: a tower (top) and a tree (bottom). The middle column shows the points in LAS files created using discrete returns, and the right column shows the much denser point clouds generated using FW data and the methods described in Parrish and Nowak (2009).

An important consideration when working with FW data is that interpolation of these detail-rich point clouds, such as those shown on the right-hand side of Figure 2.52, to a 2D raster grid (commonly referred to as a DEM or DSM) is something that should typically be avoided. Simply stated, 2D raster grids cannot adequately represent vertical structure (Stoker, 2004), so the additional information about vertical structure gained through the use of full waveforms is lost in the gridding step. If the end-user's application requires a gridded data set, a 3D grid (i.e., voxels, not pixels) works significantly better for representing the dataset generated from the FW lidar. To illustrate this concept, Figure 2.53 shows a comparison of 2D and 3D gridded representations of a transmission tower, generated from a high-density point cloud produced from FW lidar. The loss of information about vertical structure in the 2D grid is clear when compared with the 3D grid. Of course, the best option, when possible, is often to avoid gridding altogether and instead utilize the lidar data in its "native" point cloud format. This is gradually becoming a viable option for a variety of applications, due to the increase in software for working with lidar data in point-cloud format.

While many existing waveform processing algorithms are geared toward producing denser point clouds through the procedures outlined above, we are likely to see an increase in applications that utilize the "raw" waveforms (i.e., not derived point clouds) for information extraction. These types of applications extract meaningful information from the shape (or metrics related to the shape) of the waveforms themselves, without the intermediate step of estimating the waveform positions corresponding to individual laser reflections and georeferencing these to produce a point cloud. For example, it is possible to derived information about surface roughness, slope, and reflectance—and, possibly other surface characteristics, as well—from the waveforms themselves, without the intermediate step of producing a point cloud.

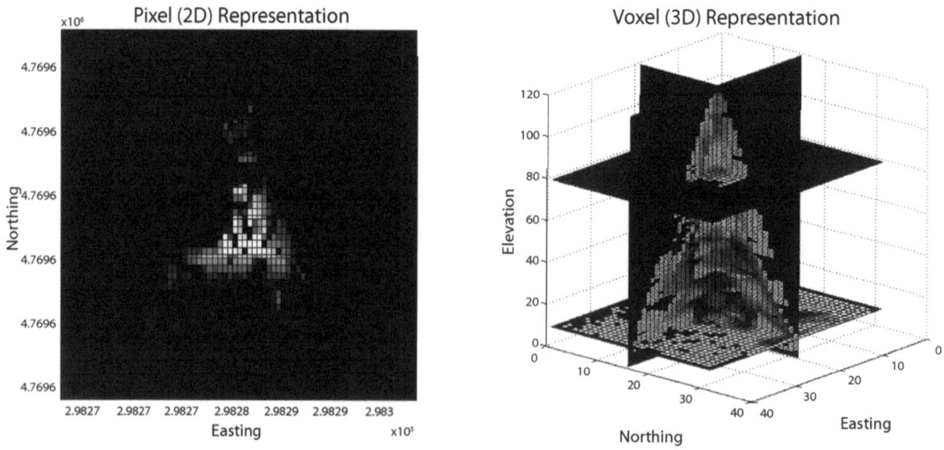

Figure 2.53 Left: 2D grid of pixels created from the lidar point cloud containing a large tower. Right: Voxel (3D) representation. The gray values on the left represent elevations, whereas the gray values on the right represent lidar intensities.

2.4.4 Summary and Outlook

Full-waveform digitization is one of the most exciting advances of the past decade in commercial, airborne topographic lidar. Based on a recent literature review, the rate of publication of scientific papers related to full waveform lidar increased steadily between 2004 and 2010, with the current publication rate being nearly three times higher than at the beginning of that period. While third-party commercial software packages for working with waveform data have lagged a bit behind the scientific community's adoption, commercial/production use of FW systems is now also increasing rapidly. Recently, ASPRS added full waveform support to the widely-used LAS format. However, there are still some challenges in FW lidar. These include lack of standards or guidelines for FW acquisition and processing, a relative dearth of third-party COTS software packages for working with the data, a need for improved efficiency in storing and accessing large waveform data files, and some lingering misunderstandings within the broad lidar community regarding both the advantages and limitations of FW lidar. Nevertheless, there is little doubt that each of these challenges will be overcome in time—perhaps quite quickly—and that future users of lidar data will increasingly benefit from the additional information provided by FW systems.

REFERENCES

Anderson, J., M.E. Martin, M-L. Smith, R.O. Dubayah, M.A. Hofton, P. Hyde, B.E. Peterson, J.B. Blair, R.G. Knox, 2006. The use of waveform lidar to measure northern temperate mixed conifer and deciduous forest structure in New Hampshire, *Remote Sensing of Environment,* 105:248–261.

Baltsavias, E.P., 1999. Airborne laser scanning: basic relations and formulas, *ISPRS Journal of Photogrammetry and Remote Sensing,* 54(2–3):199–214.

Blair, J.B., D.L. Rabine, and M.A. Hofton, 1999. The Laser Vegetation Imaging Sensor: a medium-altitude, digitisation-only, airborne laser altimeter for mapping vegetation and topography, *ISPRS Journal of Photogrammetry and Remote Sensing,* 54(2–3):115–122.

Chauve, A., C. Mallet, F. Bretar, S. Durrieu, M. Pierrot-Deseillingny, and W. Puech, 2007. Processing full-waveform lidar data: modeling raw signals, *ISPRS Workshop on Laser Scanning 2007 and SilviLaser 2007,* 12–14 September 2007, Espoo, Finland, pp. 102–107.

Drake, J.B., R.O. Dubayah, D.B. Clark, R.G. Knox, J.B. Blair, M.A. Hofton, R.L. Chazdon, J.F. Weishampel, and S.D. Prince, 2002. Estimation of tropical forest structural characteristics using large-footprint lidar, *Remote Sensing of Environment,* 79(2–3):305-319.

Goetz, S., D. Steinberg, R. Dubayah, and B. Blair, 2007. Laser remote sensing of canopy habitat heterogeneity as a predictor of bird species richness in an eastern temperate forest, USA, *Remote Sensing of Environment,* 108(3):254-263.

Guenther, G.C., 2007. Airborne Lidar Bathymetry, *Digital Eelvation Model Technologies and Applications, The DEM Users Manual,* 2nd Ed. (D.F. Maune, Editor), American Society for Photogrammetry and Remote Sensing, Bethesda, M.D., pp. 253–320.

Gutierrez, R., A. Neuenschwander, and M.M. Crawford, 2005. Development of Laser Waveform Digitization for Airborne LIDAR Topographic Mapping Instrumentation, *Proceedings of the IEEE International Geoscience and Remote Sensing Symposium (IGARSS),* 25–29 July 2005, Vol. 2, pp. 1154–1157.

Harding, D., 2009. Pulsed Laser Altimeter Ranging Techniques and Implications for Terrain Mapping, *Topographic Laser Ranging and Scanning: Principles and Processing* (J. Shan and C.K. Toth, editors), CRC Press, Taylor and Francis Group, Boca Raton, Florida, pp. 173–194.

Hofton, M.A., J.B. Minster, and J.B. Blair, 2000. Decomposition of Laser Altimeter Waveforms, *IEEE Transactions of Geoscience and Remote Sensing,* 38(4):1989–1996.

Hoge, F.E., R.N. Swift, and E.B. Frederick, 1980. Water depth measurement using an airborne pulsed neon laser system, *Applied Optics,* 19(6):871–883.

Hug, C., A. Ullrich, and A. Grimm, 2004. LITEMAPPER-5600 – a waveform digitizing lidar terrain and vegetation mapping system, *International Archives of Photogrammetry and Remote Sensing,* 36(8/W2):24–29.

Hyde, P., R. Dubayah, B. Peterson, J.B. Blair, M. Hofton, C. Hunsaker, R. Knox, and W. Walker, 2005. Mapping forest structure for wildlife habitat analysis using waveform lidar: Validation of montane ecosystems, *Remote Sensing of Environment,* 96(3–4):427–437

Jutzi, B., and U. Stilla, 2006. Range determination with waveform recording laser systems using a Wiener Filter, *ISPRS Journal of Photogrammetry and Remote Sensing,* 61(2):95–107.

Lefsky, M.A., W.B. Cohen, S.A. Acker, G.G. Parker, T.A. Spies, and D. Harding, 1999. Lidar remote sensing of the canopy structure and biophysical properties of Douglas-fir western hemlock forests, *Remote Sensing of Environment*, 70(3):339-361.

Lemmens, M., 2007. Airborne Lidar Sensors: 2007 Product Survey, GIM International, 21(2), URL: http://www.gim-international.com/files/productsurvey_v_pdfdocument_11.pdf (last date accessed: 16 September 2009).

Lemmens, M., 2009. Airborne Lidar Sensors: 2009 Product Survey, GIM International, URL: http://www.gim-international.com/files/productsurvey_v_pdfdocument_28.pdf (last date accessed: 16 September 2009).

Mallet, C., and F. Bretar, 2009. Full-waveform topographic lidar: State-of-the-art, *ISPRS Journal of Photogrammetry and Remote Sensing*, 64(1):1-16.

Means, J.E., S.A. Acker, D.J. Harding, J.B. Blair, M.A. Lefsky, W.B. Cohen, M.E. Harmon, and W.A. McKee, 1999. Use of Large-Footprint Scanning Airborne Lidar to Estimate Forest Stand Characteristics in the Western Cascades of Oregon, *Remote Sensing of Environment*, 67(3):298-308.

Nayegandhi, A., J.C. Brock, C.W. Wright, and M.J. O'Connell, 2006. Evaluating a Small Footprint, Waveform-resolving Lidar over Coastal Vegetation Communities, *Photogrammetric Engineering and Remote Sensing*, 72(12):1407–1417.

Optech, 2007. *ALTM Waveform Digitizer: Operation and Processing Manual.* Optech, Inc., Vaughan, ON, Canada.

Parker, G.G., M.A. Lefsky, and D.J. Harding, 2001. Light transmittance in forest canopies determined using airborne laser altimetry and in-canopy quantum measurements. Remote Sensing of Environment, 76(3):298–309.

Parrish, C.E., 2007. *Vertical object extraction from full-waveform lidar data using a 3D wavelet-based approach*, Ph.D. dissertation, University of Wisconsin-Madison, Madison, WI, 171p.

Parrish, C.E., and R.D. Nowak, 2009. Improved Approach to LIDAR Airport Obstruction Surveying Using Full-Waveform Data, *Journal of Surveying Engineering*, 135(2):72–82.

Parrish, C.E., I. Jeong, R.D. Nowak, and R.B. Smith, 2011. Empirical Comparison of Full-Waveform Lidar Algorithms: Range Extraction and Discrimination Performance, *Photogrammetric Engineering and Remote Sensing*, 77(8):825–838.

Persson, Å., U. Söderman, J. Töpel, and S. Ahlberg, 2005. Visualization and analysis of full-waveform airborne laser scanner data, *Proceedings of ISPRS WG III/3, V/3 Workshop "Laser scanning 2005,"* 12–14 September, Enschede, the Netherlands.

Reitberger, J., P. Krzystek, and U. Stilla, 2006. Analysis of full waveform LIDAR data for tree species classification, *Symposium of ISPRS Commission III: Photogrammetric Computer Vision (PCV06), International Archives of Photogrammetry, Remote Sensing, and Spatial Information Sciences,* 36(3):228–233.

Reitberger, J., P. Krzystek, and U. Stilla, 2008. Analysis of full waveform lidar data for the classification of deciduous and coniferous trees, *International Journal of Remote Sensing,* 29(5):1407–1431.

Roncat, A., G. Bergauer, and N. Pfeifer, 2010. Retrieval of the backscatter cross-section in full-waveform lidar data using b-splines, *International Archives of Photogrammetry, Remote Sensing and Spatial Information Sciences*, 38(3B):137–142.

Stoker, J., 2004. Voxels as a Representation of Multiple-Return Lidar Data, *Proceedings of the ASPRS Annual Conference*, 23–28 May, Denver Colorado (American Society for Photogrammetry and Remote Sensing, Bethesda, Maryland), unpaginated CD-ROM.

Tolt, G., and H. Larsson, 2007. Waveform analysis of lidar data for targets in cluttered environments, *Proceedings of Electro-Optical Remote Sensing, Detection, and Photonic Technologies and Their Applications*, 18 September 2007, Florence, Italy, SPIE Vol. 6739, 67390A.

Toth, C.K., P. Zaletnyik, S. Laky, and D. Grejner-Brzezinska, 2011. The potential of full-waveform lidar in mobile mapping applications. *Archives of Photogrammetry, Cartography and Remote Sensing*, Vol. 22, pp. 401–410.

Wagner, W., A. Ullrich, T. Melzer, C. Briese, and K. Kraus, 2004, From single-pulse to full-waveform airborne laser scanners: potential and practical applications, I*nternational Archives of Photogrammetry and Remote Sensing*, 35(B3):201–206.

Wagner, W., A. Ullrich, V. Ducic, T. Melzer, and N. Studnicka, 2006. Gaussian Decomposition and Calibration of a Novel Small-Footprint Full-Waveform Digitising Airborne Laser Scanner, *ISPRS Journal of Photogrammetry and Remote Sensing*, 60(2):100–112.

Wagner, W., M. Hollaus, C. Briese, and V. Ducic, 2008. 3D vegetation mapping using small-footprint full-waveform airborne laser scanners, *International Journal of Remote Sensing*, 29(5):1433–1452.

Wehr, A., 2009. LiDAR Systems and Calibration, *Topographic Laser Ranging and Scanning: Principles and Processing* (J. Shan and C.K. Toth, editors), CRC Press, Taylor and Francis Group, Boca Raton, Florida, pp. 129–172. 2.5 3D Flash Imaging Lidar and 3D Video Panoramic Lidar

2.5 3D FLASH IMAGING LIDAR AND 3D VIDEO PANORAMIC LIDAR

2.5.1 Why Lidar?

Photogrammetry is derived from three Greek words: *photos* meaning "light," *gramma* meaning "something drawn or written," and *metron* meaning "to measure." Photogrammetry is the art, science, and technology of obtaining reliable information about physical objects and the environment through processes of recording, measuring, and interpreting photographic imagery and patterns of electromagnetic radiant energy and other phenomena. In addition, the science of photogrammetry has been broadened by the application of *remote sensing*. In applications of remote sensing, imagery may be acquired, not only through the use of a digital camera (which in itself is a remote-sensing system), but also by recording the scene through one or more special sensors. These special sensors usually operate by other means such as electronic point (or *flying-spot*) scanning or line scanning often with active sources of illumination that commonly use wavelengths outside the normal visual range of the camera such as microwaves, radar, thermal infrared, ultraviolet, as well as multispectral combinations. Special techniques are applied in order to process and interpret remote sensing imagery for the purpose of producing conventional maps, thematic maps, resources surveys, etc. Indeed when laser LiDAR[1], (Light Detection and Ranging) was developed it became the consummate photogrammetric remote sensing instrument. It is important to note that, traditionally, lidar technology has developed along two distinct branches: continuous wave (CW) and pulsed (or direct-detection (Richmond and Cain, 2010)) which implies two distinctly different source and detection technologies. CW lidar relies on the highly coherent nature of laser light and is uniquely suited to a wide variety of measurement tasks but is less well suited to imaging due to the rapid scaling of the computational power required as resolution increases. Because of this, CW lidar will not be dealt with further in this discussion and the reader is referred to the literature for additional information.

Likewise, traditional pulsed lidar, whose characteristics are covered elsewhere in this Manual, will not be covered other than to point out key differences with a modern variant, 3D flash imaging lidar (also known as LADAR for LAser radar (Medina, 1992; Stettner et al., 2000)).

---

[1]LiDAR - an acronym like its older sibling – laser – has come into common usage as a common noun an will be denoted lidar.

As typically employed today in civil, military, scientific, and commercial applications, pulsed imaging lidar, at its core, employs a pulsed laser at a particular wavelength and pulse energy in combination with a very sensitive single point detector scanned rapidly in raster fashion over a scene to generate a sampled representation of it. A very short pulse, typically on the order of nanoseconds, originating in the laser is partially backscattered from a region, sample, or target to generate a return characteristic of the sample or target. The time-of-flight (TOF) of the returned pulse is measured by an accurate clock, which, when divided by two and multiplied by the speed of light (c), represents the range. The intensity of the return waveform over time (or its peak intensity) may also be recorded to produce contrast. The field-of-view (FOV) of the laser is typically matched to the detector with optics optimized for the scene or sample. In order to generate range imaging, the FOV must be scanned mechanically or electronically to generate a "field of regard (FOR)." The raw range imaging data is streamed to fast storage and later reconstructed using computation. Applications include sampling air columns for aerosols or contaminants, wind fields to determine wind speed, airborne terrain or littoral mapping, industrial construction validation, weapons identification and targeting etc. Notice a key distinction here that, because traditional lidar employs only a single point detector scanned over the FOR, imaging in 3D is deferred some period of time until computer-aided analysis can synthesize a scene. Unless the laser pulse repetition frequency (PRF) approaches the upper end of the newest systems at 500 kHz, full scene reconstruction will not occur on time scales typically associated with video imagery. Rather, there is typically a significant latency of seconds to minutes (or even hours) depending on the configuration, the FOR and the computing power available at the sensor. For many applications requiring essentially real-time video imaging, this latency is unacceptable.

A key aspect of lidar imagery is the ability to locate range information within an absolute frame of reference; otherwise known as geo-location, using knowledge of inertial location, range, and scan angle over time. Today, this frame of reference is conveniently defined by satellite-based global positioning systems (GPS) which are easily incorporated in to the lidar data reduction chain. This allows us to geo-locate any point in the data stream (or "point cloud"). However, despite its versatility, the uncertainty of geo-location, from the standpoint of spot-scanned lidar, depends heavily on the angular accuracy of the scanning mechanism multiplied by the enormous lever arm of a sample or scene at long range as well as platform jitter. In addition, knowledge of a regional volume must be built up over time by sampling. Thus, knowledge of the spatial content of the 3D volume element, or "voxel" produced by sampling the region with pulsed beams, is strictly limited by Nyquist sampling theory. Finally, the uncertainties associated with the reference frame itself must also be accounted for and incorporated over the time frame involved in the sampling process. In the case of an airborne platform employing a GPS and inertial measurement unit (IMU) which tracks platform location and attitude, the uncertainties in both the GPS and IMU are varying in time over the sampling interval. This is a particular issue if elements of the scene are changing over time scales short compared to the total scan time. All of these sources of uncertainty must be accounted for and de-convolved to the extent possible when producing maps of high resolution. While the imagery generated by modern scanned lidar systems is often exquisite, this takes time and is typically not well suited to the demands of video.

2.5.2 Why 3D Flash Imaging Lidar and (more recently) 3D Video Panoramic Lidar?

For applications requiring video-rate, geo-located, panoramic imagery, new technologies have been developed based on time-of-flight ranging cameras which, taken together, comprise 3D Flash Imaging Lidar (commonly shortened to flash lidar). Flash lidar actually encompasses a wide variety of technologies suitable for 3D video imagery (Stettner et al. 2000). At short range (1 m to 10 m), systems suitable for vision and autonomous navigation tasks (Halterman and Bruch, 2010; Kahlmann et al. 2007) tend to be quite different in execution from long range (10m - 10km) systems (Stettner et al., 2005; Jack et al., 2007) destined for mapping, tracking, targeting and autonomous rendezvous. At very long ranges (10 km to 1.0 Mm) or under conditions of extremely attenuated returns, a closely related technology, Geiger-mode lidar (Yuan et al., 2010; Marino et al., 2005), based on photon counting arrays is suitable for targeting and tracking. It is, however, the long range (so called "linear mode") technology and its remarkable advantages in real-time, geo-located, multi-sensor, 2D/3D fused panoramic video to which we now turn our attention. In particular, the constraints of this introductory overview force us to focus chiefly on the most widely available implementations of "flash" technology based on cameras produced by Advanced Scientific Concepts (ASC) (Stettner, 2010) and (in a somewhat different sense) Raytheon Vision Systems (RVS) (Asbrock et al., 2008). More broadly, flash imaging lidar cameras have been incorporated into a wide variety of remote sensing instruments by Ball Aerospace and Technologies Corp (BATC) and, with their video co-processing architecture, form the basis of Ball's real-time technology (Craig et al., 2007). While we will only focus on only two, Ball's implementations of imaging lidar are widely published and have demonstrated the breadth of the art (Nicks, 2010; Ramond et al., 2011).

3D flash lidar systems are analogous to a digital camera with a flash for flood illumination, but with the flash being provided instead by a pulsed laser and by the use of a flash-focal-plane array (FFPA) detector with clocks in each ROIC pixel cell to determine the time needed for the flash to depart, reflect from a target, and return. Recall that the TOF of the reflected laser pulse gives a range measurement along with intensity for each pixel in the image. Analogous to scanning lidars, this information set over the range of all detector pixels is referred to as a 3D lidar image or *point cloud* because the scene includes both spatial and range information. In the context of imaging lidar, the term image and point cloud are synonymous.

The entire scene within the sensor's field of view (FOV) is imaged with a single flash of the laser on the order of a nanosecond. This produces an image that correlates each pixel in time, which makes the sensor insensitive to any relative motion between the platform and target and eliminates the need for complex processing algorithms to account for relative motion or laser scan time over the FOV. 3D flash lidar systems are able to produce these 3D images at rates up to 30 Hz, much faster than all but the fastest flying-spot scanners. While the PRF, and hence the data rate, of scanning lidars is fundamentally limited to by the PRF of the laser source, the only limitation on the data rate of a flash lidar is the size of the focal-plane array. Consequently, there is no fundamental upper limit to the resolution as there is with raster scanning lidars[2]. The sensor frame rate and laser pulse rate can now be synchronized to produce full motion 3D video imagery. Each individual point in the point cloud can now be fused (colored) with red-green-blue (RGB) information from a visible wavelength context camera or with the intensity information of cameras sensitive to other wavelengths. In a related process, the points in the cloud can be inter-connected algorithmically to create a virtual "wire mesh" surface on which to "drape" video imagery. The context video image acquires relief when draped over the mesh surface and can be manipulated in 3D for optimal viewing during acquisition. Additional processing products, such as 3D target (or hazard) identification can be added to this processing chain for real time target ID. In this case, both 3D and 2D imagery lead to synergistic recognition. An example of a 3D image from a Ball flash lidar and the sequence of base image

[2] One only needs to look at the remarkable progression in pixel density in the digital 2D camera world to see what is possible. However, the tradeoff here is that the radiometry of higher resolution drives us to more powerful lasers.

processing steps, along with landing hazard detection, is shown in Figure 2.54. As noted previously, each pixel of the resulting 3D imagery array is correlated in time. This, and the fact that all mechanical motion is slow compared to the pulse width, makes the sensor insensitive to relative motion between the sensor and target. This eliminates the need for complex processing algorithms that account for relative motion or laser scan time of the FOV. This is in sharp contrast to a spot scanning lidar where a laser beam is raster scanned over the sensor FOV and each pixel intensity and range must be separately correlated with and filtered against platform motion. The flash lidar system has an advantage in that the platform relative motion can be applied to the entire array of pixels, greatly reducing processing requirements and enabling real time processing of the data into actionable intelligence products.

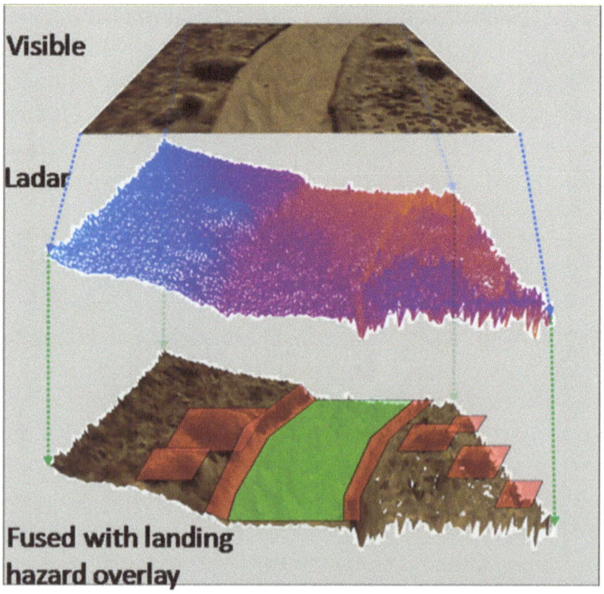

Figure 2.54 The Ball Total Sight™ Flash Lidar system combines baseline point cloud processing and image fusion with higher level processing such as landing hazard detection in near real-time.

In the ASC implementation, flash lidar data is commonly represented by "cubes" (Richmond et al., 2000; Wallace et al., 2006) (stacks of "pixel tubes") that represent wavefront time series for each pixel in the FPA, as in Figure 2.55. The pixel tube stack is comprised of voxels. A new "frame" is triggered by each laser pulse, and each corresponding time-series is initiated by the reflected wavefront following the TOF. The primary mode is one in which each pixel is triggered independently by photons in the returning wavefront and each pixel tube represents a buffered time series of that wavefront for a given pixel FOV (IFOV). Another mode, gated mode, effectively samples only a specific depth of the FOV, but the data fundamentals are the same. In the BATC system, multiple tubes are processed in parallel by powerful and power efficient field-programmable gate-arrays (FPGA). The goal is to offload as much of the parallel post-processing to FPGA firmware as possible, leaving the serial portion to be implemented in dedicated CPU's, both co-located and networked remotely. BATC developed the ability to process this rich data content at rates exceeding 30Hz (video rates). This allows the generation of "live" 3D video that is easily manipulated as a "virtual world" throgh an OpenGL, QT (or equivalent) graphical user interface as well as the new generation of 3D or holographic display systems (Klug, 2002) for unparalleled user interaction. More importantly, however, is the ability of the system to perform automatic target recognition (ATR) on large scenes stitched together from individual frames to form "superframe" (or mosaic or panoramic) maps enhanced by all manner of other context and calculated data. In addition to high speed mapping, ATR, pose, and change detection are the keys to the real potential of this approach for autonomous rendezvous and docking (ARD) (Gravseth, 2009), proximity operations (PROXOPS), autonomous hazard detection and avoidance (HLDA), autonomous robotic vision (ARV), imaging velocimetry, and differential mapping. Other functions do not require ATR but are nonetheless enhanced by the high frame rates such as altimetry, ranging velocimetry, and terrain-relative navigation (TRN) (Amzajerdian et al., 2011). This line of thinking can be extended to extra-planetary mapping through the marriage of star-tracker guidance cameras with 3D flash lidar (Earhart et al., 2011). Figure 2.56 shows a typical functional layout of a BATC flash lidar System wherein the order and capability of any element can be rearranged to suit the mission at hand. Because missions are so widely variable, trade studies are performed to optimize all aspects of system design and utilize available lasers, optics and back-end processors to their full potential.

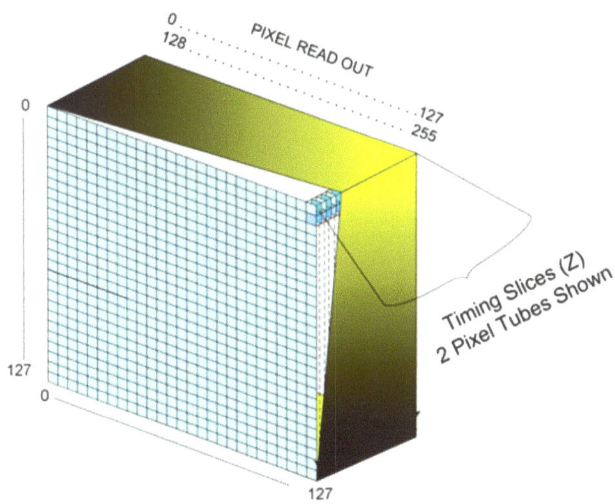

Figure 2.55 Geometric representation of the "data cube" with the X and Y coordinates represented by the array rows and columns while the tube "depth" in time represents Z as a TOF measurement.

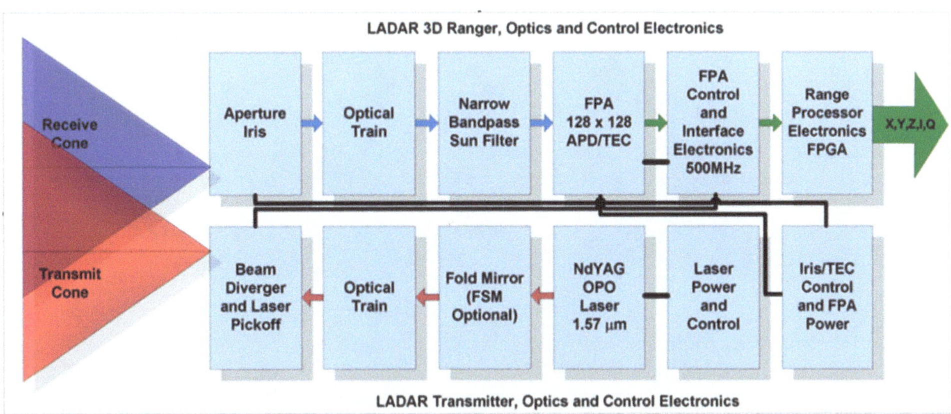

Figure 2.56 Functional block diagram of a 3D flash system in an eye-safe configuration where X,Y,Z is ortho-rectified point cloud information, I is intensity, Q is a frame quality factor.

The image capture process using 3D flash lidar employs the following sequence:

1. A short duration laser pulse with energy sufficient to illuminate the FOV of the camera. This illumination is "back-scattered" by Lambertian targets in the scene.
2. The laser aperture typically contains a diffuser to shape the laser output profile to a square (or rectangular) "top hat" pattern, increasing the efficiency and uniformity of the illumination while matching the receiver FOV. This creates a number of opportunities to enhance specific illumination attributes that result in better imaging of challenging objects at long range, such as wires, and makes more efficient use of the FFPA.
3. The camera, as master, generates a ready pulse to fire the laser. The outgoing pulse is sampled to start the TOF timer (if an optical sync pulse is used) or the output of the laser Q-switch is monitored for electronic sync.
4. The photonic energy back-scatter is collected by an optical receive (Rx) aperture and focused onto the 3D FFPA "hybrid" comprised of a readout-integrated-circuit (ROIC) and (typically) either an indium-gallium-arsenide (InGaAs) avalanche photodiode (APD) or PIN-type (non-avalanche) photodiode detector array. The choice depends on the gain required and the wavelength of the source laser.
5. The solid state APD or PIN detector pixels produce photo-electrons from the incoming photons. The gain of the detector determines how many electrons are produced per photon as determined by its quantum efficiency.
6. Each pixel in the detector array samples the wavefront independently and is connected to an amplifying and threshold circuit in the ROIC. A timer/counter in the ROIC measures the TOF of the reflected light imaged on the APD/CMOS hybrid array. The range to any object and surface in the scene is computed along with the intensity of that reflected light. Both are used to produce 3D video output at frame rates from 1 to 30 Hz or even 60 Hz.

3D flash lidar cameras from ASC or RVS sense laser wavelengths ranging from 400 nm to 1700 nm depending on the semiconductor detector material used. The prevalent use model favors InGaAs because of the high quantum efficiencies achieved at both 1064 nm to 1570 nm laser wavelengths. At shorter wavelengths, silicon (Si) is favored. Because the nature of 3D flash lidar focal planes favors dissimilar technologies (e.g., InGaAs and CMOS), 3D flash lidar focal planes are created by hybridizing the focal plane detector and read-out semiconductor material by using indium "bumps" to create the inter-chip connections. The Ball Total Sight™ Flash 3D Lidar seen in Figure 2.57 is an example of a system based on the ASC TigerEye™ camera operating at 1064 nm and designed to operate at ranges to 10,000 feet.

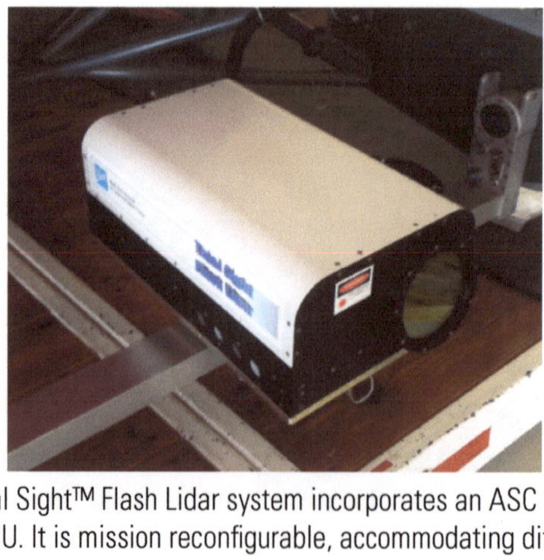

Figure 2.57 The Ball Total SightTM Flash Lidar system incorporates an ASC FFPA, Areté laser and Applanix POS 510 GPS/IMU. It is mission reconfigurable, accommodating different projection optics, receiver optics, wide-angle scanning mirror, and additional VIS/IR context cameras.

Typical laser wavelength for 3D flash lidar cameras used in proximity to humans is the eye-safe wavelength of 1570 nm; preferred because it is blocked by the cornea preserving the integrity of the retina. Note that differing objects in the FOV at the same range will backscatter differing intensities based upon their reflective characteristics resulting in a distinct contrast in intensity (the basis of 2D imagery). This intensity difference is useful when imaging diverse environments such as city streets with distinctive lane markings and street signs (e.g., road surface versus markings).

A different detection approach (known as gated-mode flash lidar (Stettner and Bailey, 1995) locks the independent pixel counters to sample the first return of the illuminating pulse in unison after a fixed period of time known as the hold-off gate. It "gates" the sampling process so that the pulse is registered only within a specific time window and is blind to early or late returns. If the range to the target is unknown, the time window is scanned until a return is registered. This can even be done using an intensified camera and laser illumination (Sturz, 1993) but requires advancing the time window gate to sample the range depth of interest. In contrast, a flash lidar focal plane such as the ASC design, samples a natural volume defined by the pixel tube depth because it captures an entire reflected waveform with 20 or 44 samples over each flash. While the gated mode approach is less versatile than the independent pixel mode in that the range window must be advanced with each pulse if target range is variable, it has the advantage of a 2× to 3× improvement in signal-to-noise ratio (SNR). This is because no threshold circuitry is involved to add noise to the sampling process. The higher sensitivity allows longer ranges for a given pulse energy. Using this mode, if there is no knowledge of object positions, the entire depth space must be gate sampled which introduces latency and hence somewhat less than real-time performance. Interrogating the entire space of interest gate-by-gate is an inefficient use of laser energy and is a time intensive process requiring multiple laser pulses. However for long ranges, limited laser power, or small Rx apertures, it offers a compelling tradeoff, particularly when combined with other information that allows for a bounding of the gate sweep. For very long ranges, the signal for each gate must be averaged over many laser pulses in order to gain the benefits of improved SNR. In general, the gating and the signal summing severely limit time resolution of this mode. By contrast, flash lidar as defined by independent pixel mode, uses only a single laser pulse to interrogate the entire depth of space defined by the FOV with each frame where only the laser/video repetition rate (as frames per second (FPS)) acts to limit the time resolution of movement within the scene. Unless called out specifically, flash lidar will refer only to independent pixel mode (or HITS mode in ASC parlance).

2.5.3 Putting It All Together

The overarching challenge of the system integrator is to get useful information to the user to support real-time or near-real-time decisions. In some scenarios, this means sufficient embedded data processing power to produce a processed digital elevation map (DEM), performed at the sensor and at the frame rate of the sensor. In other scenarios, this might mean only basic pixel processing and color fusion at the sensor to decrease the data bandwidth, followed by a radio frequency (RF) downlink, with further non-embedded processing on the ground. Pixel processing, in this case, refers to the basic embedded processing necessary to format raw data into the "cubes" mentioned earlier along with associated headers and metadata containing state information such as GPS/IMU data. In either case, the latency to produce a georeferenced DEM or other actionable intelligence products is on the order of a second. Data can be output to storage or other processing "consumers" at any intermediate point in the processing chain.

As an example, Figure 2.58 summarizes the BATC real-time processing chain, providing typical data rates at each point in the processing. For operation at 30 Hz, raw data rates from the lidar instrument and color context camera are 20.6 MB/sec and 23.6 MB/sec, respectively. The real-time processing chain performs the following processes on the data stream:

- Pixel processing immediately reduces the lidar data stream to 1.97 MB/sec in an optimized range/intensity/quality frame format. Note that the frame-wise nature of the data means that only one set of GPS/IMU state data must be saved per frame (16,384 points) of lidar data, dramatically reducing metadata requirements versus a spot scanned system. For implementations which cannot support the size, weight, and power for further onboard processing, this (or after optional color video fusion) is the optimum point for an RF downlink in this case.
- Orthorectification expands the data stream from a frame-wise format to a point cloud form. In this form it is possible to drape high-resolution color imagery over the 3D surface sampled by the lidar, and present full motion 3D video in real-time to a user on an off-the-shelf PC running the Ball Lidar Viewer, a Windows™ application for viewing lidar and context video streams.
- Geolocation maintains the point-cloud format of the data but georeferences the points, so that they are suitable for output to LAS format or for accumulation into an extended-area map or panorama. Up to this point in the processing the latency introduced to the data stream is <<1 sec.
- Scene stitching is an enhancement to geolocation that is uniquely available to flash lidar data due to its frame-wise nature. Intra-scene features can be utilized to improve the alignment of points from one frame to its neighbors.
- Re-gridding involves down-sampling the lidar data to a predefined grid resolution specified by the mission requirements. This can result in dramatic reductions in data volume. For example, in one set of Ball test flights, a multiple-pass data collection coupled with the collection geometry and airspeed resulted in significant oversampling of the area of interest. Not only did this enable improved alignment of the data frames with each other, but when the data was resampled to a 30 cm fixed grid, the total data volume was reduced by over 90 percent while at the same time realizing a significant improvement (reduction) in range noise.

2.5.3.1 Scan Mirrors on Flash Lidars?

While flash lidar does not require an aperture scanning mirror, its use is enhanced by the added capability. Examples of this enhancement include platform motion and jitter removal, staring applications from moving platforms, scanning a reduced FOV for longer range, scanning shaped beams optimized for certain applications etc. The Ball Total Sight™ Lidar is an example of such a system and is intended for the commercial and tactical military arenas. The system incorporates an ASC 128 pixel × 128 pixel flash lidar detector array and is mission reconfigurable, accommodating different laser systems, projection optics, lidar receiver optics, large-aperture wide-angle scanning mirror (LA-WASM), and additional VIS/IR context cameras. Ball has flown a number of different context cameras with the flash lidar, including visible, SWIR, and MWIR cameras, allowing fully multispectral scene analysis. These systems may be designed such that the receiver can accommodate an external platform laser system. This feature allows a host system laser on board an existing platform (such as a laser designator) to be used as the illumination source. Dual use of an existing laser substantially reduces the size, weight, and power requirements of the flash lidar. Various Ball flash lidar systems in different flight configurations are shown in Figure 2.59. Other recent configurations include the compact Topographic Mapping Flash Lidar (TMFL) as well as the new Total Sight™ commercial prototype on various platforms in Figure 2.60.

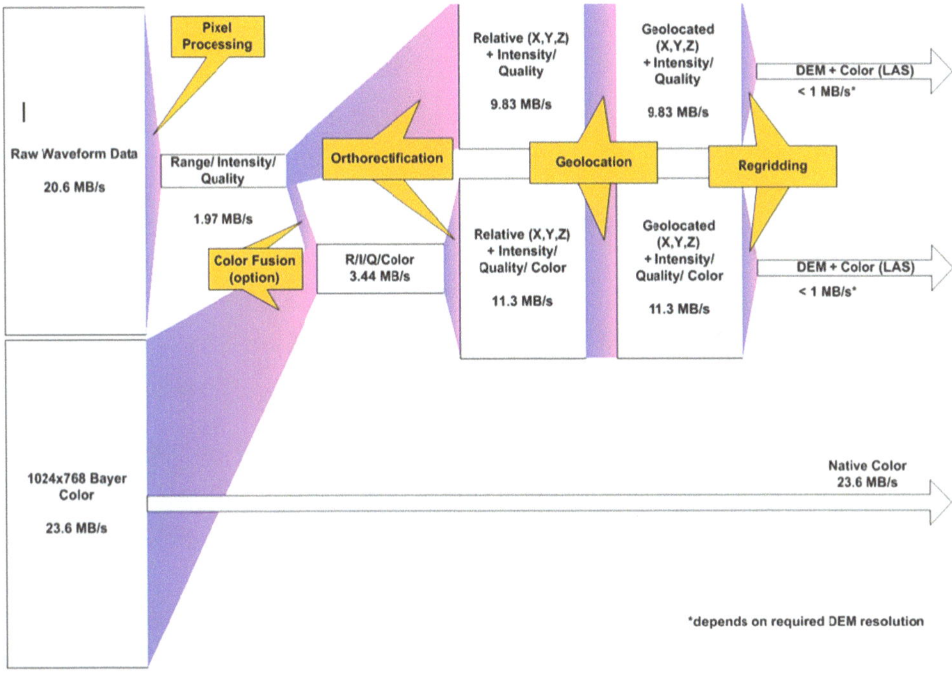

Figure 2.58 The Ball Aerospace Total Sight™ real-time processing chain enables a "map as you fly" capability and reduces data downlink bandwidth requirements for real-time delivery. The processing chain and outputs are configurable based on mission and data requirements; data rates shown are for operation at 30 Hz, uncompressed

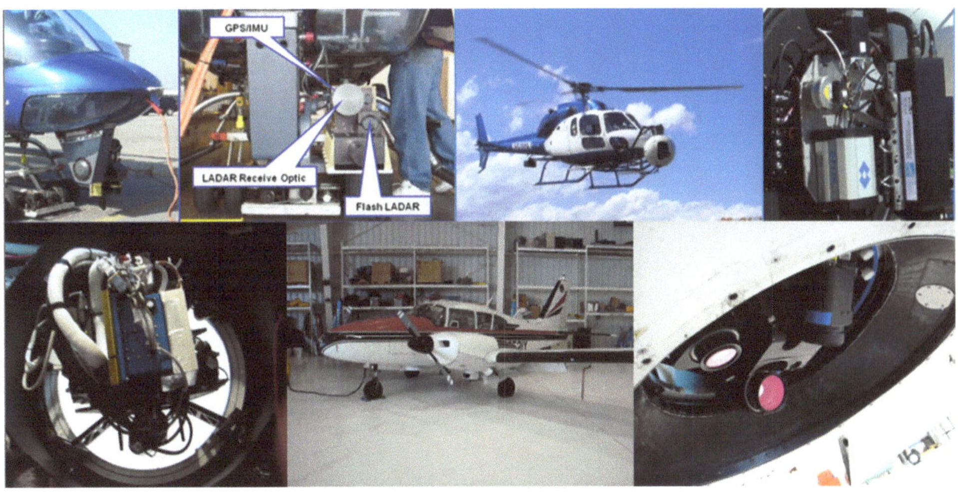

Figure 2.59 Previous generation Ball Aerospace flash Lidar systems have been flown on multiple aircraft and in different configurations. Context imagery is provided by visible, SWIR and MWIR cameras, depending on requirements. Top row are early prototypes being flown on a helicopter nose mount and a PictorVision gimbal for demonstration. Bottom row images are that of a custom prototype configured to assist an airborne Pictometry™ camera suite.

Figure 2.60 Top row: The Ball Aerospace Topographic Mapping Flash Lidar (TMFL) in its mounted configuration on a Twin Otter aircraft. Bottom row: the Ball Total Sight™ Flash Liddar with a SWIR context camera on a stabilized PictorVision™ gimbal as mounted on an AStar™ helicopter prior to its flight over the Griffith Observatory in Los Angeles, California.

Figure 2.61 The Griffith Observatory near Los Angeles, California using the Total Sight™ Flash Lidar. Left is the stitched, shaded point cloud colored by height as seen by ground control during and directly after the flight using three transects of the scene in August 2011. Upper right is the satellite view as seen on Google Earth™. Lower right is a KML depiction of all flight transects flown that day as plotted in Google Earth™.

Three-dimensional battle space characterization, intelligence, and situational awareness operations demand actionable information with minimal latency. Wide-spread use of lidar systems and data has been limited in part by complexities in data processing, limited bandwidth of ISR systems, and data latency. Ball's Total Sight™ 3D Flash Lidar is one example of a system that provides these capabilities by combining flash lidar sensors with state-of-the-art lasers and robust real-time processing systems. The combined capability enables a number of low-latency intelligence products, such as real-time 3D "map as you fly" technology, full-frame georeferenced orthorectification for high precision targeting, and 3D target detection and tracking. The same technology can be utilized in tactical applications to aid helicopter pilots in cluttered landing zones or visually degraded environments, such as helicopter "brown out." Brown-out is the dust-blinded situation that pilots of vertical takeoff and landing craft such as helicopters and V22 Osprey's find themselves in dry dusty conditions. Flash lidar operating in gated mode, is capable of selectively eliminating photons reflected from virtually all of the dust except that near the landing target or dangerous obstructions allowing a true visual look through for safer landings and takeoffs.

Table 2.8 Specific benefits of real-time 3D Flash lidar

• Suited to real-time video and "live" dem generation
• Can be extremely compact when necessary
• Insensitive to target or platform motion
• Platform jitter may be used for super resolution
• Stitched correlated frames reduce random noise
• Frame by frame feature correlation
• Insensitive to ambient and stray light
• Multi-sensor real time video symbiosis
• Suited to real-time 3D or holographic displays
• Reduced image storage requirements
• Images easily fused or draped with 2D context video
• High shock tolerance, reliability and long life
• Low computational power draw requirements
• Simple to calibrate and georectify
• Unambiguous, absolute, full-frame tof ranging
• Instantaneous point geo-location via GPS/IMU
• Range and intensity co-registered per pixel
• Pixels perfectly registered within a frame
• Precision scanning mechanisms useful but optional
• Super-resolution techniques easily implemented
• Smaller/lighter than raster scanning systems
• Ability to "see" into obscuration (range-gating)
• Full volumetric 3D scenes possible with multiple Synchronized flash lidar cameras
• Eye-safe laser capable

The key advantage of flash lidars relative to laser spot scanning systems is the ability to instantly locate pixel-correlated range information within an absolute frame of reference; otherwise known as geo-location, using knowledge of inertial location, pixel offset, pixel range, and scan angle over time all at video rates or higher. Namely, these are lidars optimized for video rate geo-located mapping, target tracking, scene stitching, etc.

Flash lidar technology is also inherently efficient with mass, power, and volume. This is because the system requires only one set of receiver optics, one detector, and no mechanisms. Using one set of optics and detector to provide both long and short range functionality not only saves mass and volume as compared to using two separate range sensors, but also provides a simpler interface with the aircraft. Because the laser is only used for flood illumination, there is no complex mechanism needed to control laser pointing nor is there a need for precision alignment between the laser and receiver optics. Use of an imaging focal plane array yields a self-aligning system, thus significantly simplifying assembly, integration and test. Elimination of both tight laser alignment requirements and scanning mechanisms also cuts parasitic mass and power, and enhances the sensor's shock tolerance, reliability, and useful life.

3D flash lidar sensors are also insensitive to ambient lighting conditions. Because the system captures a new complete image with each flash, glint and clutter outside of the expected range to the target may be filtered and discarded. This capability makes data processing extremely efficient and reliable since range, bearing, and pose algorithms do not have to deal with interpreting data from glint and clutter.

Modern digital flash lidar systems determine the time of the peak of the returned pulse, and in doing so, are able to achieve excellent range precision. This higher precision combined with the electronics and algorithms to process the image data in real time allows these systems to achieve superior real-time geo-location accuracy as compared to scanning lidar systems. Real time processing can also dramatically reduce data storage requirements. Additionally, there are published results suggesting that enhancement to 3D flash lidar data models is possible and applicable to all 3D flash lidar products. Analysis of processed flash lidar data indicates that 4x resolution enhancement (known as "super resolution" (Bulyshev et al., 2011)) is feasible through platform jitter or intentional "micro-scanning" (Watson et al., 1992). Studies also show reductions in random noise as multiple image frames are blended to create a single high resolution DEM. This has been demonstrated in BATC prototypes and has been incorporated into resident firmware. Finally, the very concept of the modulation-transfer function or MTF must be modified to account for the 3D nature of the data in relation to overall optical system design and was developed early in the Ball program (Staple et al., 2005).

3D flash lidar cameras are available from Advanced Scientific Concepts, Inc. (ASC 3D) and Raytheon Vision Systems (RVS), both based in Santa Barbara, California. While ASC has shipped various iterations of its products for Space (STS-127, STS-133, Space Exploration Corporation (SPACEX) Dragon vehicle Space Exploration Technologies, 2011)), unmanned air and ground vehicles and surveillance remains the key contributor to the technology's development. RVS has limited its public activity to the Ball built NASA Sensor Test for Orion RelNav Risk Mitigation (STORRM) Development Test Objective (DTO) tested on STS-134. ASC 3D flash lidar cameras, which are available commercially, currently have 16,384 (128 × 128) independent pixel range-finders on each sensor chip, allowing the sensor to act as a 3D video camera with functionality well beyond just range finding. These cameras are capable of streaming approximately 500K points per second, with processing done on-camera to generate the 128 × 128 range maps at 30 Hz. ASC cameras have demonstrated single-pulse 3D flash lidar imagery over physical ranges from centimeters to kilometers. Higher pixel counts are in development and will be available in the near future in various aspect-ratio formats. The cameras are available standalone in the TigerEye™ and TigerCub™ versions. RVS flash lidars used in the STORRM DTO are 256 × 256 format and exhibit similar characteristics, though without sampling full waveforms. Many other related systems and capabilities exist in the military and intelligence arenas.

Ball Aerospace has built a number of flash lidar prototypes to meet specific platform and operational requirements. In the autonomous rendezvous and docking (ARD) arena, perhaps the best known is the NASA STORRM flash lidar which demonstrated ARD capability for unmanned International Space Station (ISS) re-supply ships. This sensor incorporated a Raytheon Vision systems FFPA and an eye-safe 1570 nm Fibertek laser as part of a larger Lockheed-Martin system known as the "Vision Navigation System (VNS)." This system demonstrated ISS docking ring acquisition from beyond 5 km and provided real-time pose information capable of guiding a re-supply ship to docking at sub-meter range: see Figure 2.62.

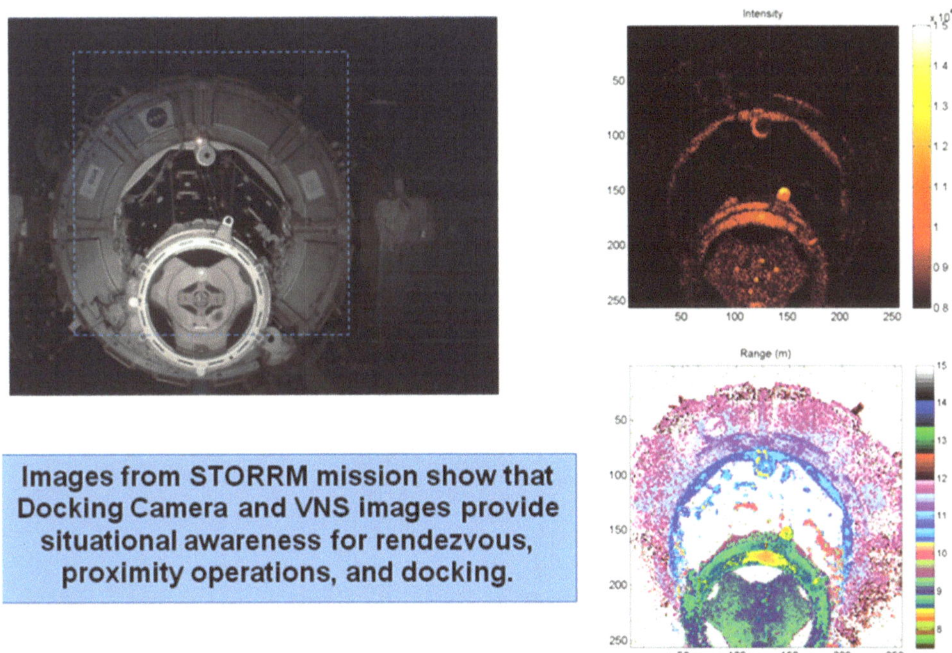

Figure 2.62 Images from STORRM mission.

2.5.4 Mapping for Civil, Commercial, and Military Uses

2.5.4.1 Map As You Fly

Building DEMs using lidar imagery is not new. The imagery is simply stitched together and often fused with visual imagery into larger images or maps of a target area. Prior to the Ball Aerospace Total Sight™ system these data products have been produced in postprocessing and involved a number of steps including data download and distribution, GPS or iterative closest point (ICP) stitching, and frame association. The time from data collection to data product dissemination can take hours to days with current technology. The Ball Total Sight™ Flash lidar system is unique in that it combines a proven processing chain with Ball proprietary real-time processing algorithms that drastically reduces the time between data collection and data product from hours to days to a matter of seconds. Total Sight™ utilizes a panoramic "super frame" concept, where sequential imagery is associated spatially. The proprietary processing system stitches the individual images together to form a 3D map of the area of interest. This process is described graphically in Figure 2.63. The upper images in this figure represent single lidar imagery frames from the Total Sight™ sensor. The real-time algorithms utilize inertial navigation sensor inputs and scene characteristics to stitch the individual images into a DEM. Subsequent frames are added to the DEM in real-time. The creation of real-time 3D mapping products is essential in applications such as battlefield management, search and rescue, disaster relief, and hazard avoidance where low lantency of information for decision making is critical.

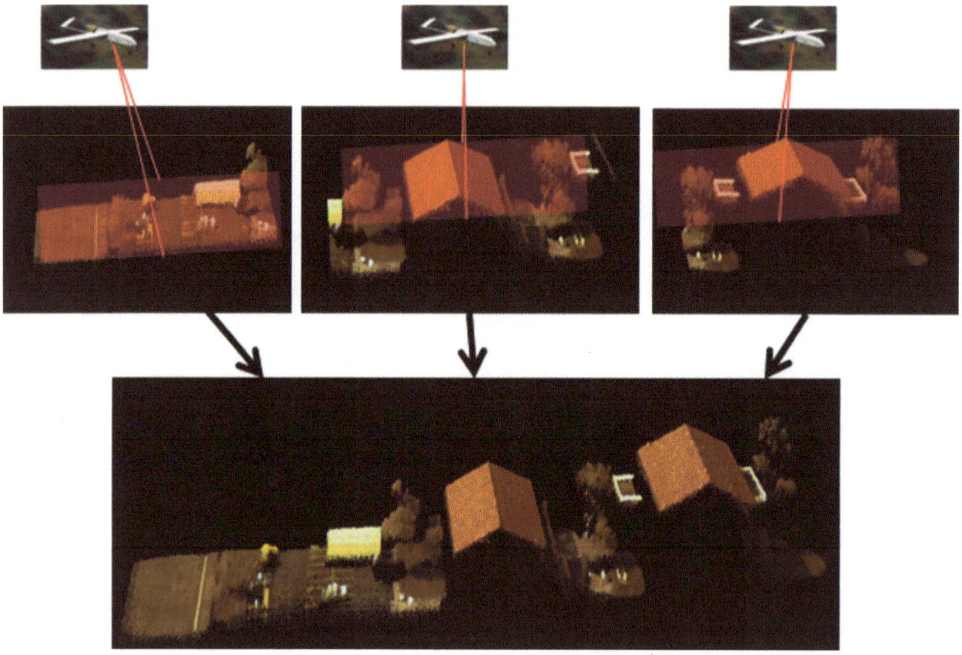

Figure 2.63 The Ball Aerospace Total Sight™ Flash Lidar associates individual image frames (top) to create integrated 3D mapping products or panoramas in real-time.

The "Map as You Fly" concept shown in Figure 2.64 is from a flight test over Denver, Colorado in May 2010. This data was used as a test case in the development of the real-time processing system. The data contains primarily urban imagery from a helicopter flying at ~20 meters/sec. The DEM shown in Figure 2.64 is of the Denver South High School. The image is created by using five separate transects of the site that are stitched together using the Ball real-time algorithms.

The availability of real-time DEMs and mapping products has obvious advantages to applications where data latency cannot be tolerated, such as battlefield characterization and disaster relief. More specialized applications include improvised explosive device detection/geographical change detection and improved military targeting using full-frame geo-rectification.

Flash lidar in an integrated system can further increase the value of a low-latency mapping mission. For example, in response to a DHS SBIR, Pictometry created their Real-time Aerial Mapping System (RAMS). The purpose of RAMS is to capture aerial imagery as a disaster or major event unfolds and to process and transmit that imagery to the ground in real-time. The RAMS system can have high-resolution aerial images rectified and on the ground in an exploitation system within 15 seconds of capture out to a 20 mile radius. An example of this imagery is shown in Figure 2.65. Pictometry and Ball Aerospace teamed to add the real-time lidar capability of the Total Sight™ system to RAMS in order to gather surface elevation data as well.

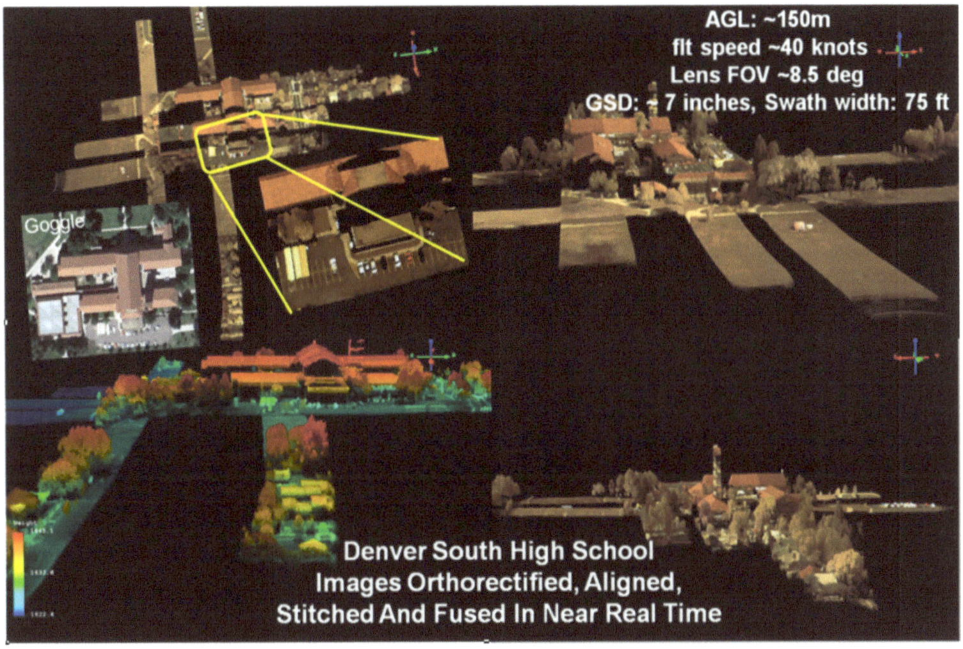

Figure 2.64 A "Map as You Fly" DEM of Denver South High School created from a helicopter data collect in May 2010. The image is the result of the stitching together of five separate transects of the site.

Figure 2.65 Images of Orleans Parish before and after Hurricane Katrina. The combined system could provide a better operational picture of water levels, structure damage, and road impassability, and support quantitative analysis of debris fields.

The integrated system would capture the lidar and imagery data simultaneously from a single aircraft and transmit that information to the ground through a high-speed point-to-point tracking radio link. This information could be used to help assess the current condition of roads, bridges, and structures after a major storm or other catastrophic event. As the event progresses, this same system can economically collect data about the size and locations of debris fields. Captured separately, it might be difficult to distinguish between debris fields and vegetation, but combined with the high resolution imagery, an operator can distinguish between the two.

(a) Lidar range + intensity only

(b) Lidar range + intensity + fused color

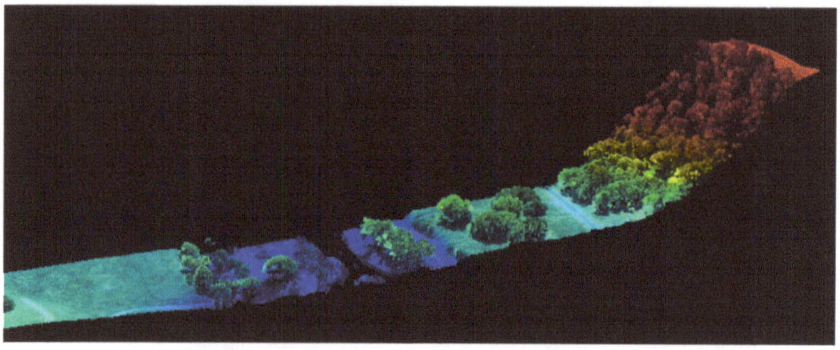

(c) Lidar range + intensity: colored by height

(d) Lidar range + intensity + fused color: colored by height

Figure 2.66 Side-by-side comparison of the variety of information available in real-time, along-track DEM fusions that can be streamed to the pilot and/or ground personnel looking for features or changes: (a) lidar range + intensity only, (b) lidar range + intensity + fused color, (c) lidar range + intensity: colored by height, and (d) lidar range + intensity + fused color: colored by height.

2.5.4.2 Geographic Change Detection

Geographic change the residual of the comparison of high fidelity DEMs acquired at different times. Geographic change detection is broadly used in applications such as improvised explosive device detection as well as in disaster relief efforts. Changes between DEMs could include flooded areas, washed out, blocked or destroyed roads, or disturbed earth along roadsides and thoroughfares. Creation of DEMs from airborne data is time and labor intensive and introduces delays in the required civil or military response. Real-time processing and creation of 3D mapping products are essential for these unique applications due to the time critical nature of the resulting product. The Ball Aerospace Total Sight™ Flash Lidar and processing system enhances the utility of these data products by getting the data into the hands of users instantaneously. The more quickly these products can be transmitted to field commanders or relief workers, the more effective they can be in responding to these situations. Ideally, the airborne imagery would not just be processed in real-time, but also be transmitted to the ground and displayed in real-time, which Ball has demonstrated (see Figure 2.67).

Figure 2.67 (a) An airfield hangar at night with Lidar point-cloud image fused with the reflected intensity return only, and (b) the same image fused with MWIR video to show thermal signatures. Note the cold metal roofs that appear in black as well as the cold ground signatures of recent cars parked there.

2.5.4.3 Full-Frame Georectification

The ability to build 3D mapping products in real-time also improves the fidelity and accuracy of battlefield targeting. Intelligence, Surveillance, and Reconnaissance (ISR) image data products are used for real-time situational awareness, battle space characterization and target tracking and strike. This data is also cataloged for intelligence analysis where future targets may be identified. Imagery is often collected by an integrated sensor turret on an aircraft that combines 2D imagery with a single range point from a laser range finder. This means that the imagery often has a single georectified point at the center pixel of the image from the sensor's laser range finder. Target geolocation accuracies degrade as the center pixel range is associated with targets on the outer parts of the image. These errors are compounded in rugged terrain or in urban settings, where scene topography can vary greatly. The Ball Aerospace Total Sight™ Flash Lidar mitigates this problem by acquiring a range grid over the entire field of view. An example of this is shown in Figure 2.68, where every point in the image has an associated range-based geolocation. This provides more accurate coordinates for targets in both real-time and in post processing of intelligence data. The more accurate target coordinates result in more precise offensive strikes and reduced collateral damage.

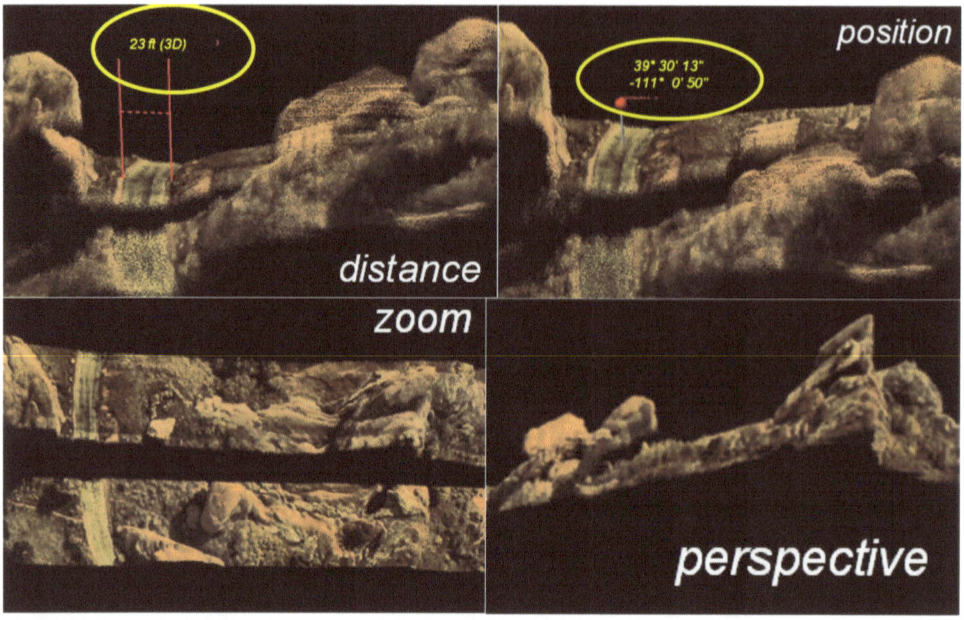

Figure 2.68 Full-frame georeferenced and orthorectified images result in higher accuracy object geolocation of targets for more precise offensive strikes and reduced collateral damage.

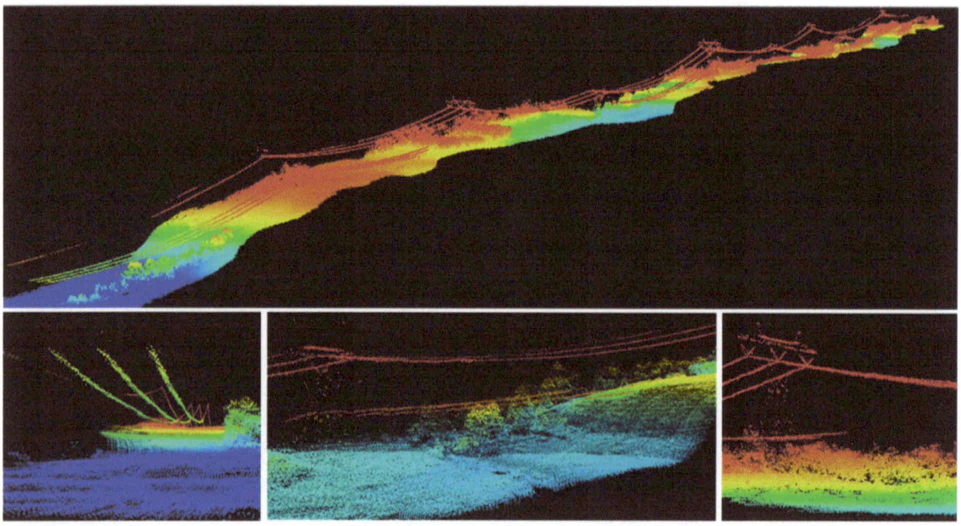

Figure 2.69 Ball Total Sight™ real-time data products from a Pictometry corridor mapping flight shows terrain features as well as power line height and sag. This data was simply downloaded from the system and ingested into QT Modeler for visualization.

2.5.4.4 Corridor Mapping

Routine mapping for corridors such as high tension power lines and pipelines is required by State and Federal agencies, which is an expensive, repetitive and processing intensive task. The Ball Aerospace Total Sight™ Flash Lidar and real-time processing system expedite this process by creating detailed 3D mapping products of these corridors in real time. Pictometry flew the Total Sight™ sensor over a power line corridor on board a Piper Aztec in the fall of 2010.

The resulting Total Sight™ data was rapidly converted into LAS format and displayed using QT Modeler™ with no additional postprocessing. The image in Figure 2.69 shows a sample of the resulting data. The larger image is over the power line corridor where the power lines and stanchions are plainly seen over the corridor terrain. Parameters such as line height and line sag are easily interpreted from the resulting DEM. It is important to note that this data was downloaded from the aircraft with no additional processing prior to LAS conversion. This capability results in more rapid, efficient, and less costly corridor mapping projects.

Figure 2.70 illustrates the same capability from a UAV platform for tactical decisions.

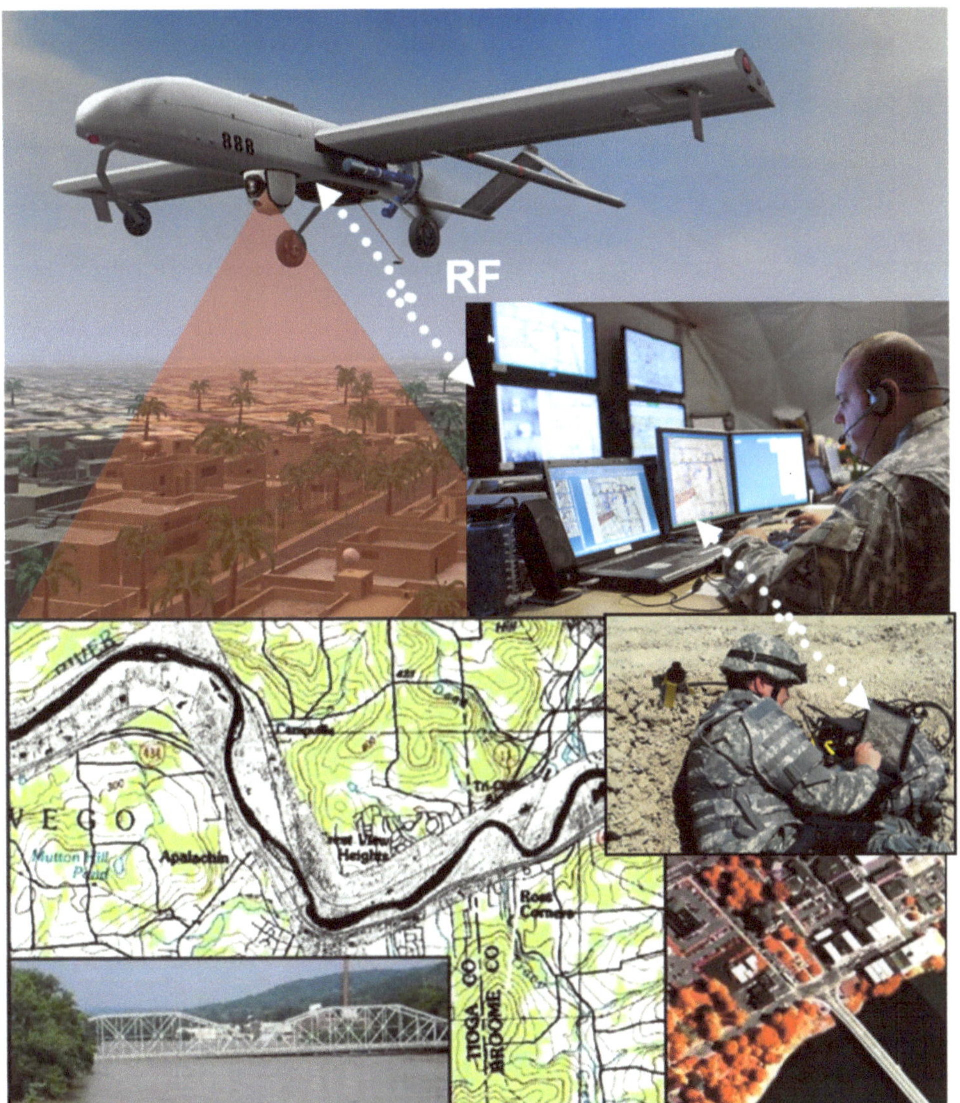

Figure 2.70 Reduced SWAP (size, weight and power) of 3D flash Lidar over scanning systems allows for compatibility with multiple platforms, real time RF downlink and autonomous processing. Real time orthorectified and geo-referenced 3D imagery enables night time ops, rapid change detection support and real time tactical decision aids.

2.5.5 A Brief Historical Aside – First Use of 3D Flash Lidar

The first major development and application of 3D flash Imaging Lidar was made under the Office of Naval Research's Organic Mine Countermeasures Future Naval Capabilities (OMCM FNC) program. At the time, Tucson based Lite Cycles, Inc. was developing an innovative and highly compact airborne active sensor system for mine and obstacle detection in the littoral zone. Known as Rapid Overt Airborne Reconnaissance (ROAR™), the system used an innovative integrated scanner, detector, and telescope (ISDT) receiver architecture. The ISD coupled all receiver components and lidar electronics to achieve the system compaction required for tactical UAV integration while providing a large aperture. It also included an advanced compact multifunction laser transmitter; an industry-first high-resolution, compact 3D camera from Advanced Scientific Concepts, and a scanning function for wide area search. The goal was a highly compact, airborne, active multi-spectral/3D imaging sensor for rapid overt mine, minefield, and obstacle detection and localization in very shallow water (VSW) and the surf-zone (SZ). The VSW is notionally defined to be 10 feet to 40 feet water depth and the SZ is notionally defined to be 10 feet water depth to the mean high tide line. The ROAR™ system was designed for deployment from a tactical UAV and adopted the Navy's vertical takeoff unmanned airborne vehicle (VTUAV) parameters and the program provided much of the early funding and development impetus behind ASC's revolutionary camera. Further development of the ASC camera continued under the direction of the Air Force Research Laboratory, Wright-Patterson AFB in areas such as ROIC design, super-resolution algorithms and automatic target recognition algorithms.

2.5.6 Adaptive Lidar for Earth Imaging from Space

The Electronically Steerable Flash Lidar (ESFL) (Weimer and Ramond, 2011) is another flash imager invented at Ball Aerospace. A demonstration unit has been built under a NASA Instrument Incubator Program (IIP) grant with the help of co-Investigators Michael Lesfky of Colorado State University, Ingrid Burke of the University of Wyoming, and Yong Hu of NASA Langley. The instrument concept is illustrated below in Figure 2.71. A 1064 nm Nd:YAG pulsed laser passes through an acousto-optic crystal. A piezoelectric transducer attached to the crystal is commanded by a custom RF driver. The transducer sets up an acoustic wave in the crystal which creates a transmissive diffraction grating that splits the input laser beam into undeflected and deflected beams. Every RF drive frequency produces its own standing wave, which in turn causes its own deflected beam at a specific deflection angle. Changing the frequency of an RF tone changes the angle of the corresponding deflected beam.

The scene is imaged onto a flash focal plane array. The flash focal plane is a pixilated array, with each pixel having its own processing electronics. The arrays are comprised of detectors bonded to a ROIC (read-out integrated circuit). The detector array, which converts the incident photons into electrons, is typically InGaAs or Si and can be APDs or PINs. The ROIC contains all the processing electronics for the collected electronics such as buffers, counters, and amplifiers.

Each laser footprint may be imaged onto a single or multiple pixels. The RF driver is commanded by the payload computer, which can be synchronized to feedback from payload attitude control as well as other data from the focal plane. The ESFL IIP instrument was flown on a Twin Otter aircraft in November 2009, May2010, and on an AeroCommander in September 2010.

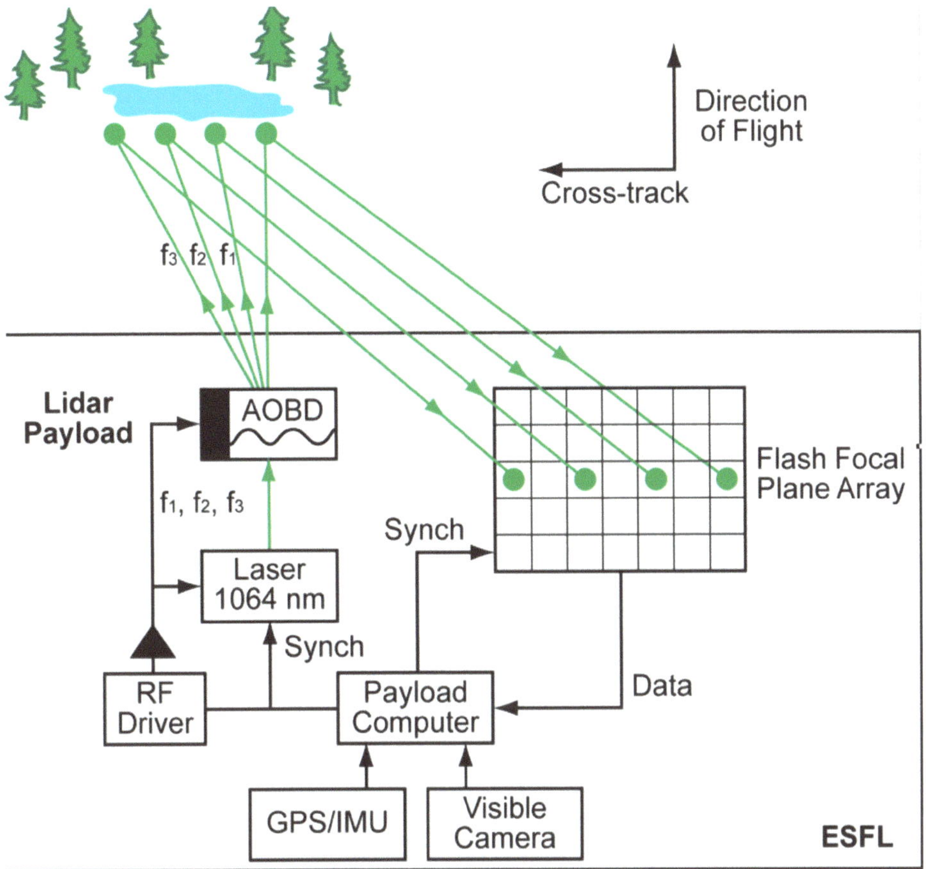

Figure 2.71 ESFL notional concept in the form of a functional block diagram.

2.5.7 Advantages of the Design

2.5.7.1 General Advantages

The general advantages of this design are:

- No scanning mechanism
- All data from a single lidar pulse is co-registered
- Eliminates many errors associated with scanning mirrors
- Sub-sampling of beam footprint
- Beam can be imaged onto one pixel, or multiple pixels to enable multiple sampling densities simultaneously
- Spatial configuration can be changed shot-to-shot
- Number of beams N can be altered at any time, limited primarily by radiometry.
- Can be reconfigured to adapt to the scene below to steer around undesirable targets and steer to desirable ones to maximize photon efficiency.
- Beams can be steered toward a pre-determined desired target (see example below).

2.5.7.2 Advantages for High Altitude Flight or Space

There are many features of the ESFL design that make it particularly advantageous for space flight, or for high altitude airborne flight as well. The ESFL IIP instrument has been designed with a path to space in mind, utilizing the lessons we learned from the CALIPSO lidar development and its on-orbit performance. Many of these advantages would apply equally to flight on a high altitude aircraft. An instrument concept for a space-based ESFL has been developed based on a 400 km orbit, each beam having a footprint with a 25 m diameter and imaged onto 3 pixels × 3 pixels on the flash focal plane array. It is important to note that other configurations are possible, limited primarily by the required signal-to-noise ratio. The current paradigm for flying lidars in space is to use a "transect" approach where M lasers are needed to project M fixed-position beams on the ground. In contrast, there are many advantages that the ESFL instrument offers:

- Single low repetition-rate laser
- Low repetition rate (40 Hz) means fewer shots required for a mission and thus higher reliability
- Fewer spacecraft resources use less power, less mass, less size than a system with greater number of lasers
- Simpler optical system offered by only using one laser means easier integration and therefore less cost
- Provides across-track coverage
- Up to ~6 km across-track deflection
- Spatial configuration can be changed shot-to-shot on orbit to steer around "undesirables"
- Cloud-avoidance capability
- Separate forward-looking camera can be used to feed to the ESFL and steer through holes in the clouds when possible
- Boresight control
- Eliminates need for a separate mechanism
- Sub-sampling of beam footprint
- 25 m beam diameter on ground spread over 3 pixels × 3 pixels
- 8.3 m sampled per pixel.

Furthermore, the flexibility of the beam steering design allows the possibility to satisfy users with very different requirements simultaneously. For example, there is the potential to satisfy sampling needs of both biodiversity and biomass communities within the vegetation science world:

- Beams can be spread out evenly across the 6 km (~ 1 km spacing) to address biomass sampling schemes
- Beams can be made contiguous to address biodiversity sampling
- ESFL configuration can switch between these two schemes on–orbit at any point
- Newer concept being developed combines cross-track sampling and contiguous nadir along-track sampling.

In order to demonstrate the potential for adaptive lidars for Earth remote sensing, a demonstration unit of ESFL for aircraft testing was funded by the NASA Earth Science Technology Office. The system was applied to the problem of remote sensing of the Earth's forests as defined in the NRC Decadal Survey of Earth Science Mission - DESDynI (Deformation, Ecosystem Structure, and Dynamics of Ice). A picture of the airborne ESFL instrument is shown in Figure 2.72. The transmitter begins with a 5 ns, 1064 nm pulsed laser beam. The laser beam crosses a beam expander and then enters the acousto-optic beam deflector (AOBD), which is a commercial TeO_2 shear wave device. A custom RF driver was designed and built for the ESFL application at Ball Aerospace which allowed us to design the direct digital synthesizer using techniques that could be easily translated to a design for space. The AOBD and custom driver allowed the output of up to 10 beams placed at any configuration within the 8.6 degree FOV, which translates to a range of 25 to 55 MHz RF frequencies.

Figure 2.72 ESFL airborne instrument.

The focal plane is a 128 pixel × 128 pixel InGaAs FFPA from Advanced Scientific Concepts a technology partner for the effort. Each of the 128^2 pixel elements provides a full lidar waveform. In contrast to detectors that output a waveform from only a single element, this focal plane is also a waveform imager. The focal plane oscillator rate is adjustable between approximately 100 to 300 MHz but is typically set to 200 MHz and can divide the returned signal into 44 time bins of 0.75 m each. The instrument operates at a 30 Hz frame rate.

A visible camera is co-boresighted to the lidar to provide context to aid in data collection and processing and to provide input for adapting the beam pointing.

ESFL flies with an Applanix POS (Model 510) global positioning system/inertial measurement unit (GPS/IMU) that is co-registered to the lidar receiver FOV prior to flight. GPS and inertial movement of the sensor are recorded in the data stream, which after postprocessing allows geolocation of every pixel in the focal plane for every shot. Typical geolocation errors are sub-meter after post-processing.

2.5.7.3 Aircraft Demonstrations of Concept

Figure 2.73 shows an example of the ESFL capability to reconfigure the projected beam pattern shot-to-shot. Three successive laser shots are shown. In the first shot, a single beam was deflected across-track. In the next, three are output. In the final frame, eight beams are shown deflected across-track. Surface albedo variation, in part, causes the change in intensity between beams. In flight, this pattern was repeated over and over again. The positions and numbers of beams were arbitrarily chosen in this example, but it serves to illustrate a major departure from the single fixed-beam "transect" approach to space-based lidars that is currently the paradigm. If instead there were some knowledge of the scene gathered real time from a separate camera, the number of beams and their positions could be positioned to maximize the observation desired.

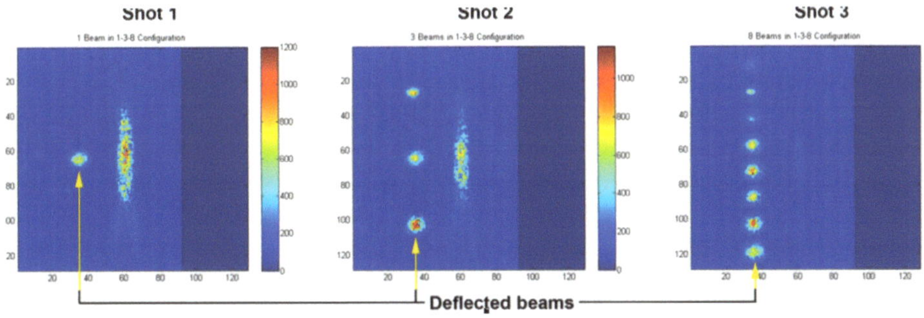

Figure 2.73 Three successive laser shots showing three different beam configurations. The transmitter cycles between 1, 3, and 8 deflected beams at a rate of 30 Hz. Data have been uncorrected for focal plane responsivity in this example. The light remaining in the undeflected laser light was spread into a line and projected ahead of the deflected beams allowing its use and maintaining eye-safety.

One example of such a closed-loop implementation can be seen in Figure 2.74. In this case, a laboratory test was performed to demonstrate how ESFL can be used to avoid clouds. Randomly-generated synthetic cloud patterns of varying density were generated in MATLAB. These were used to create a movie to simulate a dynamic cloud scene, as would be expected to be viewed in the lidar FOV on-orbit. This movie was projected on the laboratory wall and the lidar aligned to this scene. The default beam configuration was chosen by the operator; a three beam case shown in the left panel of Figure 2.74. The visible context camera co-boresighted with the ESFL lidar was used to image the cloud movie. The lidar computer controller received the data from the camera and in real-time identified the regions with and without clouds using a simple brightness-based algorithm. The AOBD was commanded to change the beam deflections of the three beams independently and in real-time, so as to avoid the clouds. The entire scene, including cloud pattern plus ESFL beams, was detected on the lidar receiver. The approach demonstrated that system latencies were sufficiently small that the multi-beam pattern could be steered around clouds at speeds that would be necessary for potential space missions. Future work with ESFL will include flying above clouds to demonstrate this from aircraft.

Figure 2.75 is an example of a swath of data taken from the May 2010 campaign over the SFAustin forest near Nacogdoches, Texas. The figure shows a river winding through the forest, which is comprised primarily of deciduous trees with a relatively closed canopy. This represents a stringent test of the instrument, as the data successfully penetrated the foliage to give a strong ground return that is essential for biomass science. The ground return together with the tree canopy can also be seen clearly in Figure 2.75.

2.5.7.4 Areas of Current and Potential Future Application

Automotive

- Autonomous navigation
- Collision avoidance

Aviation

- Helicopter brownout
- Wire detection
- Landing zone mapping

Defense

- Unmanned ground vehicles
- Unmanned air vehicles (UAV/UAS)
- Laser trackers

Marine

- Hazard identification
- Unmanned underwater vehicles (UUV)

Robotics

- Unmanned ground vehicles (UGV)

Industrial

- Welding control
- Movie set camera placement

Space

- Autonomous rendezvous & docking
- On-orbit satellite servicing
- Entry descent and landing
- Autonomous landing and hazard avoidance, collision avoidance, situational awareness / surveillance

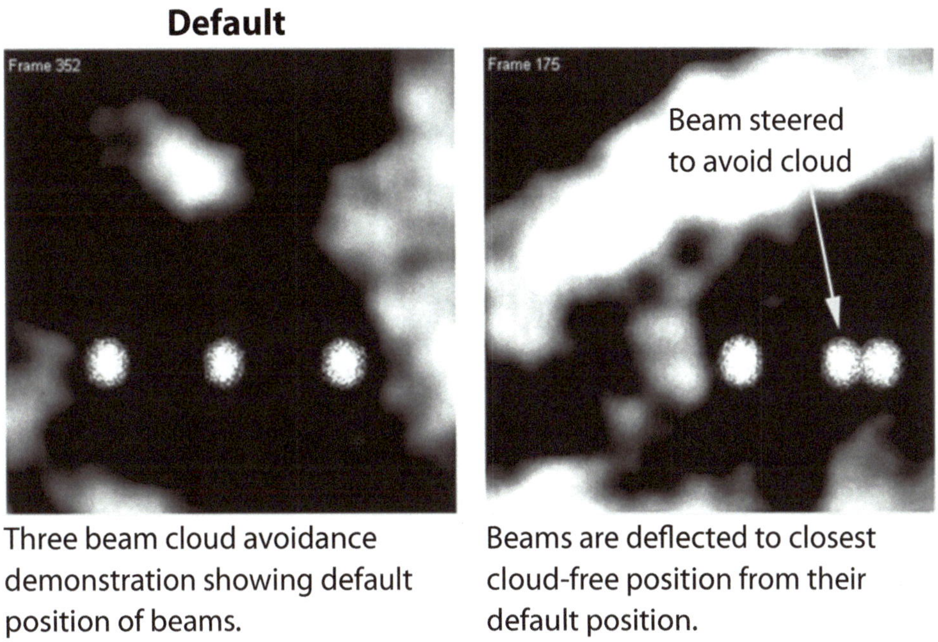

Figure 2.74 Two frames of data from the ESFL lidar receiver during a laboratory cloud-avoidance test. A simulated cloud pattern movie as would be seen from space was projected on a wall. The ESFL visible camera detects the pattern, and then deflects its beams as needed to avoid the cloud.

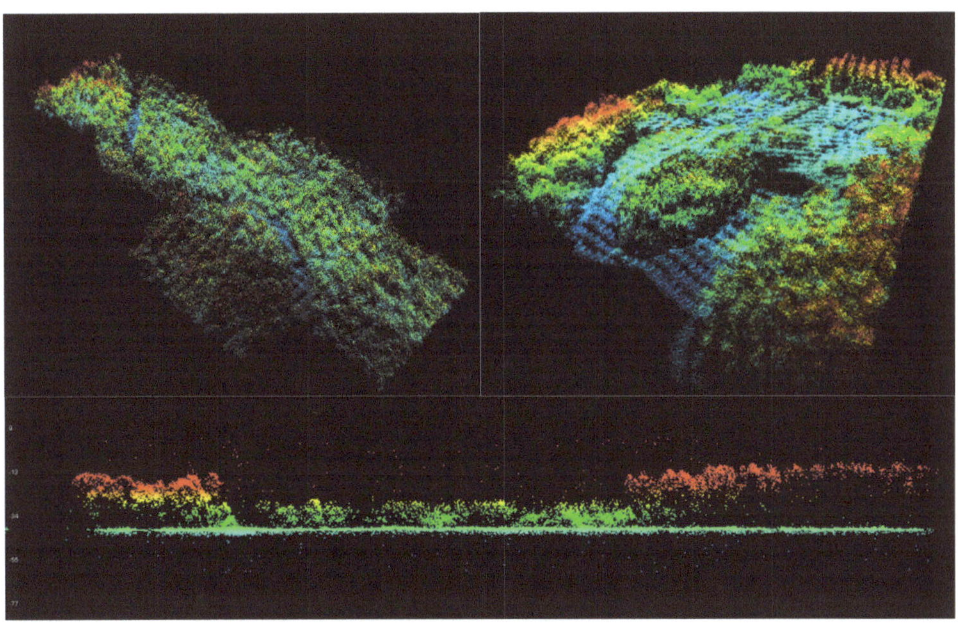

Figure 2.75 Upper Left: SFAustin Forest near Nacogdoches, Texas in May 2010. The Angelina River is also visible. Upper Right: ESFL data taken over the SERC forest in Maryland. Lower: SFAustin Forest data showing tree structure and ground returns

- Threat detection
- Object ID

Topographical mapping

- Change detection
- Hazard mapping

Transportation

- Train collision avoidance (through fog/obscuration), track profiling
- Airport bird detection
- Traffic control and monitoring system

Emergency services

- Firefighting smoke and dust penetration
- Flood mapping and monitoring
- Debris field mapping

2.5.8 Conclusion

The use of lidar data has been limited due to the cost and complexity of lidar systems, the magnitude of data and the intense processing that can be required to transform raw lidar data into point clouds and eventually DEMs and other data products. The Ball Aerospace Total Sight™ Flash lidar system combines the benefits of flash lidar with real-time processing hardware and algorithms to provide real-time 3D mapping and analysis products. This new system enhances the effectiveness of airborne lidar data where time critical information is required and streamlines production mapping product delivery. Real-time DEMs are critical to applications where data latency is problematic, such as rapid response in disaster relief efforts, improved battlefield characterization, and timely and accurate military targeting. The real-time capability can also enhance and streamline production mapping operations such as corridor mapping. The Total Sight™ system has been field tested in a number of aircraft environments and can be modified to suit customer and mission requirements.

The ESFL instrument provides a glimpse of the dramatic advantages of actively controlled remote sensing. This capability can be used for all manner of resource monitoring from both airborne and space platforms and provides an active remote sensing capability beyond anything that is currently available.

This overview provides only the briefest introduction to flash lidar as demonstrated by the engineers and technologists at Ball Aerospace, Advanced Scientific Concepts, and Raytheon Vision Systems in the long range remote sensing arena and makes no attempt to address the vast universe of other possible scientific, commercial, and military applications. This universe includes the many other types of short and very long range sensors that are becoming available in the commercial marketplace that will, no doubt, revolutionize the way we see and interact with the world around us but will find only limited use in the photogrammetric remote-sensing arena.

Our hope is that the reader will be left with the distinct impression that these instruments are something of a paradigm shift in real-time imaging and provide an exciting new avenue for the development of data products that allow for rapid, cost effective, decision making when only dimensional relationships will answer the need for information. Many times this will replace the need for individuals or teams to be in harm's way in order to ascertain these relationships. These technologies are evolving rapidly and you are invited to consider how they could be applied to solving your next real-time photogrammetric remote-sensing problem.

ACKNOWLEDGMENTS

The author would like to acknowledge the vast contributions of the airborne civil, airborne tactical, and spaceborne lidar groups of Ball Aerospace. These include but are not limited to Carl Weimer, Patrick Earhart, Eric Coppock, Tanya Ramond, Jeff Applegate, Jeremy Craner, Donald Hall, Tom Delker, Roy Nelson, Jack Hollister, Jim Medbery, Dennis Nicks, Kevin Miller, and Jim Mascarielli.

REFERENCES

Amzajerdian, F., M. Vanek, L. Petway, D. Pierrottet, G. Busch, and A. Bulyshev, 2011. Lidar systems for precision navigation and safe landing on planetary bodies, *Proceedings of SPIE: International Symposium on Photoelectronic Detection and Imaging 2011: Laser Sensing and Imaging; and Biological and Medical Applications of Photonics Sensing and Imaging 8192, 81920*, 24 May, Beijing, China, http://link.aip.org/link/doi/10.1117/12.904062.

Asbrock, J., S. Bailey, D. Baley, J. Boisvert, G. Chapman, G. Crawford, T. de Lyon, B. Drafahl, J. Edwards, E. Herrin, C. Hoyt, M. Jack, R. Kvaas, K. Liu, W. McKeag, R. Rajavel, V. Randall, S. Rengarajan, and J. Riker, 2008. Ultra-high sensitivity APD based 3D LADAR sensors: Linear mode photon counting LADAR camera for the ultra-sensitive detector program, *Proceedings of SPIE: Infrared Technology and Applications XXXIV 6940, 69402O* (B.F. Andresen, G.F. Fulop, and P.R. Norton, editors), 17 March, Orlando, Florida, http://link.aip.org/link/doi/10.1117/12.783940.

Bulyshev, A., M. Vanek, F. Amzajerdian, D. Pierrottet, G. Hines, and R. Reisse, 2011. A super- resolution algorithm for enhancement of FLASH LiDAR data, *Proceedings of SPIE: Electronic Imaging Computational Imaging IX 7873, 78730F* (C.A. Bouman, I. Pollak, and P. J. Wolfe, San Francisco, California, http://link.aip.org/link/doi/10.1117/12.876283.

Cook, D., 2007. MBE based HgCdTe APDs and 3D LADAR sensors, *Proceedings of SPIE: Infrared Technology and Applications XXXIII* 65429: 65421A (B.F. Andressen, G.F. Fulop, and P.R. Norton, editors), 05 June, Orlando, Florida, http://link.aip.org/link/doi/10.1117/12.724347.

Craig, R., I. Gravseth, R.P. Earhart, J. Bladt, S. Barnhill, L. Ruppert, and C. Centamore, 2007. Processing 3D flash LiDAR point-clouds in real time for flight applications, *Proceedings of SPIE: Sensors and Systems for Space Applications 6555, 65550D* (R.T. Howard and D. Richards, editors), 09 April, Orlando, Florida, http://link.aip.org/link/doi/10.1117/12.720143.

Earhart, R.P., R.M. Craig, C.J. Grund, I.J. Gravseth, X. Huang, and C.S. Weimer, 2011. Flash LADAR System, United States Patent 7961301.

Gravseth, I., 2009. Testing of a Flash lidar system for use as a navigation sensor in exploration missions, *Proceedings of the 32nd Annual Guidance Navigation and Control Conference*, AAS 09-09-0, 02 February, Breckenridge, Colorado.

Halterman, R., and M. Bruch, 2010. Velodyne HDL-64E lidar for unmanned surface vehicle obstacle detection, *Proceedings of SPIE: Unmanned Systems Technology XII 7692, 76920D* (G.R. Gerhart, D.W. Gage, and C.M. Shoemaker, 06 April, Orlando, Florida, http://link.aip.org/link/doi/10.1117/12.850611.

Jack, M., J. Asbrock, S. Bailey, D. Baley, G. Chapman, G. Crawford, B. Drafahl, E. Herrin, R. Kvaas, W. McKeag, V. Randall, T. De Lyon, A. Hunter, J. Jensen, T. Roberts, P. Trotta, T.P. Yuan, R. Sudharsanan, X. Bai, J. Boisvert, P. McDonald, E. Labios, M.S. Salisbury, G.M. Stuart, H. Danny, A.A. Portillo, A.B. Roybal, S. Van Duyne, G. Pauls, and S. Gaalema, 2010. 32 × 32 Geiger-mode LADAR camera, *Proceedings of SPIE: Laser Radar*

Technology and Applications XV 7684, 76840C (M.D. Turner and G.W. Kamerman, 06 April, Orlando, Florida, http://link.aip.org/link/doi/10.1117/12.851525.

Kahlmann, T., F. Remondino, and S. Guillaume, 2007. Range imaging technology: New developments and applications for people identification and tracking, *Proceedings of SPIE: Videometrics IX 6491, 64910C* (J. Angelo Beraldin, F. Remondino, and M.R. Shortis, 29 January, San Jose, California, http://link.aip.org/link/doi/10.1117/12.702512.

Klug, M.A., 2002. Display applications of large-scale digital holography, *Proceedings of SPIE: Holography: A Tribute to Yuri Denisyuk and Emmett Leith 4737, 142* (H. J. Caulfield, editor), 03 April, Orlando, Florida, http://link.aip.org/link/doi/10.1117/12.474953.

Marino, R.M., W.R. Davis, G.C. Rich, J.L. McLaughlin, E.I. Lee, B.M. Stanley, J.W. Burnside, G.S. Rowe, R.E. Hatch, T.E. Square, L.J. Skelly, M. O'Brien, A. Vasile, and R.M. Heinrichs, 2005. High-resolution 3D imaging laser radar flight test experiments, *Proceedings of SPIE: Laser Radar Technology and Applications X 5791, 138* (G.W. Kamerman, 30 March, Orlando, Florida, http://dx.doi.org/10.1117/12.609679.

Medina, A., 1992, *Three Dimensional Camera and Rangefinder*, United States Patent 5081530.

Nicks, D. 2010. Flash LADAR flight testing and pathway to UAV deployment, *Proceedings of AIAA: Infotech@Aerospace 2010*, AIAA-2010-3443, Held online.

Ramond, T., E. Saiki, C. Weimer, J. Applegate, Y. Hu, T. Delker, L. Ruppert, and B. Donley, 2011. Topographic mapping flash lidar for multiple scattering, terrain, and forest mapping, *Proceedings of SPIE: Laser Radar Technology and Applications XVI 8037, 803710* (M.D. Turner and G.W. Kamerman, editors), 27 April, Orlando, Florida, http://link.aip.org/link/doi/10.1117/12.884145.

Richmond, R., R. Stettner, and J. Glessner, 2000. Eye-safe laser radar focal plane array for three- dimensional imaging, *Proceedings of SPIE: Laser Radar Technology and Applications V 4035, 172* (G.W. Kamerman, U.N. Singh, C. Werner, and V.V. Molebny, editors), 26 April, Orlando, Florida, http://link.aip.org/link/doi/10.1117/12.397790.

Richmond, R., and S. C. Cain, 2010. Direct-Detection LADAR Systems, Bellingham, Washington, SPIE Press, ISBN 978-0-8194-8072-9.

Space Exploration Technologies, 2011. SPACEX's Dragoneye navigation sensor successfully demonstated on Space Shuttle, SpaceX, URL: http://www.spacex.com/press.php?page=20090925 (last date accessed: 11 April 2012).

Staple, B., R.P. Earhart, P.A. Slaymaker, T.F. Drouillard II, and T. Mahony, 2005. A range/depth modulation transfer function (RMTF) framework for characterizing 3D imaging LADAR performance, *Proceedings of SPIE: Laser Radar Technology and Applications X 5791, 226* (G.W. Kamerman, editor), 30 March, Orlando, Florida, http://link.aip.org/link/doi/10.1117/12.602773.

Stettner, R., and H.W. Bailey, 1995. *3D Imaging Underwater Laser Radar*, United States Patent 5446529.

Stettner, R., H. Bailey, and R. Richmond, 2000. Eye safe laser radar focal plane array for three- dimensional imaging, *Proceedings of SPIE: Laser Radar Technology and Applications V, 4035, 172* (G.W. Kamerman, U.N. Singh, C. Werner, and V.V. Molebny, editors), 26 April, Orlando, Florida, http://link.aip.org/link/doi/10.1117/12.397790.

Stettner, R., H. Bailey, and R.D. Richmond, 2004. Eye-safe laser radar 3D imaging, *Proceedings of SPIE 5412, 11, Laser Radar Technology and Applications IX* (G. W. Kamerman, editor), 12 April, Orlando, Florida, http://dx.doi.org/10.1117/12.553992.

Stettner, R., H. Bailey, and S. Silverman, 2005. Large format time of flight focal plane detector development, *Proceedings of SPIE: Laser Radar Technology and Applications X 5791* (G. W. Kamerman, editor), 30 March, Orlando, Florida, http://link.aip.org/link/doi/10.1117/12.609692.

Stettner, R., 2010. Compact 3D flash lidar video cameras and applications, *Proceedings of SPIE: Laser Radar Technology and Applications XV 7684, 768405* (M.D. Turner and G.W. Kamerman, editors), 06 April, Orlando, Florida, http://link.aip.org/link/doi/10.1117/12.851831.

Sturz, R., 1993. Laser-illuminated range-gated video, *Proceedings of SPIE: Airborne Reconnaissance XVI 1763, 70* (T.W. Augustyn and P.A. Henkel, editors),20 July, San Diego, California, http://link.aip.org/link/doi/10.1117/12.140823.

Wallace, A., R. Sung, G. Buller, R.D. Harkins, R.E. Warburton, and R.A. Lamb. 2006. Detecting and characterizing returns in a pulsed LADAR system, *Proceedings of IEEE: Vision, Image and Signal Processing 153, Issue 2,* 06 April, http://dx.doi.org/10.1049/ip- vis:20045023.

Watson, E.A., R. Muse, and F.P. Blommel, 1992. Aliasing and blurring in microscanned imagery, *Proceedings of SPIE: Infrared Imaging Systems: Design, Analysis, Modeling, and Testing III 1689, 242* (G. C. Holst, editor), 21 April, Orlando, Florida.

Weimer, C., and T. Ramond, 2011. Adaptive lidar for Earth imaging from space, *Proceedings of SPIE: Lidar Remote Sensing for Environmental Monitoring XII 8159, 815907* (U.N. Singh, editor), 21 August, San Diego, California, http://link.aip.org/link/doi/10.1117/12.893042.

2.6 GEIGER MODE LIDAR

The demand for airborne lidar elevation data and derived products in diverse fields, including applied geologic and biological sciences; industrial inspection and planning; and mapping/surveying, has grown significantly over the past two decades. To continue to meet this increasing need in a cost-effective manner, sensor developers and data collection specialists continuously look to improve the efficiency and speed of their sensors and collection processes.

Originally developed for military applications in the United States of America (Albota et. al., 2002), lidar systems built with Geiger Mode Avalanche Photodiode (GmAPD) detector arrays provide significant efficiency gains over current linear mode time-of-flight sensors available in the commercial market. Because GmAPD detectors are single photon sensitive, they provide highly efficient detection given minimal return signals, reducing the laser pulse energy requirements by a couple of orders of magnitude (Itzler et al. 2010), reducing the overall size, weight, and power (SWaP) of the lidar sensor. Additionally, the photon counting process is digital and noiseless, requiring only basic digital circuitry to detect the avalanche current required for time of flight range measurements. Finally, the digital detection and counter circuitry of a GmAPD focal plane array is compact allowing for the production of larger detector arrays compared to linear mode detectors. This ability to include more detector elements in a given area results in higher area coverage rates compared to linear airborne systems.

In recent years, commercial sources for high-resolution GmAPD arrays have become readily available, making it possible to transition this technology from the military market to the commercial/non-defense sector, where it is likely to be rapidly adopted for numerous applications where low power and/or increased area of coverage are critical to cost-effective lidar collection operations. In the following sections, we describe the technical fundamentals of GmAPDs, provide an operational comparison with linear mode flash and coherent detection methodologies, discuss the processing workflow for Geiger mode data, and present a few examples of airborne Geiger mode products.

2.6.1 Geiger Mode Avalanche Photodiode Fundamentals

Originally developed as a more power efficient alternative to traditional linear mode 3D lidar systems in the late 1990s (Heinrichs et al., 2001; Albota et al., 2002; Aull et al., 2002), Geiger mode detectors rely on an array of avalanche photodiodes operating in an over-biased scenario. If the reverse-bias voltage across an avalanche photodiode is maintained below a critical level, the output current is linearly proportional to the incident photocurrent. Above this critical level, or avalanche breakdown voltage, the absorption of a single photon can create an avalanche current pulse large enough to be detectable by simple digital timing circuitry. These Geiger mode avalanche photodiodes (GmAPD) are readily integrated into a large-format focal plane array (FPA) with integral CMOS ROIC (readout integrated circuitry) as shown in Figure 2.76. Each detector in the FPA consists of an APD detector and a dedicated time-of-flight (TOF) CMOS timer. Also, since the active area of the detector is relatively small compared to the spacing between adjacent detectors a microlens array with a dedicated lens for each detector is typically added to the FPA. This ensures that every photon incident on the FPA is coupled to a detector as efficiently as possible.

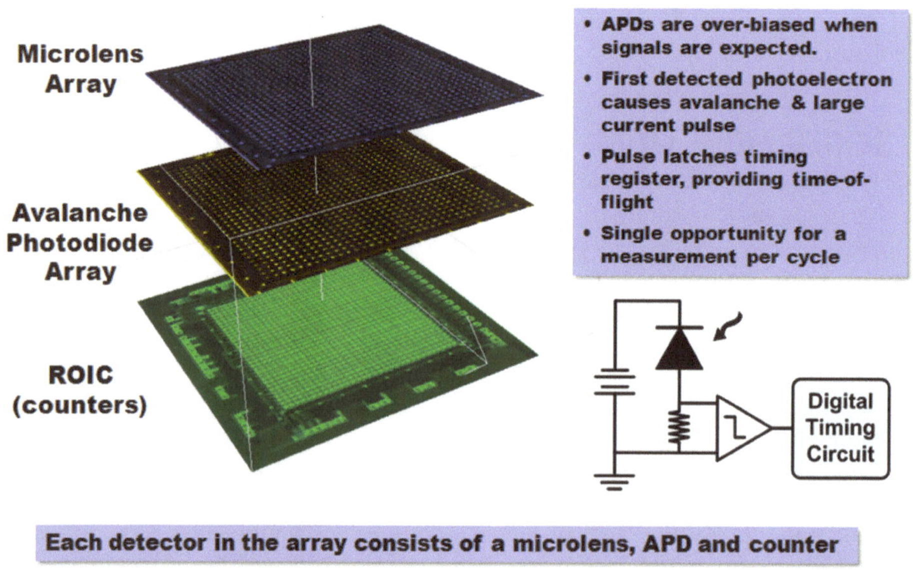

Figure 2.76 GmAPD elements are integrated into a package with microlenses and digital timing circuitry to create a compact lidar detector array.

GmAPD detectors arrays make excellent lidar receivers for the following reasons:

1. *Low Timing Jitter* – Since light travels about one foot per nanosecond, range precision in time of flight lidar systems requires quick temporal measurements. In GmAPDs simple digital timing circuits are used that respond to the leading edge of the avalanche current within 100 picoseconds, allowing for excellent range precision (Itzler et al., 2010).
2. *High Detection Sensitivity and Efficiency* – Ideal GmAPDs produce avalanche currents for single photons. Currently available GmAPD elements, however, have photon detection efficiencies of approximately 40 percent at room temperature (Itzler et al., 2010). This means that a reflected pulse containing 5 to 10 photons is required for high probability detection: an order of magnitude less than are required by linear detectors (Fouche, 2003).
3. *Compact Detection Circuitry* – Because the individual detection elements require no analog gain circuits, compact lidar detector arrays can be produced allowing for increased

point densities. This density improvement enables airborne systems to map larger surface areas faster.

4. *Low Noise Detection* – As the detection process is natively discrete, GmAPD detectors do not incur noise penalties from analog gain stages.

Although analog gain noise is not an issue, GmAPDs arrays are susceptible to noise from other sources (Itzler et al., 2010). The three primary sources of noise are the dark count rate of the device (i.e., the number of detections with no incident light), cross talk between detectors (stray photons generated due to the avalanche process), and the amount of solar background signal present in the scene, a noise source common to time-of-flight lidar systems. Of the three, the total noise rate is dominated by the background rate during the day which is typically two orders of magnitude higher than the dark count rate or level of cross talk. Solar background noise is reduced by using a narrow band interference filter centered on the transmit laser wavelength on the receiver and by reducing the range gate duration to the minimum necessary to image the scene. Cross talk noise results from the fact that when an avalanche occurs, a photon is produced. Occasionally, this may cause a nearby detector to also avalanche. Cross talk is controlled by building the FPA so the detectors are optically isolated from one another and by minimizing the bias level. The dark count rate is a result of thermal noise and is controlled by both minimizing the level to which the FPA is over biased and/or by cooling the FPA through the use of thermoelectric coolers.

2.6.2 Geiger Mode Sensor Operations

Traditional linear mode flash or spotlight lidar sensors calculate range based on the time-of-flight of a laser pulse as shown in Figure 2.77. For each acquisition frame, a single high energy pulse is used to illuminate the scene and the light reflected back to the receiver is converted and amplified into a current-driven signal proportional to the incident light intensity. This intensity waveform is then either processed in real-time using deconvolution and time-of flight calculations to recover a discrete set of range and intensity returns; or digitized and stored for later analysis and range processing. To overcome the additive shot noise present in the photoelectric conversion process, linear mode detection requires around 100 photons hit the receiver element to provide adequate signal-to-noise levels for range measurements (Heinrichs et al., unpublished data, 2001). This relatively high energy return requirement forces sensor designers to use relatively high energy laser pulses, resulting in pulse widths on the order of 10 nanoseconds or larger to maintain reasonable instantaneous energy limits with today's lasers, limiting the base resolving power of these systems to the 1 meter range.

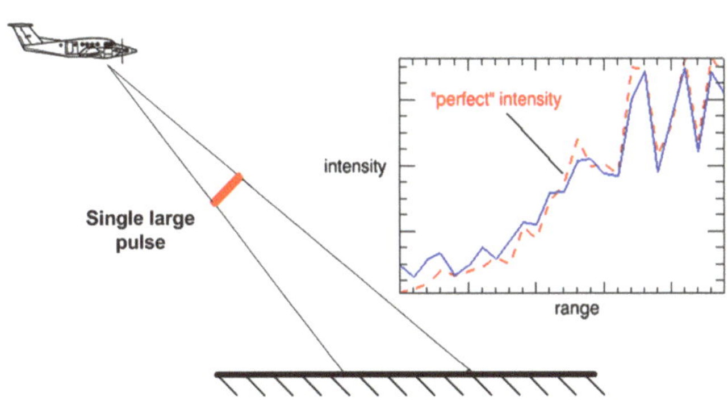

Figure 2.77 Traditional linear mode lidar sensors illuminate the scene with a single, high energy pulse and record the intensity return over time to produce range measurements.

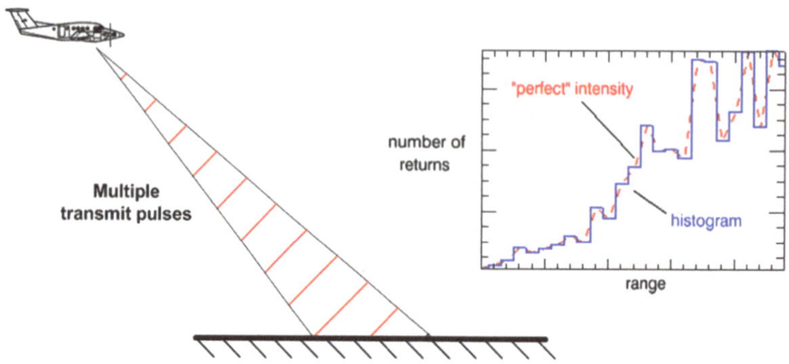

Figure 2.78 Geiger mode lidar sensors utilize multiple, low energy pulses with high repetition rates measure range and estimate intensity from reflective surfaces.

GmAPD lidar systems also compute range based on precisely measuring the time-of-flight of a laser pulse. However, because GmAPD elements can only respond to a single photon per pulse, Geiger mode sensors operate by sending out multiple, lower energy, smaller width pulses at a high PRF rate, and accumulating the number of detections at each APD element over a specified number of pulses as shown in Figure 2.78. Range measurements are determined by the time at which the GmAPD element fires while the relative intensity of the return is measured by the detection count at a given time/range bin over multiple pulses. Conceptually, the large energy contained in a linear mode pulse is distributed over a set of lower energy pulses fired more rapidly, significantly lowering the average power requirement of the laser and decreasing the base resolving power of the system to the 50 to 75 cm range. Because the GmAPD detection process is binary (on/off), there is no noise penalty incurred by "splitting" the energy into small pulses. For maximum efficiency, GmAPD sensors are typically designed so that each pulse provides no more than one possible photoelectron per detector element with pulse widths between 200 to 700 picoseconds (Marino et al., 2003) at pulse repetition rates (PRFs) of up to 20 kHz.

Lidar systems utilizing GmAPDs hold many advantages over traditional linear mode systems:

1. *Lower Laser Power* – Because GmAPD elements are single photon sensitive, lower power transmit lasers operating in more efficient high PRF conditions can be used to produce the same measurements. Thus, the overall system SWaP can be reduced significantly.
2. *Inherent Full Waveform Digitization* – Since the detection process is inherently binary, there is no need for expensive, low noise analog-to-digital converters for full waveform capture. A digital representation of the intensity waveform is naturally produced by the detection process.
3. *High Bandwidth Readout* – Because GmAPD detection requires only simple timing circuits, arrays can be constructed which support high bandwidth data readout. The fundamental limit on the readout rate is the reset time of the avalanche photodiodes. During this interval, which is less than 50 nanoseconds for today's devices, the sensor is not able to record data. This limit is not an issue, however, since the pulse repetition rate for today's lasers is measured in microseconds.

While Geiger mode sensors provide significant SWaP advantages compared to linear mode systems, they require more accurate sensor position information due to their multi-pulse mode of operation. The precise placement of a range measurement in 3D geodetic coordinates requires accurate knowledge of the line of site of the sensor. This data is provided by an inertial navigation system (INS) with an integral GPS receiver for most airborne lidar sensors. This process is particularly challenging for Geiger mode sensing from moving platforms, since the return signal is collected over multiple pulses. Therefore, a more accurate INS is required for a GmAPD system when compared with a linear mode system.

2.6.3 Geiger Mode Lidar Processing Flow

Because GmAPDs integrate range measurements collected over multiple laser pulses to recover a complete return waveform along a line of sight, producing elevation and other derived products requires a unique, four-stage workflow as shown in Figure 2.79.

In the initial processing phase, referred to as L0 to L1 processing in Figure 2.79, the sensor position and line of sight orientation metadata is combined with the time of flight data producing returns in angle-angle-range format which is then converted to 3D geodetic coordinates using conventional coordinate transformation methods. During this stage, timeouts, where no return signal is detected, are culled from the data set to reduce the volume of data to be processed in subsequent stages. As can be seen in the image in Figure 2.79, L1 data has a significant amount of noise which is dominated by random solar background noise and to a lesser extent, dark count and cross talk noise. This unfiltered point cloud is then passed to the second stage, referred to as L1 to L3 processing, where advanced noise filtering and point registration algorithms are applied. In addition to the 3D point cloud data, relative intensity images are also created from the same data, enhancing the interpretability of the final product. These output products can then be used to create standard elevation and derived products using the same techniques employed by traditional linear mode sensors (L3 to L4 processing) such as digital first return surface and bare earth models. Figure 2.80 shows an example of a wide area, 30 cm mapping product created using data from an operational airborne GmAPD lidar sensor in support of the humanitarian operations following the 2010 earthquake in Haiti.

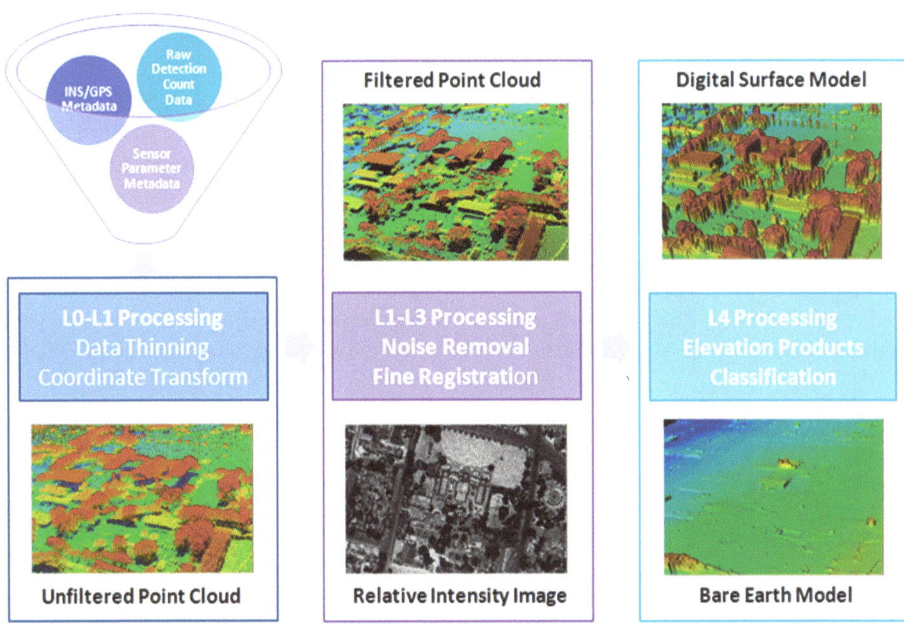

Figure 2.79 GmAPD sensors require a unique processing flow due to their fused multi-pulse method of operation. (ALIRT LIDAR data of the Republic of Haiti, collected and processed by MIT Lincoln Laboratory and NGA for US SOUTHCOM in support of Operation Unified Response)

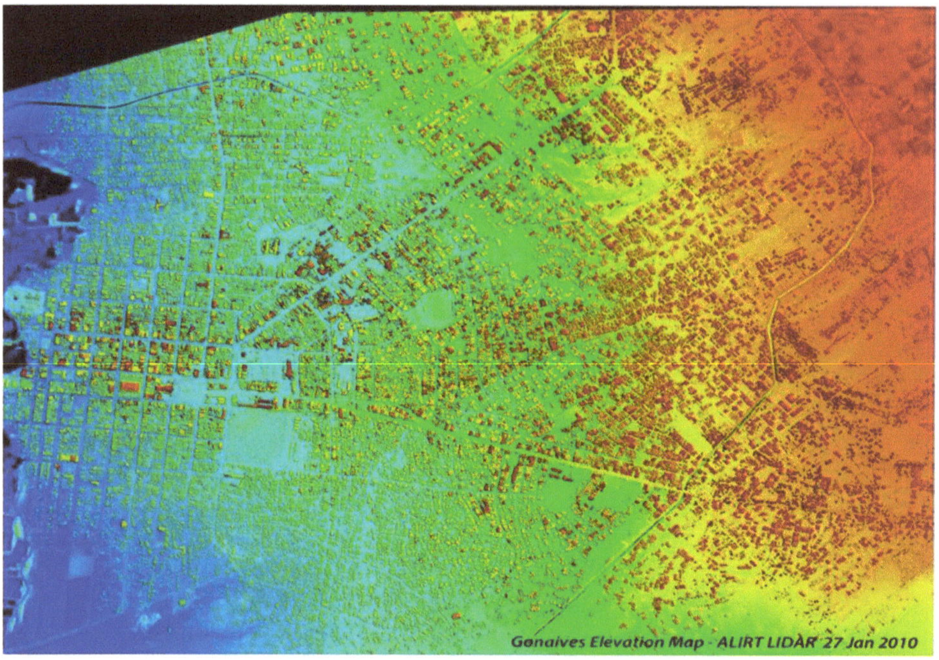

Figure 2.80 A 30 cm mapping product produced using a GmAPD sensor following the 2010 earthquake. (ALIRT LIDAR data of the Republic of Haiti, collected and processed by MIT Lincoln Laboratory and NGA for US SOUTHCOM in support of Operation Unified Response)

2.6.4 Summary

Lidar sensors built using Geiger Mode Avalanche Photodiode (GmAPD) arrays provide significant advantages over traditional linear mode devices. Because the photodiode elements are capable of detecting a single photon, a less powerful transmitter is required compared to a conventional system, allowing GmAPD systems to operate in environments with stringent size, weight, and power requirements. The primary advantage, however, is the ability to provide high area coverage rates due to the ability to form large arrays of detectors. As more Geiger mode Avalanche Photodiode focal plane arrays become commercially available over the next few years, we expect to see GmAPD sensors deployed much more often across a wide range of airborne, terrestrial, and maritime applications.

REFERENCES

Aull, B.F., A.H. Loomis, D.J. Young, R.M. Heinrichs, B.J. Felton, P.J. Daniels, and D.J. Landers, 2002, Geiger-mode avalanche photodiodes for three-dimensional imaging, *Lincoln Laboratory Journal*, Volume 13.

Albota, M.A., R.M. Heinrichs, D.G. Kocher, D.G. Fouche, B.E. Player, M.E. O'Brien, B.F. Aull, J.J Zayhowski, J.J. Mooney, B.C. Willard, and R.R. Carlson, 2002, Three-dimensional imaging laser radar with photon-counting avalanche photodiode array and microchip laser, *Applied Optics* 41:7671–7678.

Fouche, D.G. 2003, Detection and false-alarm probabilities for laser radars that use Geiger-mode detectors, *Applied Optics*, 42:5388–5398.

Heinrichs, R.M., B.F. Aull, R.M. Marino, D.G. Fouche, A.K. McIntosh, J.J. Zayhowski, T. Stephens, M.E. O'Brien, M.A. Albota, 2001. Three-dimensional laser radar with APD arrays, *Proceedings of SPIE Laser Radar Technology and Applications VI*, Volume 4377.

Itzler, M.A., M. Entwistle, M. Owens, K. Patel, X. Jiang, K. Slomkowski, S. Rangwala, P.F. Zalud, T. Senko, J. Tower, J. Ferraro 2010. Geiger-mode avalanche photodiode focal plane arrays for three-dimensional imaging LADAR, *Proceedings of the SPIE*, 7808, 78080C.

Marino, R.M., T. Stevens, R.E. Hatch, J.L. McLaughlin, J.G. Mooney, M.E. O'Brien, G.S. Rowe, J.S. Adams, L. Skelly, R.C. Knowlton, S.E. Forman, and W.R. Davis, 2003, A compact 3D imaging laser radar system using Geiger-mode APD arrays: System and measurements, *Proceedings of SPIE Laser Radar Technology and Applications VIII*, Volume 5086.

AUTHORS

Section 2.1.1 was written by Robert T. Pack, Associate Professor at the Utah State University Department of Civil & Environmental & Engineering. Sections 2.1.2–5 were written by Valerie Brooks, Manager, Airborne Survey Productsat Optech Incorporated. Section 2.2 was written by Jamie Young, Senior Manager – LIDAR Solutions for AeroMetric, Inc. Section 2.3 was written by Nuno Vilaça, Svein Vatslid, Peter Rindle, and Sven Kurz of Trimble. Section 2.4 was written by Christopher E. Parrish, Lead Physical Scientist, NOAA, National Geodetic Survey. Section 2.5 was written by Rex Craig, Principal Optical Engineer, Ball Aerospace & Technologies Corp. Section 2.6 was written by Philip W. Smith of Harris Corporation.

3
ENABLING TECHNOLOGIES

BRUNO SCHERZINGER, JOE HUTTON, and BECKY MORTON

3.1 LIDAR GEOREFERENCING

This chapter describes the technologies and methods used to georeference the data obtained from mobile lidar systems:

- Precise Satellite-based positioning
- GNSS-aided inertial navigation system
- Direct Georeferencing system for mobile lidar

3.1.1 Precise GNSS Positioning

A satellite-based positioning system is referred to as a Global Navigation Satellite System (GNSS). Today, there are four such systems either in development or in operation: the Global Positioning System (GPS) (United States), GLONASS (Russian Federation), GALILEO (European Union), and Compass (Peoples Republic of China). At the time of publication only GPS and GLONASS are fully operational.

Most high-accuracy survey grade applications use a combination of the GPS and GLONASS signals to improve the reliability and accuracy of the overall solution, primarily through enhanced satellite coverage.

3.1.1.1 Technology Overview

A GNSS can be succinctly described as a constellation of satellites carrying atomic clocks that broadcast time and an arbitrary number of receivers each of which computes its position on the Earth from measured signal propagation times from the visible satellites to the receiver. The current operational GNSS's, GPS, and GLONASS, have been extensively documented (see for example, AIAA, 1996; Hoffmann-Wellendorf et al., 1994; Wells et al., 1986). This overview describes their basic characteristics with particular emphasis on how they are used to compute a position fix with centimeter-level accuracy.

GPS comprises three major segments: space, control, and ground. The GPS space segment comprises 24 to 32 satellites that orbit the earth in six orbital planes with an approximate 12-hour orbital period (see Figure 3.1). A complete constellation comprises 24 satellites for global coverage. Up to 32 satellites may be active if older satellites and their newer replacements both are active, as has been the case during the past GPS replenishment program and current GPS modernization program. The orbital planes are inclined 55 degrees with respect to the equatorial plane. The design of the space segment with the minimum 24 satellites puts at least four satellites in view at any time anywhere on the Earth. Typical visibility in most parts of the world ranges from five to 12 satellites. Satellites transmit spread-spectrum modulated signals centered at 1575.42 MHz, known as the L1 signal, and 1227.6 MHz, known as the L2 signal. The code division multiple access (CDMA) modulation used by GPS (also used in cellular telephony) is a form of spread-spectrum modulation that allows multiple satellites to transmit signals on the common L1 and L2 frequencies. The modulated signal contains one or two pseudo-random codes that are unique to each satellite. GPS satellites are identified by the pseudo-random number (PRN) in the range 1 to 32 used to initialize a GPS satellite's pseudo-random code generators. The Clear Access or CA code has a repetition period of one millisecond and a bit or chip length of 1 microsecond. The Precise or P code has a repetition period of 267 days and a chip length of 100 nanoseconds. It is encrypted to form a Y code, the purpose of which is to deny unauthorized access to precise positioning and to prevent spoofing of the P-code by an adversary. Each satellite broadcasts a unique P(Y) code on the L1 and L2 signals that enables authorized users to calibrate and remove signal delays induced by the ionosphere.

The GPS modernization program was started in 2008 and is currently ongoing with the replacement of older satellites with Block IIM and Block III satellites. It adds several new signals to the traditional L1 and L2 signals described previously. The civilian L2 or L2C signal is a clear access signal modulated on the L2 carrier that permits commercial users to remove ionosphere-induced signal delays. The L5 signal is a civilian-use signal with a center frequency of 1176.45 MHz, and is intended for safety-of-life applications such as civilian aircraft terminal guidance and landing.

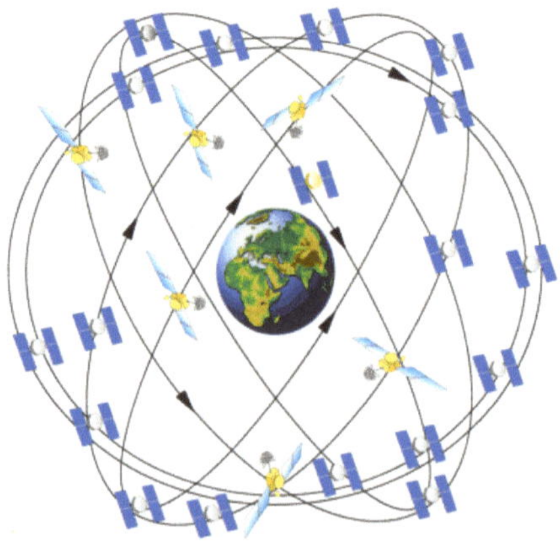

Figure 3.1 GPS Satellite Constellation (Courtesy of www.gps.gov).

The GPS control segment comprises the master control station, located at Falcon Air Force Base in Golden, Colorado, plus four tracking stations located at Hawaii, Ascension Island, Diego Garcia in the Indian Ocean, and Kwajalein Island in the West Pacific. The control segment has the task of controlling the operation of the satellites and calibrating their orbits. The control station periodically uploads navigation and clock correction parameters that the satellites then broadcast in their navigation messages. GPS time is coordinated universal time (UTC) time plus the number of leap seconds since 06 January 1980 when GPS time was initialized. GPS time increments continuously without leap second correction, and consequently has as of 2011 advanced 15 seconds against UTC since initialization. It is maintained by the Master Clock at the United States Naval Observatory (USNO) in Washington, D.C.

The GPS user segment comprises all user equipment, more commonly called GPS receivers, which perform positioning or time transfer using the GPS signals. A GPS receiver performs two fundamental functions: observables generation and navigation solution computation. The observables for a given frequency (L1, L2, L5) and a given signal type (CA, P(Y), L1C, L2C) from a given GPS satellite comprise the pseudorange, the carrier phase and the Doppler frequency. The GPS receiver contains an RF section that converts the received signal from the antenna to a digital sample stream and a signal processing section that processes the digitized signal. The signal processing section contains high-speed hardware organized into multiple tracking loops each of which locks onto one of the modulated codes in the GPS signal. The tracking loops are organized into channels, each of which generates the observables from one satellite. An all-in-view GPS receiver typically has 12 channels to track all GPS satellites in view and generate their observables. A single frequency GPS receiver generates the CA pseudoranges and carrier phases on the L1 carrier from each satellite. A high-performance multi-frequency receiver generates the pseudoranges and carrier phases from the CA, P(Y), and L2C signals modulated on the L1 and L2 frequencies. This requires significantly more signal processing hardware; consequently, such a receiver is more expensive than a single frequency receiver.

The Russian GLONASS is structured similarly to GPS, also comprising space, control, and user segments. The space segment comprises a constellation of 24 satellites in three orbital planes separated by 120 degrees. The orbital period is 11 hours-15 minutes, which implies the GLONASS satellites have slightly lower orbit than the GPS satellites. Each orbital plane has eight designated orbital slots, each having a unique slot number that is used to identify the satellite occupying the slot. The satellites use frequency domain multiple access (FDMA) modulation to transmit a common CA code and P code on one of 20 dual-frequency channels. The L1 frequencies comprise 1602 MHz plus or minus an integer multiple of 562.5 KHz. The L2 frequencies comprise 1246 MHz plus or minus an integer multiple of 437.5 KHz. At the time of publication, the Russian Federation has announced a modernization plan to transition to CDMA modulation so as to allow for better inter-operability with GPS.

The GLONASS control segment is located in Moscow. GLONASS time comprises Moscow UTC, which differs from USNO-UTC by tens of nanoseconds. It also differs from GPS time by the accumulated leap seconds since GPS time was initialized. Consequently the signal times of transmission contained in the GLONASS navigation message will contain a common bias against GPS time that needs to be accounted for in a GNSS receiver that tracks both GPS and GLONASS signals.

The GLONASS user segment comprises GLONASS receivers that track GLONASS signals and compute the position of the antenna. The hardware configuration of a GLONASS receiver is very similar to that of a GPS receiver, the only difference being the demodulation and tracking of signals on multiple frequencies. A GLONASS receiver can contain frequency-dependent inter-channel biases that do not appear in a GPS receiver in which all channels operate on the same L1 and L2 frequencies. Consequently GLONASS receivers are less accurate than GPS receivers, which is another reason for the intended transition to CDMA modulation.

A GNSS receiver is designed to receive, demodulate, and track signals from GPS and GLONASS satellites, and may also handle Galileo signals when these become available. The signal processing hardware contains over 40 channels that can be configured for either GNSS satellite and possibly Galileo as well. Such a receiver becomes capable of generating observables for all GPS and GLONASS satellites in view. This becomes advantageous when visible sky occlusion by buildings or trees blocks the signal paths to some GPS satellites. A GNSS receiver typically uses GPS time to schedule observables sampling and therefore needs to handle the differences between GPS and GLONASS times.

The pseudorange to each GNSS (i.e., GPS or GLONASS) satellite is the computed time of signal propagation from the satellite to the GNSS antenna multiplied by the speed of light. Each GNSS signal includes the navigation message that contains the parameters required by the receiver to compute the satellite position and system (GPS or GLONASS) time when the received signal left the satellite. The receiver maintains a local inaccurate estimate of GPS time in a crystal clock in order to compute the time of propagation. A computed pseudoranges thus comprises the true range between a satellite and the GPS antenna plus the receiver clock error plus several smaller range errors due to atmosphere induced signal propagation delay and satellite orbit and time errors. The receiver clock error is common in all pseudoranges and can be as large as one millisecond, which is an equivalent range error of 300,000 meters. Other range errors include ionosphere delay, troposphere delay, orbital error, multipath error, and receiver noise. These are typically less than 10 meters, and hence negligibly small compared with the receiver clock error.

Figure 3.2 shows the vector position relationship between a GNSS satellite and a receiver on the Earth. If the receiver has access to the satellite position and the relative satellite to receiver position, then it can compute its position. In reality, the receiver measures the pseudoranges to the satellites and computes the satellite positions from the navigation data messages that the satellite includes in its transmitted signal. From these the receiver computes a navigation solution using trilateration of the pseudoranges. Figure 3.3 shows the basic geometry between a GNSS antenna and four satellites. The lines between the satellites and the antenna indicate the line-of-sight ranges. The receiver requires a minimum of four pseudoranges to solve for four unknowns: three components of position on the Earth and the receiver clock error. The geometry of the satellites impacts the position and time solution error. A position and time solution will have good accuracy in all dimensions if the satellites used in the solution are uniformly distributed across the visible sky. It will have poor accuracy in some dimensions if the satellites are located in one sector of the sky. This is captured in a figure of merit called the Geometric Dilution of Position (GDOP). In simplest terms, the position and time solution error standard deviation is given by the pseudorange error standard deviation multiplied by GDOP. GDOP can be decomposed into the Position Dilution of Position (PDOP) and the Time Dilution of Position (TDOP). PDOP is typically used as a figure of merit for coverage of a position on the Earth by the GPS satellites. A position solution obtained from CA pseudoranges exhibits an error of 1.5 to 3 meters RMSE, depending on the satellite geometry.

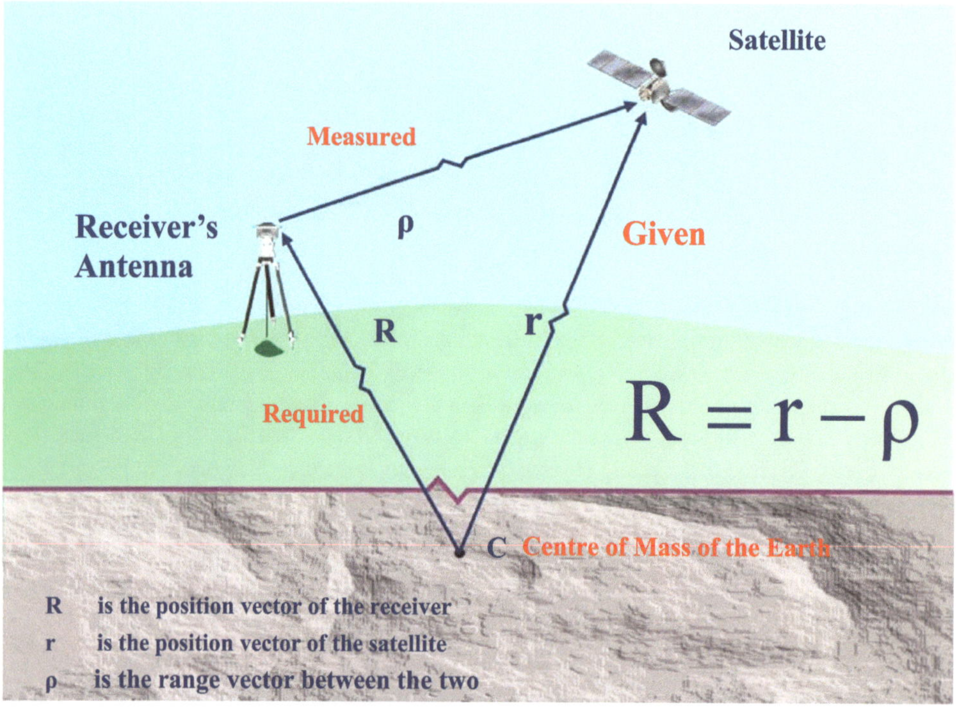

Figure 3.2 Basic Concept of GNSS positioning.

3.1.1.2 Differential GNSS

The key technology advancement that moved GNSS into the commercial surveying and mapping market was differential GNSS. A DGNSS system comprises one or more base receivers (also called reference receivers) at known positions and any number of roving receivers that move within an area covered by the base receivers. Figure 3.4 shows the geometry between two GNSS antennas and four satellites. The lines between the satellites and each of the two antennas indicate the ranges. The baseline vector between the antennas is the DGNSS solution. If the position of one antenna is known, then the position of the other antenna can be computed from the baseline vector. The base receivers both individually or collectively provide observations of the errors in the observables and broadcast navigation data. In a real-time DGNSS system, the base receiver station broadcasts the corrections to the roving receiver over a data channel so that the roving receiver can correct its observables and achieve positioning accuracies on the order of one meter. DGNSS corrections services that provide these corrections typically have continental coverage. In North America the DGNSS services include OmniSTAR, Navcom, and the US and Canadian Coast Guard differential beacon networks. OmniSTAR and Navcom use a continental array of base receivers to measure the atmospheric signal delays and broadcast orbital errors, and then transmit the differential corrections using satellite links. The US and Canadian Coast Guard networks modulate the differential corrections on existing 500 kHz marine direction-finding beacons along coastal regions and inland waterways. Other parts of the world have their own services. The accuracy of a DGPS position solution obtained with broadcast correction is on the order of 1 to 3 meters.

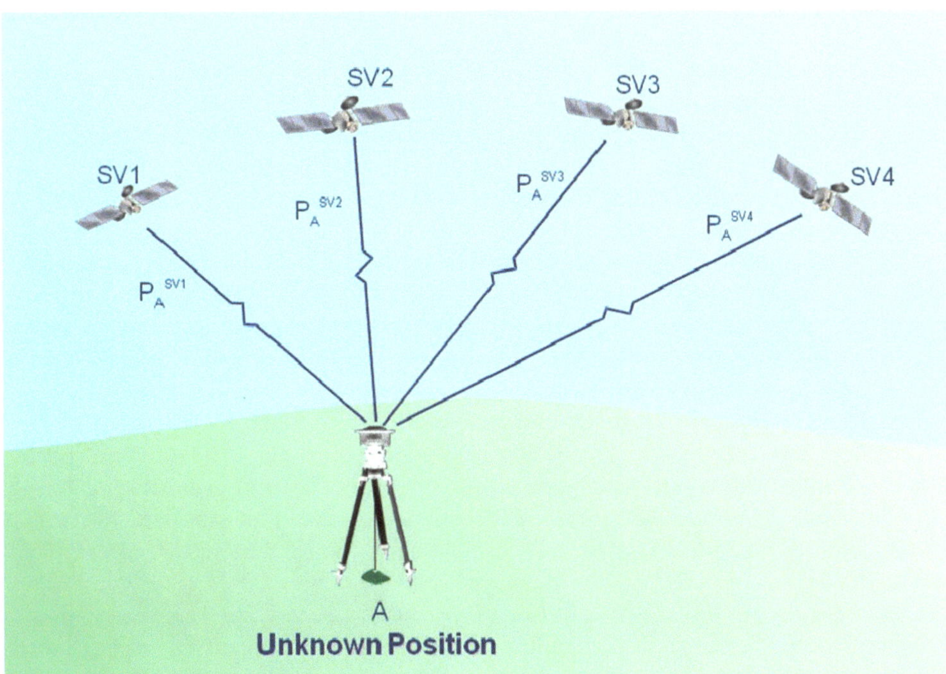

Figure 3.3 GNSS positioning by trilateration.

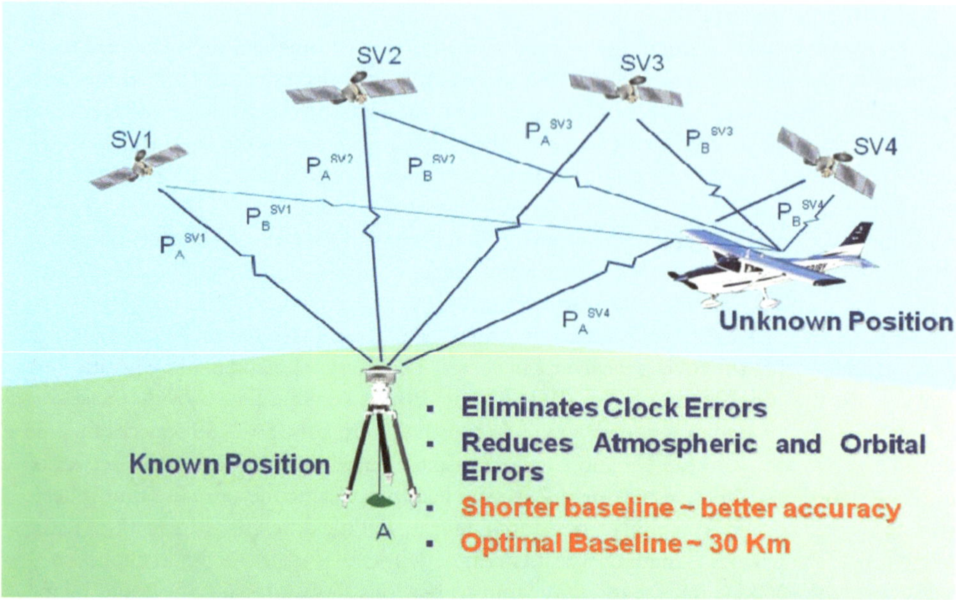

Figure 3.4 Differential positioning.

A more recent advancement in GNSS positioning is integer ambiguity resolution. This technique uses the fact that a GNSS receiver can track the L1/G1 and L2/G2 carrier phases modulo one wavelength with high accuracy. The receiver typically contributes L1/G1 phase noise on the order of a few millimeters. Each reported L1/G1 and L2/G2 carrier phase contains among other errors an unknown integer number of wavelengths. A DGNSS system with integer ambiguity resolution combines the phase data from two receivers so as to eliminate all significant errors other than the integer ambiguities. An integer ambiguity resolution algorithm is a statistical search for the set of unknown integer ambiguities among all feasible combinations of integers. Figure 3.5 shows the concept in a 2D plane. In this simple example, the intersecting carrier phases from two satellites define a grid of feasible positions in a search space. Each feasible point corresponds to one candidate set of two integer ambiguities. The carrier phase from a third satellite allows one of the feasible positions to be selected. Redundant L1/G1 and L2/G2 phase observations from two or more receivers each tracking five or more satellites provide the information for rapid unambiguous resolution. Successful ambiguity resolution for baseline lengths between roving and base receivers of 10 kilometers or less typically requires 30 to 120 seconds of signal processing. Newer algorithms are providing single epoch or instantaneous ambiguity resolution under conditions of good satellite geometry (de Jonge et al., 2000). Once the ambiguities are resolved, the corrected phases for each observed satellites become precise ranges that allow the computation of the baseline vector between the receivers with a typical accuracy of 2 to 10 centimeters.

Integer ambiguity resolution was first used to compute baseline vectors between stationary receivers as an alternative to conventional theodolite or total station survey methods. The receivers recorded the phase data, and the surveyor computed the baseline vectors in post-processing. Each receiver was required to occupy a station for 10 to 30 minutes so that phase errors could be averaged. The achievable baseline accuracy is less than one centimeter, which is within the requirements for network surveys. A more recent advancement in this methodology is kinematic ambiguity resolution (KAR) resolution while one or both receivers are moving. This allows the relative positioning of one receiver with respect to the other with accuracies on the order of a few centimeters. If a fixed base receiver broadcasts the data requirement for KAR, then a roving receiver can compute its position in real time with decimeter accuracy. This is called real-time kinematic (RTK) positioning. It has become a preferred method of positioning for applications where up to 10 centimeters position error is tolerable. These include seismic surveying and construction site surveying.

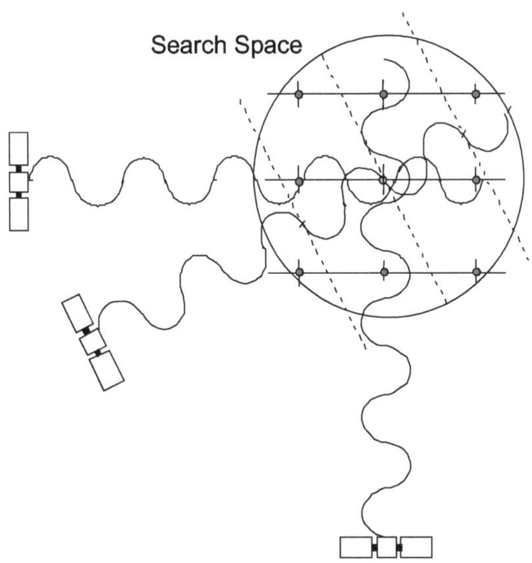

Figure 3.5 Example of integer ambiguity resolution in a 2D plane.

3.1.1.3 GNSS Attitude

Another application of RTK is attitude determination. The relative position vector between two antennas rigidly mounted with respect to each other in a horizontal plane is given in North, East, and Down components. The heading of the baseline vector can then be computed as the arctangent of the ration of East and North components. This concept can be extended to an array of three or four antennas to compute a complete roll, pitch, and heading solution of the array. A GNSS attitude measurement system thus becomes a possibly cost-effective alternative to an inertial system. The accuracy of the attitude array is dependent on the separation between antennas in the array. The attitude error can be computed as the RTK position error between antennas divided by the baseline length. Thus, a one-centimeter position error divided by a one-meter baseline length yields a 10-milliradian or 0.5-degree attitude error. GNSS attitude arrays by themselves are typically not suitable for continuous and accurate attitude measurement. However they can be used to shorten the alignment time and enhance the performance of a GNSS aided inertial navigation system.

3.1.1.3 Network Differential GNSS

Differential GNSS in its simplest form relies on a strong correlation of atmospheric and satellite orbital and clock errors between the roving receiver and a single reference receiver at a known position. This allows the cancellation of these common-mode errors at various levels of accuracy, the most accurate being the phase interferometry methods described previously. Almost exact cancellation of common mode errors can be had on baseline lengths of 10 kilometers or less. Such short baselines may however be inconvenient or impossible to implement, for example in survey applications that cover larger distances. Examples are any type of airborne survey, large area road mapping, and marine hydrography. An alternative approach to single baseline DGNSS is network DGNSS. It uses multiple reference receivers to calibrate spatial models of the atmospheric signal delay errors that are applicable over large areas (see Figure 3.6).

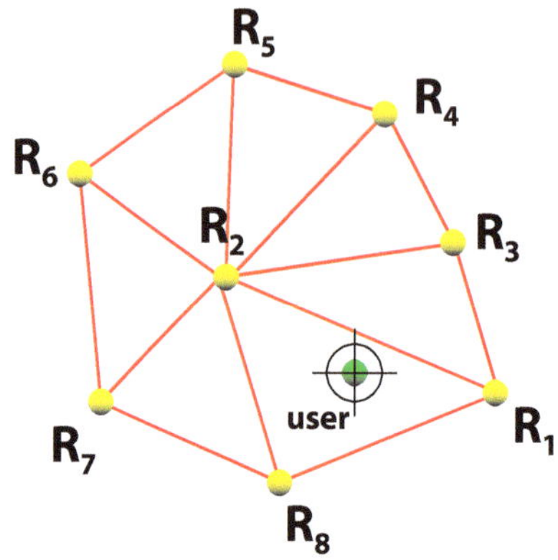

Figure 3.6 Network DGNSS Concept.

As a simple example to illustrate the method, assume two reference receivers separated by 100 kilometers form a simple network. They respectively experience pseudorange delay errors of 3 and 5 meters due to the ionosphere. A simple model for the ionosphere delay along a line between the reference receivers is a linear function of distance from the first reference receiver with slope 0.02-meters per kilometer of distance. A network DGNSS system will estimate a pseudorange delay of 4 meters for a roving receiver located between the reference receivers at 50 kilometers from the first reference receiver. Actual network DGNSS systems for precise positioning typically use four or more reference receivers separated by up to 50 kilometers to provide accurate spatial error models capable of calibrating atmospheric delay errors in the GNSS observables to the decimeter-level accuracy needed for integer ambiguity resolution. An example of a network DGNSS product is the Trimble Virtual Reference Station (VRS). It comprises a dedicated set of four or more reference receivers and a server that computes and transmits differential correction information to a roving receiver using a radio modem. It allows the roving receiver to implement RTK positioning accuracy over a 50 km2 operational area instrumented by the VRS reference receivers as if a single reference receiver were always close by (hence the word virtual in VRS).

3.1.1.4 Satellite-based Augmentation Systems

A satellite based augmentation system (SBAS) is a network DGPS system with continental or global coverage, and uses satellites to relay the differential correction data to the roving receivers. The SBAS uses a network of reference receivers across the area of coverage to compute large scale spatial error models for atmospheric signal delay errors. It then relays the correction parameters to roving receivers operating in its area of coverage using L-band channels on geostationary communication satellites. OmniSTAR, a division of the Trimble, provides subscription DGPS services to users with two levels of accuracy. The OmniSTAR VBS service provides global meter-level positioning accuracy using standard global network DGPS techniques. The OmniSTAR XP (GPS only) and G2 (GPS plus GLONASS) services provide service provides global decimeter-level positioning accuracy using a methodology called precise point positioning described later in this chapter. The wide area augmentation system (WAAS) was implemented and is maintained by the US Federal Aviation administration and the Department of Transportation for the purpose of en route and terminal guidance navigation for commercial aviation. It provides differential corrections with 1 to 3 meter positioning accuracy to users on an L-band channel that is compatible with the GPS signal structure so that a GPS receiver can receive and decode the signal without requiring a separate antenna and RF section. It uses 25 reference receivers located across the United States and two master stations that collect reference receiver data and compute the corrections. Europe has an equivalent SBAS called the European Geostationary Navigation Overlay System (EGNOS).

3.1.1.5 Development History

GPS first started development in 1973 with the formation of the US Department of Defense NAVSTAR Joint Program Office (JPO) under the direction of Dr. (Col.) Brad Parkinson. An experimental constellation of eleven Block I navigation development satellites was used to test the concept and explore operational issues. The first Block I development satellite was launched in February 1978, the last on 09 September 1985. Thereafter, the Block II and Block IIA operational satellites were launched between 1989 and 1994 to replace the Block I satellites and achieve a fully operational constellation having a minimum of 24 satellites.

GLONASS was developed between 1976 and 1995, culminating in the deployment of the first complete constellation of 24 satellites now known as legacy satellites. This first version of GLONASS was capable of 20 meter position accuracy for civilian usage. This was at first significantly better than GPS CA accuracy of 100 meters due to selective availability (SA), but was significantly worse than the 5 to 10 meters accuracy that GPS delivered after SA was switched off on 01 May 2000. GLONASS fell into disrepair following the collapse of the Soviet Union, with a usable constellation diminishing to six satellites before restorative plans became activated in 2001 under President Vladimir Putin's directive. In 2010, a full constellation of 24 satellites was again achieved, now consisting of replacement type M-satellites having long lifespan and better accuracy than the legacy satellites. The first type K-satellites with yet further improvements in lifespan and accuracy were launched in 2011.

Commercial GPS activity began in 1984 when Trimble Navigation released its first GPS product (Trimble Corporation, 2012). Since then Trimble and other companies have delivered GNSS receivers and GNSS-based products into the marketplace for a variety of applications. These include precision surveying, aircraft and ship navigation, commercial positioning for hiking and fishing, and time transfer. In fact, most commercial transactions over electronic media such as wire transfers or the Internet are tied to GPS time.

Councilman first developed precise GPS positioning using carrier phase processing in 1980 (Councelman and Gourevitch, 1981). Hatch (1981) and Remondi (1985) conducted further developments in ambiguity resolution and decimeter level positioning. Their positioning methods required a static initialization of the roving receiver to identify the phase ambiguities. The first on-the-fly (OTF) ambiguity resolution was demonstrated in 1990. The first RTK receivers from Trimble, Leica, and Novatel emerged during the early 1990's, while as of 2011 multi-frequency receivers capable of tracking GPS, GLONASS, Galileo, and COMPASS are quickly becoming standard for precision survey.

3.1.1.6 Types of Sensors

GNSS receivers can be categorized as low cost and moderate performance for mass-market applications, and high performance for precision positioning applications that include surveying. The low-end receivers are single frequency (L1) receivers configured as OEM components or chip sets for use in applications such as cell phones, car navigation systems, and hand-held personal navigators. They are designed to deliver a position and time solution at minimum cost.

Figure 3.7 High-performance, Multi-frequency GNSS Receiver Board (Courtesy of Trimble Navigation).

High performance receivers are multi-frequency GNSS receivers with over 200 tracking loops configured to track a combination of GPS, GLONASS, COMPASS, and Galileo signals (see Figure 3.7). These receivers are available either as OEM boards for integration into high-end positioning and navigation systems or as self-contained instruments for a particular application. The first such application-specific receivers, exemplified by the Trimble 4000S series receivers and Ashtech P12 and Z12 receivers, were designed for static survey applications. These receivers were designed for static occupation of survey points, and included a keypad and LCD screen for control and display, and an internal data logging capability. The network adjustment was performed post-mission using companion postprocessing software from the data recorded by the receivers. More recent receivers are designed to interface with a separate control, display, and data-logging unit or to a laptop PC. They offer RTK functionality as an option, thereby allowing point-to-point surveys to be conducted in real time with occupation time per point on the order of seconds, though with reduced accuracy. This capability meets the requirements of applications such as seismic, construction and pipeline surveying.

3.1.1.7 Postprocessing

An alternative to real-time processing is to postprocess the GNSS observables and satellite ephemerides recorded by GNSS receivers during a mission. Several GNSS manufacturers offer GNSS postprocessing software, which typically runs on a PC-compatible computer. It takes advantage of the non-causal postprocessing environment to compute a more accurate and more consistent ambiguity resolution solution than is possible in real-time. The simplest such technique is to compute position time histories forward and backwards in time, and then combine them in some fashion to ensure maximum position accuracy at each time point. A simple data quality assurance method compares the separation between the forward and backward position time histories and thereby identifies ambiguity resolution failures or wrong fixes in either solution when the difference exceeds a few centimeters.

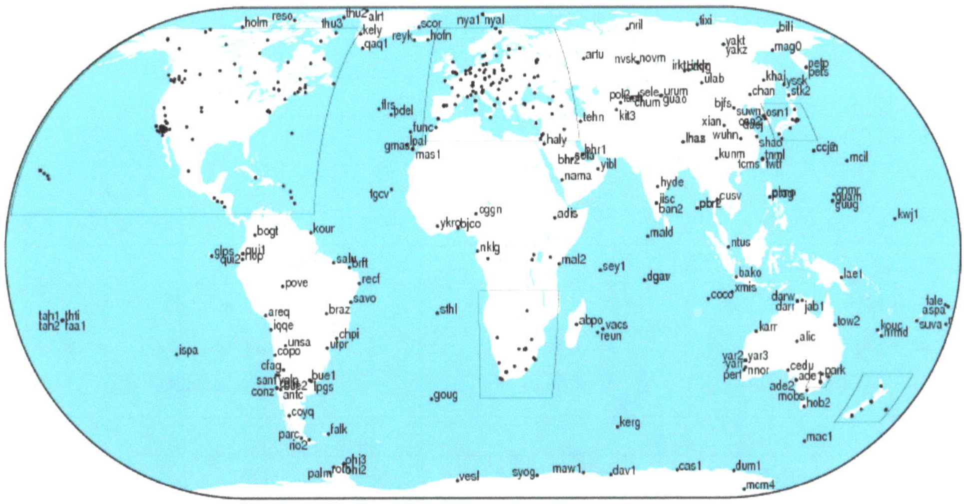

GMT 2011 Oct 25 16:48:30

Figure 3.8 The IGS Network (Courtesy of the International GNSS Service Website).

Precise point positioning (PPP) refers to the post-mission computation of a position using the observables from a single receiver and precise satellite orbit and atmospheric calibration parameters. The precise parameters can be downloaded over the Internet from several sources such as the International GNSS Service (IGS) typically, several days after the mission. The IGS is a service established by the International Association of Geodesy (IAG), officially started its activities on 01 January 1994 after a successful pilot phase of more than a year. It collects, archives, and distributes GNSS observation data sets for a wide range of applications. For details on the IGS, see Dow et al. (2009) or NASA-JPL (2012) (Figure 3.8). Other sources of precise post-mission data are JPL and the Continuously Operating Reference System (CORS).

Among a number of GNSS and geodetic products, the IGS produces high accuracy GNSS satellite ephemeris, satellite clock corrections, and ionospheric and tropospheric information. These products can be used in mobile surveys to refine the GNSS data. GNSS data refinement can be done on different levels. First, if GNSS is operated in DGNSS mode while the base station is more than 100 km away from the mapping area (a common occurrence for airborne mapping), using the IGS data minimizes orbital and atmospheric errors, and therefore, improves the positioning accuracy. Second, GNSS can be operated in PPP mode (no base station). In this case, using the IGS data improves the positioning accuracy down to a decimeter level and makes it possible to use GNSS without a base station in some mapping applications.

The IGS accomplishes its mission through continuous GNSS tracking stations, data centers, and analysis centers. Network stations are currently some 250 globally distributed stations continuously collecting GNSS data using quality dual-frequency receivers. Data centers are responsible for collecting data from tracking stations, reformatting and archiving data, and submission of data to analysis centers. Analysis centers are responsible for data processing and analysis. The products are then delivered to the Global Data Centers using designated standards on a regular basis. The IGS products are available after different periods of time and with different accuracy. Table 3.1 sows the IGS products, their precision and their timelines.

Table 3.1 IGS Combined Product Precision and Timelines (after Dow et al., 2009)

Product	Availability	Accuracy
Precise Ephemeris		
Predicted (Ultra Rapid)	Near Real Time	0.20 meters
Rapid	17 hours	0.10 meters
Final	10 days	0.05 meters
Satellite Clock Corrections		
Predicted (Ultra Rapid)	Near Real Time	30 nanoseconds
Rapid	17 hours	0.5 nanoseconds
Final	10 days	0.3 nanoseconds
Atmospheric Parameters		
Tropospheric Zenith Path Delay	Less than 4 Weeks	0.004 mm
Ionospheric Grid TEC	Less than 4 Weeks	0.20 meters

3.1.1.8 Capabilities and Limitations

GNSS has become a significant utility to surveyors and navigators. It is capable of providing baseline position accuracies comparable with traditional survey methods, and requires significantly less expertise and training to operate. It is however sensitive to its environment, and can without proper planning and quality control give poor or misleading data. The two key limitations that a user of precise GNSS positioning must plan for are signal shading and multipath reflection and residual GNSS errors. These are described briefly below. For further details on the subject, see AIAA (1996) and Wells et al. (1980).

3.1.1.9 Signal Shading and Multipath Reflections

The 1.5 GHz GPS signal propagates along the line of sight from a satellite to a GNSS antenna, and thus is prone to shading by obstacles. Also, the signal can be reflected off of fixed surfaces or refracted by foliage, both causing signals with multiple path lengths to impinge on the antenna. This is caused "multipath error." Multipath signals are always delayed compared to line-of-sight signals because of the longer travel paths caused by reflection. Code phase multipath can be as large as several meters if the antenna is stationary or slowly moving. For kinematic applications, the GNSS antenna motion with respect to the signal reflectors tends to randomize the multipath error and thereby diminish its impact on positioning accuracy. Typical phase multipath does not exceed 0.05 centimeters. Consequently, a GNSS receiver must be operated with a clear view of the sky and away from reflecting surfaces. This is the key limitation to performance that operators of GNSS equipment must be aware of when installing a GNSS receiver or planning a survey mission.

3.1.1.10 Residual GNSS Errors

GNSS errors can be classified into three major groups as shown in Figure 3.9. DGNSS processing can either eliminate or attenuate these errors provided that a GNSS mission has been planned to account for these error sources. Such planning requires the separation between the base and roving receivers to be less than a maximum distance over which the residual common errors are small.

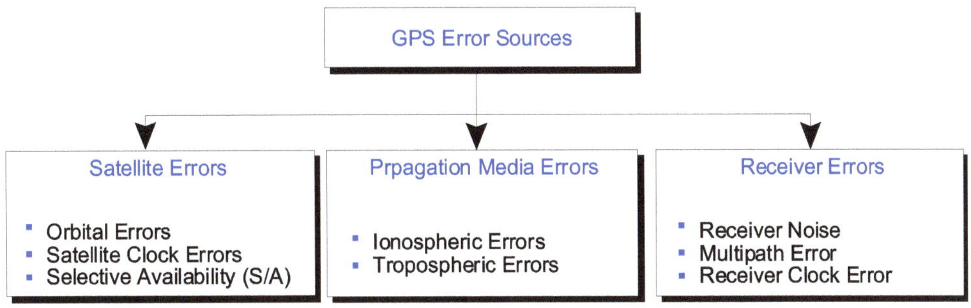

Figure 3.9 GNSS errors.

3.1.1.11 Satellite Orbital Errors

A number of gravitational and non-gravitational forces affect each satellite in its orbit. The actual satellite orbit therefore deviates from its predicted elliptical form. The navigation message received by a GNSS receiver includes the predicted (broadcast) ephemeris, which is a number of parameters from which the satellite trajectory is computed with respect to an Earth-Centered Earth-Fixed (ECEF) frame such as the World Geodetic System 1984 (WGS84). Satellite trajectory is required to be able to compute the position of the GNSS receiver's antenna. Currently, the broadcast ephemerides allow for satellite position computation with 2 to 10 m accuracy.

3.1.1.12 Satellite Clock Errors

Although GNSS satellite clocks are very precise, small biases or drifts result in large GNSS positioning errors (0.3-meters per nanosecond). Therefore, each satellite-transmitted navigation message carries three coefficients that a GNSS receiver or GNSS processing software uses to compensate for satellite clock errors. In post-mission, DGNSS processing removes the satellite clock error. When using single point positioning (without a base station) in either static or kinematic positioning modes, the broadcast clock correction parameters are used with a residual error of 10 nanoseconds or 3 meters. Using single point GNSS positioning together with the IGS-derived satellite clock corrections (accurate to 0.3 to 0.5 nanoseconds) improves the satellite-to-receiver remaining range error to about 10 to 15 cm.

3.1.1.13 Selective Availability (S/A)

Selective Availability is an intentional error imposed by the US Department of Defense to limit the highest achievable autonomous (i.e., non-differential) GPS positioning accuracy to authorized military users. This is done by intentionally degrading the quality of the broadcast transmitted parameters in the CA signal that are required to compute each satellite position and clock corrections. S/A limited the achievable autonomous positioning accuracy to 100 meters horizontal and 150 meters vertical. S/A was turned off on 02 May 2000, and as a result the autonomous positioning accuracy improved to 1.5 meters horizontal and 3 meters vertical RMSE.

3.1.1.14 Propagation Media Errors

Another limitation that must be considered in planning GNSS operations is the effect of the atmosphere. The ionosphere and troposphere introduce errors in the observables, which can contribute several meters of position error. GNSS satellites transmit spread spectrum signals that propagate through about 26,000 km of media to be received by a GNSS receiver. Signal transmission-to-reception time is about 0.07 seconds, during which the satellite signals pass through different layers of the Earth's atmosphere. Two major atmospheric layers affect the propagation of GNSS satellite signals, namely the ionosphere and the troposphere.

3.1.1.14.1 Ionosphere Errors

The ionosphere has lower and upper limits of approximately 50 and 2,000 km, within which ions and electrons are available in quantities sufficient to affect the propagation of GNSS signals. The ionosphere distortion is due to the total electron count (TEC) along the signal path between a satellite and the receiver. It is affected by the daily radiation of the sun and by the 11-year solar cycle. The ionosphere is particularly active at equatorial and northern latitudes. The year 2000 was the most recent "solar max" at which the sun's electromagnetic radiation reached a peak and thereby caused a corresponding peak in ionosphere activity. The ionosphere's effect on the GNSS signal, among other things, is a bending of the signal path, time delay of arriving code signals, time advance of arriving carrier phase signals, and scintillation. In some extreme cases of high solar activity, ionosphere scintillation causes rapid fluctuation of the GNSS signal strength, which may prevent a GNSS receiver from computing a position fix. Ionosphere errors can be dealt with using different approaches. Using DGNSS significantly reduces the effect of the ionosphere error, especially when using dual-frequency receivers. In single point GNSS positioning, the IGS-derived ionosphere grid help reduce the ionosphere errors. In real time applications, either modeling or compensating for the errors using a differential GNSS service is necessary.

3.1.1.14.2 Troposphere Errors

The troposphere is the non-ionized portion of the atmosphere, and extends from the Earth's surface to about 50 km. It introduces a non-dispersive (i.e., frequency-independent) propagation delay of the GNSS signals. The total effect can be divided into two main effects, the effect of the dry and the wet components. The dry component contributes by about 90 percent of the total error and can be modeled to about 2 percent accuracy using meteorological surface measurements. The wet component is more difficult to deal with because of the variations of the water vapor content that cannot be correlated over time and space, and thus, can only be modeled to about 10 to 20 percent accuracy. Troposphere errors can be reduced using DGNSS. Modeling troposphere errors is also used in either real time or post-mission.

3.1.1.14.3 Receiver Noise

A GNSS receiver noise comprises the CA code noise and the carrier phase noise. CA code noise is currently at the level of 0.1 to 0.5 meters. Carrier phase noise is usually at the level of one centimeter or less. The extremely low carrier phase noise is a key reason for the high accuracy of a kinematic ambiguity resolution position solution. DGNSS processing combines the observables from two receivers, and typically increases the single receiver noise variance by a factor of 2.

3.1.1.14.4 DGNSS Residual Errors

Generally, DGNSS reduces atmospheric and orbital errors, eliminates satellite and receiver clock errors, and increases the single receiver noise. Atmospheric and orbital errors are correlated with distance; the shorter the separation baseline between receivers, the greater the correlation between these errors at each receiver. These errors are almost the same at two receivers separated by short baselines (1 to 30 km), so that differential processing of the receiver data will affect the almost complete cancellation of the errors. Since these errors also distort phase data used in kinematic ambiguity resolution, the reliability of ambiguity resolution on short baselines is better than on longer baselines. For an overview on the subject, see AIAA (1996). For details on the ionosphere errors, see Tiberius (1998); for the troposphere errors, see Collins and Langley (1997). Table 3.2 briefly describes the DGNSS remaining errors while Table 3.3 shows a summary for typical errors for GNSS airborne differential positioning by carrier phase signal (c.f., Tiberius (1998), Shi and Cannon, (1994)).

Table 3.2 DGNSS Residual Errors.

Error	Error Characteristics
Orbital	• Correlated between satellites • Significantly reduced by between-satellite differencing (DGNSS) • Using precise orbits and satellite clock corrections improves positioning accuracy for long baselines
Ionosphere	• Frequency-dependent, thus, dual frequency data eliminates the error for long baselines. • Broadcast model reduces the error by 50 percent • In double difference airborne kinematic case error is typically 1 to 2 PPM for mid-latitudes between sunspot highs
Troposphere	• Frequency-independent, thus, cannot be removed by dual-frequency data • Dry component can be modeled and removed • Wet component needs meteorological data and more difficult to model because of variable nature of water vapor • Over long baselines the wet component effect on positioning can be estimated for airborne applications
Multipath	• Site-dependent and, thus, cannot be removed using differential GNSS • In kinematic applications, the multipath signature has a strong correlation with vehicle speed. Therefore, multipath gets random (and less) for higher speed

Table 3.3 Typical Dgnss Airborne Relative Positioning Errors.

GNSS Error Source	Typical Relative Positioning Error (PPM)	Relative Positioning Error For a 50 km baseline (m)
Orbital (SA is on)	1	0.05
Ionosphere	1 to 10	0.05 to 0.50
Troposphere	2	0.10
Signal Multipath	0.01	0.05
Receiver Noise	0.001	<0.025
Total Error	2.5 to 10.25	0.1 to 0.5

3.1.2 Planning Considerations and Quality Control

Use of GNSS for precision positioning requires mission planning to avoid or diminish its limitations discussed previously, and thereby ensure the quality of the data. The following aspects of a mission must be planned:

- the time of day of the mission,
- the expected mission trajectory, and
- the baseline length(s) between receivers.

The ever-changing GPS and GLONASS satellite constellations provide different levels of coverage throughout a 12-hour period. GPS coverage can range from 5 to 12 satellites depending on time of day and location. GLONASS can provide supplementary coverage of up to 10 satellites. The time of day of the mission should be planned so that good satellite coverage is available throughout the mission with allowances for overrun of the expected mission duration. Mission planning software or services over the Internet (for example, Trimble, Corporation (2012)) from some GNSS manufacturers can be used to predict satellite coverage at a specified project area during a specified mission time. Ionosphere activity during years of high solar activity can also compromise the quality of GNSS data if not planned for. A mission-planning option is where feasible to conduct a survey at night when the ionosphere is quieter.

The mission trajectory should be assessed for sources of signal shading and multipath reflections. The 1.5 GHz GNSS signal travels along its line of sight, and is easily shaded by structures and vegetation. Buildings and trees also can reflect the signal so as to generate multipath reflections to the antenna. Multipath error is the hardest error to control because it is local to one of the two antennas and therefore cannot be attenuated in a differential GNSS solution. The best method to date of attenuating multipath is with an antenna ground plane or choke ring.

In order to achieve a KAR solution with 2 to 5 centimeters accuracy with low risk, the baseline length between two receivers used to generate the KAR solution should be limited to a maximum of 10 kilometers during high solar activity and 50 kilometers during low solar activity. This ensures that common mode errors in the observables from the two receivers, most notably ionosphere and troposphere delays, cancel almost exactly in the differential solution. Reference or base receiver placement also affects the solution accuracy. The base antenna should be located well away from large vertical surfaces or structures that can generate multipath reflections.

3.1.3 GNSS-Aided Inertial Navigation System

3.1.3.1 Technology Overview

Direct georeferencing is the direct measurement of the position and orientation parameters of a remote sensor. It is an enabling technology for quantitative data acquisition and mapping applications where precise orientation of the sensor in addition to its position is required. Such mapping applications include aerial photogrammetry, airborne multi-spectral scanners, lidar, and multi-beam sonar bathymetry. In all of these applications, the sensors have a wide field of view of the terrain being mapped. A direct georeferencing system provides the position and orientation of the sensor that is required to stabilize and register the acquired data in geographic coordinates.

The state-of-the-art technology in DG systems is the GNSS-Aided inertial navigation system. The following discussion first describes the inertial navigation system, and then the concept of GNSS-aiding.

3.1.3.1.1 Inertial Navigation System

An inertial navigation system (INS) comprises two main components. The first is the inertial measurement unit (IMU; Figure 3.10), which comprises three accelerometers, three gyros, and the electronics to provide digitally encoded samples of the accelerometer and gyro data on a serial interface. The accelerometers are arranged into an orthogonal triad that measures the specific force vector experienced by the IMU. The specific force vector is the vector sum of the local gravity vector and the acceleration vector experienced by the IMU. The gyros are likewise arranged into an orthogonal triad that measures the angular rate vector experienced by the IMU. The angular rate vector comprises the sum of the earth rotation vector plus the angular rate vector experienced by the IMU with respect to the earth. A typical IMU outputs digitized samples at a 50 to 1000 Hz sampling rate of the total specific force and angular rate of the IMU with respect to a stationary inertial reference typically referred to as "the stars." High data rates are typically required to adequately sample short-term vibration dynamics as well as long-term vehicle dynamics.

The second component is the navigation processor (NP). It solves Newton's equations of motion of the IMU on the rotating Earth-based on the measured accelerations and angular rates. In order to do so, the NP must establish the local North, East, and Down directions of a navigation coordinate frame using a procedure called alignment. Determining the Down direction is called leveling, and is based on the assumption that the local gravity vector is the vertical reference. Once leveling is completed, the horizontal axes of the navigation frame define a plane that is tangent to the local geoidal surface, and hence said to be locally level. Aligning a locally level navigation frame to the true North direction is called heading alignment. A mechanism called gyrocompassing uses the horizontal component of the earth angular rate vector as the heading reference. A stationary or otherwise non-accelerating INS uses the earth rate vector measured by the gyros to gyrocompass its navigation frame to true North. A typical aircraft INS must undergo a stationary alignment lasting 10 to 30 minutes to accurately align its navigation frame. Thereafter, it navigates free-inertially, which implies that it computes its navigation solution using only the acceleration and angular rate data from the IMU without any subsequent corrections to its alignment or position solution.

Figure 3.10 Inertial Measurement Unit (courtesy of Applanix).

The INS technology from the early-1950's to the late-1970's comprised the gimbaled platform. In a gimbaled platform INS, the accelerometer and gyro triads were located on a mechanical four-axis gimbaled platform. The alignment process rotated the platform into alignment with the North, East, and Down directions. Thereafter during navigation, the gyros measured the platform angular rate and provided the feedback signal to the gimbal axis motors to null the platform rotation with respect to the Earth surface, thus maintaining the platform locally level and aligned to true North in spite of the vehicle's angular rate. The accelerometers thus reported North, East, and Down accelerations at all times, which the NP integrated to compute North, East, and Down velocity and from this updated latitude, longitude, and altitude. The gyro technology used in these mechanical INS's was the floated spinning mass gyro. The platform gimbal axes were instrumented with synchro-resolvers that measured the gimbal shaft angles. The mechanical design of the gimbaled platform caused the shaft angles to be the vehicle roll, pitch, and heading, which formed a complete orientation solution for the vehicle. The INS thus computed a complete navigation solution for the vehicle comprising geographic position, North-East-Down velocity, and orientation with respect to the North-East-Down directions.

The strapdown INS is the current state-of-the-art in inertial navigation. It emerged as the dominant INS technology during the late-1970's with the development of the ring laser gyro (RLG) and a class of avionics computers that could perform the numerically intensive strapdown navigation computations. It is so named because the accelerometers and gyros are hard-mounted, or "strapped down" to the INS case, and hence rotate with the vehicle. The accelerometers and gyros thus report the acceleration and angular rate of the INS resolved in the INS body frame. Using the angular rate data from the gyros, the NP updates the direction cosine matrix (DCM) that describes the orientation of the INS body frame with respect to a mathematical navigation frame that is locally level and aligned to true North, as shown in Figure 3.11. The alignment process in the strapdown INS initializes this DCM and thereby establishes the alignment of the mathematical navigation frame. The body-to-navigation DCM is thus the mathematical equivalent of the mechanical gimbals in the platform INS. The strapdown INS then transforms the measured acceleration vector to the navigation frame and thereafter computes velocity and position in the same way as does the platform INS. The vehicle orientation parameters are derived from the components of the body-to-navigation DCM.

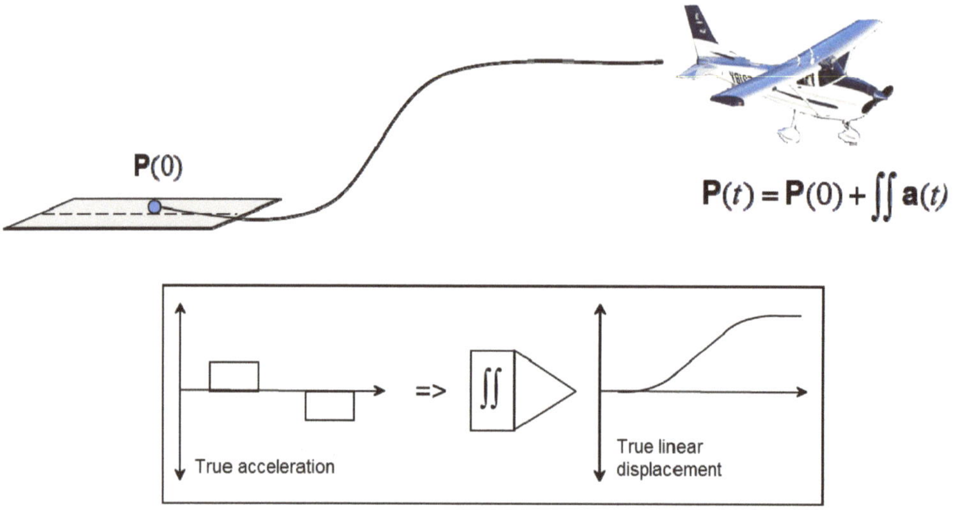

Figure 3.11 Inertial Navigation.

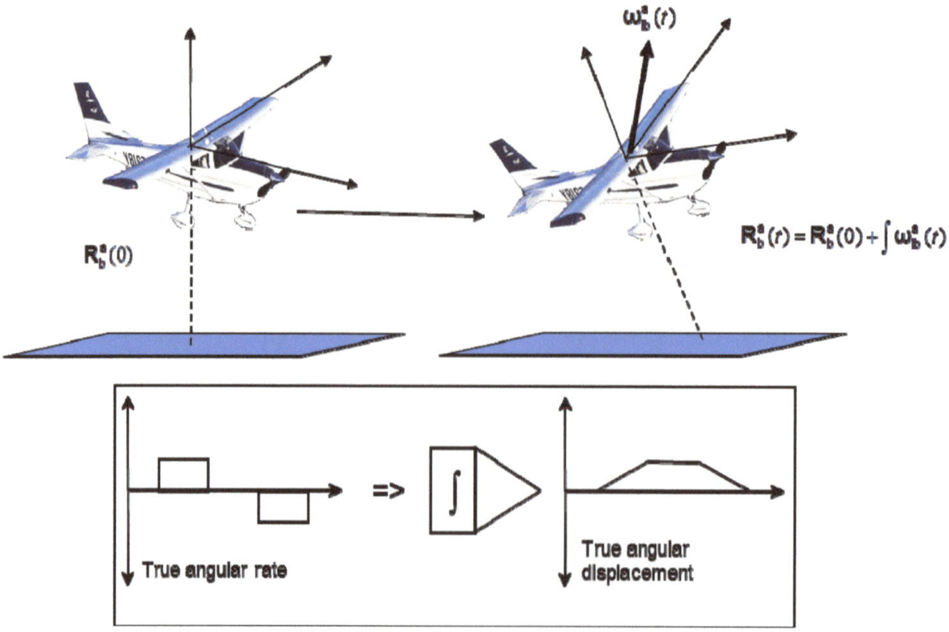

Figure 3.12 Inertial Navigation.

An INS is thus a sophisticated dead reckoning navigation system that navigates from an initial known position to its current position using only the measured accelerations and angular rates as measured by the inertial sensors (Figure 3.12). As with any dead reckoning method, it is prone to position drift, which in this case is caused by errors in the measured inertial data. A free-inertial navigator exhibits a position drift that on average grows linearly with time. An INS is typically classified in terms of the number of nautical miles that it drifts in one hour following a full ground alignment. A so-called medium accuracy INS that is used for aircraft navigation uses RLG's and typically exhibits an approximate one nautical mile per hour drift. Superimposed on this average position drift is the Schuler oscillation. This is a sinusoidal position error with an approximate period of 84 minutes, and is a consequence of the INS being constrained to navigate on the Earth surface. It was first documented by Schuler (Schuler, 1923) in 1923. Higher accuracy INS's such as are used on ships and submarines are capable of better long-term position accuracies.

The dominant errors in an INS that cause a position drift are the alignment errors and the inertial sensor errors. The alignment errors are the deviation of the navigation frame alignment from locally level and North azimuth. A navigation frame tilt error from locally level causes the gravity vector to be projected into the horizontal components of the acceleration vector transformed into the navigation frame. The inertial navigation algorithm integrates these projected specific forces assuming they are real accelerations, and outputs an erroneous time-linear velocity and a position that grows as time squared. The dominant inertial sensor errors are the accelerometer and gyro biases. These are the non-zero outputs of the accelerometers and gyros when the INS is in fact inertially stationary (for example floating in deep space with no specific forces acting on it). The other major inertial errors are the accelerometer and gyro random noises. These are uncorrelated noises caused by a variety of mechanisms that includes analog amplifier noise, digital quantization error, and in the case of RLG's lock-in error. The inertial navigator algorithm continuously integrates these inertial sensor errors into INS velocity and position errors. In particular, the horizontal gyro biases translate to constant latitude and longitude error rates, which are the major component of the INS's position error rate (see Figure 3.13). Smaller inertial sensor errors include scale factor error, input axis non-orthogonality, and vibration rectification.

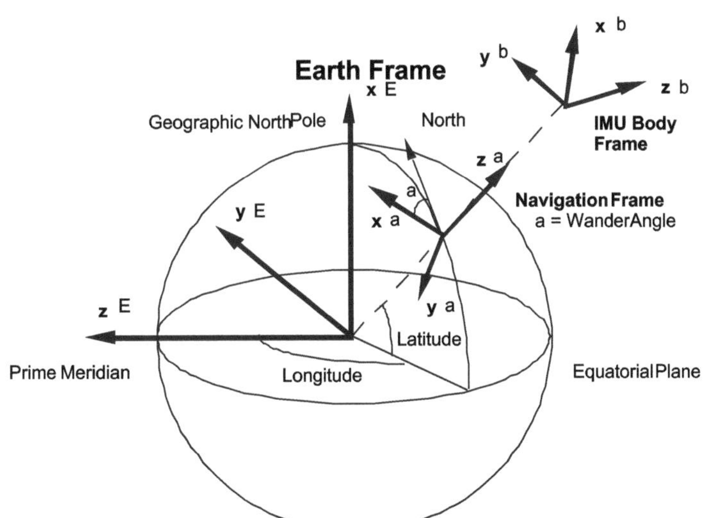

Figure 3.13 Frames of reference for Inertial Navigation.

The inertial sensor errors are in fact the residual errors following a factory calibration of the IMU. An uncalibrated IMU will exhibit large biases, scale factor errors and input axis non-orthogonalities that vary with the internal temperatures of the inertial sensors. The calibration firmware in an IMU implements temperature-dependent models of these errors. A typical model is a 3rd to 6th order polynomial interpolation of an error as a function of temperature. An IMU calibration procedure estimates the model parameters at different test temperatures across the specified operational temperature range. Ideally, an IMU calibration will model the mean inertial sensor errors at any temperature, so that the residual errors are random from one IMU power-up to the next. A typical IMU performance specification thus quotes the standard deviations of these inertial sensor errors about their calibrated mean values.

3.1.4 GNSS-Aided Inertial Navigation System

An aided INS (AINS) combines the components of an INS with navigation aiding sensors for the purpose of regulating the INS errors to those of the aiding sensors during vehicle navigation. The typical AINS uses a Kalman filter to estimate the errors in the inertial navigation solution and the inertial sensor errors. The Kalman filter uses measurements of the current inertial navigation solution differenced with aiding sensor navigation data to observe and estimate the INS errors. This is possible when the aiding sensor error characteristics are different from the INS errors. A GNSS position solution contains errors that can be characterized as noisy and statistically bounded within an error ellipse of fixed volume. By contrast, the INS position errors are characterized as smooth but unbounded. These error characteristics are said to be complementary, and hence amenable to Kalman filter error estimation. The Kalman filter processes measurements that are the differences of the inertial navigator and GNSS positions, and thus comprises the noisy but bounded GNSS position errors plus the growing INS position error. The Kalman filter contains a model for these errors, and thereby is able to suppress the GNSS short-term errors and estimate the INS errors. The accuracy of the INS error estimates depends on the magnitude of the GNSS position error. It will obtain a better error estimate when the GNSS solution is a kinematic ambiguity resolution solution with decimeter accuracy than when the GNSS solution is an uncorrected CA solution (see Figure 3.14).

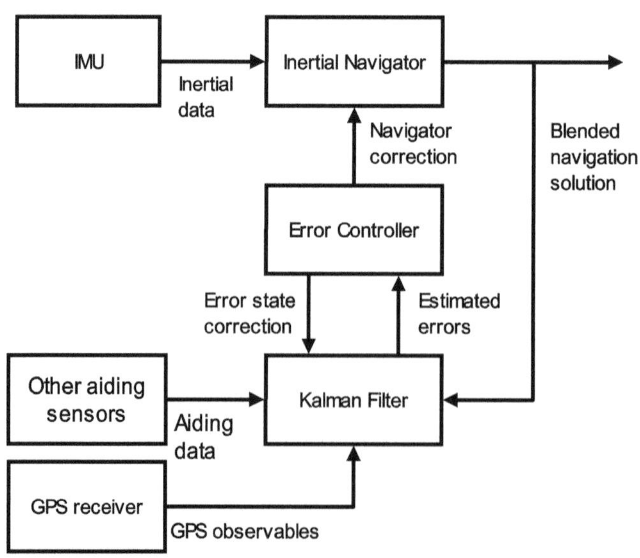

Figure 3.14 GNSS-aided INS architecture.

The AINS uses the Kalman filter's estimated inertial navigation errors to correct the position, velocity, and attitude computations in the inertial navigation algorithm. This error correction closes the loop around the INS and thereby regulates the INS errors to be consistent with the aiding sensor errors. In a GNSS-AINS, it regulates the INS position and velocity errors to be consistent with the smoothed GNSS position and velocity errors. It furthermore regulates the INS attitude errors and thereby continuously improves the INS alignment. Thus, a GNSS-AINS is able to align from a cold initialization while the vehicle is moving. This is called a mobile alignment, or in the case of an airborne AINS an "air-start."

A key capability of the GNSS-AINS is an attitude solution whose accuracy is better than a free-inertial attitude solution. This admits the use of lower quality inertial sensors in an AINS than would be required in a free-inertial INS to achieve the same attitude accuracy. The AINS achieves this because its Kalman filter continuously calibrates the inertial sensor errors as part of its overall error estimation function. For example, the INS will integrate a constant accelerometer bias to a time-linear INS velocity error. A gyro bias will rotate the navigation frame out of locally level at a constant angular rate. The gravity vector will project increasingly into the horizontal accelerometer input axes, and the INS will integrate the resulting time-linear acceleration error to a time-quadratic velocity error. The AINS Kalman filter recognizes these INS velocity errors when it compares the INS and GNSS velocities. The Kalman filter has a mathematical model that describes the velocity error as a function of the various sources of acceleration error, including the accelerometer and gyro biases. Using this error model, the Kalman filter is able to estimate these biases from the observed velocity error. The Kalman filter maintains the inertial sensor error estimates even as they undergo thermal drift. This ongoing calibration of the IMU provides for roll, pitch, and heading accuracies that are significantly better than the INS could achieve without aiding. A GNSS-AINS that uses a relatively inexpensive IMU can achieve roll and pitch accuracies on the order of a few arc-minutes and a heading accuracy on the order of 0.1 degrees or better.

3.1.5 Components of a GNSS-AINS for Direct Georeferencing

A GNSS-AINS system for survey and remote sensing applications typically comprises the following major components:

- IMU
- Roving GNSS receiver and antenna
- Navigation computer system (NCS).

Optional components include:

- RTCM differential corrections receiver
- RTK subsystem: base receiver, radio modem transmitter and receiver
- Distance Measurement Indicator (DMI)
- Secondary GNSS receiver and antenna for computing heading from GNSS to aid the inertial solution.

Those applications requiring the best achievable accuracy in post-processing will also include the following components:

- Mobile data acquisition system (DAS)
- Reference (base) receivers with self-contained data logging
- Data retrieval and post-processing computer.

The IMU containing the accelerometers and gyros is ideally compact and unobtrusive so that it can be located at or close to the desired instrumentation point, for example an aerial camera's perspective center. At the same time, its inertial sensors must be sufficiently accurate so that the overall system delivers the required orientation accuracy. These requirements typically conflict; high quality inertial systems are large, whereas tactical grade IMU's having the desired forma factor are not sufficiently accurate. A best compromise is achieved by using a small IMU with low noise inertial sensors in a GNSS-AINS (see Figure 3.15).

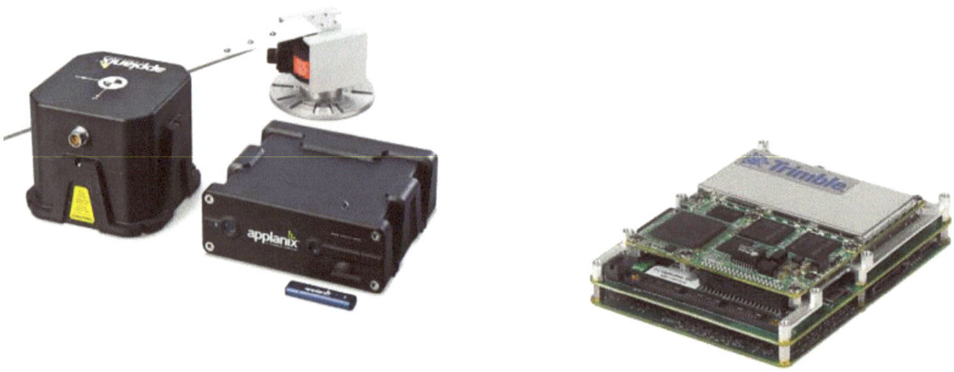

(a) **(b)**

Figure 3.15 Examples of DG Systems for lidar: (a) Stand Alone, and (b) OEM Board Set.

The NCS runs a software mechanization of the GNSS-AINS algorithm described previously and computes a position and orientation solution in real-time. For some applications, such as sonar, bathymetry is the primary data product of interest. For others requiring a postprocessed solution, a real-time solution that meets its accuracy requirements provides quality assurance of the recorded data. A properly designed system will initialize and align while the vehicle is moving. The alignment and transition to full navigation performance is automatic and hence transparent to the operator pushing the power switch.

The roving GNSS receiver is typically embedded in the NCS. It must be capable of outputting L1 and L2 observables of sufficient accuracy and quality for computing a KAR position solution. The antenna must also be dual frequency capable and of sufficient quality to reject most multipath interference.

Without any source of differential corrections, the computed position accuracy is expected to be on the order of 10 to 15 meters (CEP). The optional RTCM differential corrections receiver provides GNSS differential corrections in RTCM SC-104 format from a differential corrections broadcast service such as OmniSTAR or the US Coast Guard differential beacon network. With these corrections, the NCS computes position with a typical DGNSS accuracy of 1 to 5 meters (CEP). The RTK subsystem comprises a base receiver that outputs RTK data and a radio modem transmitter and receiver that relay the data to the roving receiver. With the RTK data, the NCS computes position with an accuracy of 0.02 to 0.1 meters (CEP).

The data acquisition system (DAS) is also typically embedded in the NCS. It records time-stamped IMU and GNSS data to a non-volatile storage media such as a hard disk or tape. The IMU time-stamp accuracy must be 50 microseconds or better to ensure that the position and orientation solution is accurate at the sensor sampling time on a possibly fast moving platform such as an airplane. It is also required to ensure a good inertial-GNSS integration in the GNSS-AINS. A few hundred microseconds of time misalignment between the IMU and GNSS data can cause the GNSS-AINS to estimate incorrect accelerometer biases and hence compute inaccurate orientation data.

The base receivers with data logging are required to compute a precise position solution in postprocessing. These are also dual frequency receivers, and are located around or within a project area so that the distance from any one base receiver to the roving receiver is within 10 to 50 kilometers. This tolerable distance depends on the ionosphere activity in the project area at the time of the survey. Ionosphere interference with the GNSS signals causes signal delay and scintillation. The data logging capability is either built into the receiver or is provided by a personal computer connected to the receiver.

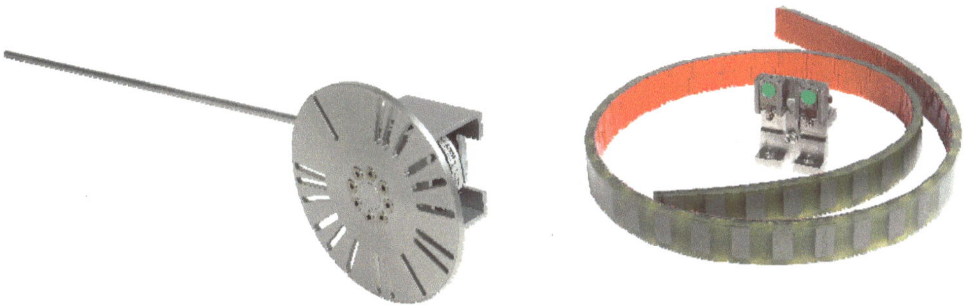

(a) (b)

Figure 3.16 Examples of wheel mounted DMI's: (a) external, and (b) rim mounted.

A Distance Measurement Indicator or DMI provides along track velocity measurements when the GNSS-AINS is mounted on a vehicle. The most common DMI's are wheel mounted, and produce a "pulse" each time the wheel rotates a specific angle (Figure 3.16). The pulses are summed to compute the distance travelled, or converted into a measurement of vehicle speed. When processed in the Kalman Filter, the DMI measurements help to constrain the position error growth of the GNS-AINS solution, especially when the GNSS signals are blocked when in a tunnel or under a bridge, or by other obstructions such as buildings or trees.

The data retrieval and postprocessing computer is used to retrieve and process the recorded data. The end product of postprocessing is either the desired position and orientation solution formatted to the particular application, or parameters derived from the solution. For example, the end product for aerial photogrammetry is the set of exterior orientation parameters for each photo. These are essentially the position and orientation solution at the time of exposure in a particular industry standard format.

3.1.5.1 Development History

The aided INS concept emerged shortly after the development of the Kalman filter, a core component of the AINS. This technology was first developed for and deployed in military aircraft, beginning in 1968. More recently, the US Air Force Embedded GNSS-Inertial (EGI) program has spawned a standard GNSS-AINS configuration.

The direct georeferencing concept for remote sensing has been postulated in technical research publications for several decades (for example, Schwarz et al., 1993). It has been put into practice in several experimental systems such as airborne multi-spectral scanners and laser altimeters. The orientation sensor typically was an aircraft INS that navigated free-inertially during the data acquisition. These systems demonstrated the DG concept, but were impractical because of the large size and high cost of the INS. An example is the early work performed by the Canada Centre for Remote Sensing (CCRS) led to the development of the Multi-spectral Electro-optic Imaging Sensor (MEIS).

A practical DG system became feasible with the advent of GNSS as a commercially available precise positioning system and the arrival of strapdown IMU's for tactical and motion sensing applications in military weapons systems. A "tactical" IMU is designed for short-term missile guidance. It is required to be small and light in order to fit into an air-to-air missile case, and relatively inexpensive in order to be affordable in an expendable weapon. It uses low cost inertial sensors such as fiber-optic gyros (FOG) to meet a 1 to 10 degree/hour performance requirement. Such an IMU is impractical in a free-inertial navigation application. Its gyro errors would drive the position error drift to upwards of 100 nautical miles per hour. It is however suitable in a GNSS-AINS, which calibrates the inertial sensor errors on an ongoing basis. A "motion sensing" IMU is designed to measure the short-term motion of a platform such as a radar antenna for the purpose of signal error compensation. A typical application is synthetic aperture radar (SAR) motion compensation, which is required to measure the radar phase center displacement along the line of sight with an accuracy standard deviation of less than one centimeter over a 10 to 30 second integration time. Its inertial sensors exhibit stable and low noise errors that can be accurately calibrated on-line. These same characteristics become important for high accuracy orientation measurement such as required for the aerial photogrammetry application.

Applanix Corporation was the first organization to offer for sale a GNSS-AINS specifically for commercial surveying applications. The Position and Orientation System for Land Vehicles (POS LV) first appeared in 1993. The POS/MV for multibeam sonar bathymetric survey was made available in 1994 (Scherzinger et al., 1997. Shortly thereafter, the POS AV for airborne remote sensing and POS TG for railroad track geometry measurement were released (Hutton et al., 1997).

3.1.6 Direct Georeferencing Systems for Mobile Lidar

In a lidar installation, a GNSS-AINS DG system is used to measure the position of the laser reference point and the orientation of the laser range at the exact time of measurement. In a scanning lidar system, the laser reference point is the mirror; in a fiber optic system it is the fiber optic bundle. The orientation is given in as Euler angles with respect to the North, East, and Down directions. The DG INS or IMU is rigidly mounted to the lidar housing so that it is rigid with respect to the laser reference point.

During a mission, the DG system records the IMU and GNSS data and the time of each laser scan, all in a common time base such as GNSS time. The DG postprocessing software computes the time tagged position and orientation of the laser reference point at a high data rate, typically 200 Hz. The lidar postprocessing software then interpolates the DG position and orientation data to the exact time of scan. With this data and the range measured by the laser, it computes the three-dimensional ground spot coordinates of each laser range. Typically the software first computes the earth-centered-earth-fixed (ECEF, also called Cartesian) coordinates and then converts these to the desired mapping frame. The resulting three-dimensional elevation map contains the data required for a DEM. This computation is illustrated in the following equation:

$$\begin{bmatrix} X_p \\ Y_p \\ Z_p \end{bmatrix}^E = \begin{bmatrix} X_m \\ Y_m \\ Z_m \end{bmatrix}^E + \begin{bmatrix} dx_p \\ dy_p \\ dz_p \end{bmatrix}^E \tag{3.1}$$

where X_p, Y_p, Z_p are the coordinates of where the laser hits the ground (point p) resolved in the Earth Center Earth Fixed (ECEF) frame (E), X_m, Y_m, Z_m are the coordinates of the reference point (m) for the lidar ranges resolved in the ECEF frame, and dx_p, dy_p, dz_p are the relative displacements from the lidar reference point to point p on the ground resolved in the ECEF frame.

The coordinates X_m, Y_m, Z_m of the lidar reference point resolved in the ECEF frame (E) are computed directly from the latitude, longitude and altitude output by the DG System.

The relative displacements from the lidar reference point to the point p on the ground resolved in the ECEF frame are computed from the lidar ranges and the orientation from the DG system. In the case of system that uses a scanning mirror they are computed as follows:

$$\begin{bmatrix} dx_p \\ dy_p \\ dz_p \end{bmatrix}^E = C_g^E(\lambda, l)\, C_b^g(\Phi,\Theta,\Psi)\, C_r^b(\theta_x, \theta_y, \theta_z)\, C_m^r(\theta, 0, 0) \begin{bmatrix} 0 \\ 0 \\ \rho \end{bmatrix}^m \tag{3.2}$$

where ρ is the range from the center of the lidar scanning mirror to point p on the ground, C_m^r is the rotation matrix defining the transformation from the mirror frame (m) to the mechanical reference frame of the lidar (r) defined by the mirror roll angle θ, C_r^b is the rotation matrix defining the orientation of the mechanical reference frame (r) with respect to the INS or IMU body frame (b) defined by the misalignment (boresight) angles (θ_x, θ_y, θ_z), C_b^g is the rotation matrix defining the orientation of the INS or IMU body frame (b) with respect to the local level geographic frame (g) defined by the roll, pitch and heading rotations (Φ, Θ, Ψ), and C_g^E is the rotation matrix defining the transformation from the local level geographic frame (g) to the ECEF frame (E) defined by the latitude and longitude (λ, l).

3.1.6.1 Boresight Calibration Requirements

The IMU usually does not provide an orientation reference for its inertial sensor input axes to the accuracy required by DG applications, hence the offset angles between the IMU and image sensor reference frame axes must be calibrated. This procedure is commonly called boresighting of the IMU.

The mounting misalignment or boresight angles between the IMU and lidar reference frame are usually computed though a combination of a laboratory and *in-situ* calibration, or sometimes solely through an *in-situ* calibration.

The laboratory calibration typically involves collimating the laser beam to achieve a perfect vertical reference, against which the roll and pitch output of the DG system are compared to determine the roll and pitch misalignment angles. Such a calibration requires a true horizontal reflecting surface. The laser beam is directed towards the horizontal reflecting surface, and its scan angle is set to nadir. The complete lidar instrument is then rotated in roll and pitch until the incident and reflected beams are coincident and hence perpendicular to the horizontal reflecting surface. The roll and pitch angles reported by the DG system are then the roll and pitch misalignment angles of the DG IMU with respect to the laser beam. This procedure is repeated a number of times to average down the errors in the DG orientation.

The *in-situ* calibration can be done using a reference surface test. Here a series of opposing scan lines with 100 percent overlap are collected of a flat bare surface (usually a runway or parking lot). Cross sectional heights of opposing lines are then compared with each other. A roll misalignment angle will cause the cross sections of horizontal surfaces to look like an "X" when they are overlaid, and hence can be estimated directly from the height differences. The pitch misalignment can similarly be determined by collecting opposing lines over a bare sloped surface (i.e., a treeless hill, sloped roof). Along track sections are then compared to determine the pitch misalignment angle, which has the effect of offsetting the slope horizontally. Finally the heading misalignment is determined by collecting opposing lines with 100 percent overlap over the corner of a building. The lines are then overlaid, and any heading misalignment will have the effect of the corner of the building to be translated horizontally between the opposing lines.

A variation of the reference surface test for airborne applications is to collect a block of opposing flight lines with 50 percent overlap and several cross strips. The overlapping surfaces can then be adjusted in a least-squares fashion to solve for the boresight angles that minimize the height separation of the surfaces.

Some lidar systems also have the ability to provide intensity for each return. This allows specific points to be identified in a lidar "image" for which the ground coordinates can be surveyed, thus allowing traditional photogrammetric techniques to be used to determine the boresight angles. Such a technique is often called "lidargrammetry."

No matter which method is used, the boresight calibration must be designed such that the misalignment angles can be measured to accuracies that do not contribute significantly to the geographic position error. Usually this means calibrating them to values smaller than the GNSS-AINS orientation accuracy, which can often be challenging, especially for the high-altitude airborne systems that require better than 0.005-degrees roll and pitch error.

3.1.6.2 Lever Arm Calibration Requirement

A lever arm is the relative position vector of one sensor with respect to another. A DG system uses the lever arms to correctly translate the aiding measurements to the IMU location and to the lidar reference point. It is usually convenient to define a single coordinate frame whose origin is the lidar reference point, and whose axes are aligned with the mechanical reference frame of the lidar. There are typically four sets of lever arm vectors that must be measured: the IMU lever arm, the primary GNSS lever arm, and if a DMI or GNSS heading are used, a DMI lever arm and secondary GNSS lever arm. The IMU lever arm is the relative position of the inertial sensing center of the IMU with respect to the lidar reference point, the primary and secondary GNSS lever arms are the relative position of the phase center of the primary and secondary GNSS antennas with respect to the lidar reference point, while the DMI lever arm is the relative position of the center of the wheel to the lidar reference point. The lever arm vectors characterize the installation of the DG system with respect to the lidar, and by vector subtraction provide the relative position of the GNSS antenna and DMI wheel with respect to the IMU. These are required by the inertial-GNSS integration software which must translate all position and velocity aiding measurements to the point of navigation (i.e., the IMU). As with the boresight angles, these lever arms must be measured to a better accuracy than the GNSS positioning accuracy in order for it not to contribute to the overall positioning error budget.

3.1.6.3 Time-tagging Requirement

In order to accurately compute the position and orientation of the lidar reference frame at the time the laser range measurement is made, the lidar and DG system must be synchronized to a common time base with a relative accuracy to better than 10 microseconds. This is commonly done with one of two methods:

1. The DG system outputs a one pulse per second (PPS) signal and associated GPS time. The lidar system records its internal system time at the receipt of the one PPS signal and then computes the difference between its system time and GPS time. This difference is then used to compute the GPS time of the laser range measurements
2. The lidar system sends an event pulse to the DG system synchronized to its internal system time. The DG system records the GPS time at the time it received the event pulse, and then sends this back to the lidar system.

3.1.6.4 Postprocessing

The final product of an airborne sensor such as a lidar or camera is generated by postprocessing software. This implies that the airborne sensor essentially stores its raw data in some form during a mission. Likewise, a DG system for airborne applications logs its raw IMU and GNSS data during a mission, and the position and orientation solution is computed post-mission with specialized GNSS-AINS processing software. The software first implements a Kalman filter based GNSS-AINS algorithm (see Figure 3.6) that generates the position and orientation time history. It then performs a smoothing operation backward in time on the data to generate a time history with improved accuracy. This method of post-mission processing generates the best achievable accuracy that can be had from the recorded inertial and GNSS data. It is an important component of a high accuracy DG system capable of 20 arc-seconds orientation accuracy for aerial photogrammetry.

3.1.6.5 Planning Considerations

The absolute accuracy of the blended position of a GNSS-AINS DG system is limited to the absolute positional accuracy of the GNSS. Hence it is important that proper mission planning be conducted to ensure that the best possible GNSS accuracy is achieved. To this end, the generic GNSS mission-planning considerations described previously apply here.

The best GNSS positioning accuracy (2 to 15 cm) is achieved using carrier phase DGNSS techniques described at the beginning of this chapter. To obtain this accuracy, a mission must be planned to provide conditions for reliable ambiguity resolution throughout the mission. Error sources that can prevent maintenance or re-fixing of integer ambiguities include ionospheric delays, multipath, and poor satellite geometry. Even if the correct ambiguities are found and maintained for the entire mission, these error sources that can, if not properly managed, still degrade the accuracy of the solution.

In the case of land-based mobile lidar collection, it is important to know the GNSS environment where the vehicle will be travelling and determine ahead of time where there will be significant signal blockages. In these cases it may be necessary to do multiple passes from different directions and do an adjustment on the final point cloud to remove the effects of inertial drift during the satellite blockages.

3.1.6.5.1 Static Data Collection

If practical, a mission should begin and end with a static data acquisition each lasting a minimum of 5 minutes. The static data allows the GNSS post-processing software uses the constant position information to obtain the correct initial and final ambiguities with high probability of success.

3.1.6.5.2 Minimizing Multipath

Multipath reflections can be a major source of position error and cause for integer ambiguity resolution failures. All base receivers should use antenna choke rings or ground planes to attenuate low elevation signals, and should be mounted at least 100 m away or above all reflecting surfaces.

3.1.6.5.3 Limiting Baseline Separation

If the mission requires the 2 to 10 centimeter positional accuracy that a kinematic ambiguity resolution solution can provide, then the maximum baseline separation must be limited to 20 to 50 km depending on the diurnal and seasonal solar activity. This allows the GNSS processing software to recover fixed integer ambiguities following cycle slips or loss of phase lock at any time during the mission. For missions with flight lines greater than 100 kilometers, multiple base receivers must be used to ensure the maximum separation between the aircraft and any base receiver is less than 50 kilometers. Current GNSS processing software packages are able to combine data from the multiple base receivers to produce the optimal solution with the least amount of error.

3.1.6.5.4 Planning for PDOP

To obtain the highest positioning accuracy, the collections should be planned during times of good satellite coverage and distribution so that PDOP is 3 or less throughout the mission. Simple satellite prediction software tools are available from various vendors that provide the information needed to plan for PDOP. An example from Trimble Navigation is shown in Figure 3.17.

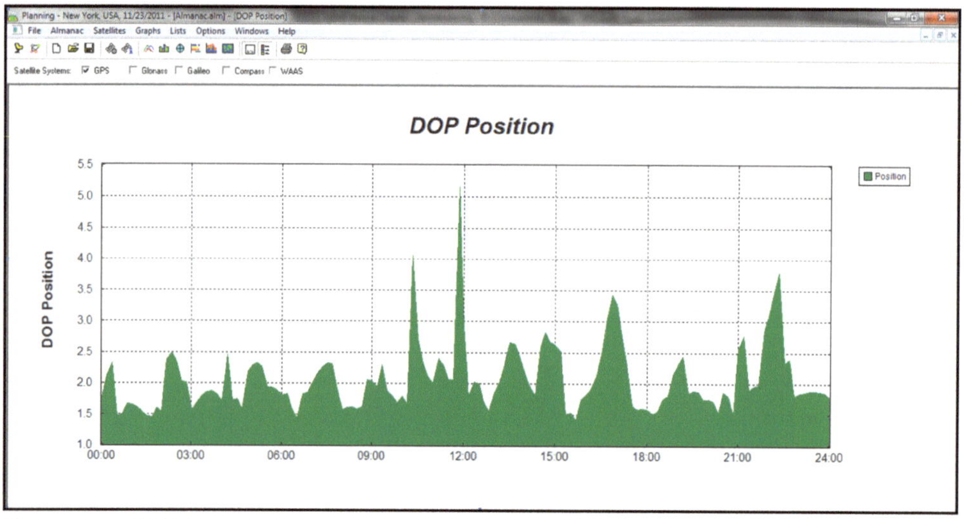

Figure 3.17 PDOP Planning SW (courtesy of Trimble Navigation).

Of course, when significant satellite signal blockages are predicted for vehicle-based applications, there may be no choice but to adjust the final point cloud to remove the effects of high PDOP and inertial position drift in the case of full signal outages.

3.1.6.5.5 Inertial Navigator Alignment

If a stand-alone INS is used as part of a DG system, then the INS must undergo a stationary ground alignment prior to free-inertial navigation. The ground alignment lasts between 5 and 30 minutes, depending on the latitude. If the aircraft is buffeted by wind or by persons moving about its interior during this phase, constant tilt and heading errors outside of the specification for the INS can result. If a power interruption occurs in the air, the INS will lose its alignment and the mission must be aborted.

In contrast, a GNSS-AINS can align itself while stationary or in motion. In fact, the in-motion alignment is accelerated and the quality of the alignment improved if the aircraft performs an accelerating maneuver such as take-off or a turn. An in-air alignment requires about three minutes of nominally straight and level flight to allow the GNSS-AINS to compute an initial roll and pitch, followed by a series of turns to align the heading. Thereafter, the GNSS-AINS improves its alignment with every maneuver. A typical zigzag survey pattern provides the maneuvers required by the GNSS-AINS to maintain a high quality alignment. For highest heading accuracy during a survey, it is recommended that the first survey line be started after a turn of 90 degrees or more, and then after the final line the aircraft do a turn of at least 90 degrees (for heading alignment in backwards processing).

For vehicle-based applications the start-and-stop motions of the survey vehicle are usually enough to keep the system aligned, or a secondary GNSS receiver can be used to provide GNSS heading aiding.

3.1.6.6 Quality Control

The quality of the data generated by a DG system does not become directly apparent until it is combined with the imaging/lidar system data. Consequently quality control for a DG system becomes a process of managing each step in the data acquisition and post-mission processing phases to achieve a consistent and reliable quality assessment.

Proper mission planning as described in the previous section goes a long way towards obtaining repeatable results. Once the mission begins, the DG system must be monitored frequently for GNSS dropouts or other data acquisition failures. A severe failure such as loss of GNSS data for an extended time period may be grounds for aborting the mission. Once the aircraft has landed, the recorded data should be checked at the hanger for outages and other immediate indications of bad or missing data. This allows the mission to be re-flown possibly the same day.

If the recorded data are deemed to be acceptable, then the data are handed over to post-mission processing. The DG post-processing software typically has several quality assessment indicators. The most basic of these are the inertial-GNSS residuals. These are the corrected differences between the inertial and GNSS position solutions at each GNSS epoch, and indicate the consistency between the solutions. The residuals will appear to be random in a successful inertial-GNSS integration, indicating that the integration process has removed all sources of bias errors in the data. The processing software will typically perform a statistical analysis on the residuals and report a simple quality indicator to the user.

Once all the data have been processed and the georeferenced data assembled, the final quality is given by the separation between computed and true positions of ground control points in the georeferenced data. This is the first direct measurement of quality, and comes at the end of the mapping process. If the previous steps have been managed properly, then the final quality measurement will be consistent, reliable and reproducible.

REFERENCES

Ackermann, F., and H. Schade, 1993. Application of GPS for aerial triangulation, *Photogrammetric Engineering & Remote Sensing*, 59(11):1625-1632.

AIAA, 1996. *Global Positioning System Theory and Applications, Volume 1, Progress in Astronautics and Aeronautics,* Volume 163, American Institute of Aeronautics and Astronautics, 1996.

Britting, K.R., 1971. *Inertial Navigation System Analysis,* Wiley-Interscience, New York.

Brown, R.G., and P.Y.C. Hwang, 1992. *Introduction to Random Signals and Applied Kalman Filtering*, Second edition, Wiley, New York.

Chatfield, A.B., 1997. Fundamentals of high accuracy inertial navigation, *Progress in Astronautics and Aeronautics*, American Institute of Astronautics and Aeronautics, Volume 74.

Collins, J.P., and R.B. Langley, 1997. Estimating the residual tropospheric delay for airborne differential GPS positioning (A Summary), *Proceedings of the 1997 Scientific Assembly of the International Association of Geodesy* (F.K. Brunner, editor), Rio de Janeiro, Brazil, 03-09 September, IAG Symposia, Springer-Verlag, Vol. 118, pp. 331-336.

Councelman, C.C., and S.A. Gourevitch, 1981. Miniature interferometer terminals for Earth surveying: Ambiguity and multipath for the Global Positioning System, *IEEE Transactions on Geoscience and Remote Sensing*, GE-19(4):244–252.

de Jonge, P.J., Y. Bock, and M. Bevis, 2000. Epoch-by-epoch TM positioning and navigation, *Proceedings of ION-GPS-2000*, September, Salt Lake City, Utah.

Dow, J.M., R.E. Neilan, and C. Rizos, 2009. The International GNSS Service in a changing landscape of Global Navigation Satellite Systems, *Journal of Geodesy*, 83:191–198.

Farrell, J.A., and M. Barth, 1999. *The Global Positioning System and Inertial Navigation*, McGraw-Hill, New York.

Gelb, A. (editor), 1984. *Applied Optimal Estimation,* MIT Press, Cambridge, Massachusetts.

Hatch, R., 1981. The synergism of GPS code and carrier measurements, *Proceedings of the 3rd International Geodetic Symposium on Satellite Doppler Positioning,* Las Cruces, New Mexico.

Hoffmann-Wellendorf, B., H. Lichtenegger, and J. Collins, 1994. *Global Positioning System Theory and Practice,* Third edition, Springer-Verlag, New York.

Hutton, J., T. Savina, and L. Lithopoulos, 1997. Photogrammetric applications of Applanix's Position and Orientation System (POS), *Proceedings of the ASPRS/MAPPS Softcopy Conference,* 27–30 July, Arlington, Virginia.

IEEE, 1983. Special Issue on Global Navigation Systems, *Proceedings of the IEEE,* October.

Kayton, M., and W.R. Fried, 1997. *Avionics Navigation Systems,* Second Edition, Wiley and Sons, New York.

IHO, 1987. *Standards for Hydrographic Surveys,* Third edition, International Hydrographic Organization Special Publication No. 44.

Kayton, M., and W.R. Fried, 1997. *Avionics Navigation Systems,* Second Edition, Wiley and Sons, New York.

Landstar, 2012. URL: www.racal-landstar.com (last date accessed: 08 January 2012).

Langley, M. Nakiboglu, K.P. Schwarz, J.M. Tranquilla, and P. Vanícek, 1987. *Guide To GPS Positioning,* Canadian GPS Associates.

Loncarevic, B.D., and B.M. Scherzinger, 1994. Compensation of ship attitude for multibeam sonar surveys, *Sea Technology,* June.

Moffitt, F., and E.M. Mikhail, 1980. *Photogrammetry,* Harper and Row, Inc., New York.

Mostafa, M.M.R., and K.P. Schwarz, 2000. A multi-sensor system for airborne image capture and georeferencing, *Photogrammetric Engineering & Remote Sensing,* 66(12):1417–1424.

Mostafa, M.M.R., J. Hutton, and E. Lithopoulos, 2001. Direct georeferencing of frame imagery - An error budget, *Proceedings of the Third International Mobile Mapping Symposium,* 03–05 January, Cairo, Egypt.

NASA-JPL, 2012. URL: http://exep.jpl.nasa.gov/files/exep/TPFIswgReport2007.pdf (last date accessed: 08 January 2012).

Omnistar, 2012. URL: www.omnistar.com (last date accessed: 08 January 2012)

Reid, D.B., E. Lithopoulos, and J. Hutton, 1998. Position and Orientation System for Direct Georeferencing (POS/DG), *Proceedings of the Institute of Navigation 54th Annual Meeting,* 01–03 June, Denver, Colorado, pp. 445–449.

Remondi, B.W., 1985. Performing centimeter-level surveys in seconds with GPS carrier phase: Initial results, *Journal of the Institute of Navigation,* Winter 1985–1986, 32(4).

Rogers, R.M., 2000. *Applied Mathematics in Integrated Navigation Systems,* AIAA Education Series.

Scherzinger, B., 1997. A Position and Orientation Post-Processing software package for inertial/GPS integration (POSProc), *Proceedings of the International Symposium on Kinematic Systems in Geodesy, Geomatics and Navigation (KISS 97),* June, Banff, Canada, pp. 197–204.

Schuler, M., 1923. The disturbance of pendulum and gyroscopic apparatus by the acceleration of the vertical, *Physikalische Zeitschrifte,* July, Vol. 24, pp. 334–350.

Schwarz, K.P., M.A. Chapman, M.E. Cannon, and P. Gong, 1993. An integrated INS/GPS approach to the georeferencing of remotely sensed data, *Photogrammetric Engineering & Remote Sensing,* 59(11):1167–1674.

Shi, J., and E.M. Cannon, 1994. Precise airborne DGPS positioning with a multi-receiver configuration: Data processing and accuracy evaluation, *Proceedings of the International Symposium on Kinematic Systems in Geodesy, Geomatics and Navigation,* Banff, Canada, pp. 393–402.

Siouris, G.M. 1993. *Aerospace Avionics Systems, A Modern Synthesis*, Academic Press, San Diego, California.
Tiberius, C.C.J.M., 1998. Recursive data processing for kinematic GPS surveying, *NCG-Nederlandse Commissie voor Geodesie*, Delft, Netherlands.
Trimble, Corporation, 2012. URL: www.trimble.com (last date 08 January 2012).
Wells, D., N. Beck, D. Delikaraoglou, A. Kleusberg, E.J. Kraklwsky, G. Lachapelle, R.B. Langley, M. Nakiboglu, K.-P. Schwartz, J.M. Tranquilla, and P. Vanicek, 1986. *Guide to GPS Positioning*, Canadian GPS Association, Fredericton, New Brunswick, 1986.
Woolven, S., and B.M. Scherzinger, 1996.An integrated position, attitude and heading solution for seafloor mapping with multibeam sonars, *Proceedings of the IEEE Oceans 1996 Conference*, Fort Lauderdale, Florida, September.

3.2 GROUND SEGMENT

3.2.1 Ground Reference Stations (Base Stations)

Ground reference stations are established on known surveyed locations. These stations are used to compute the difference between the known location and the positional information provided by GPS/GNSS satellites. A GPS receiver must acquire signals from at least four satellites to reliably calculate a three-dimensional position. By computing the difference between the known location and the GPS-derived location one derives the correction needed to minimize GPS error.

3.2.1.1 GPS Error

There are several factors which contribute to errors in the calculation of distance from GPS signals. Table 3.4 presents the four most common factors:

Table 3.4 Factors contributing to error in distance calculations from GPS signals

Error Factor	Effect
Propagation Error	Signal slowed by interaction with charged particles and water vapor in the atmosphere
Satellite Error	Orbit deviations and satellite clock errors
Receiver Error	Receiver clock errors and noise
Multipath Error	Signal interference from bouncing off nearby objects

3.2.1.1.1 Mitigation of GPS Error - Differential GPS (DGPS)

DGPS is used to correct for the known sources of error from GPS satellite broadcasts. The process computes the correction between the known location of the ground reference station and the location predicted from the GPS signal. Ground reference station receivers collect GPS observations concurrently with a receiver on the aircraft. Observations on both ground and airborne receivers are typically recorded in one-half second or one second epochs. Since the ground and airborne receivers are in close proximity and the errors in the observations are considered common to both receivers, the correction calculated from the ground reference receiver can be used to more accurately locate the GPS receiver on the aircraft.

3.2.1.2 Real-Time DGPS versus Post-Processed DGPS

The corrective factors derived from the ground reference station can be applied to the roving (aircraft) receiver in real-time or through post-processing. Real-time processing has the obvious advantage of immediacy but comes with disadvantages in the level of accuracy that can be obtained as well as physical line-of-sight limitations.

There are several sources of continuously available real-time GPS data available throughout the world. The Wide Area Augmentation System, or WAAS, is a system under development by the Federal Aviation Administration (FAA) to provide real-time DGPS through correction to geostationary satellites. Commercial companies, such as OmniSTAR and LandStar, offer real-time DGPS signals that are computed and rebroadcast over satellite or radio signals to subscribers. DGPS corrections are broadcast using a standard format known as RTCM, named after the Radio Technical Commission for Maritime Services which created the format.

The most typical method of differential processing for lidar mapping projects involves post-processing of the GPS data. The GPS recordings taken by the ground reference station and the aircraft receiver are transferred to a computer system for processing. The standard form for GPS data exchange is RINEX (Receiver Independent Exchange Format).

In addition to DGPS data from a project-specific ground reference station, users within the United States can incorporate DGPS data from a network of continuously operating reference stations (CORS; http://www.ngs.noaa.gov/CORS/) that are located on surveyed NGS benchmarks and made available by the National Geodetic Survey (NGS). NGS offers an on-line service called OPUS that provides DGPS data from post-processing of the GPS to at least 3 CORS stations.

3.2.1.3 Location of the Lidar Instrument in Space

The use of DGPS provides an accurate position of the phase center of the airborne GPS antenna; however it is the location of the lidar instrument that is required for mapping. Therefore, careful measurement between the GPS antenna and the lidar instrument is required. This is referred to as the lever arm calibration which is covered in detail in Chapter 4 - Airborne Installation and Integration of ALS Systems.

3.2.2 Typical Configurations and Requirements of Ground

3.2.2.1 Reference Stations

Ground reference stations serve as the basis for georeferencing the lidar data. Since differential correction of the GPS data is derived from the known location of the ground reference station, selection and survey on this point is critical. Ground reference stations are typically established on surveyed monuments such as those published by NGS or USGS. On projects requiring ground reference stations on new locations, the position of the GPS receiver is tied into an existing network of published control.

3.2.2.2 Survey-Grade Receivers

Ground reference stations for lidar projects require the use of high-end GPS receivers that operate with dual-frequency and multi-channel capability. The dual-frequency component will allow the GPS unit to pick up data from both the L1 and L2 frequencies providing critical information to limit the effects of atmospheric delay and allow for greater separation between GPS receivers (longer baselines). GPS receivers must recover all of components of the signal from the GPS satellites including the carrier wave, the satellite clock correction, ephemeris, and other information.

3.2.2.3 Static Observations

Ground reference stations require static survey GPS observation. According to NGS guidelines, GPS receivers should collect data continuously and simultaneously for at least three, five-hour sessions on three different days during the project.

3.2.2.4 Accuracy

Survey-grade GPS receivers as described above are able to provide accurate solutions of lidar trajectories with 5 cm RMS accuracy or better when base lines are kept within 20 to 50 km.

AUTHORS

Section 3.1 was written by Bruno Scherzinger, Chief Technology Officer, and Joe Hutton, Director for Airborne Products, both with the Applanix Corporation. Becky Morton, Senior Program Director at Towill, Inc., wrote section 3.2.

4
AIRBORNE INSTALLATION AND INTEGRATION OF ALS SYSTEMS

BOBBY TUCK, JAMIE YOUNG, and CHRIS GUY

4.1 ROTARY WING PLATFORM

4.1.1 APPLICATION OF TECHNOLOGY

Lidar has evolved from a basic planning tool to a tool for the engineering design. With the improved accuracy and increased point density, the lidar product can now match the requirements of most engineering projects. Helicopter lidar is an airborne product; there are areas that cannot be collected from the air such as under sides of bridges, box culverts, and inside of other structures. The swaths can be adjusted by the flying height and the point density can be controlled by flying heights and speed of the aircraft. Accuracies of down to 0.03 feet are possible with attention paid to detail of the mission and with high accuracy ground control. The maneuverability of the helicopter allows it to be a good platform for the corridor mapping where the corridor may have a lot of meanders and curves. The helicopter is able to fly around the curves with flat turns and eliminate the need for lost time due to turns onto various flight lines. The helicopter is also able to land at the project area and initialize off of the control points on the project, therefore reducing the baseline length of the airborne GPS. The helicopter systems are very good for corridor mapping such as power lines, railroads, gas lines, airport obstruction surveys, flood studies, and highway planning and design.

Tuck Mapping Solutions, Inc. has used *RIEGL* scanners for over ten years and all of our research has gone into providing feedback that was later used in the development of subsequent *RIEGL* systems. Reference is made to the *RIEGL* scanners and the associated software because that is where Tuck Mapping has the majority of their experience. There are other manufacturers such as Leica and Optech that also manufacture the high accuracy scanners, and I am not in a position to say how they should be mounted and their performance. I will be describing the results of my research and experience in this subchapter.

4.1.2 INSTALLATION OF THE SYSTEM

The installation of a lidar system in a helicopter is much more challenging than the installation for fixed wing aircraft. In an airplane, we would typically install the antenna over the center of the mapping system to ensure close proximity of all sensor components, and therefore, the antenna is generally mounted on the top of the fuselage. On a helicopter, it is impossible to mount the antenna over the system due to the rotor breaking GPS lock on each revolution. As a remedy to this problem, we are mounting the antenna on the vertical tail fin of the helicopter and surveying a horizontal and vertical offset to the mapping system. This installation provides some challenges, but also some distinct advantages. The challenge is to survey the proper offset in both the vertical and horizontal. The vertical tail fin on a helicopter is not on the centerline of the fuselage and this offset must be surveyed. When we measure the offsets, the helicopter is on the ground in a basic level configuration. When the helicopter flies, the center of gravity will move as the fuel is burned, and this changes the attitude of the helicopter during the mission. When the helicopter is on the ground, it is basically at the same attitude each time with only slight tipping due to loading. The IMU will be initialized at approximately the same attitude on each mission.

To ensure a well distributed point cloud, we mount the system as close to the center of gravity to reduce the amount of tip and tilt during the mission. For this, Tuck Mapping designed a pod (see Figure 4.1). If the system is mounted on the skids instead of a pod, the system will experience high yaw during the mission. Also, because the skids are made from spring material for cushioning landings, this results in a lot of vibration of the skids during the flight. This induced vibration does not allow the IMU (Figure 4.2) to reach its full potential.

Figure 4.1 Helicopter Platform with Pod

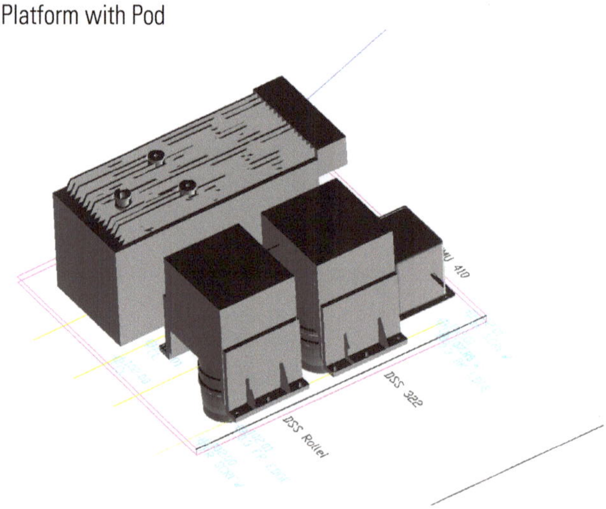

Figure 4.2 Inertial Measuring Unit (IMU).

The offsets of the antenna to the system need to be determined before the mission and surveyed very accurately. The offset needs to be surveyed to the phase center of the antenna. The postprocessing is better achieved if these offset values are in the system before the mission is undertaken. The system needs to be mounted on the aircraft in a location that has the least amount of vibration and yaw during the mission. The IMU cannot function to its ultimate if it is exposed to high levels of vibration. Every effort needs to be made to reduce vibration of the IMU during the mission.

4.1.2.1 Calibration of the Lidar System

Although some manufacturers will advise that the system calibration is not critical and that it can be calculated during the mission with the quality control (QC) postprocessing, I do not believe that this is the best way of performing a calibration. At Tuck Mapping, we will calibrate our system each time the system is removed from the helicopter pods and any time maintenance has been performed on the scanner, camera, or the IMU. Our calibration site is a thirty-six block area where we have surveyed the roofs of the houses and have run digital levels on all the streets see Figure 4.3.

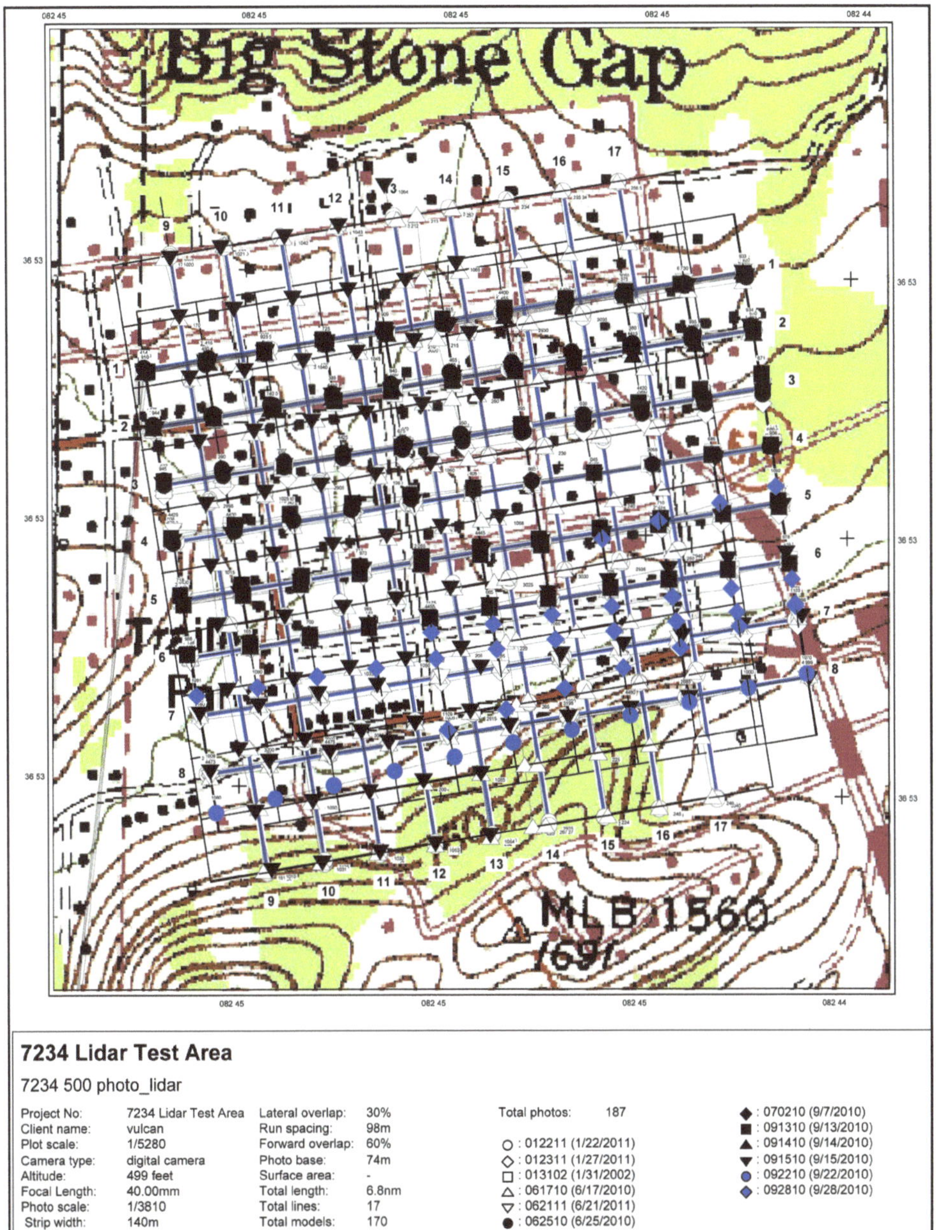

Figure 4.3 Map showing Calibration Site.

Targets are in the streets, and they will be refreshed before each calibration mission and inspected to determine where the control points have been disturbed. Our calibration site is over a housing development where we use the rooftops for boresight misalignment calculations. RiProcess, which is a *RIEGL* post-processing software, utilizes planimetric surface extraction to calculate misalignment between IMU and lidar scans. This is the reason that we fly the calibrations with adjacent lines flown in opposite directions. In addition, we will fly lines that are perpendicular to the first lines so we will have generally four swaths that will cover a single structure and each of the swaths are flown in different directions. This procedure allows us to calculate roll, pitch, yaw, or errors in the system. Our technicians ensure that each point in the point cloud that hits a common structure point will have the same coordinates. This calibration in RiProcess is completed before any LAS files are created. It is imperative that the coordinates of the LAS are calculated properly before the LAS is created, so there is no need to change the coordinates of the point cloud after the survey has been calculated. As a general rule, the person that is completing the postprocessing of the raw data has a better understanding of the coordinate geometry than the technician that is classifying the point cloud. We use a highly trained surveyor to complete the post processing of the lidar mission because of their background in surveying. The point cloud may not be calibrated to the ground at this point, but the x and y coordinates of the point cloud will not change. The calibration of the point cloud to the ground takes the point cloud as a unit and moves it vertically to reduce the root mean square of the vertical errors. This calibration process may be completed using the RiProcess software.

In addition to the system calibration, it is important to complete an on-site QC calibration of the system during each mission. Since the airborne system consists of multiple components such as cameras, IMU, and lidar scanners, the base plate can have slight movement due to differences in temperature, which can result in system error. Generally, the QC calibration does not move the sensors very much. However, in rare cases, it can expose major problems if the platform has been damaged during hard landings or rough roads when tailoring the system.

4.1.2.2 Flight Planning for eagleeye™ Mapping System

A thorough understanding of the client specifications is important for the flight planning before the mission execution. Since our system is operated with a digital camera and also lidar scanners, the system must be flown at the proper speed, altitude, sidelap, and power to meet or exceed the specifications of the client.

The performance of the lidar scanner must be understood to calculate the accuracy that can be expected from the acquisition. Depending on the type of camera that is used on the system, the pixel resolution and the horizontal accuracy requirements will dictate the altitude for the acquisition. Many times the GSD (ground space distance) required for the final product will dictate the flying height. With the higher mega-pixel camera, we can generally fly the lidar scanner at the altitude for the specified vertical and the horizontal accuracy as well as the pixel resolution that will exceed the client specifications. The lens that is installed on the camera will also be a factor in the flying height. For example, with the Rollei 60MP camera, we can use the 47 mm lens or the 35 mm lens. We map with the 47 mm lens, which has a narrower Point-of-View (POV) that allows for superior ground resolution in builtup areas of cities; the trade-off is that the base-height ratio is not as good. With the same camera and the 35 mm lens, the base height ratio will be approximately the same as the Leica RC-30, and the imagery can be used to obtain good vertical accuracy for stereo photogrammetry.

If the camera is not the determining factor for the altitude, the flying height can be determined by the vertical accuracy required. In order to make this determination, the operator must have a good understanding of the accuracies that can be acquired with their particular

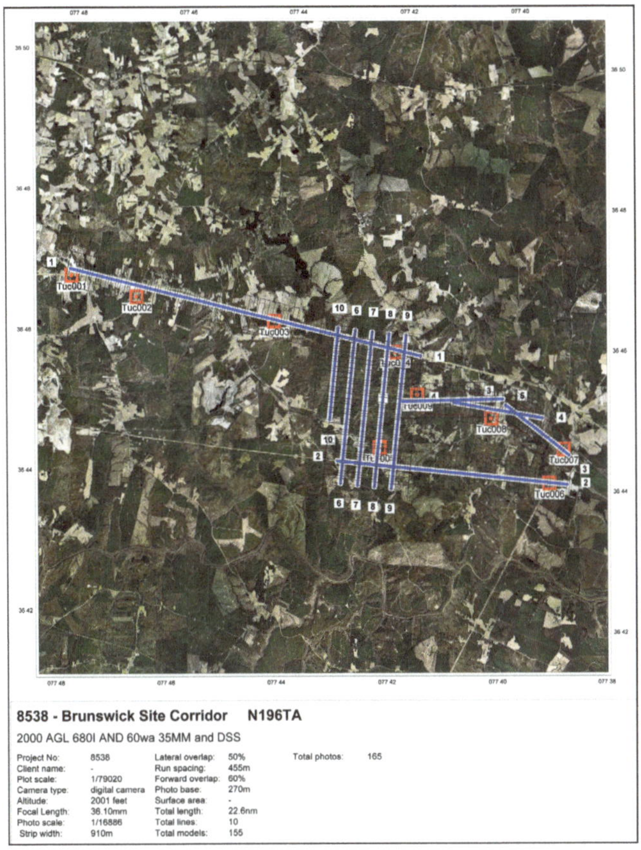

Figure 4.4 Flight Plan.

Figure 4.5 Graphic of Actual Flight Plan.

sensor at various altitudes. This can only be determined by field testing at a calibration site; the sensor must be calibrated at multiple altitudes to determine performance accuracies. By providing a QC survey for each mission, the operator will also be able to correlate his flight parameters to the final results and gain a better understanding of future mission planning.

The flight plan needs to be designed to provide the correct altitude to achieve the required specifications. Also, the flight plan needs to be designed with the surveyor in mind so that the ground control points can be accessed. Often, the flight plan needs to be laid out with extended lines into areas that can be targeted and surveyed. The person laying out the flight lines needs to design the ground control plan so the project can be triangulated and/or calibrated for lidar accuracy. Flight lines should be laid out with 50 percent sidelap so that all areas are scanned from two directions to prevent shadowing of the point cloud. Cross-flight lines will help tying all flight lines together and will ensure that the point cloud has met the relative accuracy. The cross-flight lines will also ensure that the full extent of the project area is covered by the lidar points by including a buffer around the project area. It is a good practice to put the center of the outside line along the edge of the mapping area which will provide a buffer that is approximately half of the flying height. The ends of the lines need to be extended past the ends of the mapping area.

The characteristics of the mapping area are also important so the direction of flight and the sidelap can be designed to best represent the results of detail. For most of our helicopter projects, we fly at 50 percent sidelap, so we have two opportunities to capture lidar reflections on structures and vegetation for complete coverage. We fly both sides of a power line ROW so we are viewing the line features from two positions, and also we will get more point density under the tree canopy on the opposite side of the ROW from the flight line. Features such as insulators will be depicted better and will not be shadowed from the sensor during the acquisition.

If we are flying the helicopter sensor at higher altitudes to get wide area coverage of the ground DEM, we generally relax the large sidelap requirement. Generally a 25 percent to 30 percent sidelap is sufficient depending on the amount of terrain changes in each flight line. These overlaps provide for full coverage of the ground, even when flight plan deviations occur. Since the lidar swath on the ground is directly proportional to the flying height above the ground, particular care must be taken during the planning so the ground coverage is known when there are extreme changes in the terrain, such as flying in mountainous areas. We will generally fly terrain following with the helicopter, but there can be depressions or ridges that cannot be accounted for. We will use the lidar feedback from the nadir of the sensor to determine the actual height above the ground, and we will also use radar altimeters for this indication. A problem arises when flying parallel to a valley or a ridge or both. If the helicopter is flying down the center between the ridge and the valley, you may have a ridge that reduces the swath width on one side and the valley on the other side will increase the swath on that side.

The locations of the base stations are important to ensure high trajectory accuracy. There are several methods of obtaining base station data: conventional base stations, a Smart Base Network of receivers using a combination of receivers and CORS stations, or a CORS station for single baseline Post Processed Kinematic (PPK) data.

The Smart Base Network is basically a network of stations that provides at least four stations that encircle the project area. This network will give the same results as having an OPUS (Online Positioning User Service) solution computed at the center of the helicopter for each GPS epoch during the mission. If you are using four CORS stations to encompass your project and one of the stations malfunctions, you may either lose the data or have a degradation of the accuracy of the collection because some of the data will be outside the control network.

Figure 4.6 Base Station Setup.

Figure 4.7 Control panel on Ground.

Figure 4.8 Control point on Imagery.

A helicopter survey is frequently used to produce very high accuracy DEM's. This accuracy is based on the lidar sensor, the installation, and the flight planning, but also on the quality of the GPS satellite data. Flying through spikes on the Dilution of Precision (DOP) charts will produce large GPS errors in the trajectories of the mapping system. A wise policy is to use a PDOP of 3.5 or less for helicopter lidar surveys. The PDOP (Precision Dilution of Precision) must be calculated and the missions planned to take advantage of the good GPS satellite geometry.

4.1.2.3 Ground Control Surveys

The survey control requires planning to meet the project's specifications. The accuracy of the survey needs to be twice the required accuracy of lidar points. The targets need to be seen from the air for the triangulation of the imagery. Any monuments need to be placed in a unique location so that they can be properly referenced and identified in the office. Any control monuments driven below the surface or left higher than the surface are sources of error (see Figure 4.7). The target material needs to be large enough to be seen from the aircraft camera system (see Figure 4.8). The target material needs to be set flat against the pin, and it is even better to put a hole in the material so the pin is visible in the middle. Many states require that the monuments have an identification mark with the name of the surveyor that can be contacted.

4.1.2.4 Acquisition of Data

The cutoff angle for the base stations needs to take into account the lay of the land and the obstacles around the base station. The base stations need to be set up in the clearest location on or near the project that provide the best open sky coverage. Base stations need to be set in pairs in case the unit has a malfunction during the mission. The base station is best set up on or near the project since the accuracy is directly proportional to the base line distance to the project. The coordinates of the base station point need to be surveyed in advance with a high accuracy survey so the coordinates for the project can be based on the correct coordinate values. The elevations of the base stations need to have digital levels run from a first order vertical benchmark. If a first order vertical benchmark is not available, the best elevation possible needs to be established on the base station, and then used as the project reference point. If accuracies in the range of 0.05 to 0.1 feet are desired, the digital levels need to be established on points the entire length of the project. Once the project is postprocessed, the point cloud and imagery elevations will be calibrated to the control point solution throughout the project

Before the flight, the best policy is to collect five minutes of static trajectory data to increase the trajectory accuracy. Airborne initialization can be accomplished, but the results can vary depending on the direction of flight in relation to the base stations. The best results will be achieved if the aircraft is initialized at the project in close proximity to the base station (Figure 4.9). If the helicopter is remote to the project, it is a good idea to fly towards the base station and then circle the base station before the mission. Depending on the placement of the antenna on the aircraft, the turns may need to be shallow to prevent multi-pathing of the GPS signal. The GDOP (Geometric Dilution of Precision) needs to be monitored during the mission to double check these values with printed charts.

Figure 4.9 Platform Initialization Process.

Figure 4.10 GPS Base Station at Project Site.

The locations of the base stations and the launching of the helicopter are important for the acquisition of the data. Under ideal conditions, the operator has the helicopter initialized on the project area and the GPS base would be located in the center of the project (Figure 4.10). This provides for the shortest base lines during the data acquisition and better accuracy. Typically, this scenario is possible when flying military bases, airports, or areas where permission can be obtained for operation of the helicopter from the project area. Our required procedure is to place base stations to guarantee the recordings of the base station data. The receivers sometimes fail to operate or an operator sometimes errs during base setup, and the mission will be lost without the second base. A failed mission due to the GPS is very costly and is not recoverable unless there is a reliable CORS nearby.

If the operator cannot set up the base stations on the project area, the base stations can be set up at the airport where the helicopter is based. Most airports will have a Primary and Secondary Airport Control station, and these are very good stations for base station operations. These control stations have been established by the NGS and are tied to a number of very high order stations.

Because of the cost for the acquisition of the data, it is imperative that the operator know and understand that the data has been collected properly before they leave the project area. The data needs to be viewed and a preliminary trajectory can be determined to assure that the satellite data is of the quality that will meet or exceed the client's specifications. Also, the imagery needs to be reviewed to assure that all the project area has been covered properly and the scanning data has design overlap and no gaps.

4.1.2.5 Processing of Data

The acquired data needs to be verified before the system leaves the project area. During the post processing, the trajectory needs to be determined forward and in reverse with a comparison of the delta in the position between the two trajectories. The health of the individual satellites needs to be evaluated if trajectory values are not satisfactory. If the trajectory delta values are high, the final results of the mission can be compromised with high accuracy errors. When the IMU and the lidar angles and distances are applied to the trajectory, the point cloud will be tied to the trajectory on the ellipsoid, but not to the control network. The point clouds from each swath needs to be calibrated to each other so every point in each swath that hit the same ground point have the same coordinate value. Many software programs will not use planar surfaces such as house roofs to adjust the lines to each other.

The raw data from the system needs to be processed in order to get it into the format required by the client. Each project has a required set of deliverables that are determined by the client to achieve the final results to satisfy the requirements of the specific project. Many projects will have a wide range of deliverables that will provide for many users to complete their individual requirements.

The scanner raw data from the system is generally in a proprietary format from the manufacturer. This data must be postprocessed to combine the trajectory of the helicopter with the IMU angles and the scan angle and ranges of the lidar points.

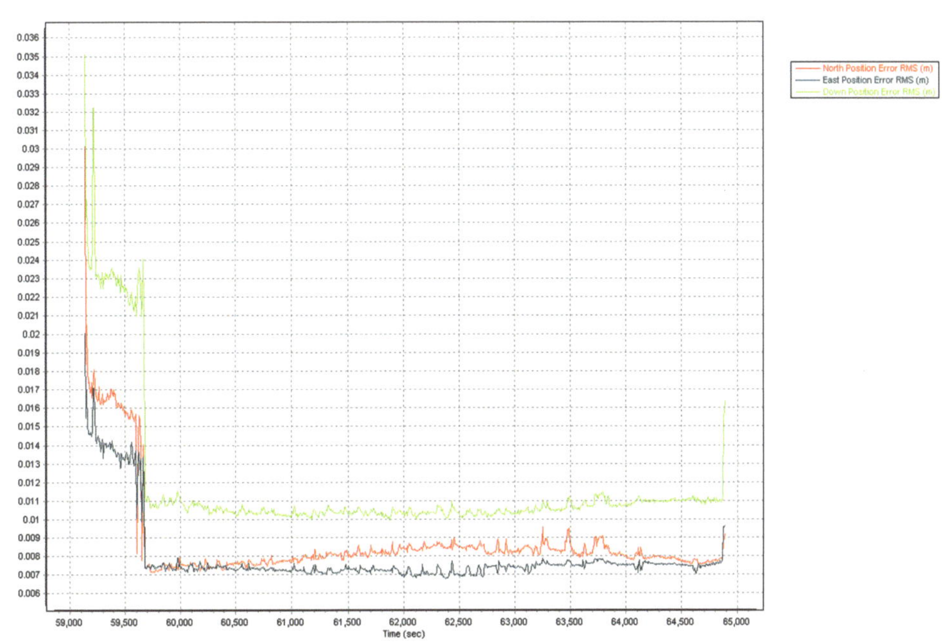

Figure 4.11 Graphic of Lidar System Performance.

The first process is to process the airborne GPS from the base station and the GPS in the system. This will provide for GPS trajectory of the helicopter during the mission. This trajectory is generally processed starting with initialization location and calculating the trajectory in the same direction as the flight of the helicopter. The trajectory is then processed starting with the end of the mission and processing the flight in reverse. Next, a comparison of the two trajectories is completed to determine the differential between the trajectories. If there is a large differential in the two trajectories, the technician must then start to review the satellite data to determine if there is missing data or if there is bad satellite geometry that will prevent a good solution from being determined. Sometimes individual satellites are broadcasting inferior data, and the data from these satellites must be removed from the solution. There may be some gaps in the epochs from the satellites that must be repaired before a good solution can be determined. The post-processing trajectory software provides many tools for correcting problems with the satellite data in order to provide the operator with an acceptable solution. In rare occurrences, an acceptable solution cannot be calculated and the mission must be repeated. Generally this is the result when proper planning has not taken place and the operator has not known or paid attention to the satellite geometry. The flight may have been performed during a spike in the satellite ephemeris. It is not recommended to avoid acquiring data when the DOP (Dilution of Precision) is greater than 3.0 to 3.5. A lower DOP will result in a better the geometry of the satellites and a stronger geometric solution of the trajectory.

Once an acceptable trajectory has been calculated, it is combined with lidar scanner data and a point cloud formed. When the trajectory, IMU, and the angles and ranges of the scanner have been applied, the technician must then begin to compare the different swaths from the mission. The software will apply the system's bore sight (calibration offsets and angles) to the data to bring the point in the individual swaths together to ensure that a feature in one flight line will match the corresponding point in all the flight lines. Points such as structure roofs will be compared to ensure that the peaks of the roofs have the same coordinate values from one line to another. The software will compare roof lines that show up in each swath and will determine a revised offset and angle that will bring all the points that are common to a common coordinate value. The software will allow the technician to run iterations of the software until the angles and offsets bring the data together. The ground control is then applied to the point cloud to take out any systematic errors of the system. Each swath of the point cloud is very well calibrated to each other and can be brought together to about one centimeter, but it will still not have true coordinates. The ground control will allow the technician to make minor adjustments to the horizontal and vertical positions of the point cloud to make it fit the true geodetic position of the points. Since the vertical accuracy of the point cloud is very good from line to line, the entire point cloud is adjusted vertically to match the vertical elevation of the ground control.

If the surveyor has surveyed to plus or minus three inches, that is as good a result as the lidar data will produce. When very high accuracy surveys are required, a surveyor must utilize digital levels for this vertical control. The control becomes a QC for the lidar and a QC control must be at least twice as accurate as the survey being checked for quality. If the client wants 0.05-feet vertical accuracy, the ground control survey must be determined within .025 or less. This is a similar rule of been used of control for photogrammetric missions for many years.

When the data is calibrated and all points in each flight line will provide for the same coordinate values for the same feature in the adjacent lines, the operator will then export the geo-referenced point cloud as a LAS file. This LAS file is then imported into one of many different software packages to classify the data according to the feature that the point represents. These feature classes are determined by the clients and will vary from client to client and project to project.

The imagery will then be processed in a similar fashion. The raw imagery from the camera will be processed to create an image that has proper color and is geo-referenced. The trajectory and the IMU is applied to the principal points of each image to provide the first rough estimate of the coordinate value of the principal point and ultimately of each pixel of the image. The ground control is applied to provide accurate horizontal and vertical values to the targets in the imagery.

4.1.2.6 Quality Control

During quality control processing, we use inclined surfaces to assess the quality of the point cloud. The more the surfaces are matched and refined, the more accurate the relative accuracy between the swaths and ultimately the final product. When the point cloud is created, it may need to be adjusted so the elevations match the ground control on the project (commonly termed a "z-bump"). Errors will generally occur because the surveyor could not carry the proper elevations for the full extent of the project. Once the point cloud is established, technicians that are classifying the point cloud do not need to adjust the coordinate values. If the point cloud is properly calibrated, each of the swaths will come together to accuracy better than required by the project specifications. We always check control points to properly assess the degree of accuracy in the project.

L:\8411-Caltrans Camelia City Viaduct\Ground Control\8411-Caltran GC CA Z 2 NGVD29 PENZ.txt

Number	Easting	Northing	Known Z	Laser Z	Dz
103	6706525.049	1967274.708	14.731	14.740	+0.009
104	6708852.389	1966559.491	16.875	16.910	+0.035
105	6710759.116	1965913.624	18.937	18.960	+0.023
106	6710933.800	1966292.135	16.232	16.220	−0.012
107	6708980.725	1966942.960	23.457	23.450	−0.007
108	6706734.255	1967690.544	13.623	13.560	−0.063

Average dz	−0.003
Minimum dz	−0.063
Maximum dz	+0.035
Average magnitude	0.025
Root mean square	0.032
Standard deviation	0.035

Figure 4.12 Example Quality Control Report.

REFERENCES

Riegl Laser Measurement Systems, 2012. Airborne Laser Scanning, URL: http://www.riegl.com/nc/products/airborne-scanning/ (last date accessed: 14 December 2011).

4.2 FIXED-WING PLATFORM

The mounting of the all ALS systems is critical to the performance of the sensor. The type of aircraft selected for the ALS system is just as important to the operation of the lidar sensor as the sensor itself. The aircraft needs to have the operating space and weight capacity to carry the sensor, crew, fuel, and additional equipment as applicable for a complete lidar collection operation. Different types of sensors can fit in different aircrafts, so determining the type of aircraft that is correct for your sensor is important. It should be understood that a vast number of aircraft, for example, a Cessna 206, 207, 210, 310, 410, Partenavia, Piper Navaho, Twin Aero-Commander, Cessna Caravan, and many other aircraft are suitable for most lidar sensors.

Typically, lidar collection missions last between four to six hours in duration. So, it is important to consider the amount of fuel required for this duration considering the weight of the lidar sensor and electronics rack, crew, and fuel. The early commercial lidar sensors weighted as much as 300 pounds with ancillary equipment, so the required weight capacity of the aircraft was substantially more when including the weight of the crew and fuel. Also, the comfort of the crew is important given this is their "office" for 8 to 12 hours a day.

4.2.1 MOUNTING OF ALS SYSTEM

There are several considerations when mounting an ALS system into a fixed wing aircraft. The first item to address is the electrical requirements for the system. Typically, ALS sensors require aproximately 30 amps to operate properly at peak power. Most current ALS sensors also required a 24-volt aircraft power which is due to the power requirements for primarily the laser. Although, 12-to-24 volt converters can be used, it would not be recommended because significant fluctuation issues can arise using a 12 volt aircraft system. The electronics for the rest of the aircraft need to be considered, i.e., if the aircraft is a single engine aircraft or a twin engine needs to be considered.

Typically, the electronics and/or avionics in the aircraft have power requirements, and the entire "power budget" needs to be evaluated. Typically, if the aircraft can provide a total of 100 amps then there should not be a problem with the power requirements, but again this depends on the specific aircraft.

If the aircraft has not been specifically electrically wired for a high-performance aerial camera or lidar system, then an FAA approved Certified Aircraft Engineer will have to inspect the system and have to "sign off" on the electrical modifications made to the aircraft. Typically, a Standard Technical Conversion (STC) can be used for the modifications, but a Licensed Aircraft Mechanic is required to supervise and approve the work. All such aircraft modifications are noted in the specific aircraft's official "log book." The exact process can be obtained through the FAA or from an aircraft mechanic or avionics service supplier.

Once it is determined that the power requirements are satisfied and the electronics have been wired for the ALS system, the camera/lidar hole requirements in the floor of the aircraft will have to be examined to confirm that the ALS sensor can fully function in the camera hole. The different manufactures and different types of lidar sensors from a manufacture have different requirements regarding the camera holes, but generally speaking if you obtain an aircraft with a traditional camera hole there almost is never a problem with installing an ALS system. If an aircraft is purchased without a camera/sensor port, then make certain that the specific aircraft has an STC to install the required port. Creating an STC for an installation that does not currently exist is extremely expensive and time consuming. The most important element when understanding the installation of the sensor head into the port is making sure you have the specified FOV clearances. The hole needs to be large enough to accommodate the sensor's scan angle, any forward or back angles (if applicable), and a web-cam or video system if desired (Figure 4.13).

Figure 4.13 Optech Gemini including web-cam as seen from the from bottom of Aircraft.

Typically, most camera holes are re-enforced (as per the STC), but it is imperative to make sure that there is no possibility of aircraft vibration that will affect the operation of the sensor. The solution for this situation is to mount the sensor head using the manufacture's provided mount and constructing a plate to install the manufactures mount. Then, set this assembly within the aircraft with as many attachment points as possible. The incorporation of shock mounts into the installation will further limit any issues caused by aircraft vibration and these are highly recommended. In addition, the mounting of the sensor head should be designed to take a minimum of a 9G impact, which requires the use of aircraft mounting hardware. The following are examples of sensor head installations for a two types of lidar sensors. In addition all manufactures will provide you with the necessary assistance and equipment plans for a proper lidar installation including the sensor head (see Figures 4.14, 4.15, and 4.16). The proper installation of the sensor head is paramount in the proper function of the ALS and the resulting data.

The electronics equipment used to power and operate the laser head often referred to as the "rack" is the next component of the sensor that needs to be installed in the aircraft. The rack needs to be located for easy access by the sensor operator. Typically, the manufactures of the ALS systems provide long length cables between the components, so that the rack can be mounted in a convenient location or any location that has ample room. The electronics rack contains the data logger, galvo controller, GPS card, laser power supply, and supporting electronics required to record the information coming from the components supporting the sensor head.

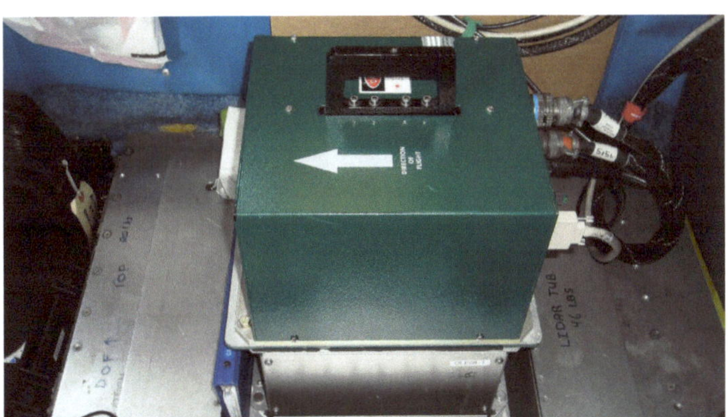

Figure 4.14 Optech ALTM Gemini Sensor Head Installation (courtesy of AeroMetric, Inc.).

Figure 4.15 Leica ALS-50 MPIA Sensor Head Installation (courtesy of Merrick, Inc.).

Figure 4.16 Leica ALS-70 Sensor Head Installation (courtesy of AeroMetric, Inc.).

The ALS manufactures have greatly reduced the size of the sensor rack over the past several years (with improved modular electronics) which has made the installation of the racks much easier and has reduced the weight and balance solution for the aircraft. As specified for the sensor head, the rack needs to be mounted to withstand a 9G impact as well. Figures 4.17 and 4.18 are examples of rack installations in an aircraft.

Figure 4.17 Optech Gemini rack installation (courtesy of AeroMetric, Inc.).

Figure 4.18 ALS-70 rack installation (courtesy of AeroMetric, Inc.).

The GPS antenna needs to be mounted on the top of the aircraft. The antenna needs to be mounted away from other transmitting antennas, such as radio antennas, because the radio antennas can interfere with the ALS antenna. The GPS antenna is an L1/L2 antenna that is specifically designed for the ALS system. The older system antennas operated at a frequency of 26db but newer antennas operate at 43db. It is critical to confirm that the ALS system has an antenna that is specifically designed for the ALS system being used.

The next critical step to measure the lever arm offsets of the antenna after mounting it to the aircraft. This measurement can be accomplished by using a total station survey instrument. The lever arm offsets are the relationship of the phase center of the antenna and its reference to a known point on the ALS sensor. This location on the sensor is typically a point on the scan mirror, also called the "laser zero" which is a point to which all components of the sensor are referenced. The lever arm offset is the location of the antenna in the X, Y, and Z direction as it relates to the laser zero. Since the antenna is located above the laser zero, the Z component of the lever arm offset is negative. The X and Y for the lever arm offsets will be in relationship to the forward or back direction of the antenna as it relates to the laser zero and the wing-to-wing direction of the antenna as it relates to the laser zero (see Figure 4.19).

It is imperative to measure the location of the antenna as it relates to the laser zero on all aircraft in which the ALS is installed. This process should always be done using proven surveying techniques, such as GPS observations and/or a total station. It should be noted that the Position and Orientation System (POS) software will estimate these locations based on the specific flight, but this solution alone can result in errors from mission to mission or cause additional problems with the resulting data; it best advised to initially survey in the location of the antenna as it relates to the sensor head known zero point.

Figure 4.19 The antenna in the center of the picture is the ALS antenna.

4.2.2 INTEGRATION WITH POS/NAV EQUIPMENT

The POS/NAV equipment is used to derive the location and attitude of the sensor while the lidar collection is occurring in real time. The GPS data is collected at up to 10 Hz (i.e., ten times per second) and the IMU data is collected at 200 Hz (i.e., 200 times per second). At a minimum the GPS data should be collected at 1 Hz, but most current system collect at higher rates. The attitude information is the roll, pitch and heading (yaw) of the aircraft. Aircraft roll during collection of lidar data and POS/NAV data should not exceed 20 degrees If possible. Some of the current POS/NAV manufactures indicate that acceptable solutions can be achieved if the aircraft banks more than 20 degrees in roll, but this procedure is not advised given the degradation of the quality of the POS data.

Typically, the ALS manufacturer will provide the POS/NAV system for your ASL system, so special consideration as to what POS/NAV system should be used is not necessary. As a user, it is important to understand the capabilities the POS/NAV when considering the addition of supplementary equipment such as a camera, hyperspectral sensor, thermal sensor, or other sensor systems to an existing ALS system configuration.

4.2.3 INTEGRATION WITH SUPPLEMENTARY EQUIPMENT

Additional sensors can be integrated with most ALS systems which is becoming more commonplace as the various technologies and requirements have advanced. Multiple sensors with lidar is more common with helicopter platforms, but is becoming more frequent with fixed wing platforms. It should be noted that ALS sensors are active systems and most supplementary systems are passive, multiple data collection is limited to reflected light (daytime) when multiple systems with an can be operated.

Most POS/NAV systems have the ability to record a host of information in addition to the information that is required for the lidar sensor. Typically, the most common supplementary equipment added to an ALS is a medium format, digital camera, but recently multiple cameras are being added to the lidar system. The important elements of this process are to understand the POS/NAV systems and their ability to record the necessary information for the supplementary component(s). Providing a "tag" to reference information from the supplementary component(s) is vital. Additionally understanding how to reference the supplementary component(s) physical location in relationship to the IMU and GPS and how it relates to the ALS sensor is also critical. Typically, this information can be deduced from the directions/plans provided by the ALS vendor and the component vendor. Typically, measurements are required to confirm the positional location of the supplementary sensors in relation to the lidar system's "laser zero."

Last, there needs to be verification with the equipment providers that each component has the ability to record data and communicate positional information. Typically, most manufactures will try to facilitate your needs and component communication and integration can be accomplished.

4.2.4 CALIBRATION OF A WHOLE SYSTEM

The calibration process is described in detail in Chapter 5.3, and the significance needs to be re-iterated that calibration is the single most critical component in the collection and processing of lidar data. The ALS needs to be calibrated every time the system is installed into the aircraft; this sophisticated hardware has limitations, and the installation can dramatically change the system calibration; any change in the location of the sensor head will considerably change the resulting data. The calibration determines the fundamental parameters that affect the resulting data, and although the ALS manufacturers provide calibration offsets for their systems, the repeatability of these parameters vary regardless of manufacturer's calibration. It should also be noted that each ALS system retains its calibration differently. Some sensors hold calibration very well and other not so well although they can be the same brand and model of sensor. Regardless of this fact, the calibration corrections change from installation to installation even if it is in the same aircraft; there is no way for the manufacturer to prevent this.

If an ALS is installed in an aircraft and remains in that aircraft, typically the lidar provider will discover (in the postprocessing phases) a trend to the calibration, which means that the calibration parameters do not deviate beyond a certain threshold and additionally can remain pretty consistent in some cases and again it depends on the system. Some systems can hold calibration over several missions; some sensors need to be calibrated on every lift so it would be necessary to perform a series of calibration passes for each mission. This process varies from provider to provider, but the best (safest) practice is to perform a "mission calibration" regardless of sensor stability because it provides backup information in the event that something unique occurred during a given mission. Typically, almost all issues that could occur during a mission can be resolved following this rule.

AUTHORS

Section 4.1 was written by Bobby Tuck of Tuck Mapping Solutions, Inc., and Section 4.2 was written by Jamie Young and Chris Guy of AeroMetric.

5
GUIDELINES FOR LIDAR DATA COLLECTION

BECKY MORTON and JAMIE YOUNG

5.1 FLIGHT MISSION PLANNING

Flight layout and mission planning are designed and implemented based upon key project components. These key components are often established by the client or in combination with recommendations from the data provider.

5.1.1 Key Project Components

Key components include:

- density of lidar points (post spacing)
- accuracy of lidar points
- window for flight operations
- project boundary
- elevation model format (DTM/DEM/breaklines)
- imagery requirements (if any)
- lidar data deliverables and formats
- accuracy of data deliverables

In addition to the key components listed above, additional factors such as the shape of the project boundary, the distribution of vegetation on the project, the impact of terrain relief, the presence of tall buildings or other features of height, etc., are brought into the evaluation for flight planning.

5.1.2 Platform

One of the first decisions involved in flight mission planning is determination of the type of aircraft to be used.

Helicopters are often selected for projects with the following conditions:

- Need for maneuverability such as along a utility corridor or river.
- Requirement for high point density (obtained from slower aircraft speed).
- High vertical accuracy (obtained from lower flight heights).

Fixed-wing aircraft are selected based upon the following factors:

- Ability to acquire wide areas of coverage at higher flight altitudes and speeds (cost effectiveness).
- Option for higher point densities with overlapping flight lines.
- Vertical accuracy sufficient for most large to medium scale mapping applications.

5.1.3 Planning the Flight Window

Planning the flight window requires careful consideration of the goals of the mapping project, project location, vegetation, ground conditions, weather, and additional site-specific factors. Many considerations are obvious such as acquiring lidar with "leaf-off" conditions when the goal is to obtain ground returns, and avoiding conditions of cloud cover, snow, rain, and forest fire. Planning for the flight window must include the forecast of windows of acceptable position dilution of precision (PDOP) to ensure adequate GPS functionality.

5.1.4 Flight Plan

Developing a flight plan starts with analysis of the project extents, topography, and land characteristics in relation to targeted specifications of point spacing. Topography and land characteristics have a large effect on point spacing and therefore require adjustments to the flight plan in order to maintain the targeted spacing. For example, mountainous regions require large variations in flight height which in turn affects the post spacing, swath width, laser footprint size, and accuracy. Forested project areas require consideration of lidar instrument settings to maximize canopy penetration and/or maximize identification of tree crowns or understory. Urban areas require flight planning with significant overlap to ensure adequate sampling of the ground between tall buildings. Corridor projects (such as rivers, highways, and transmission lines) present flight plan options such as flying with a helicopter, breaking flight lines, and/or flying multiple parallel lines to cover the alignment.

Data providers typically work from spreadsheets or tables with pre-determined flying heights and lidar system settings to achieve standard post spacing and accuracy settings given the particular options of their lidar system. These settings are fine-tuned to achieve the project-specific requirements. Each lidar system is unique and data providers will take into account those strengths and weaknesses of their particular system when designing the final flight plan.

5.1.5 Flight Height

The flight height is determined in combination with aircraft speed, scan angle, scan frequency, and pulse rate, in order to achieve the targeted post spacing. Typically data providers develop tables of standard settings for their unique lidar system and aircraft to achieve standard post spacing.

Table 5.1 Example Lidar Project Planning Table.

Post Spacing (meters)	Vertical Accuracy (cm)	Horizontal Accuracy (cm)	Flight Altitude AMT (meters)	Scan Angle from Nadir (degrees)	Aircraft Speed (knots)	Pulse Repetition Rate (kHz)	Mirror Scan Rate (Hz)	Flight Line Overlap (%)	Full Swath Width (meters)
1	15	14	1067	10	130	76.5	57	50	376
1.5	15	18	1524	12.5	130	61.5	50	40	676
2	15	20	1829	15	130	54.0	44	30	980
3	15-16	26	2438	15	130	48.5	30	30	1307
4	15-19	30	3048	20	130	44.1	20	25	2219
5	16-22	40	3658	22.5	145	30.8	15	20	3030

5.1.6 Beam Divergence and Footprint Size

The footprint of the laser upon the ground is affected by flight height, scan angle, and beam divergence along with terrain factors. In general, as the ground slopes, or as the laser pulse is directed toward the ground at an oblique angle, the shape of the laser footprint is elongated from a circle to an ellipse (see Figure 5.1). As the footprint size expands more horizontal and vertical uncertainty is introduced and accuracy is decreased. Also, as the footprint size expands the pulse energy is reduced and it is possible that there may not be sufficient power to return the pulse to the sensor. These negative factors are mitigated through lowering of the pulse rate to increase the pulse energy and/or reduction of the scan angle to reduce the angle of incidence of the pulse.

Some lidar instruments have the ability to modify the beam divergence to change the concentration of laser energy between a wider or narrower setting. This feature provides some control over footprint size and pulse energy which can be an advantage particularly in forested, mountainous, and urban locations. Even though pulse energy is reduced with a wider beam divergence, this may be the desired setting when the goal is to increase the chances of laser returns from multiple height features within the footprint. Conversely, a narrow beam divergence increases the chance of obtaining a ground sample in heavy vegetation due to the increased pulse energy.

5.1.7 Overlap of Flight Lines

Overlap of flight lines is a requirement to ensure that no data gaps occur. The amount of overlap is dependent upon the amount of relief in the terrain. Areas with a lot of relief will require a higher overlap. Overlap may also be increased to provide higher point density. Often, flights are designed with 50 percent overlap or greater in order to achieve a twofold increase in point spacing.

5.1.8 Point Density

Flight planning of lidar missions requires consideration of not only the density of the lidar points as defined by the number of points per square meter, but also consideration of the distribution of the points in both the along-track (between scan lines) and cross-track (between points on the scan line). The optimal project design will produce similar post spacing both along-track and cross-track.

Several variables are manipulated in order to achieve a uniform spacing of points – flight speed, pulse rate, and scan rate. These variables affect point spacing in the following ways:

Table 5.2 Adjusting Along-Track and Cross-Track Point Spacing.

Variable	Affect on Along-Track (between scan lines)	Affect on Cross-Track (on scan lines)
Increasing Flight Speed	Increases the spacing between points.	None
Increasing Pulse Rate	None	Decreases the spacing between points.
Increasing Scan Rate (without modifying the scan angle)	Decreases the spacing between points.	Increases the spacing between points.

5.1.9 Flight Plan Map

After the determinations have been made for flight and instrument settings, a digital flight map is prepared. The styles and tools used for preparation of flight maps are numerous but the common characteristics include the overlay of the flight lines on a visual map of the project. Flight maps should provide a visual background of the project area (for example, a USGS topographic map, Google image, orthophoto, etc.) with the overlay of the project boundary, flight lines, base station locations, and ground survey locations.

Digital flight maps provide a method of communication to the client as well as serving to establish the precise start and stop of flight lines in a format compatible with flight operations (i.e., latitude, longitude).

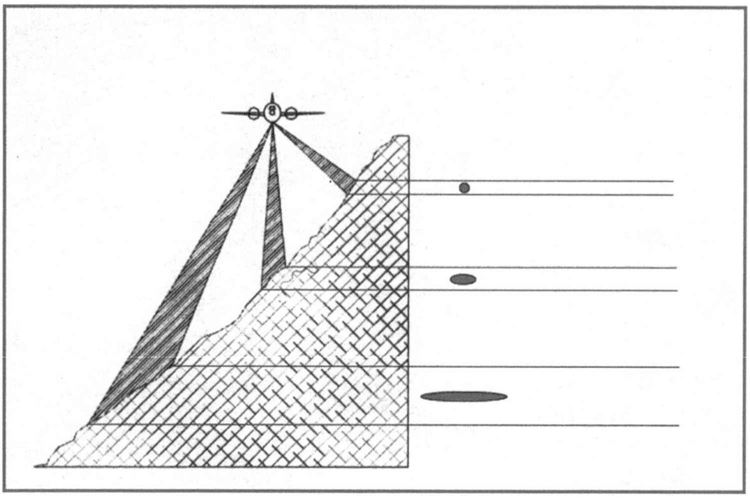

Figure 5.1 Example of lidar footprints and shape off-nadir and on sloped terrain (courtesy of EarthData, 2003).

5.2 PLANNING OF GROUND REFERENCE STATIONS

Planning of ground reference stations requires the consideration of several factors, including:

- Targeted accuracy of the project
- Availability and location of existing high-order survey locations
- Expected time of flight and satellite coverage (PDOP)
- Shape of the boundary
- Flight constraints such as restricted fly zones, terrain, obstacles

5.2.1 Location of the Ground Reference Station

Establishment of the location for a ground reference station requires consideration of the view of the satellites. Ideally the location will have no obstructions above an inclination angle of 15 to 20 degrees and will be located at least 50-feet from tall trees, buildings, and walls. High-power transmitters in the area can also cause interference and ideally the ground reference station should be located at 1-kilometer or further from such sources.

Typical lidar mapping projects will establish a ground reference station on a high-order survey location within the project area. This station, referred to as the base station, will be brought into the solution along with additional ground survey locations as part of the control network for the project.

5.2.2 Distance between Ground Reference Stations and the Airborne Receiver

In general, shorter distances between ground reference stations and the roving receiver on the aircraft improve the reliability of the solution. FEMA guidelines for floodplain mapping using lidar technology call for the roving GPS receiver (on the aircraft) to be located within 20 km of a ground reference station. Typical configurations for lidar missions establish the maximum distance between the airborne receiver and the ground reference station within the range of 10 to 50 km (maximum).

The use of at least two ground reference stations during flight operations is recommended to provide redundancy in case of a failure and also to effectively eliminate the atmospheric and orbital errors.

Table 5.3 Guidelines of distance from ground reference station to roving receiver.

Targeted Accuracy	Maximum Distance
2-foot or 1-meter contours	25 to 40 kilometers
5-foot or 2-meter contours	40 to 100 kilometers
10-foot or 5-meter contours	100 to 250 kilometers

5.2.3 Airport Ground Reference Stations

Airport locations for ground reference stations have several distinct advantages for lidar acquisition projects:

- Proximity of the control location to flight operations, often allowing the receiver to be left within a secure environment during flight operations without the need for a person to stay with the equipment.
- There are typically many high-order survey locations already established on airport facilities.
- Allows for recording of GPS data from the aircraft and the ground reference station prior to movement of the aircraft and while the recording units are in close proximity.
- Allows for recording of stationary GPS data upon landing which provides the ability to post-process the lidar trajectory with both a forward and reverse computation.

5.2.4 Setting up the Ground Reference Station

Once the location of the ground reference stations have been established, other considerations must be taken into account when establishing the station(s):

- The station should be set to log data at the same frequency as the airborne GPS and IMU systems (typically 1 Hz or 0.5 Hz).
- The ground reference station requires the ability to log data directly to a computer (ideally equipped with a removable hard drive).

5.3 IN FLIGHT CALIBRATION OF AN ALS SYSTEM

Calibration (also referenced to as bore-sighting) of an airborne lidar is one of the most critical processes contributing to an accurate and reliable data set. There are several ways to calibrate a lidar sensor including recommendations and procedures from the manufacturers of these systems. Lidar data collectors also have varying calibration processes based on their experiences which are a modified version of the manufacturers' recommendations. All airborne lidar sensors should be calibrated by the manufacturer, then again, every time the sensor is installed in an aircraft. Additionally, the systems should be routinely flown for calibration purposes to check for any variations as a result for aircraft vibration; although, if the system is properly installed, these variations seldom occur. Lastly, some form of calibration should be required for every project if not every mission, but this depends on the stability of a given system and the requirements of the project. Lidar data collectors vary in the degree of their calibration process. The following will detail the process from installation calibration to project calibration and requirements of each process as it relates to the collection of the calibration data. The actual calibration processing will be summarized but cannot be detailed because most providers and manufacturers have their own methodology of processing the calibration data, and some of these processes are proprietary. The general nature of this process and the associated variables will be detailed in this section, as it is commonly understood by the lidar practitioners.

Installation of the system as discussed in Chapter 4 is vital to ensuring the calibration results will be usable for the installation calibration process. Also, having a stable platform is critical to maintaining the computed calibration values. Typically, the following variables would be checked and validated in all of the different calibrations including the installation calibration:

- Roll: Roll is the rising or dipping of the aircraft's wings.
- Pitch: Pitch is the movement of the aircraft's nose either up or down.
- Yaw: Yaw is the aircraft moving towards the left or right while in flight.
- Pitch Slope Error: When the pitch is correct at nadir of a scan and offset at the edge of a scan (see Figure 5.2).
- Torsion or Scan Factor Error: This error is represented but an upward or downward arc appearance when profiling perpendicular to flight direction within scans (swath data); the correction for this error will be consistent for all scans of a given mission.
- Z-bias: This error is an elevation offset between the lidar data set and the control used to check the lidar data. This offset is usually caused by not having the elevation offset correctly measured in the calibration file. It can be confused with the range correction variable, so the technician needs to be careful to make the correct adjustment and confirm that the range correction is properly adjusted.

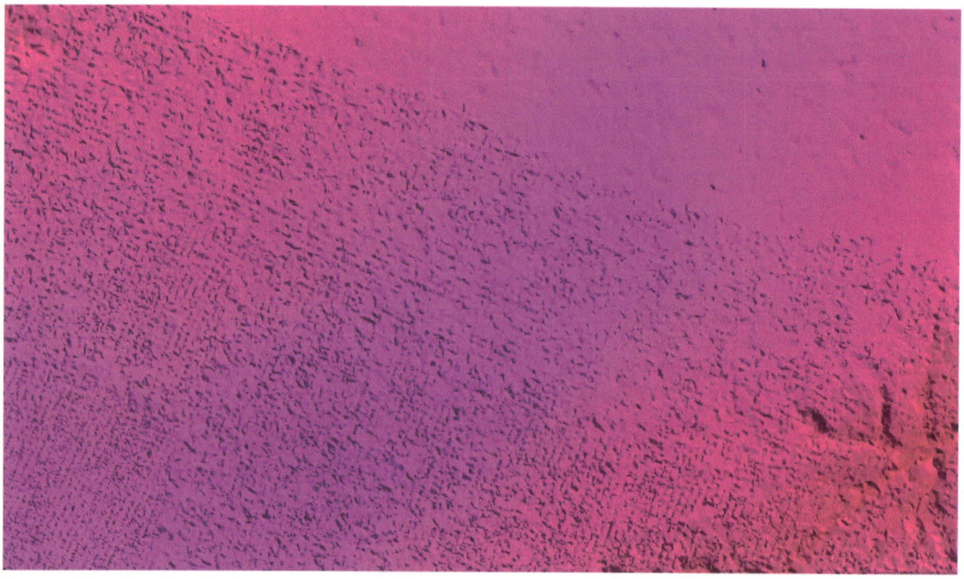

Figure 5.2 Graphic of pitch slope error (courtesy of AeroMetric, Inc., 2011).

The relative accuracy of the data is also a critical component, as well as determining the repeatability of a laser point location and elevation as it relates to other points; this has proven to be difficult at times and is more apparent in flat terrain. This scenario is addressed in most calibrations, and software tools have been developed to validate the flight line matching or repeatability of overlapping data. This process is commonly referred to as "stair-stepping" or flight line edge mis-matching, which is one of the most common indications of a calibration problem or difficulty with a data set. From the list above, there are a couple of variables that cause these problem results; Roll is the most predominate variable. Scan Factor or Torsion Errors can also impact the repeatability of the flight lines. In addition, inaccurate GPS or IMU (see Chapter 3) which will impact the lidar data repeatability.

The installation calibration should consist of several flight lines in varying directions over a calibration test range (Figure 5.3). Most lidar data collectors have established these sites at their local base of operation. The lidar operators perform extensive surveying of the test site; these surveys include building surveys for horizontal positioning, kinematic, RTK, and static surveys of points for 3D vertical positioning.

The installation calibration is conducted at several different altitudes to properly understand the affect of the errors as they relate to altitude, because at lower altitudes, the same errors will not be detectable. In addition, collection of the same set of flight lines would be flown in opposing directions. Another set of flight lines perpendicular to the initial flight lines would be flown again in opposing direction, and these lines would intersect the first set of flight lines roughly in the middle. Figure 5.3 shows the location of these flights occurring over a surveyed test range. Ideally, the calibration site would be as flat a surface as possible (for example, at an airport), because most errors are very apparent on flat surfaces. Also, these test ranges should have a series of buildings or distinct features to detect positional (horizontal) errors. The buildings can be surveyed, which will very accurately define the positional errors; four flight lines, two in opposing directions with an additional two in opposing direction perpendicular to the initial lines are required. The location of the buildings or detectable features should be in the overlapping area of all four flight lines. It is not absolutely critical for this flight line pattern to occur, but it makes the calibration process much easier and reliable. If the buildings or detectable features are not located in the overlapping areas, it is necessary to have buildings or detectable features in each of the areas of the full set of opposing flight lines in both directions. Figure 5.4 is an example of an installation calibration flight at one of the altitudes.

Figure 5.3 Visualization of calibration test flight; cross-flight shown in white (courtesy of AeroMetric, Inc., 2011).

Figure 5.4 Calibration at one altitude (courtesy of the Sanborn Map Co., 2009).

Operationally, the flight altitudes for the installation calibrations should be flown based on the characteristics of the specific system. The systems designed for corridor mapping or flown at lower operational altitudes would base the calibration flight heights based on the expected lowest altitude and highest altitude that an operator is expecting to fly their system with an altitude in between those two. This process would be the same for a system that is flown at higher altitudes or systems that typically fly a broad range of altitudes. An example for this would be if an operator with a system that can fly from 6,000 m above ground level (AGL) down to 400 m AGL. The operator typically flies the system between 2,500 m and 800 m AGL, with a typical mission altitude at 1,800 m AGL; therefore, the installation calibration would be flown at 2,500 m, 1,800 m, and 800 m, respectively. An additional calibration could be flown if the system was required to be flown at lower or higher altitudes, and this could be done during the routine calibration or as required.

In addition, to the airport runway calibration discussed, a building calibration is sometimes preformed. This is another method to detect any horizontal positional differences that could occur with a lidar sensor. This step would normally be performed during the installation calibration as well. The rooftop of a large, flat, rectangular building is surveyed on the ground using conventional survey methods as discussed above, and be used as the lidar calibration target. The aircraft flies several specified passes over the building with the lidar system set first in scan mode, then in profile mode, and finally in both scan and profile modes with the scan angle set to zero degrees.

Figure 5.5 (Calibration Pass 1) shows a pass over the center of the building. The purpose of this pass is to identify a systematic bias in the scale of the system.

Figure 5.6 (Calibration Pass 2) demonstrates a pass along a distinct edge of the building to verify the roll compensation performed by the Inertial Navigation System, INS.

Additionally, a pass is made in profile mode across the middle of the building to compensate for any bias in pitch.

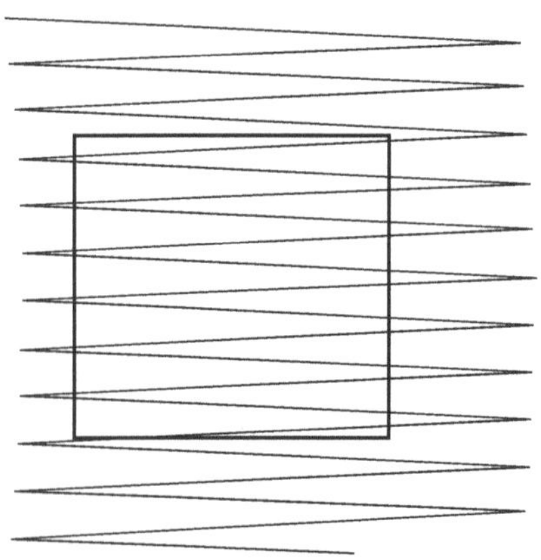

Figure 5.5 Calibration Pass 1.

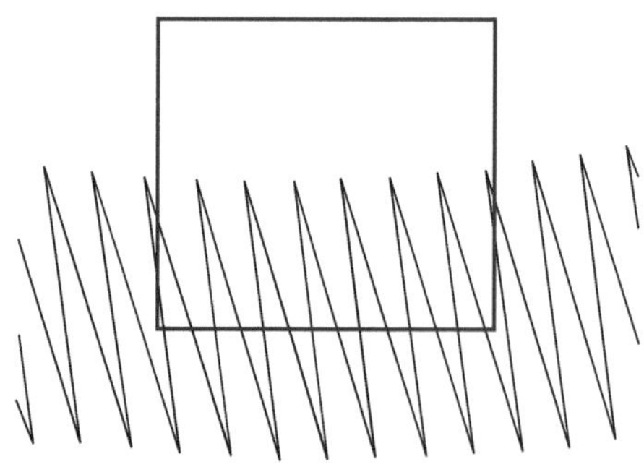

Figure 5.6 Calibration Pass 2.

The routine calibrations to check for aircraft vibrations or changes in the system characteristics would be the same procedure as the installation calibration or an abbreviated version of the installation calibration. The recommended flying height of this flight would be at the highest altitude as described above. The reason for this procedure is to identify the largest errors in lidar systems at the height farthest from the ground. Typically, these flights should occur about every three months. If the lidar operator is flying routine project or mission calibrations, these procedures will detect any changes in the system.

Typically, most lidar operators fly project or mission calibrations. The following will describe the different variations of these mission calibrations. It should be stated that there are variations (lidar system requirements, manufacturer's guidelines or software, contract specifications, vendor-specific procedures) to this process and all these variations may not be detailed in this section.

The first calibration discussed is a very detailed mission calibration process that will provide a very high probability of successful collection. The process takes about 20 minutes for the entire mission and has proven to solve most data quality issues within the project. The process is an abbreviated version of the installation calibration. If an operator performs a mission calibration, it is not necessary to perform a project calibration.

All lidar projects require an airport base of operation. In some cases there will be several bases of operation. Given that lidar programs have grown dramatically in the last five to eight years and several different lidar sensors can be used on a single, large project, this procedure needs to be considered in the calibration process, which provides the strength of redundancy over a large area.

Once a base of operation is established, the operator will perform a ground survey for one of the runways or taxiways at the base. This can be a kinematic survey providing that the aircraft has been "surveyed in" to determine the elevation of the GPS antenna and its relationship to the sensor and the ground. It should be noted that this kinematic survey of the runway usually has an accuracy of 3 to 5 centimeters based on the weight of the aircraft fully loaded including the weight of the operator(s) and pilot. This survey can be done in a vehicle other than an airplane, but given that airports are designed for aircraft, the most effective way to perform this survey is using the aircraft.

Similar to the installation calibration, the runway or taxiway is surveyed by using the GPS data from the lidar system. The runway or taxiway is taxied three times along the runway once during the project collection for this project location. If the lidar operator has more than one sensor in this location, it is recommended that all sensors taxi the same runway or taxiway. The reason for this procedure is to check the repeatability of the each aircraft's survey data for the given runway or taxiway and not induce variations over the final deliverable project area.

Once the survey is preformed, the calibration will be done at the project collection altitude. For example, if an operator is collecting 1.4-meter data for flood plain mapping and the required altitude is 1,400 meters, then the calibration over the surveyed runway would be 1,400 meters. Similar to the installation calibration, the lidar sensor would collect two passes over the runway, and two passes perpendicular to the runway. These two lines would be collected at the beginning of the mission, and then the operator would collect lidar data in the project area. After the project mission is complete, the operator would return to the base of operation, and then complete two additional calibration lines over the runway or taxiway, which would be one line perpendicular to the runway and one line parallel to the runway. The reason for this cross flight is to provide data that gives the operator the ability to detect all the variables required to properly calibrate the lidar data. In addition, this procedure provides the operator the ability to detect any drift in the lidar system over the flight mission time. If the data shows drift or differences as a result of the GPS and IMU data, the data can be corrected by normalizing the data differences of the entire flight. Consistently, over the past 15 years of commercial lidar operations, this procedure has yielded significantly better vertical and horizontal RMSE than as stated by the manufacturers of these lidar systems.

It should be stated that although this process is extremely detailed and has consistently proven to be the best way to perform calibration for project collection, several other versions of this process and the project calibration method still yield acceptable results for most lidar collections. It is important to keep in mind that the calibration for a project is a function of the required accuracies for a given project. If the manufacturer states that their lidar systems can achieve 15-centimeter vertical accuracy (z) and better than 0.5-meter horizontal accuracies (x and y), and the operator is collecting a project to meet 18.5-centimeter accuracy and a 1-meter horizontal accuracy, then it may be assumed that the calibration process may not be as detailed or rigorous, but the intention of the process described above is to state the best practice for all collections.

The most common project calibration will be detailed below, but there are variations for this process and the details of those processes will be described in the metadata by the lidar operator performing a particular project collection. Depending on the operator and the project size and configuration, this process will vary. Typically, a cross-flight perpendicular to the collection lines will be flown across the entire set of lines. Keep in mind that this cross flight distance should be limited to the ability of the IMU not to drift, because the IMU requires movement to "find itself"; if the IMU drifts, the calibration line could be in error. Also, it should be noted that the GPS collection should conform to the recommended tolerances for collection of GPS data for lidar collection for this-cross flight and all other collection lines.

There are several variations of the collection of the cross-flight lines. Some lidar operators will fly one cross line roughly in the middle of the project area perpendicular to the collection lines. This line will be flown over several days of collection within the project area. Others will fly two cross lines over several days of collection. One line being at the lower extent of the project and one line at the high extend of the collection. In addition, some operators will fly cross lines for each mission, either one in the middle of the mission's collection or two as described above. Although this is proven to be effective, the concern with this is determining the best location for these lines to be flown so that the resulting information can yield the desired results. The most important consideration is determining what are the required results of the project being flown, the terrain being flown, and the vegetation characteristics of the area being flown.

One element of calibration that has not been discussed is the calibration of the lidar intensity information (see Chapter 10.1). Usually, the manufacturer of the lidar sensor will develop the intensity correction table for the sensor purchased. Over the years, several operators have reported issues with the intensity tables provided by the manufacturer. In most of the cases, the intensity correction table needs to be changed. Over time, loss of laser power can occur, and the electronics circum to use, specific instrument malfunctions can occur and only the manufacturer can comprehend and resolve problems. Given these facts, it is recommended that the intensity of the system be checked which can be done several ways. Manufacturers of lidar sensors are very good about checking lidar intensity data collection for the operators of their systems, which usually occurs as a result of the operator's service plans. Typically, intensity collection is checked on a yearly basis and/or if the operator notices something unusual in the data, most often in the calibration flights.

The five most-common issues that have been observed when the intensity deviates from the proper result are as follows. The most common occurrence is raised white paint strips (for example, airport runway striping) in the data. Although this is not always as a result of an error in the intensity table, it seems that nearly every time it is; the other occurrence it is a result of improper operation of the system as a result of not setting the attenuation of the laser properly. Basically, what happens is that the paint strips on a runway or where there are predominate, white paint strips on a roadway, the data user will experience the elevations of these painted features to higher in elevation than the surrounding areas (see Figure 5.7). This phenomenon is caused by saturation of the energy of the lidar. This phenomenon can be re-classified (edited) through postprocessing, but it is a potential concern for the operator leading to the second problem.

The second problem is much more concerning, and the error is that the specified accuracy of the data in most areas (paved roads, dirt roads, parking lots, and man-made features) will pass the requirements, but the flat, agricultural fields will be vertically exaggerated, e.g., elevated striping. This error is almost always detected and quality checked in the data quality control phase as a result of collection of validation point surveys or provided checkpoint data. If the processing technician is observing elevated stripes in the open field data, it should be noted that this problem requires checking.

The third problem is much less frequent. At the edges of flight line lidar data from a mirror-scanning sensor there will be finger-line artifacts in the data, and the elevations will be higher in elevation than the adjoining points from overlapping flight lines (see Figure 5.8). These phenomena should not be confused with "corn-rowing" in the data which is caused by the repeatability of points in a flight line or in adjoining flight lines.

The fourth problem is pitting of the data which in a flight line. Two adjoining points within a scan will have different elevations outside a standard deviation for the given scan. This "pitting" usually occurs when there is a problem with the electronics; the manufacturers are very responsive about fixing this problem. If the problem is minor, it can be reclassified as "noise."

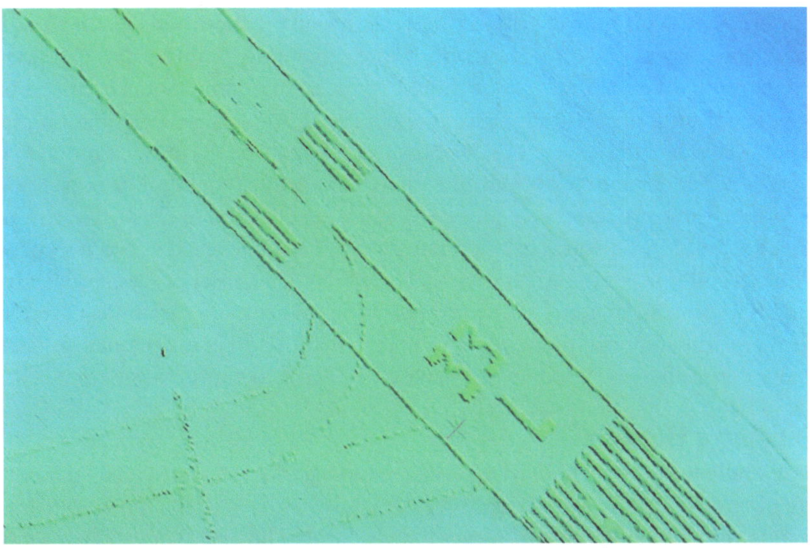

Figure 5.7 Raised paint strips (courtesy of the Sanborn Map Co., 2012).

Figure 5.8 Edge of scan fingers (courtesy of the Sanborn Map Co., 2008).

The final problem is not always an intensity issue but has been attributed to it in some cases, which is the loss of points on a very dark surface such as a recently tarred parking lot, coal piles, and composite along railways. Historically, lidar sensors have always had a difficulty with these types of surfaces, but recently, the manufactures have addressing this issue with refined filtering processes.

The lidar operator can check their resulting intensity data, and most routinely do this during the preprocessing phase. The easiest way to check the intensity calibration is to survey points in a broad range of reflective surfaces making sure to get as broad a sampling of these surfaces from dark to light surfaces. The lidar values are then checked against the surveyed values and an intensity correction algorithm is calculated based on a best fit curve which is applied to the data set. As stated before, it is extremely important to get a broad sampling, because without the full range of reflections, the new correction can induce errors into the data that did not exist before. Although, all lidar operators can check their intensity data, it is recommended that unless the lidar operator is well versed in the process, the manufacturers of the systems should perform this step for all sensors.

Nearly all lidar operators have developed their calibration practices based on the fundamentals of the sensors they operate. All lidar sensors are different, yet most lidar operators yield similar results using different practices. It should be noted that same model lidar sensors can yield very different result because all sensors are not equal. Although, all calibration practices may have not been stated in this section, the intent was to describe the most common practices and to give a better understanding of this process as it is a critical function for project success, and some form of these processes is required to yield valid data and to quantify the results.

Mapping different terrain and vegetation terrain types can be considered when calibrating a system, but several years of experience has indicated that if a conservative approach is taken in the calibration process, operators can yield favorable results in all variations of terrain and vegetation types. The important element is to realize that the differences and errors become very apparent depending on the complexity and type of terrain and/or vegetation.

5.3.1 Mapping a Flat Area

Nearly all calibration errors will be detected when mapping a flat area. Most lidar operators have vast experience mapping flat areas and are very familiar with detecting calibration issues in flat areas. The most difficult calibration difference to detect would be a pitch or yaw (heading) error in a flat area, which can be solved by the presence of man-made features within the flat area. It is important to realize that some differences of the variables of a calibration could technically be within specification of a given accuracy requirement, and these differences could be detected in a flat area. For example, if the project has a stated vertical accuracy of 15-centimeters for a data set, and a 7-centimeter difference between flight lines in this flat area is detected, by specification this data is within requirements of the data set. It is for this important reason to state a relative accuracy for all data products; the relative accuracy will be better in flat areas because it is easier to obtain these tighter variables as they are more easily detected in flat areas.

In very flat areas (e.g., tidal mudflats), it is easy to measure differences between flight lines as low as 5-centimeters or even less in very isolated cases. The difficulty with flat areas is that a user of the data could have a very accurate data set, but the integrity of the data is compromised as a result of detection of differences much lower than the stated accuracy specifications. In addition, some attention needs to be given to the accuracy of the overlapping data because of the systematic difference of the two flight lines providing the data for the overlaying data. Assuming that a lidar system operator is capable of achieving a 9-centimeter RMSEz in a project; this result means that the resulting data when checked has to be with 9-centimeters RMSEz, but it is possible to actually have differences greater than that in areas of overlap because one point could be 9-centimeters below the surface (absolute location), and one point 9-centimeters above the surface (absolute location); in the same case there could be two adjoining flight lines representing the same effect. It is important to make sure that the resulting calibration accounts for the relative accuracy of the data making sure that it is as tight as possible relative to itself making the calibration process more critical in flat areas. This process will not compromise the integrity of the data

Given that a very flat area is being mapped, the horizontal (pitch, pitch slope error, yaw) is no less important but would be less detectable in a flat area, in particular with the absence of any man-made features. If the operator is mapping a field that is very flat, realistically how many points are needed to map that field? If the field was shown as if a flat table, only need four points would be needed at the four corners of the table. If an operator mapped that table with two million points and they had a horizontal difference as a result of (pitch, pitch slope error, and/ or yaw), would the error be detectable? Thus, there is a need to evaluate all components of a project within the calibration process.

5.3.2 Mapping a Sloped Area

Calibration errors are detectable within sloped areas but are not as easily detected as in flat areas. As the slope is increased, the errors still exist but become more "hidden." In most cases these errors are filtered out and reclassified from the lidar data point cloud in the postprocessing. Given this circumstance, it is recommended that a calibration be preformed on the system as defined above. The relative accuracy of the data is less apparent in sloped area based on the components that make up the visualization of the data, but can be detected in the raw data easier when viewing by flight line. At a minimum, the project calibration cross-flights can be used to quantify the repeatability of the data. Linear features such as ridge lines could be evaluated for location repeatability and drainage features absent of vegetation should be evaluated, which has proven to be somewhat difficult in most areas. The location of man-made features becomes more important in these cases further supporting a more thorough initial calibration process for sloped areas.

The repeatability of sloped building roof tops in lidar data can be very useful in calibrating in sloped areas. The sloped roof tops have very defined structure, and if calibrated correctly can carry over to the slopes of terrain. In the case of system installation and mission or project calibration, if the repeatability and relative accuracy of these roof tops holds over a given area, the calibration corrections will carry over to the slope (see Figure 5.9). Again, the ridges lines can be checked for additional validation.

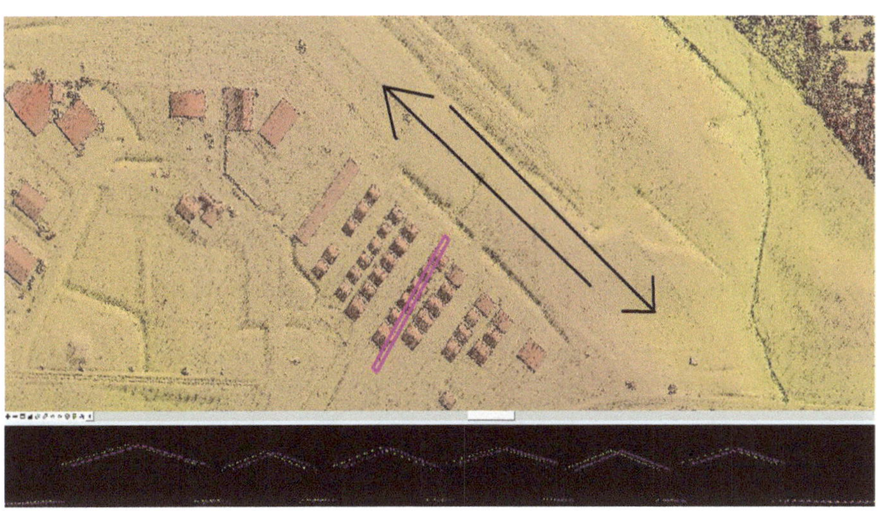

Figure 5.9 Roof ridge lines (courtesy of AeroMetric, Inc., 2012).

Roof top calibration brings up an additional query. If a lidar operator is mapping rolling hills with few distinct natural features, the manmade features and the initial calibrations become more important, and the resulting data can be validated by the calibration of cultural features. As stated above, most errors are detected in very flat areas with the existence of some manmade features, such as an airport base of operation.

The calibration process used when mapping a sloped area is further complicated as a result of the presence of vegetation. The operator must take into consideration the magnitude of vegetation and the relationship of that vegetation to the terrain complexity where the vegetation exists.

5.3.3 Mapping a Vegetated Area

Vegetated areas present a unique situation for calibration. Given that the lidar sensor is a random point sampling system, the probability of hitting an exact location twice is minimal at best. Usually, when mapping vegetated areas as in most lidar programs, there is a requirement to collect the data during the "leaf off" season unless the requirement is to collect the canopy as well. The repeatability of the vegetation features between flight passes is good but probably not within the required accuracy specifications for the vegetation features as a whole. Although, the resulting bare earth model results are very good, usually the users of the data and operators have a good understanding of the desired outcome. This type of lidar mapping is complicated when there is the existence of complex understory or riparian vegetation. These types of conditions require much higher point sample densities and the accuracy specifications are stated to better fit the types of vegetation that exist.

In heavy vegetation it is very difficult to quantify the repeatability of vegetation between two flight lines or a cross-flight making it more important to define the calibration process required to validate the data. It is important to have a repeatable calibration process for mapping vegetated areas, and even more so when having a vegetated area with the presence of major relief, such as areas in the Pacific Northwest.

A significant amount of heavily vegetated areas exist with the absence of man-made features further complicating the calibration, and additionally supporting the need to calibrate the data independently of the project area or finding an area within the project area that supports the ability to detect calibration errors. As in sloped areas, the detection of flight line areas is masked by the vegetation and slope. These errors can be seen, but are much more difficult to differentiate as an error or vegetation. Large errors can be detected by viewing the data by flight line, but this analysis would be a function of the sample density and the resulting detail of the bare earth model.

As lidar technology continues to develop the calibration process will continue to improve resulting in better processes for calibration. Calibration is a key component in the resulting lidar data processes. Although, the results of the calibration process are usually only seen in the resulting data, it is important to understand this process, as it relates to mapping different natural characteristics. The more thorough the calibration process, the higher likelihood of yielding suitable results.

5.4 FLIGHT OPERATION GUIDELINES FOR PILOTS

The flight-line guidelines for pilots as they pertain to collection are similar in some aspects to traditional aerial mapping but vary based on the requirements of the lidar data products. The constraints of the operation of the lidar system and the technology of the system have improved dramatically over the years, but lidar operators are still required to take in account the limitations of the different components of the lidar system independently. Although there is continuing discussion in the vendor community, the main limitation is the resulting GPS and IMU (direct georeferencing) data during collection. All operators have variation guidelines for the pilot operations during lidar collection. The following is a general assessment of this process and recommended practices for this process.

5.4.1 Performing a Lidar Flight Mission versus Traditional Aerial Mapping

The requirements for lidar flight mission are much tighter than those for traditional aerial mapping. These requirements are based on the need for more accurate IMU and GPS (direct georeferencing) data. In some cases, traditional aerial mapping imagery is collected without GPS and IMU data (which is becoming less the case with the more wide use of digital imaging sensors), but that being said, the requirements for lidar collection are much more rigorous.

Aside from the calibration process when using lidar, it is recommended that static initialization on all lidar collections be conducted at the beginning and the end of every lidar mission. This process does not necessarily mean that these initializations need to be preformed at the same location. For example if a very low altitude mission is being conducted over a corridor with several GPS base stations along the corridor over a long distance, the static initializations would be done at the start of the corridor and at the end of the corridor. If a county-wide program is being conducted it is usually preferable for the static initialization be done in the same location. Using the same base station helps reduce the possibility of errors in the GPS data and provides additional data in the event that the data has unresolved ambiguities. This process is becoming less common as a result of additional satellites in the GPS constellation, i.e., the Global Navigation Satellite System (GNSS). In addition, even though the IMU can be initialized while airborne, it seldom yields an acceptable IMU solution.

Typically, most lidar missions are performed at lower altitudes than most aerial missions because of the point densities required for the lidar collections. Also, given that a lidar sensor is an active sensor, lidar missions can be performed during the day or night. The lidar sensor can be operated during the presence of clouds and haze as long as those conditions do not exist between the lidar sensor and the ground. Lidar sensors cannot penetrate clouds, smoke, or rain.

5.4.2 Limitations of Airborne Flight Maneuvers

Airborne lidar collection has limitations for maneuvering in the air; there are several reasons for these limitations. As technology improves some of these the limitations could be omitted, and some data providers could argue that some of these limitations do not exist anymore. The following is the conservative approach to these limitations, and as a result, this approach continues to yield the best results to date. This being said, it is recommended at all approaches be properly evaluated and conclusions be drawn independent to these following limitations.

The main limitation to maneuvering in the air is as a result of the requirement of the accuracy of the GPS data. The GPS data is the largest contributor of error in lidar data. Typically, it is recommended that the angle of bank of the aircraft does not exceed 15 to 20 degrees (some will suggest that this should be 25 degrees). Recently, some software vendors have released software that eliminates this requirement, but the results vary from project to project.

The primary reason for this limitation is the orientation of GPS satellites within the satellite constellation. Depending on the aircraft's location in the world, the availability of satellites within a constellation differs. It should also be noted that there are different GPS satellites for different sets of satellite constellations available for GPS collection. If the aircraft banks beyond limitations, the GPS can "lose lock" on one or more satellites. Usually, most companies require a minimum of six available satellites for any lidar mission. Also, there is a requirement to have a minimum PDOP (Positional Delusion of Precision) value for collection. This PDOP value between vendors varies and is a function of the required accuracy, but it is usually between 3.2 and 3.5 maximum; the lower the value the better the PDOP. Some additional lidar operators may have other values depending on there requirements for operation.

The IMU (Internal Measurement Unit) requires movement to find itself, and the more movement the better the attitude values of the aircraft/sensor is recorded from the instrument. Typically, it is suggested that the aircraft does not continue in a straight line for more than 40 minutes; the distance flown for during this 40 minutes varies by aircraft and achievable aircraft speed. As commercial IMU technology improves, this limitation will become less and some may argue that they can exceed this time and distance, but to date, this continues to yield very favorable results. Also, it should be noted that transit from a base of operation to the project site should be considered when taking in consideration the IMU measurements. If the project area is further than 40 minutes from the base of operation, a procedural "S-turn" should be conducted to help the IMU find itself again. In the case of corridor mapping and particular in a rotary aircraft, the operators can account for these requirements along the collection route. Lastly, all adjacent collections lines within a mission should be flown in opposing directions one after another, or as the lines are flown, variations in direction should occur so that the lines are not being flown in a circular pattern. Often, weather prevents collecting all lines adjacent in a uniform manner.

The air speed at which the lidar is collected is very debatable. Lidar system manufactures and operators differ in the recommended speed. Fixed wing operators typically fly faster than rotary wing operators. What should be considered is the target accuracies that can be achieved at what speed; the slower the movement, the better the accuracy. This statement should be coveted by stating that a certain speed, you will produce a certain result and accuracies. Also, there are aircraft stall speeds that need to be accounted for, and most lidar operators have applied their own requirement as it relates to their aircraft, sensors, and safety.

Consideration should be given to flight line spacing, altitude, overlay, and the requirement to be able to stay on a flight line. The most important component of this is how is the flight mission is planned. Most gaps are not a result of a pilot's inability to stay on a flight line. Typically, all aerial pilots have the ability to be within five feet of a flight line, but this depends on the flight conditions in which they are flying. Consideration for terrain and the resulting flight line overlay, as it relates to terrain and project requirements, defines whether there will be flight line gaps. Pilots will sometimes be required to change altitude to accommodate the varying terrain in a project. It is also important for a pilot to maintain altitude to make sure the required point density is achieved. The lidar mission planning is usually conservative to ensure the flight parameters are maintained, but if the pilot is not aware of them, flying out of specification can occur.

Flight conditions can alter the result of quality lidar data, and again different lidar operators have differing opinions on this subject, which is also lidar system and aircraft type dependent. Rotary-wing requirements will be different than fixed-wing requirements. Also, the size of the aircraft can play a role in determining these parameters. The important ingredient is to note that the "smoother the ride, the better the data." Typically, sharp movements will impact the IMU data. Excessive aircraft crab could result in more "corn rowing" and gaps in the data, but all lidar operators have determined what these parameters should be, and they are open to share their experiences and services.

Lastly, restricted airspace is of significant planning and concern for pilots. There are several versions of restricted airspace and most of these issues are addressed prior to flying because commercial pilots are trained to address these restrictions. Most common is Class B airspace which are areas of high traffic, such as, around large airports. Pilots have to submit plans for approval in these areas and often will get approval in these areas only during the night. In addition, military and government installations require special permits and/or access. There are altitude restrictions that they need to considered because there are guidelines for VFR flight rules that require flying in a certain direction at a specific altitudes. Most pilots work with local air traffic control to work out most all issues in regard to flying issues that can occur during the mission.

5.4.3 Typical Pilot Error and Solutions

Typical pilot errors are a result of flight performance that are not natural to a pilot. Commercial pilots are trained and ingrained to fly a certain way, and it takes a unique set of skills pilot to be a survey pilot. The most common error made by an inexperienced survey pilot is exceeding the bank angle for the GPS. This is not natural maneuver for a pilot, and nearly all lidar systems have an internal alarm for this to alert the system operator and the pilot. The pilot will sometimes exceed the recommended speed for lidar collection, but the missions are usually planned to account for the projected variation in speed. Aircraft speed errors occur mostly when the pilot has a higher than expected head wind or tail wind which requires drift and speed adjustments. Although, most lidar systems have roll compensation, the pilot is required to keep the plane as level as possible from wing to wing (termed "flying trim"). If flying trim is not accomplished, the result could be in gaps in the data. Typically, gaps are detected in the swath coverage during the flight (real time), between mission lifts, or after the data is processed. Project coverage is almost always checked in the field, so that these missing areas can be flown in situ prior to leaving the project locations or during the same flight depending on how the data is checked. Typically, three degrees of variability from nadir is recommended for this component.

The angle of bank problem used to be more of a concern in the lidar data, but this usually occurs outside the project area, and in most cases the data can be processed without any resulting affects of the data. If there is a problem with the data, it is usually identified on site and can be re-flown.

5.5 IN-FLIGHT QUALITY CONTROL

The in-flight quality control is managed by following a set of procedures to optimize the potential for good data. This includes monitoring the system and reviewing data outputs within a system's operating system. Key components of this process include the IMU, the GPS, scanning information, and laser data information.

All systems have the ability to output IMU information during flight. The lidar system operator will monitor and record the critical components of the IMU data. The system also will indicate any errors during flight regarding the IMU data. If the system operator determines the IMU notifications are just a warning message, the message usually gets recording on the flight log for future use. Also, the operator will review the data in the field to verify that the warning will not affect the resulting data. Components monitored include the alignment of the IMU during flight, recording of the data, the fitness of the data, and that the attitude information (see Figure 5.10).

The GPS information is monitored and recorded because prior to the flight, the PDOP and satellite forecast is reviewed (see FAA, 2011). In addition, the KP index is forecasted but not monitored; the resulting GPS information may be of poor quality and can be the result of extreme solar activity. The KP-index is the effect of geomagnetic activity from solar flares on the Earth's surface, and it can greatly impact the resulting GPS data (i.e., radio transmission). Although the forecasted PDOP is planned, it does not always match the actual PDOP, thus, the requirement to monitor and record PDOP for each flight line. In some case, although less common now with a better GPS constellation, the collection of data will be halted for a specified time period until the PDOP returns to acceptable levels.

The laser information similar to the scanner information is recorded and monitored to ensure that they are functioning properly throughout every, and that the amount of returns received by the sensor is acceptable. Occassionally, the system operator will observe dropouts in the laser performance typically occurring over water which is a function of the characteristics of a mapping class laser with typically operate in the near infra-red (1,064 nm) on the electromagnetic spectrum (i.e., most of the lidar pulses are absorbed by water). Typical indications of a laser problem or incorrect laser setting would usually be high dropouts over land. The scanner information has several warning indicators to detect a scanner problem. Also, in the event that the resulting scanner information is outside the planned settings, the lidar system will detect and display the problem on the system's computer monitor (see Figure 5.11).

5.5.1 A Visual Control of a Flying Trajectory

Different lidar systems have different methods to manage the flying trajectory or flight lines. The flight lines are managed by both the pilot and system operator. In most cases the operator will select the flight line to be flown. Both the pilot and system operator have displays to monitor these flight lines. Typically, all flight planning is done prior to collection and often includes the calibration flight lines depending on how these lines will be flown. The flight management system (FMS) and system operator will record which lines have been flown by marking them complete in some accepted form (manually or digitally). In some cases, the system records the extents of the swatch coverage that has been flown. This is very effective in providing all the necessary information pertaining to potential gaps. Older systems generalized the swath coverage, but newer systems now take in account the actual elevation (automatic altitude calculation between the sensor and the Earth's surface with scan angle information) of the swath, which has proved to be a very effective way to detect potential gaps during flight. The advantage of these systems is that if you locate a gap during flight, you can adjust to capture the gap during that mission. Some systems do not provide this information, and the operator and pilot have to trust the design of the plan and the constraints planned to prevent gap problems. Most lidar operations will verify coverage after the flight and indicate any re-flights post mission. See Figure 5.12.

The GPS track recorded during flight provides some indication of how the flight was flown, and if the GPS track is too far off the planned line, the pilot and system operator have a good indication that the flight line should be flown again.

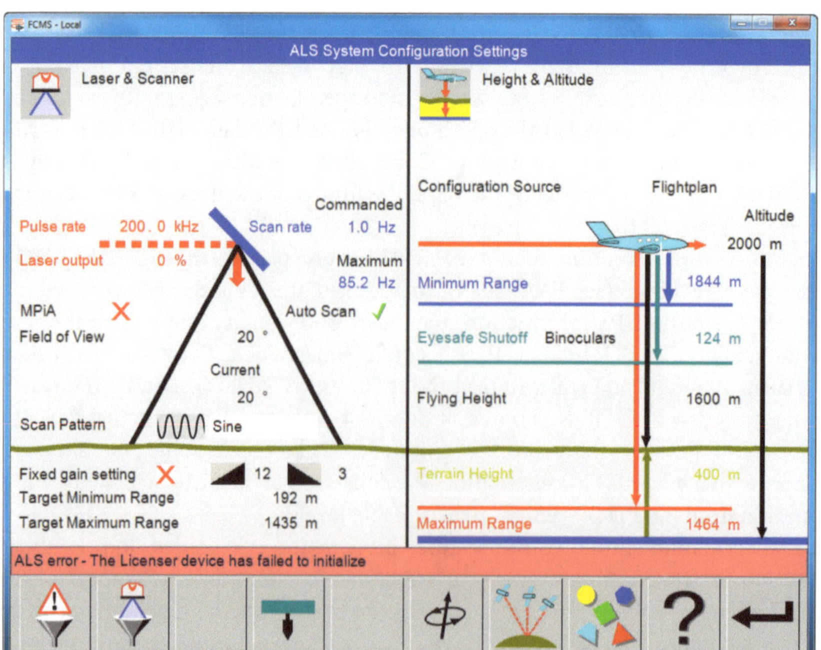

Figure 5.10 Screen shot of TracGUI ALTM –NAV (Courtsey of Lecia Geosystems, 2012).

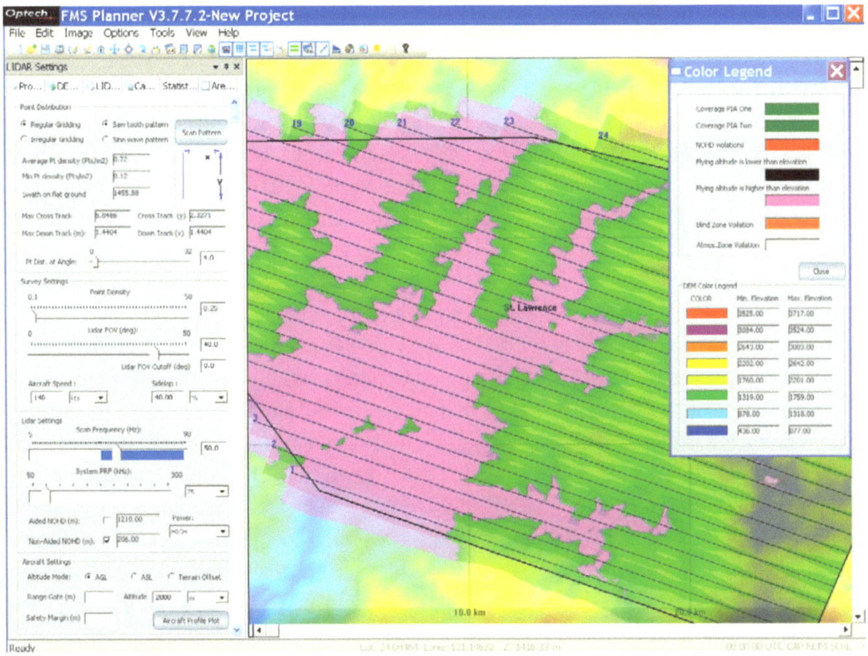

Figure 5.11 Scanner information (courtesy of Optech, Inc. 2012)

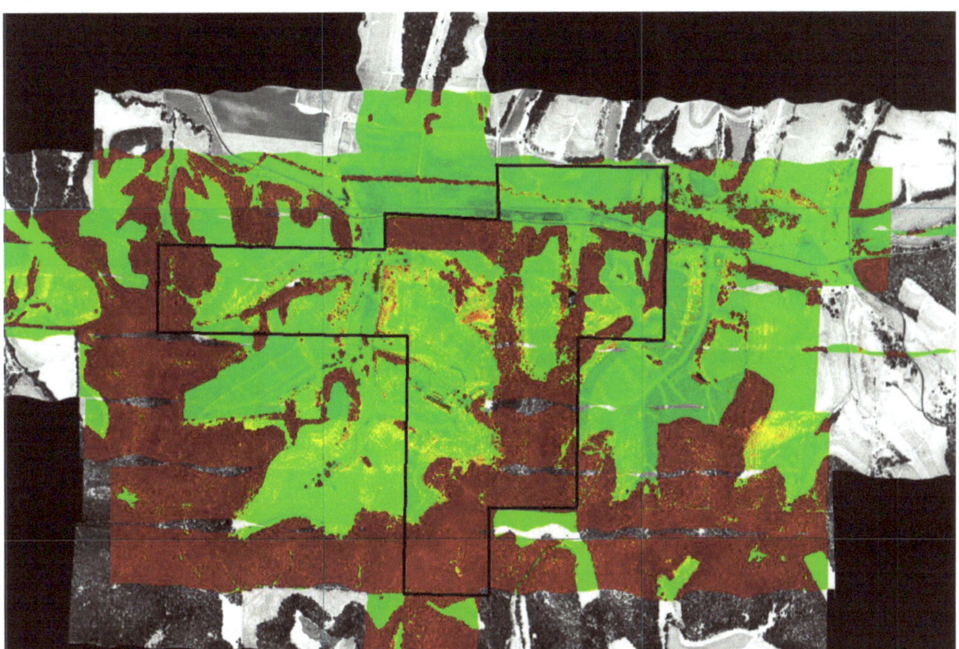

Figure 5.12 Coverage check (courtesy of AeroMetric, Inc. using GeoCue Software, 2012).

5.5.2 Power Control of ALS System

Current lidar systems require about 30 to 35 amps at peak power when the laser is turned on and level out at about 5 less amps once the system is fully operational. New systems require much less power based on technological advances. Typically, most aircrafts operating lidar systems will have a minimum of 100 amps available for all electrical systems including the lidar system. Typical aircraft electrical configurations require between 30 to 40 amps. Newer lidar systems only require about half the power, but this should be verified by the system manufacturer. The power controls for most lidar system are very simple to operate. Typically, there is a main power switch activated by the pilot; there is a main switch or circuit breakers turned on by the operator. Then, there is a sequence of switches on the system or different components of the system that are activated to fully power up the system. On the laser there are typically two switches and an interlock device to activate the laser for safety precautions. In addition, the laser on most lidar systems has an electronic firing switch with is controlled by the lidar operating system. This switch is activated only when the laser is required to fire and the operating system indicates that the laser is firing on both the pilot's and system operator's monitor or read out.

Given the class of laser used for lidar, there is also a switch in the lasers to turn the laser off immediately if the range is too close. Lower altitude operating systems also account for this issue by attenuating the power of the laser. Usually the closer you are to the ground the less power from the laser you need.

5.5.3 Typical Operational Mistakes, Faults and Solutions

The Manufactures of all lidar systems have invested significant effort in understand the potential for errors and the operation of these systems. They have developed sophisticated software monitoring tools to indicate a performance issue when it arises. In most cases issues with operating the system and recording data will be detected by the system itself.

The lidar operators have also spent many hours with their lidar system components and understand their system. They have developed tools and procedures to prevent problems while operating the system. The most typical faults with a system is the result of GPS information, but a conservative approach to calibration and collection usually mitigates this issues while in the field during QC processes or in the processing of the GPS data. The most common mistake made by a system operator is not properly detecting gaps in the data because they are very small. Recently, manufactures and operators have invested significantly in developing software to detect the smallest of gaps. In addition, careful planning and even over planning overlap has help eliminate the potential for gaps.

Figure 5.13 is a graphic visualization of the operational workings of a functioning airborne lidar system.

REFERENCES

FAA, 2011. Global Positioning System (GPS) Standard Positioning Services (SPS) Performance Analysis Report, NSTB/WASS T&E Team, URL: http://www.nstb.tc.faa.gov/REPORTS/PAN73_0411.pdf (last data accessed: 14 May 2012).

AUTHORS

Chapter 5 was written by Becky Morton, Senior Program Director at Towill, Inc., and Jamie Young, Senior Manager- LIDAR Solutions for AeroMetric, Inc.

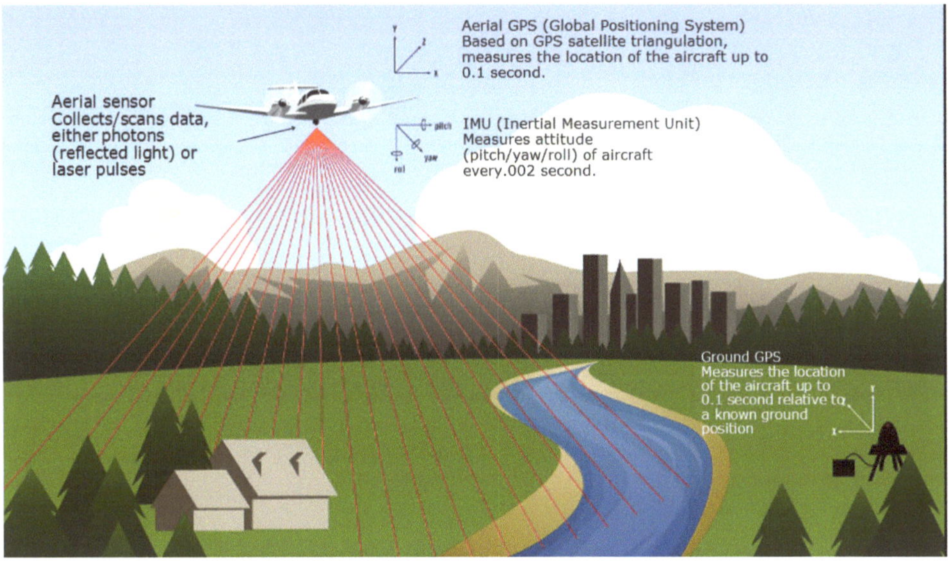

Figure 5.13 Lidar component workings (courtesy of the Sanborn Map Co., 2009).

6
DATA PROCESSING

JAMIE YOUNG and QASSIM ABDULLAH

6.1 PROCESSING WORK FLOW FOR LIDAR DATA

The following is a recommended lidar processing work flow which starts with the data coming from the field following collection. This work flow does not include the qualitative checks and validation of the data in the field. The work flow may vary slightly depending on the type of lidar system used for collection, field processing, and software packages used. In general, the following step-by-step outline an accepted process:

1. Copy the raw data from the hard drives received from the field. Make one copy in its raw state and another copy for processing.
2. Archive one copy.
3. Step up a directory and file structure for processing.
4. Extract RAW GPS and IMU data from the flight recorded data.
5. Extract RAW base station data collected during each mission (aka, a lift)
6. Please note that typically base station or CORS data is used but Precise Point Positioning (PPP) and TerraPOS software could be used to process the data when a base station or CORS data is not available or not collected based on specifications or requirements.
7. Combine RAW GPS and Base station data.
8. Input all GPS values including antenna heights, coordinate locations, phase center information, and any ancillary data.
9. Process GPS data; this process may take several re-iterations until an acceptable solution can be achieved, i.e., forward and reverse time calculations, Kalman filter results.
10. Visually check GPS graphs and solution of the GPS data.
11. Combine GPS solution with IMU data.
12. Input positional offsets for the sensor used based on calibration data.
13. Process IMU/positional information.
14. Visually check IMU/positional outputs using the graphs provided by the software.
15. Combine output of IMU/positional file (*.sbet and *.sol file are most common) with raw laser data.
16. Load IMU/positional file and last know calibration information and intensity files and raw laser data into the processing software.
17. Process lidar data.
18. Check lidar data for coverage and usability (no gaps, multiple returns).
19. Calibrate lidar data.
20. Step 19 may take several re-iterations and this process may vary significantly by lidar producer/vendor and make of lidar sensor.
21. Verify calibration against know ground survey data.
22. Please note the process from Step 1 through 19 will be repeated for all missions flown for a given project until all missions are processed.
23. Once all missions are processed, double check each mission against all other missions to make sure that all missions match each other spatially and within specification.
24. Generate a tiling scheme for the project or use a provided tiling scheme; tiles must join perfectly without gaps.

25. Import project boundary including any buffer information as specified.
26. Import all data into the tiling scheme. Typically, the lidar data will be imported in a raw coordinate and datum at this point in the workflow, but these parameters could have been determined in the GPS process.
27. Note that at this point, different software may very, and the process will vary depending on the processing solution chosen and the choice of software.
28. Verify all data has populated in all tiles seamless and without gaps for the project. This step can be accomplished in a GIS environment using a contrasting base color compared to the lidar tile raster imagery.
29. Check the data again for calibration against ground surveyed data.
30. Additional fine tuning the calibration may be preformed at this stage.
31. Set up classification macros and test the data on sample of tiles within the project area. The macros will be different for different vegetation and terrain types. An explanation of this process is required in the project metadata report.
32. Once quality classifications are achieved, process all tiles using the classification macros.
33. Manually edit and perform quality control on the lidar data for classification and validate that the lidar data meets classification specifications. This process is the most time consuming step in the process and takes between 40 to 60 percent of the entire processing time to generate a lidar data set
34. Run accuracy validation on the lidar data and verify that the data meets all qualitative and quantitative requirements. Document these results in the metadata report.
35. Generate products, reports, and the metadata report.
36. Check products reports and metadata. Some projects may require a third-party QA/QC review, additional accuracy validation, and authorized signature.
37. Deliver the final data products.

6.2 PREPROCESSING

Preprocessing is defined as office verification of data collected, the processing of the GPS and IMU data, and setting up for the post-processing of the data which would include an initial calibration of the data. In some cases, the entire calibration is preformed in this process. The calibration process is covered extensively in Section 5.3 but some detail in this section will elaborate on the calibration process.

Different lidar vendors will have independent methods to validate the data and set up the preprocessing. Different companies use different software packages, so the exact process may vary; this section will generalize the setup and preprocessing. Once the data is received from the field, the vendor will set up an established file structure to follow for the entire processing. This organizational step keeps track of the data and its location for use by many technicians within a company or agency.

Typically, the data received from the field, whether processed or not in the field will be again processed from scratch within a controlled office environment. This is to insure that the data is being processed properly. The first step in the process is to put the data in the file structure for a given provider, then the GPS and IMU data is extracted from its RAW format. Once this is completed, the GPS data is processed and checked for an acceptable solution. It is important to verify all variables input into the GPS processing software including lever arm offsets, antenna heights, point coordinate values, and antenna model information. Making sure these values are correct is the key to a good GPS solution. If the values are not correct, it will affect the entire solution, and the results will indicate an error. The first action is to check and confirm that all values are entered correctly. The technician will check the combined separation of the GPS to make sure the best POS solution can be achieved. Figure 6.1 is an example of a good combined separation plot.

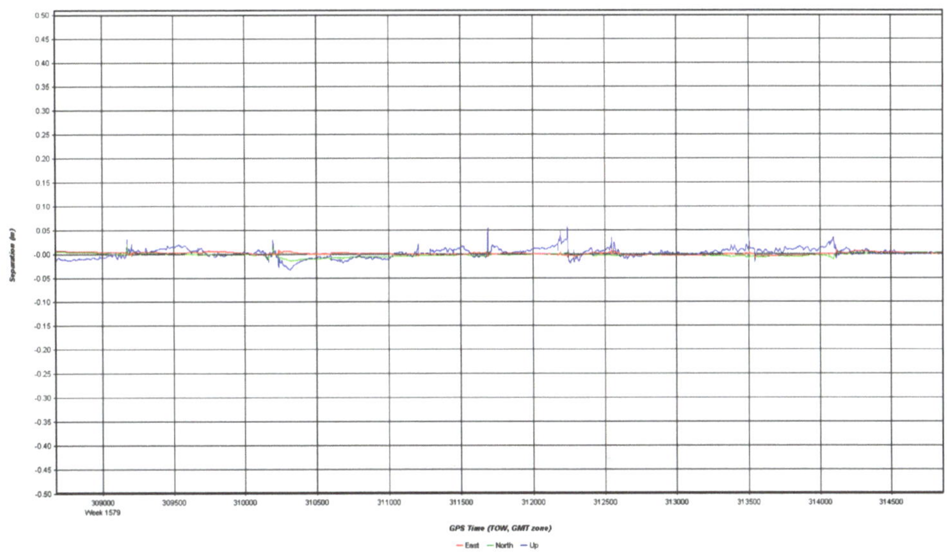

Figure 6.1 Example separation plot: x-axis is GPS time, y-axis is separation.

After the GPS data is processed and checked as suitable for use, the IMU data is than processed. The IMU data was extracted with the GPS data so it is ready to be processed at this time. Again variables that are input into the software need to be verified. These values include the roll, pitch, and heading values and the IMU correction information as it relates to a known point on the lidar sensor head. More detail of the IMU processing is provided in other sections of this manual (see Section 3.1.1). When processing the IMU information, it is extremely important to check the resulting processing graphs regardless of software used for this process. The significant values to check are accelerometer bias, gyro bias, and lever arm offsets which are shown in the following figures. Note that there are graphs for each variable to check and the figures provided is only showing one axis for each. The X, Y and Z axis for each element should be checked when processing IMU data.

Once the GPS and IMU data is processed and passes all the QC checks, it is combined with the RAW Laser data to be processed. The laser processing is processed using the lidar manufacturer's developed software. The levels of laser processing can vary by manufacture. Recently, several manufactures have developed lidar processing software that can process the data and calibrate it all at once. The important component of this software is to understand the software and how it works. It is possible to corrupt the data if you do not understand these procedures. The files required for processing the data in any manufacture's software are the calibration files, the intensity file, the trajectory file, and the raw laser information (usually referred to as strips). The calibration file contains all the software corrections for the hardware inaccuracies associated with a given system. The most common corrections include roll, pitch, and heading, but there are several other important variables that go into this file. The intensity files include intensity corrections for a given system; for different systems these files will vary. The trajectory information is the combined GPS and IMU information. The raw laser data is the data collected from the laser including times and all the return information from the laser. In the case of full waveform these files would be all the data for the waveform for each pulse captured. The time to process the laser information depends on the amount of data captured and the specific manufacturer's software you are using. The more information you have, the longer it will take to process the data.

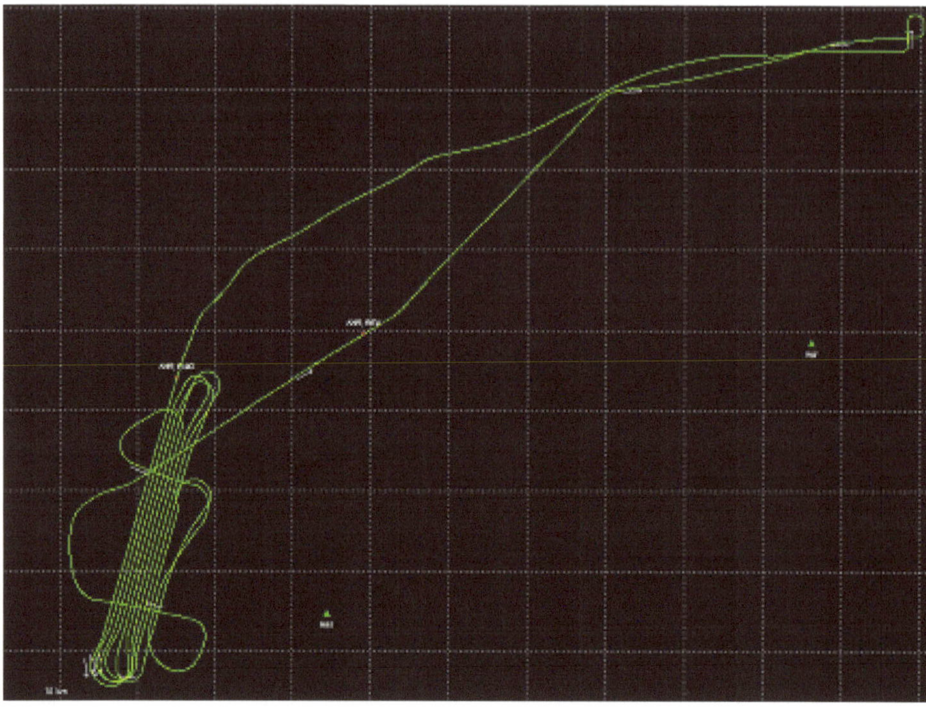

Figure 6.2 An example of a GPS trajectory from a lidar mission. The two green triangles in the image indicate the GPS base station locations.

Once the data is processed the technician will look at the data for coverage, point density achieved, and calibration. The coverage check entails checking confirm the area of interest is fully covered as required by the given specifications of a project. This would include but not be limited to point density achieved, required overlap, point distribution, and confirmation that the system did not malfunction. The calibration which is covered in Chapter 4 will be checking that the data is properly calibrated and yields a surface that is acceptable for the specified product being collected. It is important to remember that all lidar products do not yield the same accuracies and have the same specifications. It is important to review the appropriate specifications for every project and put the requirements of the project regarding the accuracy and performance specifications in a proposal for work.

6.3 POSTPROCESSING

Postprocessing is defined as the lidar processing procedure completed after the data has been calibrated and output into flight strips. Independent lidar operations have different processes and algorithms they use for postprocessing. The following will outline the most common postprocessing procedures and discuss variations in the postprocessing that may occur based on application and required project deliverables. It is important to understand and apply the best processes for a given product and application.

Once all the flight missions for a given project are calibrated and output through a given sensor manufacture's preprocessing software, these flight strips are imported into a variety of software packages depending on lidar vendor. Typically, the flight strips will be imported into a project file. This project file will contain a tiling scheme provided by the client or created by the lidar vendor.

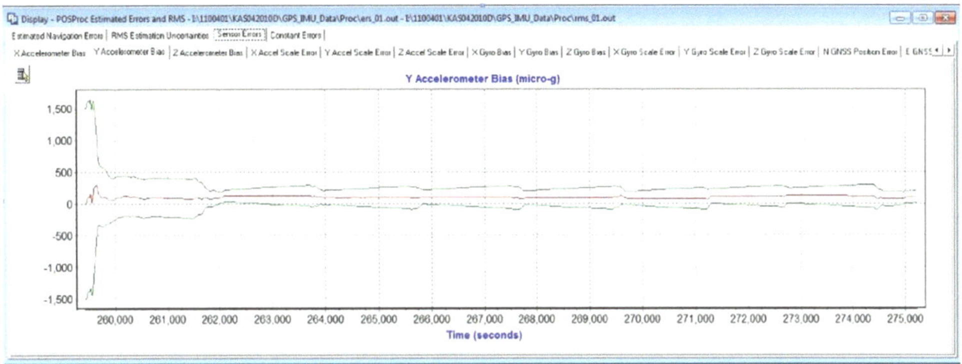

Figure 6.3 Accelerometer Bias in the Y-axis.

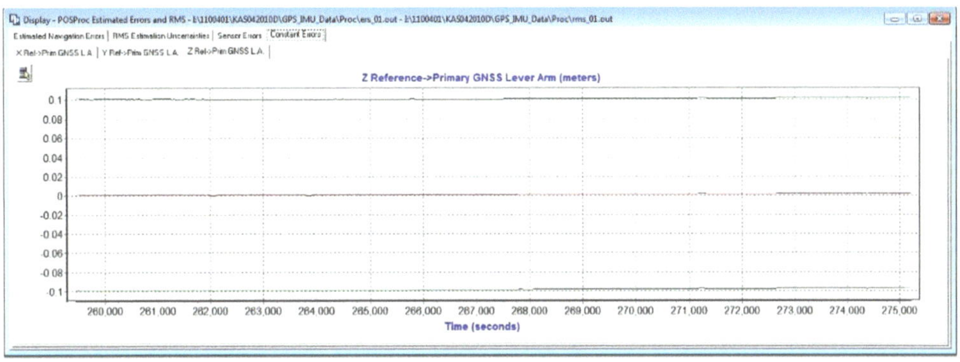

Figure 6.4 Gyro bias in the Y-axis.

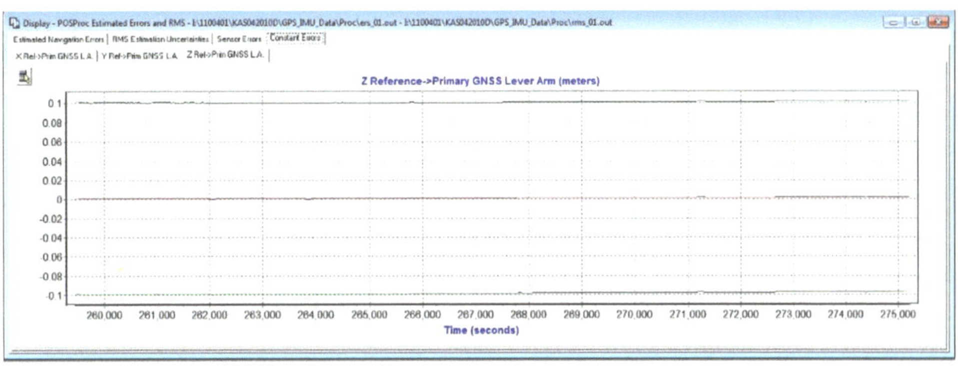

Figure 6.5 Lever arm in the Z-axis.

The tiling scheme varies greatly based on client requirements and lidar point density. As lidar sensor technology has improved, the point sample densities the tile sizes have changed (i.e., larger). The low altitude and corridor mapping tiling scheme will vary greatly from those from high altitude large area collections. Typically, tiling schemes for county and large area collections will range from 3, 280 feet × 3,280 feet to 6,561 feet × 6,561 feet. Most commonly, county and state agencies request 5,000 feet × 5,000 feet tiles. In some cases the tiling schemes will be a USGS Digital Orthophoto Quarter-Quad (DOQQ) format for a given area. The tiling scheme ultimately depends on the amount of data and preferences of the client. The lidar vendors will accommodate whatever requirement is requested but will point out the limitations if the tile size is too large or small. The tile sizes for corridor mapping could be the planimetric sheet for the corridor to match other information layers in a GIS for that given project, and in some cases be the entire corridor. Given that the amount of points collected for a corridor at a density upwards of 100 points per meter (although not as common), these tiles will be much smaller depending on the requirements of the client. The higher the point density, the smaller the tile size, and more tiles you will have in the tile scheme.

Once all of the data is imported into the tiling scheme and depending on the calibration process performed before this step occurs, the data could actually be calibrated at this point. Independent vendors perform the calibration processes differently, and even if the calibration process maybe done at this point, the results can still be favorable. It is recommended to perform calibration prior to this postprocessing stage because typically, depending on the collection of the calibration and the sensor calibration adjustment procedure, this will yield much better relative accuracy and uniform flight line matching before this stage. Although, additional correction can help eliminate some calibration artifacts at this point, if the calibration was not as precise to begin with. Prior calibration processes take in account all of the sensor variables, and the postprocessing procedure makes some assumptions about the sensor to correct for errors at this step.

It should be noted that most all the postprocessing of lidar data is done incorporating the LAS format regardless of the final outputs, and this format is a required delivery of almost all clients. Information on the current LAS format and version is available at the ASPRS website (www.asprs.org).

If the calibration process is not preformed at this stage, then a QC check will be performed to verify complete coverage of the data including the absence of data voids. Typically, this process is checked prior to this step in the field or in the preprocessing step; sometimes the importing of the data can cause data void for a variety of reasons. In addition, a process is run to validate the relative accuracy of the data, which is very important for quality control and confidence in the lidar data accuracy. Typically, with the high altitude collections, the relative accuracy should be 5 cm or less depending on the specifications of the project. Low attitude collections will have a much better relative accuracy based on the fact that these projects require much tighter accuracy requirements.

Once the coverage is checked, the trajectories will be imported into the project for a number of reasons. The first is to assign a flight line to each of the points within a flight line. This step is important to most clients for future analysis of the data and provides useful information in the QA/QC process. Given the error distribution as it relates to a given point within a scan generated from lidar systems and based on the collection parameters, some lidar vendors will trim off a portion of the scan to increase the overall accuracy of the data. This process is done based on the specifications of a specific project. As long as ample flightline overlap is collected, this "trimming" is a well accepted practice. The flightline trajectory information is required to do complete this step. Also, this information is required in the LAS metadata format and is stored imbedded within it if required by the client. It is recommended that all flight line information be deduced for future use.

The next step for postprocessing would be to create automated classification filters for the data, which is the most rigorous step in the postprocessing stage. The filters will vary depending on point sample spacing, vegetation, terrain, man-made features, and applications. All lidar vendors have developed processes that significantly vary from one another. Each variable stated above should be addressed when determining what parameters to use when classifying the data. The following table is a typical LAS point classification as required for a common project and is being used for discussion purposes in this section.

Table 6.1 LAS Classification Values Definitions (see LAS Standard; www.asprs.org)

Classification Value	Definition
0	Created, never classified
1	Unclassified*
2	Ground
3	Low Vegetation
4	Medium Vegetation
5	High Vegetation
6	Building
7	Low Point (noise)
8	Key Model points
9	water
12	Overlap

6.3.1 Point Sample Spacing (Point Density)

The point density of the lidar data determines which filters can be applied successfully. If the same filters were applied on different point density data sets, then it is likely that usable points would be omitted from the data set. The filter tolerances for different points densities should be set based on the point density. In the flight planning stage, the point density is usually determined based on the application, type of terrain, vegetation, and features in the project area. For example, heavy vegetated regions are collected with a much higher point density to increase the probability to obtain points close to or on the ground. Furthermore, highly developed urban areas are collected at higher point densities to better define the structures and ground in these areas. The goal is to maintain as many usable points within a project area; an error in setting the parameters of the filter can greatly impact the resulting data.

6.3.2 Vegetation

The vegetation classification tends to be the most difficult classification parameter to set. The classification needs to be based on the characteristics of the vegetation (species, age, size class, density, slope) within the project area. Vegetation parameters can also change within a project area because not all the vegetation within a project area will be the same. If the vegetation classification is not correct, additional removal of usable points can occur in areas where this should not occur. Typically, most vegetation occurs in proximity of drainages and water bodies complicating the classification process. In some cases the aggressive removal of vegetation can impact the resulting drainage area and remove points that define the drainage. Conversely, lack of removal of vegetation could create false information and artifacts defining the drainage area or water body. The parameters for this classification can vary within complex terrain where the vegetation exists. The vegetation classification for a very steep area will be much different for the classification in a flat area. Some regions have very little vegetation and some areas have significant vegetation. Currently, there are three typical vegetation classifications, which are low, medium, and high vegetation; the current COTS software to classify these categories are marginal at best. To properly classify vegetation into the vegetation classifications can take extensive and expensive procedures to complete. Several software providers and vendors are working on methods to improve this process.

6.3.3 Terrain

Terrain presents another unique classification challenge, particularly with the combination of vegetation and man-made structures. The classification parameters for varying elevations need to be set based on the relief. If the relief varies within a project area, then separate classification filters are needed to optimize the filter, and yield the desired result within the complex areas. In addition, the vegetation characteristics in each terrain type need to be address to make sure the vegetation is removed as required, and the definition of terrain and characteristics of the ground are maintained in that region. For example, if the filter is applied on a drainage basin with high relief and heavy vegetation, the filter needs to address the vegetation, desired ground result, and relief of the area. In addition, if there are roads in the area, the filter needs to be customized to ensure that the road structure is maintained in the region and not be too aggressive in removing parts of the road or areas of flat ground in the same location. The removal of man-made structures can impact the resulting DEM in areas of varying terrain depending on the relief. The most common problem that occurs is that "lidar-hits" on tops or ridges are removed as a result of a poor filtering. Compounding this situation is the presence of vegetation. The removal of man-made features impacts this classification process because the filter identifies the ridge, peak, or hilltop as a feature that needs to be removed. The algorithm used to filter is only as effective as the math and expertise behind the process.

6.3.4 Man-made Features

Typically, the man-made features are simpler to remove and the filters are pretty straightforward, but in is common for there to be artifacts from the buildings to be retained after the filters are applied. The most significant difficulty with building classification or the removal process is the removal or classification of additional features into the building class or the removal of these features. Typically, tree clumps or vegetation/trees close to buildings are classified or removed as part of the building . In addition, as stated above, hits on tops, peaks, and ridges can be classified or removed as buildings. Bridges, docks, and overpasses should also be considered; most projects require bridge decks to be removed. The automated filters that do this have varying results, and the filter sometimes will remove too much and/or too little of the bridge depending on the characteristics of the feature. This occurs with the overpasses as well. Typically, docks will be filtered out of data, but the process for the filter has to be set correctly to address the land base, the dock, and water to confirm their classification.

6.3.5 Other Applications

There are several lidar applications which require that the classification filters be customized to address the application. These applications may seem obvious, but the filters need to be applied for the resulting product. For example, a forestry application filter will vary greatly from a filter for flood plain mapping. A mapping project tailored to transmission line mapping will significantly vary from a transportation engineering planning corridor. The filter needs to extract the required information from the data to yield the results desired by the end user. There are several lidar vendors that specialize in different applications, and they thoroughly understand the requirements of their application. It should also be stated that several applications can be extracted from one lidar data set and the variation of filters can be run to accomplish this.

Typically, the lidar vendor will create the filter using algorithms that best fit the variables and characteristics of a given project area. In some cases several filters will be run on localized areas within a project area to maximize the outcome. The filters will be developed and tested on the project area(s), and the resulting data will be analyzed to evaluate the result. If the result is within or close to the desired specification for the project, the filter will be run on all of the data. If it is not, then the filter will be modified until the desired result is achieved.

Once the automated filtering process is complete, manual editing and QC of the data will occur. This process varies by the lidar vendor, but typically the most recommended process is to manually edit and QC the entire lidar data set. Currently, this step is the most time consuming process for lidar processing. As automated processes, software, and the technology improve, this process will become less, but a baseline level of the manual editing QC should be used. This process begins with a lidar editor who will scan each tile within a project visually to detect classification errors. His step will remove existing vegetation, man-made features, and artifacts in the data that should not be present. In addition, features will be restored into the data set that should not have been removed from the automated filtering process. This step will again verify coverage and completeness of the data.

Last, in the postprocessing procedure, accuracy assessment checks will be run on the dataset. Typically, as set of ground control points will be collected by the lidar vendor or provided to the lidar vendor from a third party to do an accuracy assessment on the data independent to the client's accuracy assessment. Most lidar vendors perform this process regardless of the requirements of the end user. The accuracy assessment is typically provided to the end user as a quality assurance report or metadata. The magnitude of information within this report varies but always supplies the required information to the client. At minimum, the report includes the location of all points in a database with supporting information of each point, standard deviation, RMSE, distribution of points, and the appropriate reporting to the required accuracy specification (NSSDA, ASPRS, NMAS, FEMA, USGS).

6.4 LIDAR DATA ACCURACY AND QUALITY ASSURANCE

6.4.1 Lidar Data Accuracy versus Quality

Lidar Data Accuracy, as defined by system manufacturers, represents the absolute horizontal and vertical accuracy of an individual lidar point, which collectively form the surface or point cloud (a term used to refer to the point-based terrain data collected with lidar).

The accuracy of lidar-based terrain surfaces are often times misunderstood by users as they are equated to the absolute accuracy of the individual lidar point defined by system manufacturers. To finalize the final accuracy for a lidar-based terrain surface, an additional dataset characteristic must also be considered in order to determine the correct accuracy. This additional characteristic is the density of the lidar points in a dataset, which reveals the fine details of a terrain surface.

The quality of the lidar data equates to how well it represents the terrain undulation and continuity. No matter how accurate an individual lidar point is, without dense lidar data to represent terrain continuity, the data cannot be classified as accurate or of high quality. The following example will help simplify the point.

Example: A user requested that lidar data be used to map berms and ditches in rolling terrain. The berms and ditches are, on average, about 1.0 meter wide and 0.60 meter deep. The data provider proposes a lidar dataset with the following specifications:

Horizontal accuracy: $RMSE_{XY} = 0.40$ m
Vertical accuracy: $RMSE_{Z} = 0.10$ m
Nominal post-spacing of lidar points: 2.0 m

Using only this dataset, can the user map berms and ditches to meet the lidar data accuracy stated by the data provider?

After careful consideration of the data provider specifications and the user requirements, one can easily conclude that the answer to this question is no. In order to map the positions of the perms and ditches with sound terrain modeling accuracy, the user needs a denser lidar dataset (better quality) with a nominal post spacing of one half meter or better to satisfy the Nyquit sampling frequency since the width of the perm is one meter. Another way to obtain the required quality from the proposed 2m post spacing lidar is to augment the lidar surface with 3D breaklines derived from another source, such as Photogrammetry. Figures 6.6 to 6.8 clearly demonstrate the effects of lidar point density on the quality of the terrain data, and eventually on the overall accuracy of the mapping products derived from this lidar dataset.

Figures 6.6 and 6.7 demonstrate, with some exaggeration, how a sparse lidar dataset affects terrain modeling. In Figure 6.7, the straight lines AB, BC, CD, and DE correspond to the shape of the terrain according to the Triangulated Irregular Network (TIN) surface that was constructed to represent the actual curve-shaped segments of the terrain between stations A, B, C, D, and E (shown with dash lines). As depicted in Figure 6.7, if one tries to measure the elevation of the hill at point F, an error of 15 feet may be introduced from using a lidar dataset with sparse points (every 50 feet for this example). Figure 6.8 shows how a dense lidar dataset can prevent the elevation discrepancy illustrated in Figure 6.7.

The above discussions emphasize the importance of introducing a new concept to describe the accuracy of lidar elevation data that combines both the absolute accuracy of an individual point and the density of the data or the post-spacing. According to this concept, accurate lidar points do not necessarily result in a good quality lidar surface. Furthermore, a high-resolution and high-definition lidar surface does not necessarily result in high absolute accuracy unless data planning, acquisition, and processing are conducted at a high standard. The subject of point density in lidar datasets and the resulting accuracy of derived products are of great importance, both to users and providers of lidar data. Unfortunately, there are no set rules to govern this topic, leaving many users to establish their own guidelines when requesting lidar data acquisitions. This fact becomes very obvious when studying the point density requirements specified by different requests for proposals (RFPs). Without direction, users tend toward the extreme, asking for sub-meter post-spacing to achieve an accuracy that is easily obtainable with less dense lidar datasets and/or without the need for breaklines. Unless the task calls for 3D modeling and above-ground, manmade or natural features, asking for highly dense lidar data may harm the budget with very little accuracy benefits, especially when the collection of breaklines is requested.

During the USGS Second National Lidar Meeting held during May 2008 at the USGS headquarters in Reston, Virginia, speakers presented a variety of views and levels of understanding as to what constitutes a reasonable and practical lidar dataset. The most misleading approach is the one calling for a lidar database to fit the broad needs of all users including those whose applications require ten points or more per square meter! An recommendation such as this not only wastes valuable taxpayer money but also makes for an impossible task, especially if there is very little capital available for such an expensive undertaking. Lidar data specifications should be tagged with user-specific needs and specifications.

AUTHORS

Sections 6.1–6.3 were written by Jamie Young, Senior Manager – LIDAR Solutions for AeroMetric, Inc.; Section 6.4 was written by Qassim Abdullah, Ph.D., PLS, CP, Chief Scientist at Fugro Earthdata, Inc.

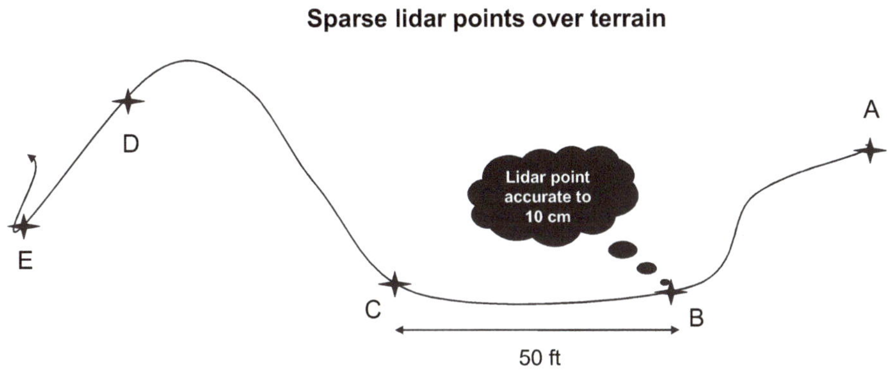

Figure 6.6 Terrain Modeling with Sparse Lidar Data.

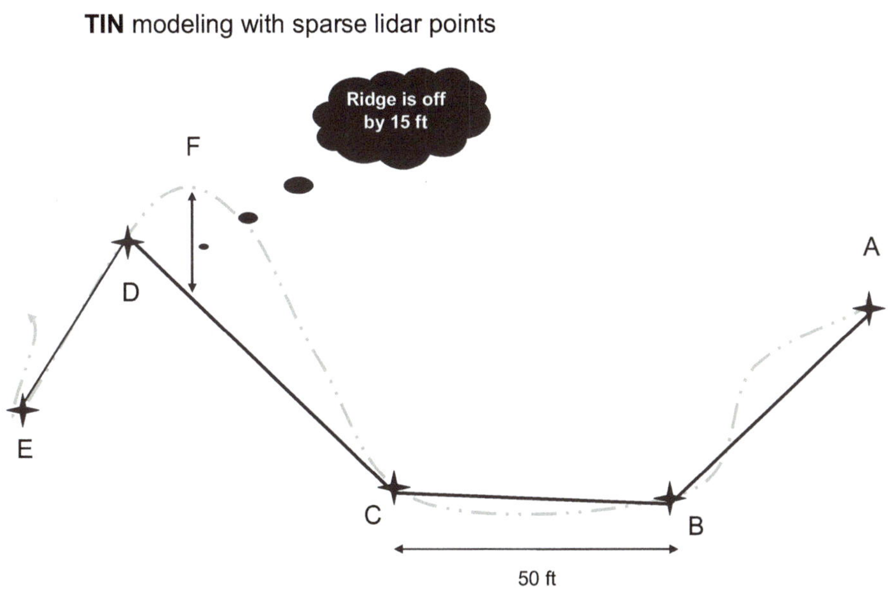

Figure 6.7 Effect of Sparse Lidar Data on Terrain Representation.

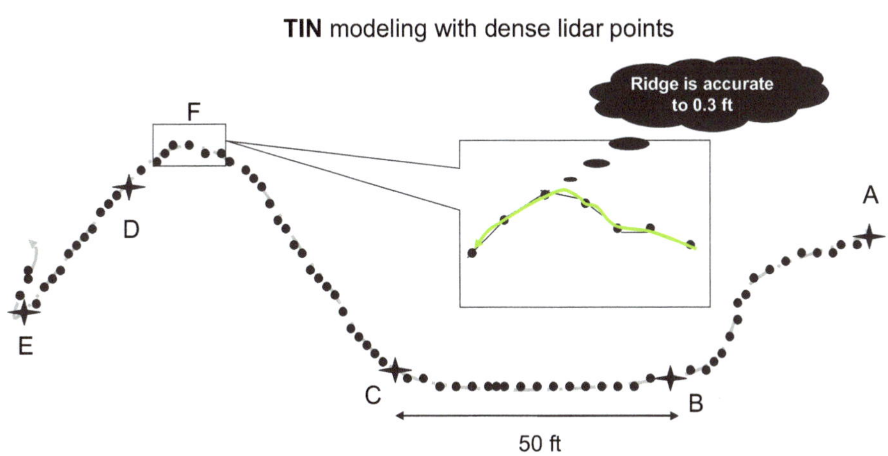

Figure 6.8 Terrain Modeling with Dense Lidar Data.

7
STATUS OF LIDAR INDUSTRY SUPPORT SYSTEMS

JAMIE YOUNG and LEWIS GRAHAM

7.1 EXISTING COMMERICAL ALS SYSTEMS

7.1.1 Fixed Wing Systems

There are several commercial fixed wing ALS systems, some of which are used in used in rotary-wing (helicopters) as well. The main producers of lidar systems are Leica Geosystems, Optech, Inc., Riegl, and Trimble. There are also bathymetric lidar systems which can be operated in a fixed wing platform. This section will detail the latest fixed wing lidar systems. Currently, client accepted lidar systems have to be able to capture at least four returns and three intensity returns. The products delivered are required to include all information as it pertains to no less than four returns and three intensity returns in most cases. There are systems that are still functioning that do not have the ability to include this extent of information, so these systems are not detailed in this section. This section will concentrate on systems that are fully operational and can meet the requirements of lidar clients.

The specifics of each system will be detailed as they pertain to collection of lidar; additional information of each system can be provided by the manufacture if additional information is required

7.1.1.1 Leica Geosystems

Leica Geosystems currently has three types of lidar models that can meet the criteria. They are the ALS-50 including Phase I, Phase II, and MPIA (multiple pulses in air) systems, ALS-60, and ALS-70HA and ALS-70HP. Currently, the most common used system from Leica Geosystems are the ALS-50 MPIA, ALS-60, and ALS-70. The ALS-70 is generally used for wide-area collections, but it can serve a variety of applications including corridor mapping from a fixed wing platform. The system consistently produces low noise data, which might be a concern based on its split beam and multi-pulse technology, but is not an issue. The specifications detailed will be for the HP model of the ALS-70 as these systems are more commonly used; although there are several HA models being used outside the United States. This ALS-70 operates at 1064 nm wavelength (near-infrared). The ALS-70 is significantly larger in physical size than the other makes and models of similar lidar sensors. The following will detail the ALS-70 specifications:

7.1.1.1.1 Standard Components

- LS70-LP Scanner Assembly
  - Scanner mirror
  - Window assembly
  - Laser transmitter – part of LC60
  - Laser receiver – part of LC60
  - IMU
- System Electronics (LC60 Laser Controller + SC70 System Controller
  - SCM70- System Controller Module
  - IPAS20- Inertial Position and Attitude System
  - GCM60- Galvanometer Controller Module

DLM65- Data Logger Module
LC60- Laser Controller
PDM60 – Power Distribution Module

- OC52 Operator Interface
- OC50 Pilot Interface
- Vibration-isolated Interface plates
- GNSS +GLONASS antenna
- Cables
- Digital Camera
- Roll Compensation

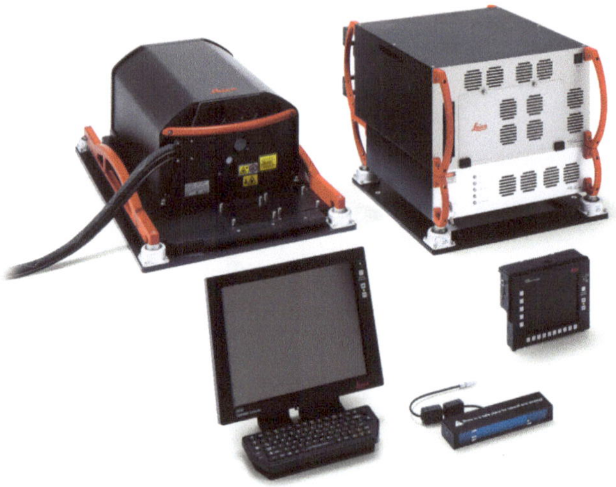

Figure 7.1 ALS-70 Lidar Sensor.

7.1.1.1.2 Notable Optional Components

- WDM65- Waveform Digitizer Module
- RGB Camera(s) integration
- Hyperspectral, Multispectral, Thermal sensor integration

7.1.1.1.3 General Specifications

Maximum Flying Height	3,500 m
Minimum flying Height	400 m
Maximum Repetition Rate	500 Khz
Maximum Scan Rate	200 hz
FOV	Up to 70 degrees
Typical Vertical accuracy (1,800 m AGL)	5 to12 cm RMSE
Total Weight	192 lbs
Beam Divergence	0.22 mr
Full Waveform	Optional
Input Power	937.2-972.4 watts
Power Required	28 volt
Number of Discrete Returns	up to 7

7.1.1.2 Optech, Inc.

Optech, Inc. has several models of lidar systems currently being used in the industry. Optech, Inc. systems currently used in the industry include the 3100EA, Gemini, Orion M200 and M300, and Pegasus. Typically, the 3100EA and Gemini are used more for wide-area collections but have also provided good results for corridor mapping from a fixed-wing platform. The Orion M series of lidar systems from Optech, Inc. provide a very good option for users wanting to do a very diverse set of applications. The Orion series operates at 1064 nm. The system provides exceptional results for corridor mapping as well as being able to collect wide areas with equal results. The system is quite compact compared to other systems, and typically produces data that is very crisp and clean. It should be noted that the Orion M is commonly used in both fixed-wing and helicopter platforms, but this section is concentrating on the M Series as it relates to fixed-wing platforms. It should be noted that the Optech M Series is the most current system be used in the industry as of June 2012. It is expected that Optech, Inc. will be releasing a new lidar system called the Orion H which is an enhancement to the M series system. The Orion H will be a multiple-pulse system that operates at 300 kHz with a beam divergence of 0.18 mr and an operating altitude up to 4,000 meters AGL. This new H-series system is expected to be released in September of 2012. The following are the specifications of the Orion M series:

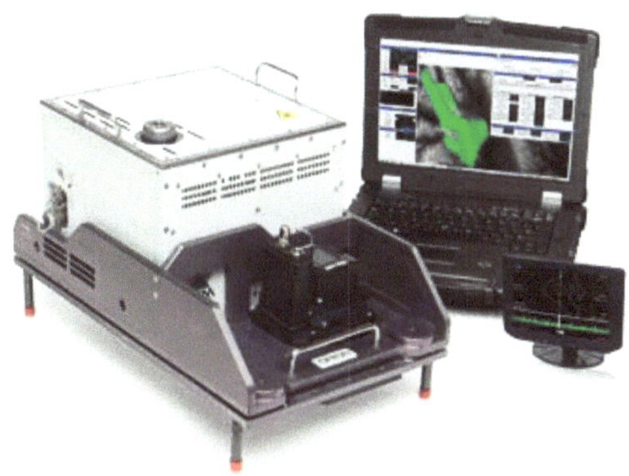

Figure 7.2 Optech Orion M Series Lidar Sensor.

7.1.1.2.1 Standard Components

- Scanner and Electronics Assembly
    - Scanner mirror
    - Window assembly
    - Laser transmitter
    - Laser receiver
    - IMU
    - System Electronics and Scanner assembly in same unit
- ALTM_NAV Operator Interface
- ALTM_NAV Pilot Interface
- Vibration-isolated Interface plates
- Applanix POS system
- GNSS +GLONASS antenna
- Cables
- Roll Compensation

7.1.1.2.2 Notable Optional Components

- Full Waveform Digitizer
- 80 MP Medium Format Camera(s) integration
- Hyperspectral, Multispectral, Thermal Sensor integration

7.1.1.2.3 General Specifications

Maximum Flying Height	2,500 m
Minimum flying Height	200 m
Maximum Repetition Rate	300 kHz
Maximum Scan Rate	90 Hz
FOV	Up to 50 degrees
Typical Vertical Accuracy (1800 m AGL)	5 to 8 cm RMSE
Total Weight	59 lbs.
Beam Divergence	0.25 mr
Full Waveform	Optional
Input Power	300 watts
Power Required	28 volt
Number of Discrete Returns	up to 4

7.1.1.3 Riegl

Riegl offers several types of lidar sensors and until recently, focused on systems tailored to specialized markets. The most common Riegl systems used in fixed-wing platforms are the LMS-Q560, LMS-Q680i, and VQ 820-G. It should be noted that the LMS-Q560 is mostly used in a helicopter, but it can be used in a fixed-wing platform. The VQ 820-G also provides the ability to operate as a bathymetric lidar system; this system operates at a 532 nm wavelength. This VQ 820-G is a very good choice for users wishing to collect both topographic at high and low altitudes and bathymetric data. This system provides exceptional data both on ground and on water floors. The following are the specifications for the VQ 820-G:

Figure 7.3 Riegl VQ-820-G Lidar Sensor.

7.1.1.3.1 System Components

- Scanner and Electronics Assembly
  - Prism mirror
  - Window Assembly
  - Laser Transmitter
  - Laser Receiver
  - IMU
  - System Electronics and Scanner Assembly in Same Unit
- Ri-Acquire Operator Interface
- Ri-Acquire/POS Track Pilot Interface
- Vibration-isolated Interface Plates
- Applanix POS System
- GNSS +GLONASS Antenna
- Cables
- Roll Compensation

7.1.1.3.2 Notable Optional Components

- RGB Camera(s) Integration
- Hyperspectral, Multispectral, Thermal Sensor Integration

7.1.1.3.3 General Specifications

Maximum Flying Height	3,000 m
Minimum flying Height	10 m (200 m typically)
Maximum Repetition Rate	550 kHz
Maximum Scan Rate	200 Hz
FOV	42 degrees
Typical Vertical accuracy (1800m AGL)	5 to 12 cm RMSE
Total Weight	60 lbs.
Beam Divergence	0.50 mr
Full Waveform	included
Input Power	185 watts
Power Required	28 volt
Number of Discrete Returns	up to 12

7.1.1.4 Trimble

Trimble currently offers the Harrier 68i lidar sensor. Although, the Harrier 68i is their current system, there have been several past models of the Harrier lidar system in use for several years. The Harrier 68i lidar sensor provides multi-pulse information and operates at 1550 nm. The Harrier 68i is generally operated in a helicopter platform and typically operates at lower altitudes; it does provide the capability to collect lidar information from a fixed-wing platform. The system provides like data accuracies and quality as the other major systems. This Harrier 68i provides the user with the option for both lower altitude collection as well as a reasonable solution for wide-area collections when it is required. The following are the specifications of the Harrier 68i:

Figure 7.4 Trimble Harrier 68i.

7.1.1.4.1 System Components

- Scanner Assembly
  - Scanner mirror
  - Window assembly
  - Laser transmitter
  - Laser receiver
  - IMU
- System Electronics
  - Harrier Control Computer
  - Applanix POS System
  - Galvanometer Controller
  - Data Logger
  - Laser Controller
  - Uninterruptable Power Supply
- Pos Track Operator Interface
- Pos Track Pilot Interface
- Vibration-isolated Interface Plates
- GNSS +GLONASS Antenna
- Cables
- Medium Format TAC P65+ 60 mp Digital Camera
- Roll Compensation

7.1.1.4.2 Notable Optional Components

- Hyperspectral, Multispectral, Thermal Sensor Integration

7.1.1.4.3 General Specifications

Maximum Flying Height	1,600 m
Minimum flying Height	30 m (200 m typically)
Maximum Repetition Rate	400 kHz
Maximum Scan Rate	200 Hz
FOV	45 to 60 degrees
Typical Vertical accuracy (1800m AGL)	5 to 15 cm RMSE
Total Weight	187 lbs.
Beam divergence	0.50 mr
Full waveform	included
Input power	616 watts max
Power required	28 volt
Number of discrete returns	Unlimited (full waveform)

The fixed wing systems detailed above are the latest systems from the four major manufactures. It is recommended that all users of lidar systems and data contact these manufactures and others that may have not been listed to make sure that the lidar system solution for any project or application is clearly understood prior to using a lidar system. Typically, this information for the lidar systems, applications, specifications and the systems' capabilities can be derived from most lidar companies. All latest model lidar systems typically will meet most applications as required by the users of data. It should also be noted that the manufactures of lidar systems continue to improve their technologies and new systems continue to be introduced on a regular basis.

7.1.2 Rotary-wing Systems

The installation and operations for typical rotary-wing lidar systems is covered in detail in Section 4.1. The lidar system manufacturers have developed lidar sensors best suited for low altitude, slow airspeed applications, i.e. rotary-wing platforms, that result in very dense lidar pulse spacing.

ACKNOWLEDGMENTS

It is important to recognize the following individuals that contributed to the gathering of this information as follows:

Leica Geosystems: Ron Roth and Dave Hull
Optech, Inc.: Jim Green and Mike Sitar
Riegl: Jim Van Rens
Trimble: Ernest Yap

REFERENCES

Leica Geosystems, 2012. URL: http://www.leica-geosystems.us/en/Leica-ALS70-Airborne-Laser-Scanner_94516.htm (last date accessed: 04 June 2012).

Optech, Inc., 2012. URL: http://www.optech.ca/ (last date accessed: 04 June 2012).

Riegl, 2012. URL: www.rieglusa.com (last date accessed: 04 June 2012).

Trimble, 2012: URL: http://www.trimble.com/geospatial/ (last date accessed: 04 June 2012).

7.2 LIDAR DATA MANAGEMENT

7.2.1 Introduction

In this section we review the general management of lidar data from data acquisition through data delivery. Note that our focus is on managing data during lidar data processing rather than techniques that might be employed for managing data for exploitation.

We first examine the overall (from a high level) generic steps involved in lidar data processing. This is followed by a discussion of the nature of lidar data (point clouds). The important yet often overlooked issue of data policies is next discussed since these policies will drive your data storage requirements. Next a detailed discussion of data flow as it relates to management is presented. The final section presents a detailed look at LAS, the industry standard format for lidar data transmission.

7.2.2 Overview of Lidar Workflows

The detailed workflow for processing lidar data is presented elsewhere in this manual. In this section we will review the high level, generic data flow as it relates to issues of data management and storage.

The general flow of lidar acquisition and processing is presented as Figure 7.5. The gross processing blocks are lettered for later reference. Of course the actual flow of data is much more complex than presented here; our goal is to provide only the detail necessary for considering issues of data management.

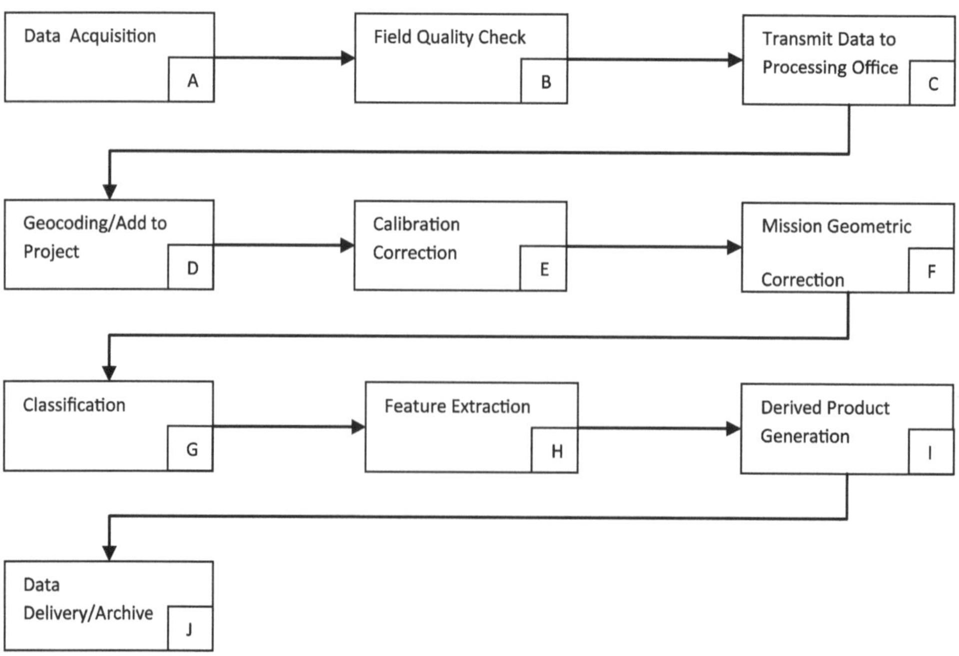

Figure 7.5 Generic Lidar Data Flows.

We use the term "mission" to mean a data acquisition that includes a single cycle of the Position and Orientation System (POS). Best practices dictate that you start the POS a few minutes prior to flying (or driving, in the case of a mobile project) and leave it operating without interruption until a few minutes after parking the sensor aircraft/vehicle. We use the term "project" to mean the amalgamation of missions into the overall deliverable area. Thus, the minimum project size is a single mission. Quite often large projects are subdivided into "blocks" to ease the logistics of managing subcontractors; we will not discuss this intermediate level since, for our purposes, it can be generalized into the project.

7.2.3 Nature of Lidar Data

Lidar data generally occur in three general forms:

- Proprietary (raw) range/timing data;
- Point cloud format;
- Point cloud/wave packet hybrid data.

Note that lidar data are often used to generate gridded elevation files (with possible optional attributes), since these data are no different from elevation data derived from other sources such as stereo image correlation. We will not deal with them here, but refer you to articles and books related to DEM processing (see, for example, Maune, 2007).

You will typically deal with raw lidar data at two points in data processing; field checks near the project site and preprocessing in the processing facility. Raw lidar data are in a vendor-specific proprietary format that is typically streaming (and streamed with respect to time). The data generally comprise a time stamp, a range (from sensor to '*declared*' *return*) and intrinsic pointing information relative to the platform reference frame. In the case of a system that provides waveform packets, the data will also contain digitized amplitude information in the vicinity of a pulse. Raw data are typically in a proprietary format that can only be used by the vendor-supplied processing software. The general content of range data is depicted in Table 7.1. Note that these data cannot be directly used to derive an X, Y, Z location in object space. For this, the time/range must be combined with system orientation data (position of mirrors and so forth) and platform information provided by a Position and Orientation System (POS).

Table 7.1 Time/Range Data.

Timing (seconds)	Range (meters)	Angles (radians)
Time 1	Range 1	Angle set 1
Time 2	Range 2	Angle set 2
Time 3	Range 3	Angle set 3

Raw timing/range data are converted to a format of X, Y, Z and attribute data in vendor-proprietary "geocoding" software (see Figure 7.6). In nearly all commercial off-the-shelf (COTS) lidar geocoding software, the output is point cloud in the ASPRS "LAS" format (discussed in detail later in this section). Newer generation systems are able to emit both point return pulses as well as digitized amplitude information in the vicinity of return pulses (waveform packets). These waveform packets can also be encoded in LAS format.

Unlike raster data, point cloud data have no regular geometric organization. A planar top-down view of 3D point cloud data is presented as Figure 7.7. Note that the data appear random in spatial organization. The major consequence of this lack of gridded organization is that all three spatial parameters (X, Y, Z) must be stored for each point. Raster data, on the other hand, are organized in a uniform grid of constant X and Y (not necessarily the same) spacing and thus only Z need be stored for each point. A wide variety of methods have been developed for transforming point cloud data to gridded rasters (see, for example, Agarwal et al, 2006). Tools for converting from point cloud to raster are included in nearly all commercial lidar processing software applications.

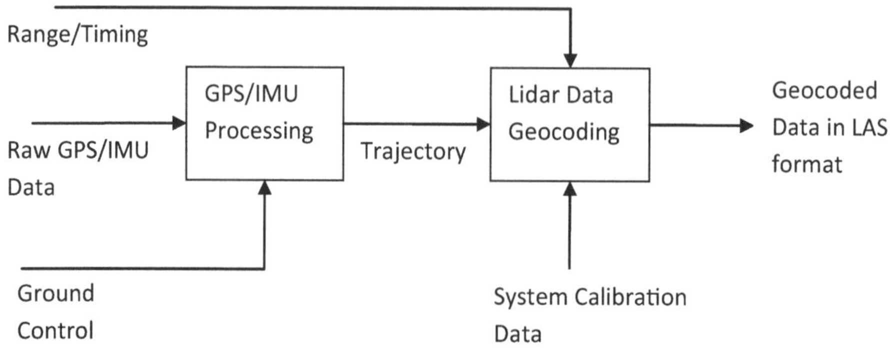

Figure 7.6 Lidar Geocoding Process

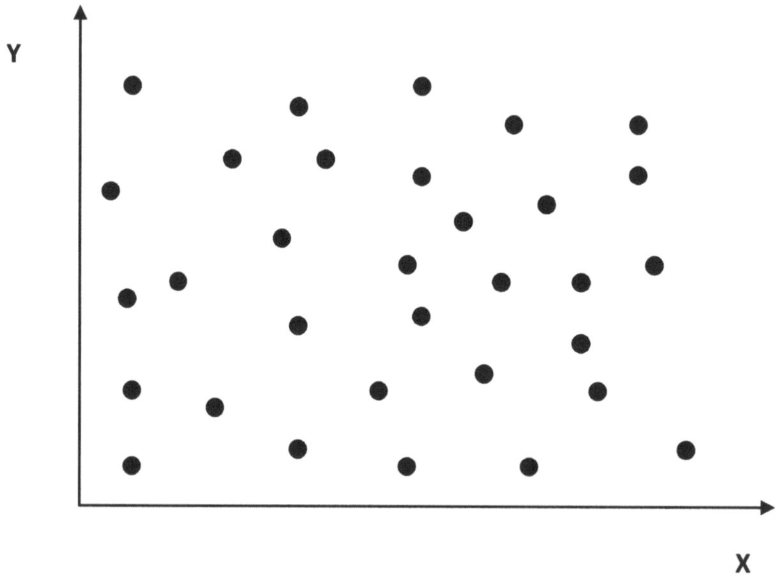

Figure 7.7 Point cloud data organization.

7.2.4 Data Organization for Processing

For most projects, the lidar data will occur in one of three general spatial forms:

- Streaming data;
- Flight/Drive lines;
- Tiles.

Streaming raw data (range data) are converted to geocoded flight line (or drive "take") data during the sensor proprietary geocoding stage. These flight lines are then divided into a tiling scheme for subsequent data processing. The date are tiled to make data processing more manageable by allowing separate segments of data to be processed by different technicians (e.g., to facilitate parallel processing). Tiling is also used to reduce the amount of data that memory-based editing tools need to manage in a session. Figure 7.8 denotes a single helicopter flight line (the bold red outline) subdivided into 804 processing tiles (brown tiles in the Figure). A detailed discussion of tiling is presented in *Library Concept and Design for Lidar Data Processing* (David et al., 2008)

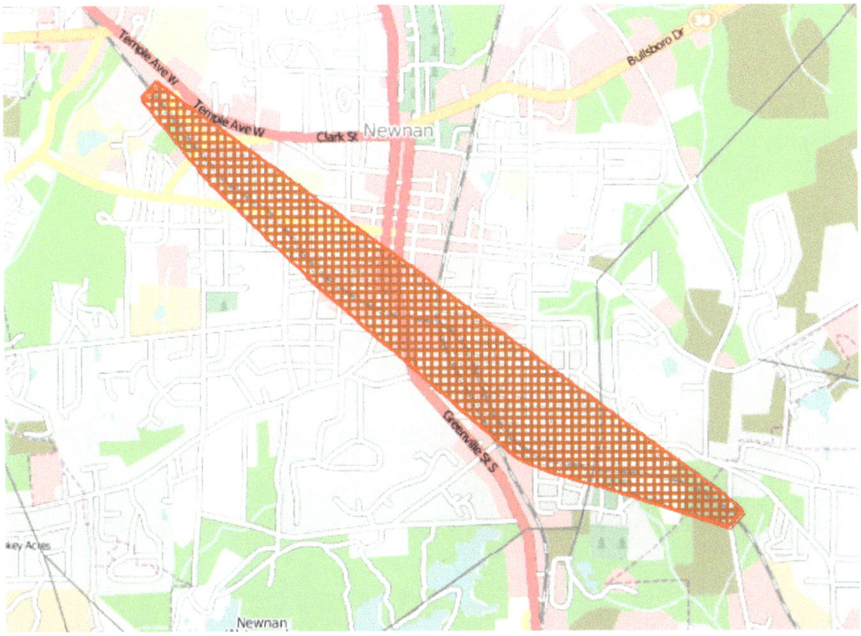

Figure 7.8 A Flight Line (red) and Lidar Tiles (brown).

An additional reason for tiling data is to merge overlapping flight-lined/drive takes. In Figure 7.9 is depicted a typical aerial project. Throughout this project, the flight line overlap is 3; that is, at most locations in the project, lidar data are contributed from three separate flight lines. Without tile merging, editing the data would be unwieldy, to say the least. The LAS file format maintains a flight line (or take, for the case of mobile) identifier that makes it possible to identify the contribution from individual flight lines in merged tiles.

Figure 7.9 Overlapping Flight Lines (shown in white).

7.2.5 Data Resolution and Precision

Precision refers to the repeatability of a process. For example, if you measure a constant quantity (say the length of a rigid metal rod at a constant temperature) repeatedly using the same technique, the range of your result is a measure of the precision of the process. Resolution, on the other hand, refers to the *fineness* with which you can make a measurement (for example, the spacing of the graticule on a rule). These two distinct properties are often confused.

The LAS format specifies the X, Y, and Z coordinates of lidar data as scaled 32-bit integers to address issues of maintaining constant resolution. It is generally desirable for data to have the same resolution over the entire project area. Floating point numbers, expressed as a fixed length mantissa and a fixed length exponent, have logarithmic resolution over their available full range. Thus, integer math maintains absolute error over the available range, whereas floating point numbers maintain constant relative error over the available range (Hamming, 1973). Unfortunately, even the names of the formats cause some confusion regarding resolution versus precision (e.g., single and double *precision* floating point).

Using LAS as the processing format generally ensures that constant resolution is maintained during standard processing. If you have direct control over the scaling of your project data, be sure to choose an appropriate value. For example, if you are expecting a project precision of 0.5 cm (for example, in a mobile mapping project) then you will need to scale to at least 0.001 in a meter system. This will limit your overall available project data range to a bit over 4 million meters (about 4,300 kilometers). These ranges are nearly always adequate for single project work.

Other areas where you need to be cautious are coordinate reference system (CRS) transformations and data compression.

Most CRS transformations are performed using floating point algorithms. This means that an X, Y, Z LAS integer coordinate is converted to floating point (hopefully double "precision"), transformations performed, and the resultant coordinate transformed back to integer. This is usually not a problem unless project scaling is such that you are running up against the edge of needed resolution. It is good practice to limit the number of times that you are transforming data.

There are a number of data compression algorithms that have been introduced to the lidar community. Several of these use lossy compression, similar to JPEG for imagery. Part of the lossy approach is to quantize either the coordinates or the GPS time. Lossy compression should *never* be used in lidar data production and associated archives. There are several common lossless compression programs available (for example, ZIP and RAR) as well as LASZip, an open source compression system designed specifically for the LAS format.

7.2.6 Lidar Geometric Correction

In this section we present a top level overview of lidar geometric correction. The intent here is not to provide technical details on this process (this is described in Chapter 6 in this manual) but to provide context to the data flow discussions.

There are two general schools of thought on lidar geometric correction. The first (and preferred method) is to segregate system calibration from project geometric deviations. System calibration involves correcting data for stable system parameters such as lever arms, local coordinate system origin, location of sensors (such as the IMU) with respect to this coordinate system, and so forth. In a high quality lidar system, the calibration will remain valid over fairly long periods of time (unless the lidar unit is dismounted from the aircraft or vehicle). The updated calibration parameters are stored in a calibration file (or database) and associated with a sensor/platform combination. System calibration can be applied by updating calibration parameters used in the sensor proprietary geocoding software, or with the appropriate third party software, as part of the project geometric correction (this latter approach will be recognized as being analogous to using self-calibration routines in photogrammetric block bundle adjustment software).

A rather common and unfortunate approach to system calibration is using visualization software to observe deviations in observed phenomenon in the data. An example would be to use overlapping flight lines that cover a building roof and observe differences between the roof line in the two passes. These observations are then used to deduce changes in system calibration parameters. The sensor geocoding software is then rerun with the adjusted parameters, and this process is repeated until the "best" solution is obtained. This approach to system calibration nearly always leads to severe problems on larger projects because it is a localized heuristic approach as opposed to a rigorous, model-based approach. The appropriate approach to system calibration is to fly/drive a signalized calibration site and use purpose-built software for deriving calibration parameters. Under a best-practices scenario, the system calibration file/database would be updated to reflect any changes in system calibration. Furthermore, trends in calibration (these parameters should be fairly stable over long periods of time) indicate a problem with the sensor that should be investigated and corrected.

Project deviations are caused primarily by inaccuracies in the GPS/IMU-derived trajectory and are not associated with system calibration (unless the system is so unstable that parameters associated with calibration are not remaining constant for the duration of the mission). These errors are caused by GPS X, Y, Z errors (particularly problematic with mobile mapping systems in urban canyons) and drift/inaccuracies in IMU parameters. These errors are non-stationary (meaning that if you return to an exact position during a mission, the errors will be different) and non-predictable. These errors are reduced in project data using a process akin to block bundle adjustment in photogrammetry. We call the process of correcting these errors *project geometric correction* or simply geometric correction. It must be accomplished using appropriate math modeling software, such as the TerraMatch product from Terrasolid (www.geocue.com/terrasolid). As detailed in a separate section of this manual, the process typically follows a sequence of:

- Mission relative adjustment;
- Mission absolute adjustment;
- Mission to mission absolute adjustment.

Lidar data should never have its geometric attributes (i.e., X, Y, Z positional values) modified after rigorous geometric correction. The LAS format (discussed later in this section) provides a mechanism for marking a data point *withheld* and for introducing new points using a *synthetic* tag.

7.2.7 Data Retention Policies

It is very easy to accumulate lidar (and imagery, for that matter) data in an open ended fashion; that is, without a clearly defined date when you will dispose of the data. Therefore, it is very important to establish overall data *policies* for your company/agency in terms of data storage and retention (the same applies, of course, to all types of data that you process and/or deliver).

We recommend that you clearly state in your delivery contracts the data retention policy that you propose. For example, if you are processing data for a prime contractor, you might specify that you will maintain a copy of the final, classified LAS data for one year from the final acceptance date. Of course, the prime contractor/client will often specify a data retention clause and this will establish the policy. It is very important to adhere to this policy or you will soon find yourself buried in data of unknown recoverability!

We use the terms on-line, near on-line, and off-line to characterize the levels of data storage typically used in production. These terms are defined in Table 7.2.

Table 7.2 Storage Modes.

Mode	Description
On Line	The data are immediately accessible from compute nodes using a connection (network, dedicated storage connection, etc.)
Near Online	The data can be brought to an on-line state using an automated mechanism under manual or software control. The most common near online storage is robotic tape devices.
Off Line	The data are not accessible using automatic means. This mode typically means data that reside on a physical media that is stored on the shelf such as optical disk, portable disk drives, non-volatile random access memory devices ("USB Memory Sticks"), tapes, and so forth. They must be physically inserted into a reader for use.

Retaining data such that it can be recovered for secondary or replacement use is rather expensive, and thus the goal is to keep data retention to a minimal level. The storage media itself are relatively inexpensive but the labor required to ensure viability can be quite high. The two biggest problems with data retention are:

- Stability of the storage media;
- Recovery of stored data from the storage media.

Stability of the storage media concerns the issue of how long the data can be expected to remain in an uncorrupted state on the media without a refresh. By *refresh*, we mean the data are read from the media, the media is checked and revitalized (if possible), and the data are written back. Some media, such as rewritable optical disk, cannot really be revitalized and must simply be replaced by new media on data refresh. Hard disks are revitalized by copying the data to a temporary media (always make two separate copies!), reformatting the media, performing an error check, and then copying the data back. Stability is the issue of the reliability of media in terms of aging and exposure to environmental conditions. For example, if you chose writable Digital Video Disk (DVD) as long term storage media (a poor choice, by the way, due to its very low data density) you would want to give careful consideration to the expected data recovery life of the media. When selecting a storage media, obtain realistic information on the reliable data shelf life. For example, we often see rewritable Digital Video Disk (DVD) specified as having a reliable data retention time of 25 or more years; however, the experiential life is more on the order of five years. An overview of the expected life of various digital media can be found at the National Archives website (National Archives, 2011) and the Optical Storage Technology Association (OSTA, 2011)

Another major consideration in establishing a storage policy is the risk of obsolescence of the storage access hardware. For example, you would be hard-pressed to find up-to-date hardware and software to recover an archive of 9-track computer tape, yet this was the standard data backup system of the 1980s. Therefore, in long term storage scenarios (for example, data that will be good source material for change analysis over long time periods), a plan must be in place to rotate the data to new types of media and possibly new data organizational formats (e.g., transforming 1980s proprietary image formats to TIFF).

As you can see, any long term off-line storage (longer than a few years) will require an active storage program. An active storage program requires periodic testing of the storage media and rotation of data to new media. If you are doing a simple media refresh, the data are written back to the same type of media (it may be replacement media such as writing data on magnetic tape to new physical tapes of the same type). If you are performing a technology upgrade, you will be transferring the data to a new type of media. This latter case is the more complex since media density is doubling every 18 months or so, and thus a repackaging of data might be involved to take advantage of the new, higher density.

You should consider service provided remote storage (i.e., "cloud-based" storage) for relatively low volume data such as corporate records, metadata, data bases, and so forth. These solutions are typically not band-width effective for high density data such as lidar and imagery. By this we mean that the possible bandwidth between your processing facility and the service provider will not economically support the transfer of the data.

In general, online and near-online data storage arrangements are much less labor intensive to maintain than offline systems. Of course, the tradeoff is that the hardware required to realize these mechanisms are more expensive than offline storage. It is very important to realize that data redundancy within a single unit such as archiving to an online RAID-5 system is not sufficient protection. If the disk hardware suffers a catastrophic failure such as a lightning strike or physical damage (fall, flood, fire, etc.) then all is lost. Thus you must maintain a duplicate system at a separate physical location. Large sample size analysis from the Carnegie Mellon Parallel Data Laboratory (Schroeder and Gibson, 2006) seems to indicate that the practical life of hard disk drives is about five years. Since this represents nearly three generations of hard disk technology, it might be a reasonable plan to completely replace online hard disk technology on a five-year cycle.

So what exactly should be archived, and when should these archived data be deleted? If not specified otherwise in the delivery contract, then we suggest you modify the base plan of Table 7.3 to suit your particular needs.

Table 7.3 Template Archival Periods.

Data	Archive Period	Notes
Raw Lidar Sensor data	1 year post-delivery	
RAW POS data	1 year post-delivery	
Survey Control	5 years post-delivery	
Geocoded Lidar data	Until geometric correction is verified	Can be reconstructed from raw data, if need be
Geometrically corrected, tiled Lidar data	1 year post-delivery	
Geometrically corrected, tiled, classified Lidar data	5 years post-delivery	If you are building a speculative Lidar data archive, these are the data you will retain
Thinned (model key-point) Lidar data	6 months post-delivery acceptance	These data are easily reconstructed from the classified data
Derived products	5 years post-delivery acceptance	

We have found that most companies tend to save too much data for too long. Additionally, these are usually of unknown quality (i.e., only portions of the data sets can be recovered) and hence are of little practical use. It takes a very disciplined approach to properly maintain data archives but the time invested is invaluable when that rare occurrence pops up of needing to restore a historic project.

7.2.8 Data Checkpointing

Data checkpoints are versions of data retained during processing to enable rapid recovery in the event of a processing error. Checkpointing begins with acquisition and occurs at checkpoints throughout production.

Most commercial kinematic lidar systems record raw data to a single data drive set during data acquisition (although metadata, such as GPS/IMU data, are often recorded to multiple devices). Thus, the first step after a data acquisition mission should be validating and copying the acquired data. You should design a copy system that is deployed with your lidar system for this purpose. We strongly recommend that duplication of data occur prior to any field quality checks (in case a catastrophic failure occurs during the QC process).

Best practices dictate that you do a field check of data prior to commencing processing in your data production facility. We recommend the basic approach depicted in Figure 7.10.

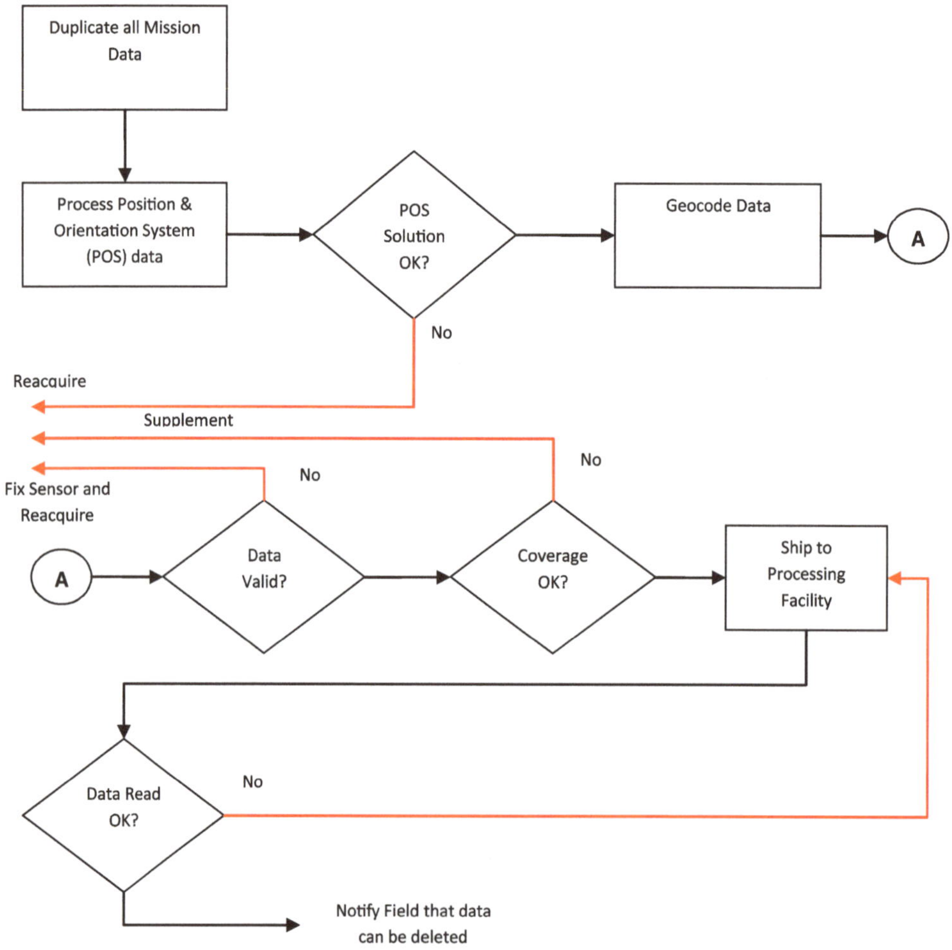

Figure 7.10 Field Data Flow.

Obviously the typical processing flow is more complex than is depicted in Figure 7.10. For example, if field personnel could not obtain a satisfactory first look solution for the POS, they would probably pass the data (through FTP) to the processing center to see if experts could obtain a satisfactory solution. The general idea is to always perform Quality Checks as close to the potential error source as possible. It is much more expensive to find an unrecoverable POS error after the collection platform has left the collection area than before. The preserving of two copies of data in the field forms the first data checkpoint. Note that this set of data should be deleted once confirmation has been received from the processing office that all data have been received and are intact. We strongly recommend that one set of field data be immediately deleted but the second set be retained until the media must be recovered for a subsequent mission. This procedure allows a bit of a safety factor in the event that the processing office has made an error in archiving the data.

We strongly recommend that two full copies of raw data be maintained at the processing facility for the duration of production of the *entire* project (block C-D in Figure 7.5). These data generally comprise:

- POS raw data;
- Lidar Range file (lidar raw data);
- Lidar platform calibration data;
- Image data (if applicable);
- Control/Checkpoint data (while this is technically not 'raw' data, it is convenient to keep it with the raw data since it would be needed for reprocessing).

We make this recommendation since the mission can be fully recovered from this data set should disaster strike during the production phase of the project (e.g., an inadvertent deletion of a top level folder containing all of the project data, and so forth). Whether these data are retained after project delivery depends on the contract and your established data retentions policies (you have established these, right?). Be extremely careful in deleting these data since one of the most common reasons for project rejection by discerning clients is the quality of geometric correction. If you retain data at this point, you will be fully able to repeat the calibration/project geometric correction steps.

Establish a temporary mission checkpoint at the conclusion of calibration (Step E in Figure 7.5). This checkpoint allows you to do a full recovery, post-calibration, should you get hopelessly lost in mission geometric correction. These checkpointed data can be deleted as soon as you have confirmed the validity of project geometric calibration.

A checkpoint should be established when geometric correction is complete (at the conclusion of Step F in Figure 7.5). Again, we recommend that you keep two copies of the data since the process of geometric correction can be quite time consuming and the checkpoint will allow you to recover to this step. After confirming the validity of the checkpointed data, you can delete the Step E checkpoint.

Classification (Step G in Figure 7.5) is generally an iterative process of applying automated processing routines (for example, "macros" in TerraScan or "point cloud tasks" in LP360) followed by interactive classification. Checkpointing during this phase of processing is done on a tile-by-tile basis and is more appropriately called "versioning." The number of versions to maintain will depend on your overall scheme for processing as well as the complexity of classification. If you employ a production model of complete automated processing followed by interactive classification (without a return to the project-wide automated processing), then create tile-based file version checkpoints at the end of automated processing. This procedure will allow you to recover a tile to the point of basic classification should an editor make an error where recovery is quicker from the automated classification (A classic example would be to inadvertently classify outside a polygon when the intent was to classify inside a polygon.).

The final step of classification is a rigorous tile-by-tile quality check. This check examines the quality of both the classification and the density of certain classes. For example, if a project requirement is to produce a ground point density of two points per square meter in open regions, this test occurs as part of the final QC during classification (a much better place for this step than final QC prior to delivery). Following successful completion of this QC step, the data are checkpointed. In many situations, this is the most valuable checkpoint within production and may be the data that you store as the long-term archive LAS for the project.

Vector data associated with the project are also carefully checkpointed. These data often are not on the same tiling scheme as the lidar data since they tend to be geometric features (e.g., roads) that span tiles. The scheme that you employ will depend on the nature of the repository used for the vector data. If you are collecting to a geodatabase using tools such as ArcGIS®, you will have versioning capability included in the collection software. If you are using a file based application, you will need to save versions of the files at critical points in collection.

7.2.9 LAS Data Format

LAS is a specification for a *point cloud* file format. It is primarily used for transmitting laser point cloud data (lidar) but can be used for any general 2D or 3D point oriented encoding.

This section provides an overview of the latest version of the specification, LAS 1.4, as well as a history of the LAS format. It is based on an article published in *Photogrammetric Engineering & Remote Sensing* (Volume 78, No. 2, February 2012, pp. 93–102.) for the same topic (Graham, 2012). Generally in this subchapter, I will use the pronoun "I" when something is primarily my personal opinion and "we" when the work was consensus work of the ASPRS LAS Working Group (LWG). The LWG is a working group within the Lidar Division of the ASPRS. It comprises a group of representatives of lidar hardware and software manufacturers as well as several representatives of end-user agencies. The LWG periodically updates the LAS specification as either hardware and/or processing standards change.

The introduction of a specification file format (LAS 1.0) for kinematic lidar data in 2002 has been very successful as evidenced by the fact that all major software vendors have adopted the specification and customers are routinely requiring that lidar data be delivered according to this specification. As with any specification, usage has made it apparent that changes need to be made both due to omissions in the original design as well as continuing maturity of the lidar industry. These updates have been occurring on an approximate two year cycle.

7.2.9.1 Highlights of LAS 1.4

As of publication of this Manual, version 1.4 is the most recent release of the specification. The update to LAS that resulted in the creation of version 1.4 was primarily motivated by the need to extend the number of possible classes from the 32 available in LAS 1.3 to 256 (ironically, LAS 1.0 did support 256 classes, but this was changed in the update to LAS 1.1).

Of course, any time one starts updating data formats, other features have to be added. The summary of major additions to LAS 1.4 is:

- Extension of offsets and field sizes to support full 64 bit addressing;
- Support for up to 15 returns per outgoing pulse;
- Extension of the Point Class field to support 256 classes;
- Definition of several new ASPRS defined classes;
- Extension of the Scan Angle field to 2 bytes to support finer angle resolution;
- Addition of a Sensor Channel bit field to support mobile mapping systems;
- Addition of Well Known Text (WKT) definitions for Coordinate Reference Systems;
- Addition of an Overlap bit to allow indicating pulses in the overlap region while maintaining the class definition;
- Addition of an (optional) Extra Byte Variable Length Record to describe "extra bytes" stored with each point.

7.2.9.2 Specifying LAS 1.4

Throughout this section, I will go in to some detail concerning the major points of the LAS 1.4 specification. This initial discussion provides information for those userss who just want to know what they should be specifying when contracting for lidar data.

The most critical features of LAS 1.4 with which most users should be concerned are support for 256 classes, an Overlap region indicator bit and specification of Coordinate Reference

Systems (CRS) by the Open Geospatial Consortium method of Well Known Text (WKT). Since "class" and "Overlap" are per-point attributes, it is necessary to use a new Point Data Record Format (PDRF) to specify these new features. LAS 1.4 adds five new formats (PDRF 6 through PDRF 10) to the specification that all include the new CRS encoding, extended classes and designation bits. The major differences between these five new formats are delineated in Table 7.4.

Table 7.4 The New Point Data Record Formats (PDRF) of LAS 1.4.

PDRF (LAS 1.4 Only)	Major Features
6	Base type
7	Red, Green, Blue (3 channel) colorization support
8	Red, Green, Blue, NIR[1] (4 channel) colorization support
9	Waveform Packet Support
10	Red, Green, Blue, NIR (4 channel) colorization and Waveform Packet support

As is evident from Table 7.4, most new deliveries will be in PDRF 6.

LAS 1.4 *requires* that the Coordinate Reference System (CRS) encoding for PDRF 6 through 10 be WKT as opposed to the old technique of GeoTIFF packets. Thus, you are assured by specifying record types 6 to 10 that your data will have proper CRS encoding. I believe that this is critical for archived data since GeoTIFF (the old encoding standard) has limited support for describing the vertical components of CRS and limited support for newly emerging CRS definitions.

I also believe that it is highly likely that the United States Geological Survey (USGS) will require point data record formats that support the designation of overlap using the new Overlap bit supported in PDRFs 6 through 10, and that this requirement will accelerate the move from prior versions of LAS to LAS 1.4.

7.2.9.3 Emulating Prior LAS Versions

During the specification process for LAS 1.4, there was a strong desire from several developers of open source software to create fields in the LAS 1.4 specification that might make it possible for a prior version of a LAS reader to ingest a LAS 1.4 file, if that file met certain conditions. These conditions are:

- The offset to data pointers must remain at or below 32 bit limitations;
- Point Data Record Format must be one of the "legacy" formats (0 to 5);
- The Coordinate Reference System (CRS) encoding must be GeoTIFF (as opposed to the new method of WKT);
- The CRS must be encoded in a Variable Length Record as opposed to an Extended Variable Length Record;
- The legacy reader software must ignore the LAS version number.

I was personally opposed to this emulation mode because it introduces (if used) some unreliability to the LAS file and relies on a software defect (namely ignoring version number) to work at all. Unfortunately, the details (outside of the LAS Working Group) became obscure with some LAS consumers thinking that if this emulation mode were supported, their existing LAS software would be enabled for the new LAS 1.4 features. This, of course, is not possible

[1]NIR - Near Infrared

since the major new features are encoded in the new point types (6 through 10) that simply are not present prior to LAS 1.4.

I think this emulation feature should be used only for experimental purposes where the exact implementation is known, *a priori*. My recommendation is that you *never* specify a data delivery of LAS 1.0 to LAS 1.3 data within the context of a LAS 1.4 file. If you need LAS 1.0 to 1.3, simply require your vendor to deliver in the proper format.

- General rule: Always specify Point Data Record Types of 6 and above and legacy fields set to zero.

7.2.9.4 The Original Specification

The original lidar binary file specification grew out of work done by EnerQuest Systems (whose software development work is now owned by Virtual Geomatics) in the late 1990s. EnerQuest had developed a proprietary point cloud binary format for laser scanned data (LASer) to make dealing with the generally unorganized data associated with lidar more efficient. Don Wicks, then president of EnerQuest, generously agreed to contribute the EnerQuest LAS work to the public domain. Leica Geosystems, Optech, and Z/I Imaging (I was then with Z/I) joined the effort and the LAS 1.0 specification was born (the bulk of the implementation work was done by John Nipper, a software developer then at EnerQuest). About a year after this industry group defined the initial specification, the ASPRS agreed to take over ownership.

The LAS specification is a relatively compact binary encoding of point location and point *attribute* data. Rather than store attributes in *referenced* records, the light-weight attribute data of LAS is stored in the same record as the point data. At the time of the definition of LAS 1.0, we did not try to anticipate how users would want to augment the format.

7.2.9.5 The Philosophy of Lidar Processing

In order to fully appreciate the design of the LAS format, it is necessary to understand the thought process that the original designers of LAS 1.0 had with respect to lidar data processing.

A time-of-flight lidar system emits a laser pulse and zero or more pulse returns[2] ("echoes[3]") are detected in the receiver. These return pulses may be reflected from nearly anything at all (e.g., birds, planes, tree leaves, buildings, ground, etc.). It is the task of the lidar data editor to *classify* these data with respect to their impinged surface. Thus, the original LAS specification supported the idea of assigning a *classification code* to an echo. This classification attribute could be set to the surface type such as Building or Ground. Indeed, much of the software challenge in lidar processing has been a drive toward developing various algorithms that can automate this processing (so-called lidar *filtering* algorithms; an unfortunate misnomer).

The original philosophy behind lidar processing was that points can never be added to or deleted from a lidar data set after acquisition. Additionally, points should never be *moved* except under rigorously modeled adjustments such as *boresight* calibration and project geometric correction (for errors introduced by inaccuracies in the positioning system). We (the original LAS design group) felt very strongly that simply *deleting* or moving data that do not fit a desired model is tantamount to fabricating data. To support the editing concept, certain classes were reserved to indicate data adjustments beyond reflectance classification. Thus, if for example, a high point (perhaps reflected from a bird) were to be omitted from the data set, it remained in the set but its classification code was set to "WITHHELD."

It was soon evident that we needed to retain the original classification of points that were being withheld from the data set (e.g., we needed to designate a point as both "Ground" and

---

[2] and, effective with LAS 1.3, possibly wave packets

[3] I use echo and return interchangeably

"Withheld") and there was a need to be able to *synthetically* insert points into an existing lidar data set (for example, digitizing a synthetic lidar point to represent the attachment point of a transmission wire to an insulator). Thus, starting with LAS 1.1, the idea of modifier bits was added to the class designation.

Today the philosophy remains that once lidar data have been geocoded in the proprietary software supplied by the hardware vendor, they should only be geometrically moved in a system calibration and/or geometric correction process and that points should never be deleted.

7.2.9.6 Extending the Lidar Edit Philosophy

The LAS 1.4 specification incorporates features that allow more robust point marking, adhering to the philosophy that points should never be physically deleted or moved during editing. For example, if points are found to be reflecting from saw grass, they should not be collected up as a group in a point editor and literally *moved* to the ground. We argue that what has really happened is that a collection of points have been marked as *withheld* and a new collection of points *added* based (one would hope!) on an accurate editing surface such as ancillary stereo imagery. This type of point marking has been available in LAS since version 1.1 by supplying appropriate attribute fields in the point records. LAS 1.4 supports four attributes related to point editing (described later).

The LAS specification is being used quite frequently as a general format for point cloud data, regardless of its source. For example, I know of at least one company who generate LAS format data from the output of a stereo image correlator and then apply lidar processing tools to the task of editing these data. We very strongly support this use of the LAS specification and have, in fact, added some fields in the version 1.4 release to expand support for facilitating these non-traditional uses (primarily the addition of a Near-Infrared channel for supported five channel 3D image data derived from line imagers).

7.2.9.7 The General LAS Layout

The format contains binary data consisting of a public header block, any number of (optional) Variable Length Records (VLRs), the Point Data Records, and any number of (optional) Extended Variable Length Records (EVLRs). This general layout is depicted in Table 7.5.

Table 7.5 LAS 1.4 Format Definition.

PUBLIC HEADER BLOCK
VARIABLE LENGTH RECORDS (VLR)
POINT DATA RECORDS
EXTENDED VARIABLE LENGTH RECORDS (EVLR)

The public header block contains generic data such as point counts and point data bounds. Interestingly, it does not contain the definition of the Coordinate Reference System (CRS).

The Variable Length Records (VLRs) contain variable types of data including projection information, metadata, waveform packet information, and user application data. They are limited to a maximum data payload size of 65,535 bytes. The continued existence of VLRs in LAS 1.4 is one of the compromise results of supporting emulation mode (discussed earlier). Software writers of LAS 1.4 files where emulation is not needed (the recommended approach) can encode all variable information into EVLRs.

The Extended Variable Length Records (EVRLs) allow a larger payload than do the VLRs and have the advantage that they can be appended to the end of a LAS file. This allows adding, for example, projection information to a LAS file without having to rewrite the entire file.

A LAS file that contains point record types 4, 5, 9, or 10 could potentially contain one block of waveform data packets that is stored as the payload of any Extended Variable Length Record (EVLR). Unlike other EVLRs, the Waveform Data Packets (if stored internally to the file) have the offset to the storage header contained within the Public Header Block ("Start of Waveform Data Packet Record").

7.2.9.8 Projects

In LAS 1.1, we defined the idea of a *Project.* A Project can be a lidar job as defined by a number of flights/drives over a project area or it could simply be a number of elevation files being used in a processing job not necessarily even related to lidar. A Project is uniquely identified by a Globally Unique Identifier (GUID). A GUID (sometimes alternatively referred to as a Universally Unique Identifier, UUID, or as a Uniform Resource Indicator, URI) is generated by a "GUID Generator" algorithm available within the Microsoft development systems or from open sources. A GUID is a 16 byte identifier that is unique across the domain of all possible GUIDs. The Project GUID existed in LAS 1.0, but we clarified its intended use in LAS 1.1.

It is a good practice to assign a Project ID at the time of creation of the project and to ensure that this same project ID is used for all files that are associated with the project.

7.2.9.9 File Source ID

The File Source ID field is used to uniquely identify *this* file within a Project. If the source of the file is a lidar flight/take[4] line, then this field is used as the flight line number. Thus, inclusion of the File Source ID solves a major problem that existed in LAS 1.0; the restriction of flight line information to 256 lines. The File Source ID field supports 65,536 individual files within a project.

It is important to note that the LAS format can embody any sort of point cloud data in addition to lidar data. This means that a project could consist of a heterogeneous collection of a variety of elevation files from any number of sources. For example, we might construct an elevation project with some files originating from lidar flights, others from United States Geological Survey (USGS) elevation files, still others from photogrammetrically-correlated data and so forth. Thus, the concept of *Project* becomes generalized.

7.2.9.10 System ID

In keeping with the notion of generalizing the LAS specification to apply to any sort of point cloud data, System ID is used to provide information about the manner in which the LAS file was generated. For a lidar system, this field should be encoded with a system identifier such as "ALTM 1210." Similarly, this field is used to indicate the method by which a file was generated when the source is not a lidar sensor. The specification provides several predefined values such as "MERGE" and "EXTRACTION" and allows the user to provide any definition (within the limits of a 32 byte description). A common example within lidar processing would be the generation of a processing tile by merging the content of several flight lines. In this case, the Project ID is set to the overall Project ID assigned to the project, a new and unique File Source ID is created, and the System ID field is set to MERGE.

7.2.9.11 Point Data Records

LAS files are homogeneous with respect to point records. The record type is indicated by the Point Data Record Format (PDRF) in the Public Header Block. All points within a single file must be the same with the format being specified by the PDRF.

---

[4] In Mobile Mapping, a single span of data collection is often referred to as a data "take."

It is important to note that the LAS specification does not impose a point ordering scheme on a LAS file. Thus the points can (and very often do) appear in any order. For example, if point n is marked as "return 2 of 4" and point n+1 is marked as "return 3 of 4", there is no guarantee that these two points are associated with the same outgoing point (you would have to use GPS time to sort the returns, as noted later). I have seen a lot of software that erroneously assumes that LAS points will be sorted by spatial proximity, by return number or other attributes. This results in, for example, poor display performance when the data are, in fact, not in the assumed order.

LAS 1.4 added 5 new PDRFs (6-10) to the specification. The salient new features of these types are discussed in a later section. While LAS 1.4 continues to support PDRFs 0–5, I strongly recommend that you always specify the new point types 6–10. The primary reason for this recommendation is that merging point types 0-5 into files using point types 6–10 will require a type 6–10 PDRF (due to expanded fields). A second reason for this recommendation is that LAS *requires* Well Known Text (WKT) encoding for the Coordinate Reference System (CRS) of PDRF 6–10. WKT is a much more reliable CRS representation than GeoTIFF (in fact, there are a number of CRSs that simply cannot be represented in GeoTIFF). Remember, if you have a LAS file and you cannot determine its CRS, the data are useless for absolutely referenced data exploitation.

It is critically important for programmers to understand that a point data record in LAS can contain fields beyond those defined by the PDRF. For example, PDRF 0 defines 20 bytes of specified information. If desired, a software implementer can add additional bytes to the end of this point type (for example, perhaps to encode an accuracy estimator for Z). Thus one must always refer to the Point Data Record Length field in the Public Header Block to determine the correct number of bytes for each point record.

LAS 1.4 (thanks to a contribution from Dr. Martin Isenburg) adds a way to formally define extensions of the standard PDRFs via a mechanism called "Extra Bytes." While nothing in the LAS specification prevents a developer from adding their own PDRF (say PDRF=21), no other software would be able to read the point data. By using the Extra Bytes facility, software that does not implement Extra Bytes will still continue to correctly read the "base data" of the point record. Those who participated in the review of the canceled LAS 2.0 effort will recognize that this was the total representation method of points in that format.

7.2.9.12 GPS Time

Most Point Data Record Formats (PDRF) support a field for the time of the pulse. This time is encoded as either GPS Week-Second time or Standard GPS time as indicated by the Global Encoding field in the Public Header Block. We recommend that all procurers of lidar data *require* that the GPS time be encoded as Standard time. This is because GPS Week Time does not allow one to determine the absolute time that a pulse was transmitted.

Although not clearly described in the LAS specification (an oversight that I just recognized while writing this article), the GPS time is the time when the pulse was *emitted* by the lidar sensor. Thus all returns (see next section) of a multiple return pulse will have the same GPS time.

When we first defined the original LAS specification, we felt that GPS time, on a per pulse basis, would be used only for certain "close to the sensor" operations. However, many algorithms for data exploitation, particularly in the area of mobile mapping, heavily rely on GPS time. Additionally, the GPS time is currently the only mechanism for correctly correlating a set of multiple returns. For this reason, GPS time was included as a field in all of the new PDRFs of LAS 1.4.

7.2.9.13 Returns

LAS is designed for discrete point data (that can optionally support wave form packets). While it can be used for any 2D or 3D point (or N dimensional with the Extra Bytes facility), the primary emphasis is on kinematic, discrete return, time of flight systems lidar. All modern lidar systems of this type are capable of multiple returns per outgoing pulse. A diagram of a potential multiple return scenario is illustrated for a five return scenario in Figure 7.11.

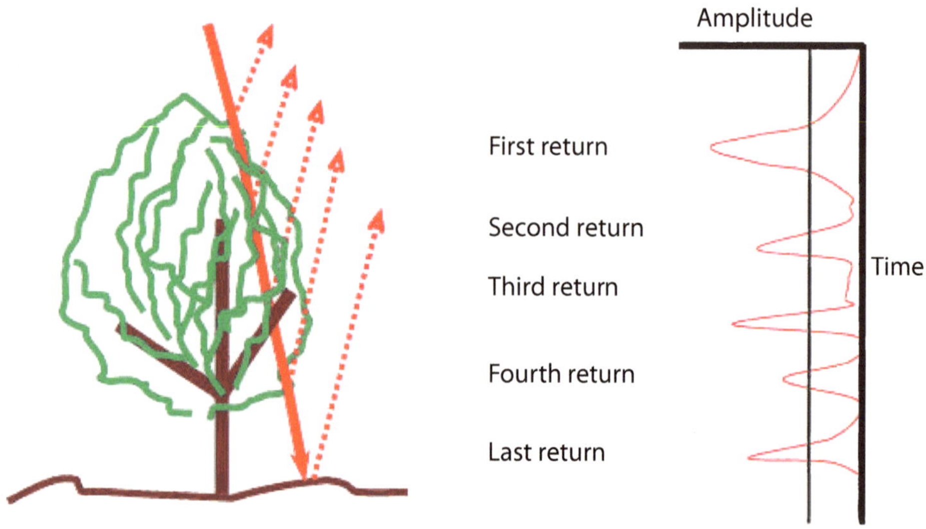

Figure 7.11 A five return system.

LAS 1.4 supports up to 15 discrete returns per outgoing pulse. The count of the number of returns per pulse is contained in the field Number of Points by Return in the Public Header Block (not the Legacy Number of Points by Return - this is used for encoding a LAS 1.1–1.3 payload within a LAS 1.4 file body - a legacy implementation that can support only 5 returns). Note that this field is an array that is a histogram of the number of returns for this file. It does not represent the physical capabilities of the system(s) that was used to acquire the data.

The return position of a point within a multiple return source pulse is such an important factor in data analysis that this position is encoded for all Point Data Record Types. PDRF 0–5 (the legacy types from prior LAS versions) support up to 5 returns whereas the new types (6–10) support up to 15.

All returns from a multiple return outgoing pulse will have the same GPS time stamp. This fact is used for correlating the returns of a multiple return pulse. It is entirely possible that returns from the same outgoing pulse will span more than one LAS file. This generally occurs when data are segregated into tiling schemes and different returns of the same outgoing pulse span a tile boundary.

7.2.9.14 Classification

The LAS 1.0 specification devoted an 8 bit field to CLASS. The purpose of the field was to allow a point to be tagged as to its reflectance surface (e.g. Ground, Low Vegetation and etc.). A second use for CLASS was to mark points that were special in some sense such as WITH-HELD or MODEL KEYPOINT. Other than reserve space in each point record for the field, the value stored in this field was entirely up to the user.

When we initially devised the LAS 1.0 specification, we did not really view the specification as a format that would be conveyed to the ultimate end-users of lidar data. It was intended to be a transport protocol between makers of lidar sensors and developers of lidar processing tools. However (and this is a very good thing), LAS rapidly became the de facto specification for all phases of lidar data processing and delivery. This holistic use necessitates specification of the values for the CLASS field. In other words, we all have to agree on which numeric value will signify the various classes.

A second shortcoming of the original CLASS field in the LAS 1.0 specification was that one could not simultaneously maintain a class attribute and a treatment attribute. For example, the field would not support the idea of a point being both Ground and a Model Key Point. Encoding a point as a Key Point destroyed its reflectance categorization.

To solve these two problems, the 8 bit CLASS field of LAS 1.0 was split into two parts; a 5 bit CLASS section and 3 bits used as Flags (see Figure 7.12). The classes ranged from zero to 31.

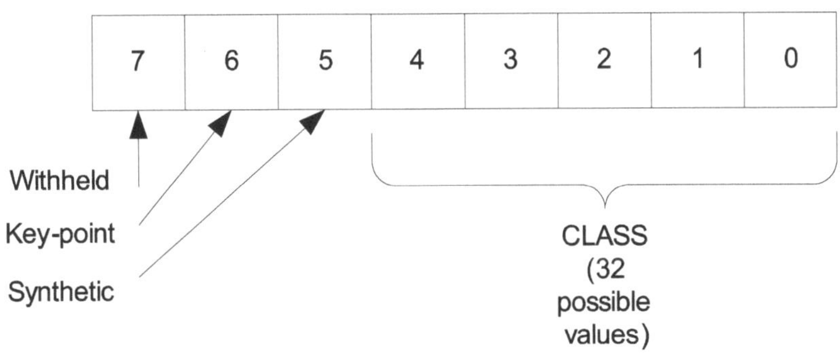

Figure 7.12 The Classification Field of LAS 1.1–1.3.

When we made this transition to LAS 1.1, there were typically only a handful of classes defined (usually less than ten). We therefore thought that 32 classes (0–31) would provide plenty of room for any classification scheme. Unfortunately, with the recent widespread use of lidar for mobile mapping applications and transmission line coding, the need for additional room for class definitions has become critical.

For this reason, LAS 1.4, for the new PDRFs 6-10, separates CLASS back in to a dedicated 8 byte field. Class IDs 0 through 63 are reserved for ASPRS use leaving classes 64 through 255 available for definition by users.

Using ASPRS defined classes allows any software application to correctly display the meaning of the class without the software vendors needing to communicate this information. Now one could argue that this could be more elegantly communicated by simply encoding the class definition table in a Variable Length Record (in fact, this is what we did in the original version of LAS) but this creates a mapping problem when merging files.

The class definitions for PDRFs 6-10 are listed in Table 7.6.

Two very important changes have occurred in this classification table; the removal of Model Key Point (class 8) and Overlap Point (class 12) as classifications. To avoid confusion with the classification of the old point data record formats 0–5, we have set these two classes to Reserved. Indicating Key Points and Overlap with PDRF 6–10 is performed by attribute bits (see next section).

Note that we have added a few new ASPRS standard class codes such as Rail and Wire types. We anticipate that the full 0–63 class slots reserved for ASPRS will be defined over time as user needs mature.

Obviously another good reason to only accept lidar data in formats 6-10 is to ensure that the classification encoding is homogeneous across all of your data.

Table 7.6 ASPRS Standard Lidar Point Classes (Point Data Record Formats 6–10).

Classification Value	Meaning
0	Created, never classified
1	Unclassified[5]
2	Ground
3	Low Vegetation
4	Medium Vegetation
5	High Vegetation
6	Building
7	Low Point (noise)
8	Reserved
9	Water
10	Rail
11	Road Surface
12	Reserved
13	Wire - Guard (Shield)
14	Wire - Conductor (Phase)
15	Transmission Tower
16	Wire-structure Connector (e.g. Insulator)
17	Bridge Deck
18	High Noise
19–63	Reserved
64–255	User definable

7.2.9.15 Point Attributes

Point attribute bits were first expanded to include classification flags in LAS version 1.1. In LAS 1.4 (for PDRF 6–10) an additional flag was added to indicate points in the overlap region. The Point Attributes of PDRF 6-10 are listed in Table 7.7. The first 4 bits are devoted to classification flags.

Table 7.7 Point Attributes for PDRF 6-10.

Classification Flags	4 bits (bits 0 - 3)	4 bits
Scanner Channel	2 bits (bits 4 - 5)	2 bits
Scan Direction Flag	1 bit (bit 6)	1 bit
Edge of Flight Line	1 bit (bit 7)	1 bit

The first flag bit is the *Synthetic* flag. If this flag is set (value 1), it indicates that the associated point was not collected by means of the method indicated in the System ID field for the file but was added via some other mechanism. The most common use of this field is moving a lidar

[5]We are using both 0 and 1 as ***Unclassified*** to maintain compatibility with current popular classification software such as TerraScan. We extend the idea of classification value 1 to include cases in which data have been subjected to a classification algorithm but emerged in an undefined state. For example, data with class 0 is sent through an algorithm to detect man-made structures – points that emerge without having been assigned as belonging to structures could be remapped from class 0 to class 1.

point. We advocate a philosophy that once "bore sighting" and project geometric correction (the equivalent of photogrammetric block bundle adjustment) have been performed, individual lidar points cannot have their X, Y or Z values adjusted. Thus if one were to move a collection of lidar points observed in an editing session to be reflecting from within low vegetation to the ground, the operation would actually proceed as a deletion of the point within the low vegetation (the deletion is actually setting the WITHHELD flag bit, not removing the point from the file) and the addition of new ground points. These newly added ground points would have their CLASS value set to GROUND and their SYNTHETIC attribute flag bit set to True (a value of 1). The Synthetic flag should be used for any *additive* operation such as digitizing points in lidargrammetry, adding a point to indicate a power line connection and so forth.

The second flag is KEY POINT. This allows the marking of a point as one that is to be kept for elevation file formation (perhaps as a contributor to a contour generation routine) while maintaining its class attribute. Thus the point could maintain a class of Water and also be flagged as a Key Point. Key point algorithms thin data to the minimum points necessary to maintain a specified accuracy criteria. A typical approach is to create a Triangulated Irregular Network (TIN) using all of the points (of the desired class such as ground) and then iteratively remove points that do not cause the regenerated TIN to violate a specified error tolerance.

The third flag is the WITHHELD flag. Points should never be deleted from an LAS file but rather flagged as deleted. Thus, for example, in the removal of water points, a point could be classified as Water (CLASS = 9) and then have the WITHHELD bit set to 1 to indicate that the point should be omitted from farther modeling.

New to LAS 1.4 is the fourth flag, OVERLAP. It is frequently desired to thin data in the overlap region of lidar flight lines (or, in the case of mobile, "takes"). This operation is used to either improve the statistical accuracy of the lidar data or to simply thin the multiple density data in these regions. Using a Class to indicate overlap (the technique prior to LAS 1.4) caused problems in that one loses the actually classification of the point (e.g. Ground). This leads to problems in analyzing accuracy in the overlap region. What exactly comprises points that should be designated as Overlap will be determined (as it is now) by the procurer of the lidar data. For example, the USGS has an extensive description of how points should be designated as Overlap.

Also new to LAS 1.4 is the ability to designate a Scanner Channel. The intended use is for mobile mapping systems with multiple laser scanners. 2 bits are used for this indicator and thus one of 4 scanners can be designated. For a single scanner system, the channel must be zero.

The final two bits of the attribute field are Scan Direction and Edge of Flight Line. Unfortunately, even though these are mandatory fields in the LAS point data, I often see them omitted. Scan Direction is defined for rotating mirror scanners and is defined to be positive for counter-clockwise rotation if the flight direction were coming out of your eye (looking at the back of the aircraft as it moves along its track). I personally do not know how this parameter might be used in data processing. The last flag, edge of flight line, is far more important for non-360° scanners. It can be used to quickly compute a graphic for the edge of a lidar flight (see Figure 7.13).

7.2.9.16 User Data

Each Point Data Record Format (PDRF) supports 1 byte of User Data. This is a data field that can be used for any purpose by processing software. There is no guarantee that data stored in this field will be preserved from one application software package to another. In fact, many software packages use this field for a scratch pad area during processing. Clients ordering lidar data may specify that this field be set to zero prior to data delivery.

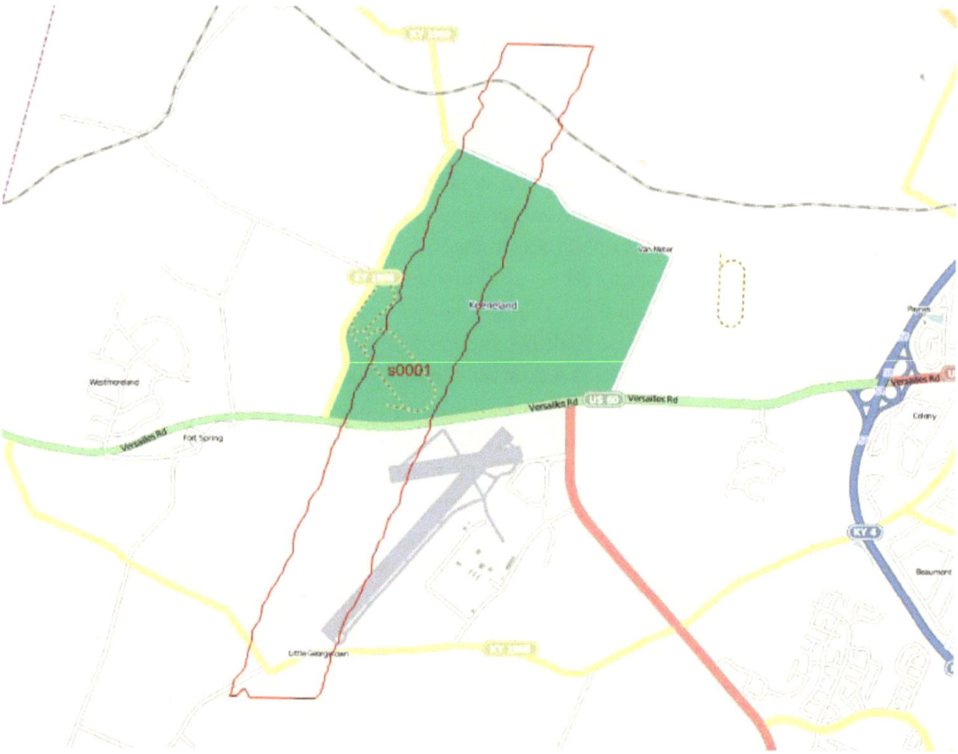

Figure 7.13 Edge of Lidar Flight Line.

7.2.9.17 Scan Angle

The Scan Angle, for PDRF 6–10, has been expanded to two bytes for better scan angle resolution. It is a signed short that represents the rotational position of the emitted laser pulse with respect to the vertical of the coordinate system of the data. Down in the data coordinate system is the 0.0 position. Each increment represents 0.006 degrees. Counter- clockwise rotation, as viewed from the rear of the sensor, facing in the along-track (positive trajectory) direction, is positive. The maximum value in the positive sense is 30,000 (180 degrees which is up in the coordinate system of the data). The maximum value in the negative direction is -30,000 which is also directly up. Note that scan angle can now, under this new definition, be used for 360° rotational sensors such as those used on mobile mapping systems.

7.2.9.18 Intensity, RGB, NIR Channels

The intensity value is the integer representation of the pulse return magnitude. This value is optional and system specific. However, it should always be included if available.

Red, Green, Blue (RGB) and near infrared (NIR) can also be stored with certain point formats. These channels are intended for storing values from colorization of the points by one or more auxiliary optical sensors (cameras). For example, colorizing lidar data with RGB reflectance data from a 360° camera is a common operation in mobile mapping.

This data, when included, is always normalized to a 16 bit, unsigned value by multiplying the value by 60,536/(intensity dynamic range of the sensor). For example, if the dynamic range of the sensor is 10 bits, the scaling value would be (65,536/1,024). This normalization is required to ensure that data from different sensors can be correctly merged. If intensity and/or color data are not included, these values must be set to zero

LAS is more and more frequently being used to represent point clouds from non-laser derived data such as 3D points from image correlators. By using Point Data Record Format 8, up to five channels of information (using the available I, R, G, B, NIR fields) can be stored.

7.2.9.19 Point Source ID

The Point Source ID field (2 bytes) is used to identify the source file in which a point was originally created within a Project data set. For example, if a project contained a source file of lidar points that originated from a flight line and that flight line had been assigned number 77, all points within that original file would keep their Point Source ID of 77 throughout all processing regardless of how many files through which that point might have propagated. This feature allows mapping back to the original source of a point even when an LAS file has resulted from composite operations such as merge and extract. The most common example is that of a processing tile formed by merging numerous flight lines into the tile. By using the Point Source ID, each point can be uniquely identified as to its original source. Thus this is the field that is used in software when performing operations such as "color by flight line."

In general, once a Point Source ID has been assigned to a point, it should not change. It should also be noted that Point Source ID is not meaningful outside the context of a single project (e.g. if points from two or more projects are blended, they cannot be segregated using Point Source ID). However, if you have control over all projects that will be merged, you can pre-plan flight lines such that back-tracking and unique flight lines are maintained.

You will note that for a pure lidar project (all data sources originated from lidar flight lines), the Point Source IDs must all map back to original flight line numbers except for SYNTHETIC points. This is illustrated in Figure 7.14. In this simple example, we have two flight lines ("strips") that have been labeled with File Source IDs of 1 and 2. The points from these two flight lines have been MERGED into a file that has been assigned a File Source ID of 500. Within this merged file, the left-most point has been 'moved' down. This is accomplished by setting the WITHHELD bit of the original point to 1 (TRUE) and then creating a new point in the desired location. Since the new point was created within Source File ID 500, it is assigned a Point Source ID of 500. In addition, this new point has its SYNTHETIC flag set to 1 (TRUE). Notice that in the merged file, each individual point can be traced back to its origin.

7.2.9.20 Waveform Data

LAS 1.3 introduced the storage of "waveform" data to the specification. This capability continues, unchanged in LAS 1.4. A waveform is essentially a digitized representation of the return intensity of a laser pulse with respect to time. The waveform nature of the return was indicated in Figure 7.11.

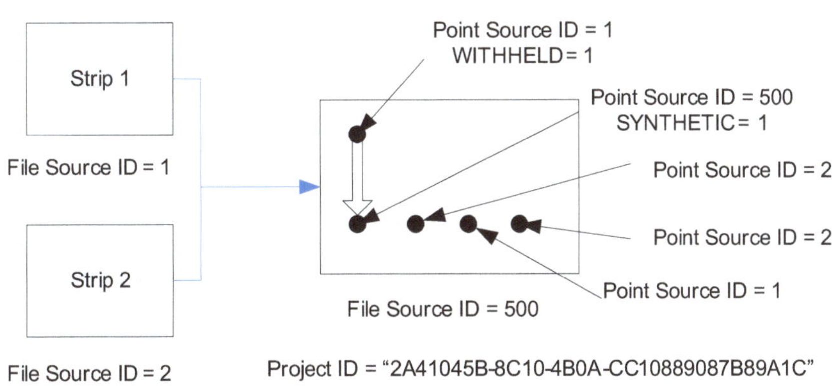

Figure 7.14 A composite file with a "moved" point.

Primarily for convenience and to reduce data storage requirements, earlier lidar systems analyzed waveform return data in front-end electronics and declared the potential location of an impact with an object space object. As digitization hardware improved and software algorithms became more sophisticated, there was a drive from both hardware vendors and data processors (particularly the academic community) to make waveform data available to processing software. Hence, this capability was introduced in LAS 1.3.

Rather than store the entire waveform, waveform packets spanning "interesting" events are stored. An interesting event could be anything the hardware vendor decides but is typically amplitude above the noise level within the region of interest range (e.g. if the nominal flying height above ground is 2,000 meters, events 200 meters from the aircraft would not normally be collected).

To allow dual use of the data (e.g. point mode and waveform mode), each waveform packet is indexed by at least one point (see Figure 7.15). This allows software based on discrete returns to continue to function. There is a misconception that this encoding scheme does not allow "returnless" waveform packets to be encoded. However, this is easily achieved by the hardware vendor (in post-processing software) tagging desired "returnless" waveform packets with a synthetic return.

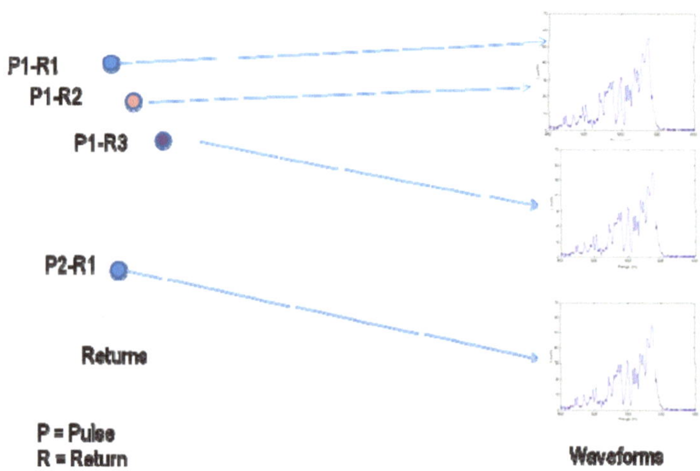

Figure 7.15 Waveform Encoding.

An inspection of the encoding of the waveform packet scheme reveals that we constructed a hybrid approach to storing waveform information. A return pulse can:

- not point to a waveform packet
- be the sole pointer into a location within a waveform
- Point into a waveform packet pointed to by one or more other returns (first example of Figure 7.15).

A single LAS file can encode up to 255 waveform packets schemes. These schemes are encoded within Wave Packet Descriptors (stored in VLRs). A waveform packet descriptor contains the fields listed in Table 7.8. As of version 1.4, compression of the waveform data is not yet supported. The temporal sample spacing is in picoseconds, providing a maximum resolution of 0.03 cm.

Table 7.8 Waveform Packet Descriptor User Defined Record.

Item	Format	Size
Bits per sample	unsigned char	1 byte
Waveform compression type	unsigned char	1 byte
Number of samples	unsigned long	4 bytes
Temporal Sample Spacing	unsigned long	4 bytes
Digitizer Gain	double	8 bytes
Digitizer Offset	double	8 bytes

Each point indexing a waveform packet includes first order coefficients for "walking the wave." This allows you to shift the location of a declared point using a first order approximation. The common operation here would be to analyze a wave packet for the location where you would like to declare a return by shifting the default return of a sensor biased toward high returns (for example, in high grass) to the first rise in amplitude. You would compute the X, Y and Z values for your new return by starting at the closest sensor declared return on the wave packet and using the first order X(t), Y(t), Z(t) coefficients to position the new pulse. Finally, this new pulse would be marked as *synthetic*.

The exploitation of waveform data in a meaningful way is just starting to appear in algorithms of commercial software (e.g. TerraScan from Terrasolid). As time goes on, we expect this to become more prevalent.

There has been some criticism of the waveform storage of LAS 1.3/LAS 1.4. These criticisms range from too much data being stored to too little data being represented. However, this is a bit of cart before the horse. As commercial software vendors add meaningful processing algorithms, we will get a better handle on what actually needs to be exchanged. Additionally, a full understanding of the nature of the problem is necessary prior to making changes. For example, we store wave walking coefficients today on a per return basis. Would a single set of coefficients suffice? A detailed analysis of the curvature of the pulse (caused primarily by atmospheric refraction) would be required prior to making this change.

7.2.9.21 (Extended) Variable Length Records

The LAS format has always provided a facility for users to store metadata that is "private" in nature (meaning that other software will simply ignore these records). With the exception of waveform data, prior to LAS 1.4, these Variable Length Records (VLR) have occurred (in the file) prior to the Point Data Records. A VLR can hold a payload of up to 65,535 bytes (because the data packet size is limited to 16 bit offsets). To legitimately use a VLR, a user must request a VLR "Key" from the LAS working group (www.lasformat.org). This Key, along with a Record ID number, allows one to uniquely identify a VLR. Note that a registration key is only needed when one intends to add a custom metadata packet to a LAS file (which, for most users, would be quite rare).

For a number of uses, the length of the VLRs proved to be a serious limitation (for example, when using this facility for storing spatial index information). Thus in LAS 1.4, we made the Extended Variable Length Record available for general use (prior to LAS 1.4, an EVLR was used only for Waveform Packet information). EVLRs are intended to be placed after the point data records. An EVLR payload size is represented by a *long long* (64 bits) so this should definitely solve size limitations!

Unfortunately, VLRs/EVLRs are designed to be sequentially placed in a LAS file (rather than daisy chained). Thus all VLRs should be sequentially placed ahead of the point data records and all EVLRs should be sequentially placed after the point data records.

There are several VLR/EVLRs reserved for LAS Specification use. An example is the encoding of coordinate reference system (CRS) information. For the legacy Point Data Record Formats (PDRF) 0-5, this CRS encoding can be either GeoTIFF or (preferred) Well Known Text (WKT). For PDRF 6-10, the CRS encoding must be WKT. An important note to software developers is that the LAS defined records can be either stored as VLRs or as EVLRs. Thus when searching a file for the CRS records, it is necessary to search the EVLRs if a CRS record was not found in the VLRs.

7.2.9.22 Extra Bytes

A new LAS Variable Length Record (VLR[6]) has been defined called "Extra Bytes." This VLR allows one to parse a Point Data Record that has been augmented with extra fields. As noted earlier in this article, a standard LAS Point Data Record can have data tacked on to the end of each point data record. Prior to LAS 1.4, a writing software would have to communicate to a reading software the content of this extra data using some previously agreed private scheme. The Extra Bytes metadata provides the ability to specify, for each field,

- Field Name
- Description
- Data Type (all standard type as well a several array types)
- A No Data signature
- Minimum permissible value
- Maximum permissible value
- Scale
- Offset

Flags are included to indicate if various attributes apply to a specific field (such as minimum and maximum values.

This facility will prove quite useful for those who wish to make use of "off the shelf" lidar software that ingests LAS format data but need to augment that data for other uses. For example, Extra Bytes will provide a canonical method of added "Sigma Z" to a point data record.

It is the responsibility of LAS software writers/readers to properly index point data records in LAS files. A software writer must *always* use the Point Data Record Length field in the Public Header Block to determine the size of the point records in a given LAS file. Simply counting bytes for a particular PDRF will result in an indexing error if extra bytes have been appended to a record. Software that is rewriting a LAS file containing Extra Bytes should simply copy these bytes to preserve the records.

It should be noted that while Extra Bytes is a tremendous improvement over no representation of augmented point data records, we do not anticipate that commercial software will attempt to build the semantic engines necessary to handle mixed mode Extra Bytes (that is, multiple files each with possibly different Extra Bytes metadata). Of course, this issue is certainly not unique to LAS!

7.2.9.23 A Well Formed LAS 1.4 File

This section is primarily my opinion so it will all be written as first person. There is a bit of a difference of opinion among designers of LAS software as to how ambiguous conditions should be handled. Commercial software purveyors tend to focus on error handling whereas the open source camp is more focused on attempting to continue to read the files. My obser-

---

[6]Note that I refer to these LAS Spec records as VLRs for convenience. They can actually be encoded as either VLRs or EVLRs.

vations, based on working the LAS format since inception and working with dozens of lidar processing companies, is that ambiguous conditions must be treated as fatal errors. This is primarily based on the observation that production managers usually do not want production technicians making decisions regarding potential error resolution. Nothing is quite as embarrassing as delivering a final lidar project in the wrong coordinate reference system! Thus the following guidelines are recommended for software authors:

1. While LAS 1.4 will support the encoding of Point Data Record Formats (PDRF) 0-5 (the legacy formats) such that, under certain conditions (described in an earlier section) software that has not been updated to LAS 1.4 could read the file, I strongly recommend that this facility never be used for production-type scenarios. There is no need at all for using this facility since one can simply write out a file in one of the legacy formats, if compatibility with these prior formats is needed.
2. In light of 1, the following fields should be always be set and maintained as zero:
   a. Legacy Number of Point Records
   b. Legacy Number of Points by Return
3. Always use Well Known Text (WKT) encoding for the Coordinate Reference System (CRS). GeoTIFF has not been maintained in terms of new CRS and vertical representation is often not supported. Note that you will need to set the WKT bit of the Global Encoding field in the Public Header Block.
4. Add an Extra Bytes VLR when augmenting Point Data Records. This eliminates the problem of trying to figure out why your Point Data Record Format 0 is longer than 20 bytes!
5. Never store both a GeoTIFF and WKT VLR in the same file. There is very often not a one-to-one mapping from GeoTIFF to WKT and thus an intrinsic error exists.
6. Try to avoid using anything other than powers of 10 for the scale values in the Public Header Block. Floating point for X, Y, Z has been purposefully avoided in the LAS definition. Non-power of ten scaling makes this a pain to avoid when writing software.

7.2.9.24 Error Conditions

While we attempted to make the LAS 1.4 format unambiguous, we were not completely successful. Again, I will resort to first person in this section since what I say should be an error, other might say to take a different path. Remember, my focus is on rigorous error detection in lidar production environments. In any of the below error conditions, the software should, of course, provide an exact message as to the reason an error has been declared.

1. While I strongly recommend against using the *legacy* payload capability of LAS 1.4, one should treat a difference between a non-zero legacy field and its equivalent standard field as an error. For example, if Legacy Number of point records is not zero and is not the same value as Number of point records, an error should be declared.
2. Global Encoding bit WKT is set (non zero) but a valid WKT VLR does not exist (even if a GeoTIFF packet does exist)
3. PDRF is 6 or greater and the WKT bit is not set (this is explicitly pointed out as an invalid LAS file in the LAS 1.4 specification). This error must be declared even if a WKT VLR is found in the file.
4. Number of point records does not equal the sum over Number of points by return
5. Return Number (in a Point Data Record) is zero

In my opinion, any data inconsistencies within a LAS file should be treated as a "hard" error, meaning the reader will exit with an error rather than continue to process the file with a warning or, even worse, no message at all. After declaration of a hard error, the file should be fixed "off-line" and reintroduced to the production workflow only after correction. This opinion is based on my experience of seeing many embarrassing data deliveries caused by improperly formed LAS files (particularly in the area of coordinate reference system issues).

7.2.9.25 What is Next for LAS

When looking back at a prior article on LAS 1.1, I notice that we have now (at LAS 1.4) added most of the features that were in the "what's missing" section of that article (RGB, waveform, etc.).

I think now our (the LAS Working Group) focus needs to be on flexibility. The LAS format "hard codes" the various Point Data Record Formats (PDRF). While, with LAS 1.4, these formats can be extended with "Extra Bytes", this is still a mechanism that requires inside knowledge of the semantics of the data fields.

Several years ago, we proposed a more flexible LAS format during the design of LAS 2.0 (a version of the specification that was ultimately withdrawn). Here the approach was similar to the "Extra Bytes" facility of LAS 1.4 but applied to every field in the PDRF (similar to an XML encoding but binary). This type of encoding will eliminate the repetitive PDRF scheme and allow us to simply specify different formats based on a set of standard fields. It will also allow for different data types to be used for some of the standard fields, thus either expanding the fields (for example, allowing X, Y, Z to be specified as 64 bit ints) or contracting fields (allowing intensity to be encoded as an 8 bit field).

Perhaps the most needed addition at the moment is spatial indexing and pyramiding. Currently a number of venders (including my own companies) perform these performance enhancing additions via VLR encoding and/or auxiliary files. It will be very beneficial to standardize these facets so that a software package from vendor B can read an indexing scheme from vendor A. Of course, there are many ways to accomplish these additions so a consensus approach will be a challenge!

As it stands, LAS 1.4 is a fairly robust format for conveying point cloud data. Based on what I see on the hardware side, this version should give us a few years of breathing room before we need to do another update to the specification.

REFERENCES

Agarwal, P. K. Arge, A. and Danner, A., 2006. From Point Cloud to grid DEM: A scalable approach, *Proceedings of the 12th International Symposium on Spatial Data Handling*, pages 771–788, Springer-Verlag.

David N., Mallet C. and Bretar F., 2008. Library Concept and Design for lidar Data Processing, *Proceedings of GEOBIA 2008, ISPRS Archives XXXVIII*, Part 4/C1, 2008, Alberta, Canada.

Graham, L., 2012. The LAS 1.4 Specification, *Photogrammetric Engineering & Remote Sensing*, Volume 78, Number 2.

Hamming, R., 1973. *Numerical Methods for Scientists and Engineers*, Second Edition, pages 19–40. Dover Publications, Inc., New York, New York

Maune, D. F. , 2007. Introduction, *The DEM User's Manual*, 2nd Edition, edited by D. F. Maune, 8–12, Bethesda, Maryland: ASPRS

National Archives, accessed November 2011, http://www.archives.gov/records-mgmt/initiatives/temp-opmedia-faq.html)

OSTA, Optical Storage Technology Association, accessed November 2011, http://www.osta.org/osta/index.htm

Schroeder B. and Gibson G., 2007. Disk Failures in the Real World, Pages 1–16 in *5th USENIX Conference on File and Storage Technology (FAST 2007)*, Feb. 13-16, 2007, San Jose, California

AUTHORS

Section 7.1 was written by Jamie Young, Senior Manager – LIDAR Solutions for AeroMetric, Inc.; Section 7.2 was written by Lewis Graham, President, GeoCue Corporation.

8

LIDAR DATA AND COMPLIMENTARY TECHNOLOGIES

MARTIN FLOOD, MICHAEL S. RENSLOW, ROBERT T. PACK,
and QASSIM ABDULLAH

8.1 LIDARGRAMMETRY – USING STEREO MODELS GENERATED DIRECTLY FROM LIDAR POINT CLOUD DATA

"The importance of lidargrammetry and its impact on the cost-benefit model for lidar mapping cannot be overstated."[1]

Since the early 1990's, laser altimetry ("lidar") has established itself as a reliable, accurate, and cost-effective method for generating high resolution, high accuracy elevation models of the Earth. It has seen wide-spread adoption in the commercial sector as both a complementary and competitive technique to older photogrammetric methods of creating terrain models. Its operational flexibility and high resolution, high accuracy data capture capabilities have also allowed it to expand into non-traditional applications outside of standard DEM/DTM generation. In parallel, new data processing and data analysis techniques are being released each year that allow users to extract greater information from the basic lidar point cloud. One such technique that has been adopted is the generation of stereo models directly from the point data for exploitation in traditional stereo compilation workflows, a technique usually referred to as "lidar-stereo" or "Lidargrammetry."

Lidargrammetry addresses a known limitation when working with lidar point cloud data; the difficulties associated with accurately extracting linear features and sharp edges whose dimensions are less than the resolution of the point data set. In other words, while it is easy to see where the road is in the point cloud, it is much harder to accurately position the edge of the curb. To compound this problem, early commercial lidar sensors provided only spatial information, the X-Y-Z locations of the returns from the target surface, with no other target object information. The ability to capture the amplitude of the return signal, usually referred to, incorrectly, as the intensity[2], was introduced by sensor manufacturers starting in 2000 and rapidly adopted by commercial providers. However, intensity data was not initially used to generate true commercial mapping products directly from the lidar data. Rather, it was predominantly used to create ancillary data, such as rasterizing the point data to create greyscale pseudo images of the project site, or as secondary data to improve automated classification of the point data. In part this was due to the lower repetition rates, 10 kHz to 25 kHz, in use by commercial sensors at the time. These repetition rates usually resulted in data sets too sparse for the intensity values to be useful for automatic feature extraction routines

[1] National Research Council, 2007. *Elevation Data for Floodplain Mapping*, Washington D.C., Committee on Floodplain Mapping Technologies, The National Academies Press, p 78.

[2] Intensity is a measurement of flux and is not synonymous with "strength" or "amplitude." It is typically expressed in power per unit area (e.g., W/m^2) (or in photometry and radiometry as power per unit solid angle). Commercial lidar "intensity" values are typically a scalar value representing the amplitude of the return signal normalized to the receiver specifics. For consistency we will maintain the colloquial usage of "intensity" here.

or for generating high-resolution rasters. With the more recent deployment of lidar sensors with repetition rates of 200 kHz to 500 kHz or more, it has become much more effective to exploit the intensity-capture capabilities of the sensors to generate more traditional imagery-based map products. Lidargrammetry is one such example of this type of exploitation. In this context we are referring to "Lidargrammetry" as the technique of using the intensity information contained in the lidar point cloud as the source data for generating stereo models of the project area by rasterizing the point data itself, rather than generating a stereo model from any ancillary imagery. However, ultimately we would expect the term lidargrammetry to evolve to also include the compilation of features from the point cloud when viewed directly in a true stereoscopic 3D environment without the need for any intermediary rasterization.

A primary driver of the development of current lidargrammetry techniques has been to create improvements to overall production efficiencies when working with lidar point data. Even with the increasing sophistication of automated classification algorithms for point data, lidar data can still require significant manual editing to finish the map product. A technique such as lidargrammetry, which reduces the manual editing time required, has an obvious impact on overall production efficiency and improved product quality. The technique is also useful in determining if an end-to-end *lidar-only* workflow is practical for a particular project, potentially eliminating the need for concurrent imagery to be captured. Finally, capturing the cost reductions lidargrammetry creates is of obvious benefit to both data producer and data end users.

While the lidar stereo approach is to a certain degree a "backward-looking" solution, after all the point cloud is already 3D, so a process that rasterizes the source point cloud to then create a traditional stereo model for 3D exploitation is arguably redundant: it does offer several advantages. By creating appropriate stereo pairs from the intensity images, it is possible to generate stereo models that can be directly exploited using existing photogrammetric workstations and the traditional photogrammetric workflow, particularly in areas such as feature extraction or breakline compilation. Using the lidar point data to generate stereo models also allows data providers to improve the overall efficiency of their lidar point data processing while enabling established mapping firms to continue to exploit their existing infrastructure when working with lidar data sets. Data providers can directly integrate proven (i.e., accepted) photogrammetric workflows with the accurate elevation data available from the lidar sensor, rather than having to rely on less well-established, and usually less well-understood, point cloud manipulation techniques. By leveraging established workflows, lidargrammetry enables the deployment of lidar technology while still exploiting an organization's existing environment (infrastructure), current skill sets (people) and preserving client satisfaction with proven and accepted processes (reputation).

8.1.1 Workflow

The workflow requirements for implementing lidargrammetry are relatively straightforward. There are no additional field data collection constraints imposed compared to a standard lidar data collection, beyond verifying intensity information will be collected[4] and maximizing point density within the limitations of the other project technical specifications. Point densities of greater than 10 points per square meter are desirable if lidargrammetry is to be used

---

[3] The techniques discussed here are based on algorithms developed by GeoCue Corporation and implemented in their GeoCue software suite. GeoCue has a patent pending on its imaging techniques, including lidar stereo or "lidargrammetry" and related "synthetic stereo" generation methods.

[4] This is almost always the case for the current generation of sensors. Older legacy data sets may not include the intensity values.

during post-processing. After the field data collection is completed, the normal preliminary processing to generate the geometrically correct point cloud, in LAS format, is followed.

At this point the workflow splits in to two separate channels. The point cloud data will be processed using the standard approach of automatic classification routines to, for example identify the bare earth points, followed by manual editing and clean-up of the classified data by editors working with specialized software designed for this task. In parallel, or even as an alternative to working with the point cloud at all, stereo models will be generated in an appropriate software tool and then these models will be passed to standard photogrammetric stereo editing workstations for feature collection and extraction of breaklines. Note this is a fundamentally different approach from importing the lidar point data into a stereo compilation environment to overlay it on a traditional stereo model. In our lidar stereo case, the stereo model itself is being generated from the point data and no separate stereo model is required. The technique used to accomplish this lidar stereo model generation is discussed in more detail in the next section. Working in a standard photogrammetric stereo editing environment, editors will compile and extract the required features from the stereo model, for example compiling breaklines (See Figure 8.1). The compiled features will then be passed back for use during the generation of surface models from the lidar point data. This step may be followed by additional product generation, such as contours or DEMs, as the project dictates. In certain cases when the point cloud data itself is not a deliverable, it may be possible to skip the point classification and editing steps complete and simply extract features from the stereo models. Simplified examples of these workflows are included below (Figures 8.2 and 8.3).

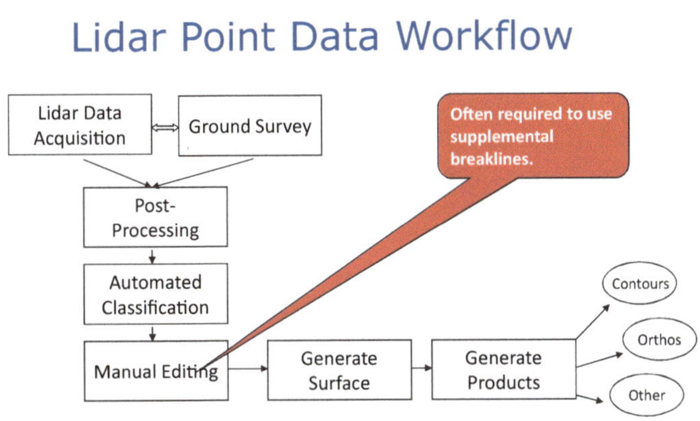

Figure 8.1 Lidar Point Data Workflow.

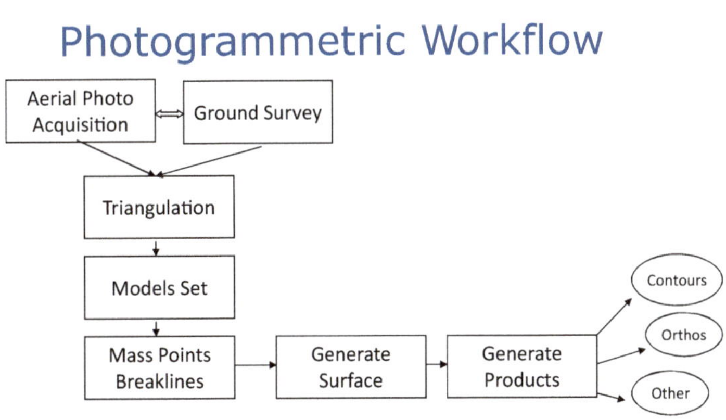

Figure 8.2 Photogrammetric Workflow

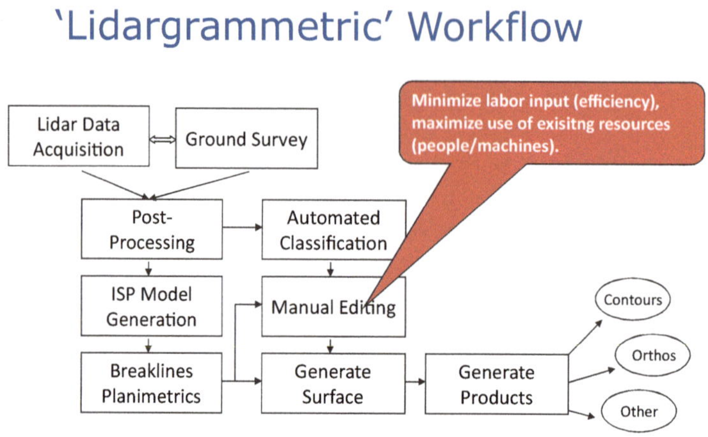

Figure 8.3 Lidargrammetric WorkflowGenerating Lidar Stereo Models

8.1.2 Generating Lidar Stereo Models

There are established techniques for rendering an orthorectified image into stereo using an ancillary elevation model. The "stereo mate" technique works by creating a second image from an orthorectified source image and a corresponding elevation model. However, with lidar point data, no true image exists. Rather the source data consists of actual 3D '"object space" points. In normal stereo vision systems, a pair of 2D images, each taken from a slightly different perspective is used to derive 3D object space points using triangulation. Technically, lidargrammetry works by reversing this process, taking 3D object space points and rendering an inferred pair of 2D images. The images are generated as if captured at positions determined using a pseudo *base-height* parameter and related parameters of the data set. As a result, the algorithms required for lidargrammetry are fundamentally different from the stereo mate process, although stereo mate techniques have been used to generate models from original 3D data. The process involves introducing parallax (displacement of the image pixels in the X direction) and generating the inferred images. The parallax introduced is proportional to the difference of the actual elevation value for a pixel from a fixed elevation value (average) across the image. The object information being used to generate the image, usually the intensity, determines the value of the pixel.

Once generated from the 3D object space points using appropriate algorithms, the inferred stereo pair (ISP) can be used directly in the traditional photogrammetric workflow. When applied to lidar data, the readily available 3D elevation values and additional co-registered object information, such as intensity, make the lidargrammetry concept relatively easy to implement. Practically, ISP generation can be applied to any set of 3D point data, ideally with associated object information suitable for generating the image. The source data is input into a software tool capable of generating the inferred images and implementing them with the appropriate mathematical model for the particular stereo viewing software to be used in downstream production. The resulting lidar stereo model inherits the level of accuracy as the lidar data itself; in other words, the lidar stereo rendering supports the same planimetric and vertical accuracy of the source lidar data.

To properly implement these ISP techniques with lidar data requires a dense data set for accurate image creation, hence the need for higher repetition rate lidar sensors to efficiently exploit the technique. It should also be noted that setting proper sensor parameters during the field data collection work is critical to intensity image quality and the resulting quality of the stereo image. As could be expected for any stereo image process, areas with low vegetation or obscured ground surfaces can cause some problems with the resulting stereo models, but these are usually far less severe than the corresponding issues with models generated from traditional imagery. While the lidargrammetry technique is currently being applied primarily to airborne lidar data sets used in the generation of map products, it is generally applicable to any 3D point cloud data set and can, for example, be used to generate a stereo model of a building façade captured using a terrestrial (ground-based) laser scanner or a mobile (vehicle-based) scan.

8.1.3 Benefits

Once generated, the intensity images can be viewed as a stereo model in standard photogrammetric software. Since the resulting images are orthographically correct, production technicians can use established viewing, measurement, and compilation techniques to extract breaklines and features from the image, exactly as if working with a standard stereo model created from traditional imagery. The extracted features are three-dimensionally correct at time of collection. This creates several significant advantages when using the technique with lidar point data. It allows for an extremely short turnaround time from sensor to exploiter (often only a matter of hours). In fact, when properly set-up and with the appropriate integrated software tools, stereo imagery can often be available ahead of the completion of automated and manual editing of the lidar point data itself. Given the large amount of point data usually collected by lidar sensors and the fact even the most sophisticated automated classification routines don't provide 100 percent accurate feature extraction, manual editing is required to finish map products generated from lidar data. The manual editing of the point data, even if just for final clean-up and QA/QC, can be extensive enough that the time required actually exceeds the time required to extract the same features directly from lidar stereo imagery using traditional photogrammetric methods.

The lidargrammetry technique also allows stereo models to be created independent of lighting conditions, including generating images from data acquired at night and/or under clouds. Obviously, the fact the production operator can exploit the images with existing stereo display hardware and software is a significant benefit to organizations with established infrastructure. Since a standard stereo model is created, *automated* 3D feature extraction techniques can be more easily implemented than in the point domain alone. Features in the lidar point cloud can be readily screened and identified using attributes such as morphology, color, and elevation in the stereo model and the information extracted or fed back into the point cloud processing environment.

It is also important to note that the collection area, the area coverage per ISP, can be much larger than for traditional imagery, so fewer set-ups are needed per project. Lidar stereo models can be generated from point clouds of 200+ million points (equivalent to a 15,000 × 15,000 pixels image). Thus, a lidar stereo model can be used to visualize huge amounts of point data in a single stereo viewing session. As an example, a pilot lidar project of 600 square miles with 1 billion lidar points was covered by only six stereo models. Thus, it becomes very efficient to inspect very large point data sets using exiting photogrammetric workstations. Since the stereo image is directly generated from the point data itself, there is no need to use hybrid techniques to import the raw lidar points into the stereo environment to, for example, find anomalies in the point classification routines. The resulting vector data set is usually very light and easily *roamable*. Qualitative assessments from early projects with ISP images generated from lidar point data also indicate that shorelines are not as easily obscured by vegetation, so actual horizontal positioning of a water body is more accurate than when compiled from corresponding imagery. Experience also indicates that the production operator can measure more accurate water levels and hydrographic breakline locations and flow direction is very easy to determine.

An added advantage when using software for ISP generation that is tightly integrated with the stereo viewing/editing software is that production operators can use iterative refinement schemes for selective improvement of specific areas of interest in the image. Since the stereo models are generated synthetically from single pass data (no "stereo overlap" required), one can set any desired base-to-height ratio. Regeneration of "problem" areas or areas of interest with greater stereo exaggeration is easily accomplished. For example, a production operator can very quickly regenerate a stereo model over a particularly portion of a project area and increase the base-to-height ratio to increase the stereo exaggeration. This can allow the operator to easily detect and follow even small relief changes in that area of the image, such as shallow ditches in relatively flat areas (see Figure 8.4).

The lidargrammetry technique also provides for enhanced interpretability of lidar point cloud features by a production operator. Feature extraction from stereo images has been demonstrated to provide a much faster manual editing environment for a trained production operator versus working with a traditional lidar point displays. While still a relatively new technique on the production floor to date, lidargrammetry has been applied to lidar data to allow rapid breakline delineation for terrain modeling (improved contours), planimetric feature capture, and for final QA/QC of results from automated point classification routines (bare earth extraction).

8.1.4 Summary

In concluding our discussion of the lidar stereo technique, it is important to note that the value of lidargrammetry has been acknowledged in independent studies by various government and professional organizations. Most notably in February 2007, the National Research Council's Committee for Floodplain Mapping Technologies presented a report to Congress that included the recommendation to adopt lidargrammetry as a key tool when acquiring elevation data for floodplain mapping[5]. The committee recommended using lidargrammetry methods to capture the required breaklines and planimetric features needed to supplement the lidar point elevation data when doing flood modeling. This approach removes the necessity to collect aerial photography in addition to lidar data, significantly reducing the overall cost of a floodplain mapping program. The committee also argued that there is still significant unrealized potential in lidar technology due to techniques such as lidargrammetry and these should be fully exploited to reduce costs and improve overall product quality. The committee did caution however that, due to the relatively recent emergence of lidar stereo techniques, the corresponding planimetric accuracy of lidargrammetry compared to photogrammetry needs further study. While there are few independent quantitative results published, qualitative evidence indicates breaklines collected from lidar stereo images generated from 2-meter posting lidar data is equivalent to 1:2400 scale imagery (horizontally).

Using lidargrammetry to exploit lidar point data sets in a traditional photogrammetric production environment has become an established technique, but much work is being done to validate the technique and look at areas for further improvement. These include working with increased resolution lidar imagery, examining "class-based" stereo images (point data classification is translated directly in to color-coding in the stereo image) and using color images created by modulating elevation values with intensity data or classification values with intensity data. Since the technique is independent of the object information used to create the *image*, the generation of stereo models from lidar imagery fused with other data, for example hyperspectral data, is also of significant interest. The increasing availability of point clouds with colorized points due to passive RGB data collection in the senor is opening-up other interesting avenues of research. In mainstream production, lidargrammetry offers obvious advantages as a supplemental technique to automated feature extraction, whether the auto-

[5] National Research Council, *Elevation Data for Floodplain Mapping*, Washington DC: The National Academies Press, 2007.

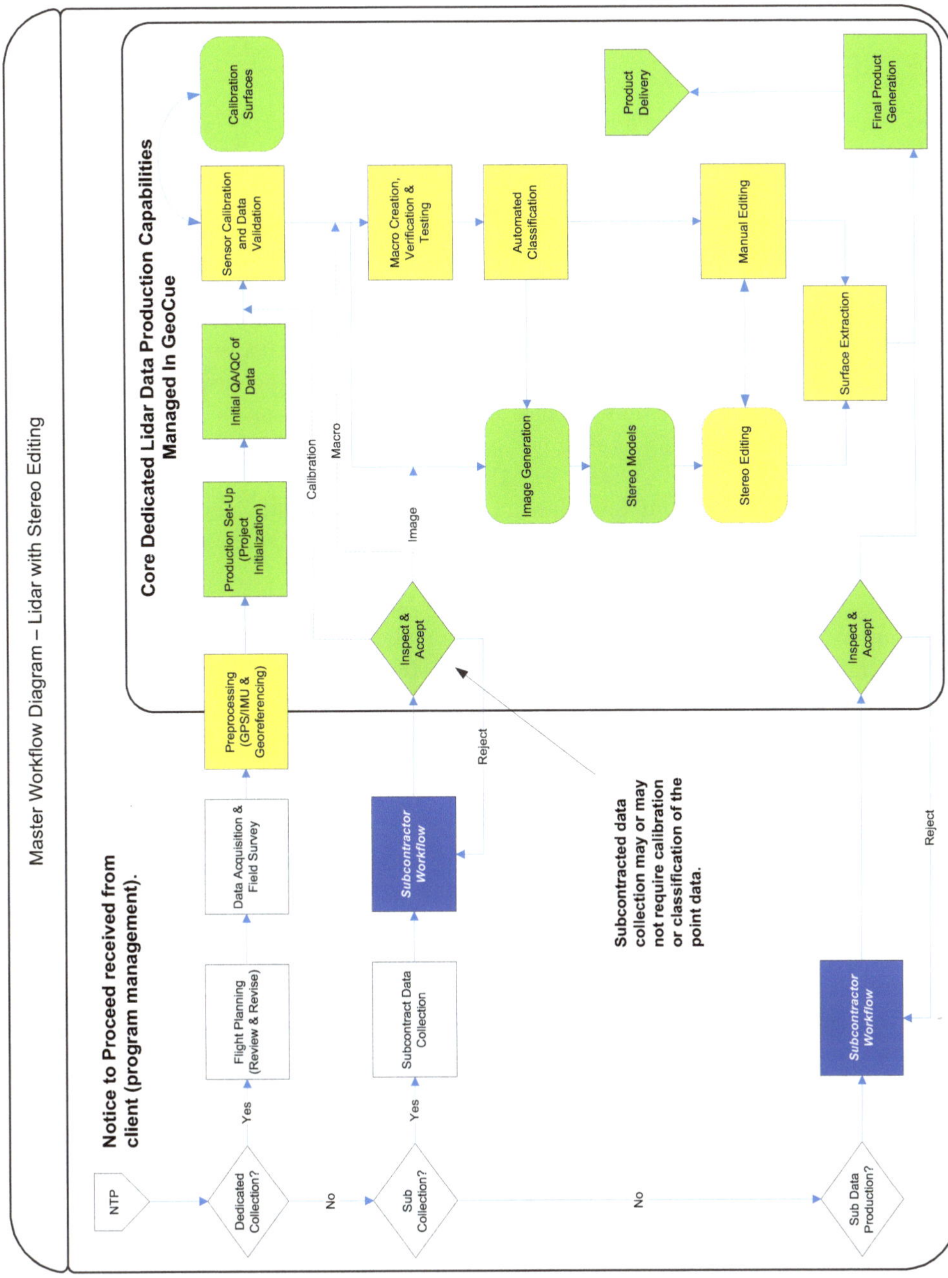

Figure 8.4 Master Workflow Diagram: Lidar with Stereo Editing.

mation is done in the point domain or the image domain. Since the technique can be applied retroactively, work with legacy/historical lidar data sets to add additional value or improve previous results is also possible.

With these lidar stereo techniques rapidly developing, and given the relatively onerous and less intuitive manual editing environments that exist in the point domain, lidargrammetry techniques offer a way for mapping firms to eliminate the difference between "lidar editors" and "stereo compilers" and deploy lidar technology with far less disruption to their established workflow.

8.2 COMBINED PHOTOGRAMMETRIC AND LIDAR MAPPING PROCESSES

For most large mapping projects, lidar and photogrammetry interact to generate geospatial data which is cost-effective, time-efficient, accurate, and often more complete. Being an active sensor technology, lidar provides the means of capturing mass points where traditional image-based stereo photogrammetry is often not able to determine the ground in vegetated or shadowed areas. Working together, photogrammetric techniques provide the means to edit lidar data, add breaklines, and validate quality.

In the last ten years, the accuracy and capability of lidar technology has increased dramatically. The ability to generate dense, high-accuracy point clouds, stable collection equipment, proven processing techniques, and speed of delivery have made lidar an attractive mapping solution. The latest airborne lidar sensors pulse at hundreds-of-thousands of points per second usually with multiple returns and pulse intensity data to create rich and very dense point clouds depicting features on the Earth's surface. The ability to process the large lidar data files has kept pace with the acquisition development, validation, and point cloud classification. The LAS lidar data standard developed and maintained by ASPRS provides a common data transfer standard and a point classification scheme (ASPRS, 2012).

Also during the past several years, the development and subsequent dominance of softcopy photogrammetry for general mapping, editing, and orthophotography altered the traditional topographic mapping approach and introduced methods that can support lidar mapping. The ability to edit lidar bare earth models while adding breaklines using lidar-derived intensity images in stereo (lidargrammetry) to support breaklines under vegetation are relevant examples of merging these two technologies.

Each technology has strengths and weaknesses, as well as fundamental differences. For example, photogrammetry is based on the stereo matching of two overlapping images, and lidar is based on a range measurement to a point from a single position. Also, the software and algorithms used in photogrammetry have been designed for smooth landscape modeling, whereas lidar depicts surface roughness and very minor elevation differences (Smith et al., 2004). Table 8.1 is a general comparison of traditional photogrammetry and lidar strengths and weaknesses (after: Habib, 2009; Renslow and Mercer, 2003). As lidar data collection is becoming very dense (8 to 20 points/m^2), some of these comparisons are changing.

8.2.1 LIDAR

Lidar is an acronym for Light Detection and Ranging (although currently "lidar" is mostly considered a word, similar to radar and sonar). A lidar system consists of the integration of two technologies, Direct Georeferencing (INS and GPS) and a laser scanner/detector. Basically, a pulsing laser ranging system is mounted in an aircraft with the INS/GPS system. Modern laser systems produce hundreds of thousands of pulses per second at a specific frequency; each pulse has a duration of a few nanoseconds (10^{-9} sec). The individual pulses are sent towards the Earth where it is reflected off a feature(s) back towards the scanner in the aircraft where it is captured by a receiver.

Figure 8.5 is an illustration of typical airborne lidar data collection. The aircraft is equipped with a sophisticated GPS for x, y, and z sensor location and an inertial system for measuring the angular orientation on the sensor with respect to the ground (combined technologies for "direct georeferencing" (Lillesand and Kiefer, 2000). The GPS in the aircraft has been initialized with a base station established at a known point, and must be 'locked in' during the data collection. Using the very accurate pulse time tags and the speed of light, the distance/range to the reflection can be accurately calculated (see Equation 1). Most lasers use an oscillating mirror which creates a sawtooth pattern of points perpendicular to the aircraft flight line. Some lidar systems use a rotating mirror which produces reflection points in a circular pattern or an evenly spaced pattern.

The essential measurement made by a lidar sensor is of time (nanoseconds): the time that elapses from the moment the pulse is emitted until it returns after being reflected by the target surface. Because the laser pulse travels at the speed of light, a known constant, time can be directly converted to distance, by the following equation:

$$R = v\left(\frac{t}{2}\right) \tag{8.1}$$

where R is the slant distance or range from the sensor to the target and back, v is the speed of light (a known value), and t is the elapsed time interval (Shan and Toth, 2009).

Table 8.1 Comparing Photogrammetry and Lidar

Photogrammetry	Lidar
Day time only photography collection (passive technology)	Day or night data collection (active technology)
Complicated and often unreliable matching procedures	Direct acquisition of 3D coordinates
Vertical accuracy is worse than planimetric accuracy	Vertical accuracy is better than planimetric accuracy
Established accuracy standards	No established accuracy standards
High Redundancy	No inherent redundancy
Dense positional information along object space breaklines	Almost no information along breaklines
Rigorous ground control required	Almost no ground control required – relies heavily on direct georeferencing
Difficult to map remote inaccessible areas, such as mud flats and extremely steep terrain	Well suited for large homogeneous areas, steep terrain, and shadowed areas
Must be able to see the ground in 3D – typically not possible in forested areas.	Penetrates vegetation to map the bare earth
Very human dependent	Most of the processes are automatic
Creates a single data layer	With multiple returns creates 3D volumetric information (i.e., vegetation structure)
Capability to create orthophotos from imagery	Requires additional imagery to create orthophotos
Typically slow and expensive	Relatively fast and cost effective for large areas.
Accurate hydrographic topology	Often difficult to capture details
All types of features visible mapped accurately	Difficult to map edges of linear and polygonal features.

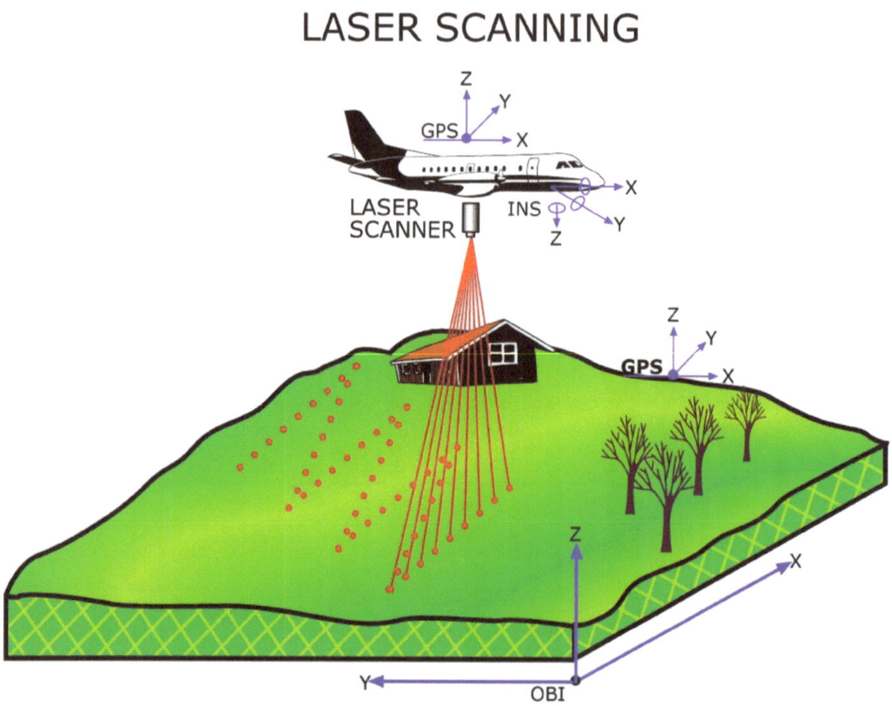

Figure 8.5 Illustration of a typical airborne lidar data collection.

Much has been written documenting the development of mature lidar systems for topographic mapping (Fowler et al., 2007). Until recently, lidar technology has focused on increased pulse rate, whereas now, the manufacturers have developed supporting solutions (for example, Multiple Pulses in the Air (MPiA)) and lidar sensors designed for specific applications (corridors, transmission lines, low-altitude/ultra-dense data, general mapping, and wide-area mapping) (Leica Geosystems, AG, 2011; Optech, Corporation, 2011; Riegl, Inc., 2011). The density of the lidar collection is customized for the specific application. Typical, large-scale projects are flown with a fixed wing aircraft with one to four points per square meter; engineering projects are flown with a helicopter at low altitudes with up to 80 points per square meter.

Lidar postprocessing requires that the raw data be validated, put into real-world coordinates, and classified to the LAS standard (see Romano et al., 2007). With a reliable point cloud of classified lidar data, the extraction of surface models can proceed. Since lidar is a collection of mass points, the user must determine if breaklines are necessary to create a meaningful digital surface model of the bare earth. This decision depends on the terrain, i.e., are there abrupt changes in the terrain that need to be identified to meet the application and accuracy requirements.

8.2.1.1 Breaklines

Breaklines are added to the lidar digital surface model using photogrammetric methods. Typical features requiring breaklines are roadway edges, hydrographic features (streams, river edges, open water), top/bottom of a steep slope, levees, large building footprints, and retaining walls. If stereo photographic data is available, then breaklines can be added in a softcopy environment with the lidar points superimposed on the 3D stereo model. The operator can then edit any observed lidar artifacts and add breaklines at the same time (see Figure 8.6). If stereo photography is not available, then another approach is to utilize lidargrammetry.

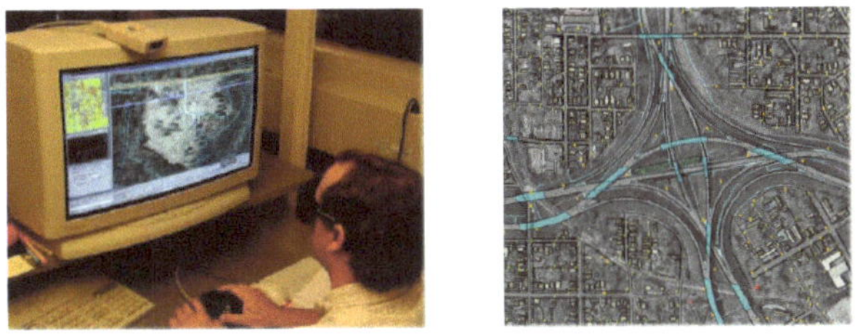

(a) (b)

Figure 8.6 (a) Compilation of breaklines from lidar-derived stereo imagery, and (b) example of breaklines, building footprints, and bridges.

8.2.1.2 Lidargrammetry

Imagery is not a direct product from lidar, although the intensity (infra-red reflection values) of each lidar 3D point can be converted to simulate a raster image. Figure 8.7 shows raw lidar points, a TIN model, and a lidar intensity raster for the same area. Lidar intensity (infra-red reflections) is quite different than normal sunlit radiation imagery collected by aerial photography. Lidar intensity (active sensor technology) can be collected at any time (day or night), and aerial imagery (passive sensor technology) requires reflected sunlight. For example, the infra-red lidar reflections (i.e., points) are nearly 100 percent absorbed by water bodies. Lidar intensity has very low reflectance off of asphalt, and high reflectance off of concrete. Depending on the reflectance angle the lidar intensity may result in variable grayscale tones. The resolution of the lidar intensity image is limited to the resolution of the lidar data. Even with these conditions, a psuedoscopic stereo pair (combination of the intensity image with 3D values) may be created from the lidar intensity and be used for compiling breaklines on the bare earth model (Flood, 2006). A first return-only lidar intensity stereo pair can be useful to map and classify general land use catagories.

In general, lidargrammetry works very well as long as there is an adequate density of points on the bare earth surface (Fowler et al., 2007, pp. 227–230). In areas of dense vegetation, the lidar mission planning must begin with very dense collection parameters (6 to 10 pts/m2) to assure adequate penetration beneath the canopy. The GIS and CAD tools within the photogrammetric workstation (especially vertical exaggeration) can be used to compile 3D features, such as breaklines, terrain features, and hydrologic feature enforcement. The operator can also make any required edits to the lidar data for final finishing of the digital surface model. Thus, it is possible to use lidar as a sole source for data extraction in a traditional photogrammetric workflow without the additional need for significant compiler staff training.

In some cases, breaklines are compiled from stereo imagery using conventional imagery and photogrammetric processes, and lidargrammetry is used to finish the editing and breaklines in areas of dense vegetation. One drawback of this approach is that another dataset is introduced to the mapping solution with a different source and likely reduced accuracy. The cost and extra time required for this approach should be considered and the process and validated accuracy well documented in the project's metadata. Lidargrammetry does bring the advantages of stereo-compilation to lidar projects, but should be considered a tool and not always the total solution.

8.2.2 Summary

The strength of lidar technology to place a very dense blanket of points on the top surface of the Earth and be processed to classify the bare earth and separate other classifications (water, buildings, vegetation, noise, etc.) has fundamentally changed the processes for topographic mapping. For many projects requiring only a DEM, lidar has become the technique of choice. Photogrammetry still offers the strength of very high accuracy, completeness in visible areas, and the benefit of established standards. Combining these two technologies and often sharing data (such as in lidargrammetry) offers the geospatial community a robust, complementary solution f.or developing useful mapping products.

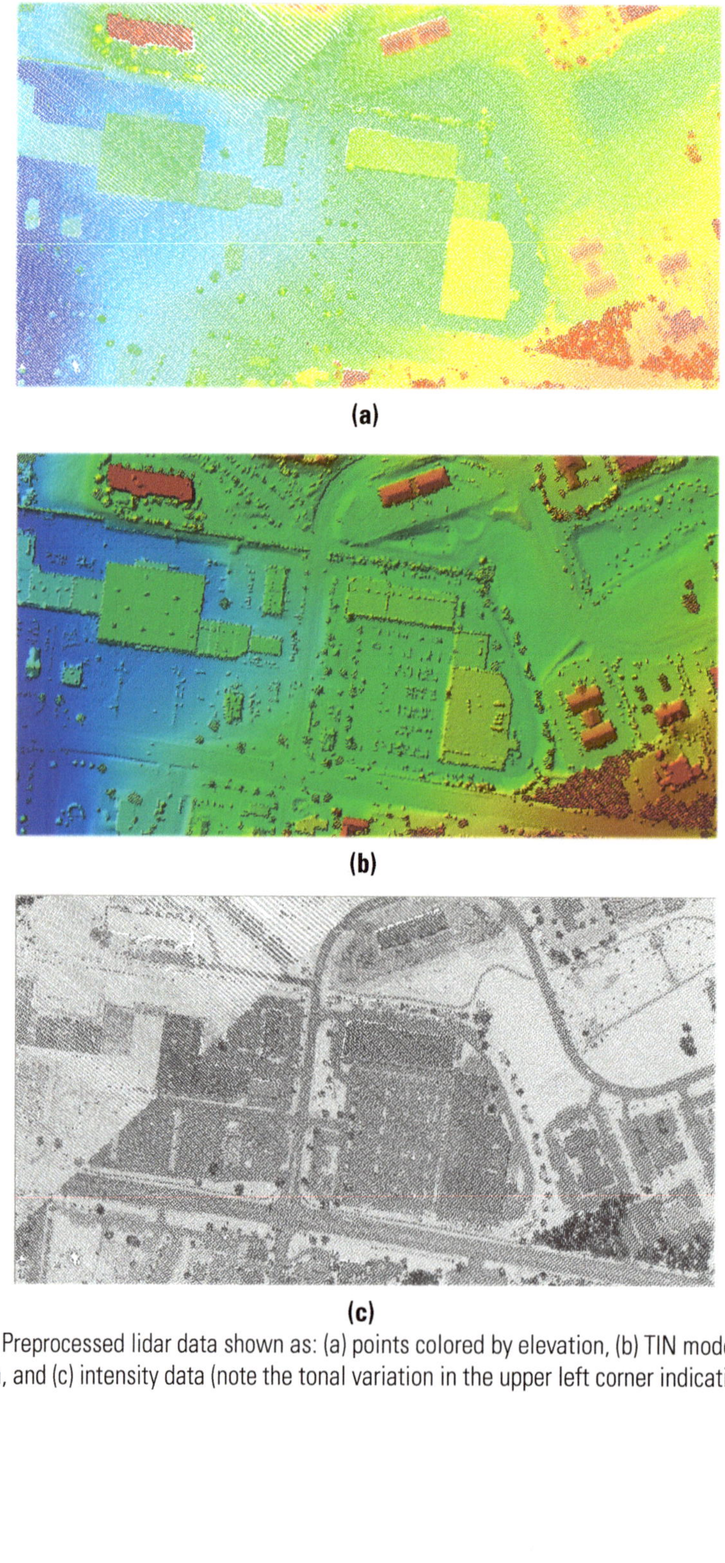

Figure 8.7 Preprocessed lidar data shown as: (a) points colored by elevation, (b) TIN model colored by elevation, and (c) intensity data (note the tonal variation in the upper left corner indicating multiple flight lines).

REFERENCES

ASPRS, 2012. The LAS Format Specification, v. 1.4, American Society for Photogrammetry and Remote Sensing, 10 June, URL: http://www.asprs.org/a/society/committees/standards/LAS_1_4_r12.pdf (last date accessed: 14 June 2012).

Flood, M., 2006. "Lidargrammetry" combines the best of two worlds, GeoPlace.Com, URL:http://www.geocue.com/resources/geocue_LidargrammetryBestofTwoWorlds.pdf (last date accessed: 06 February 2011).

Fowler, F.A., A. Samberg, M.J. Flood, and T.J. Greaves, Topographic and terrestrial lidar, *Digital Elevation Model Technologies and Applications: The DEM Users Manual*, Second edition (D.F. Maune, editor), American Society for Photogrammetry and Remote Sensing, Bethesda, Maryland, pp. 199–252.

Habib, A.F., 2009. Integration of photogrammetric and LiDAR data for accurate reconstruction & realistic visualization of urban environments, *Proceedings of FIG Commission 3 Conference*, 4 February, Mainz, Germany, URL: http://www.fig.net/commission3/mainz_2009/programme2-Dateien/TS%208.1_Habib_AKAM_FIG_GERMANY.pdf (last date accessed: 6 February 2011).

Leica Geosystems, AG, 2011. URL: http://www.leica-geosystems.com/en/Airborne-LIDAR_86814.htm (last date accessed: 6 February 2011).

Lillesand, T.M., and R.W. Kiefer, 2000. *Remote Sensing and Image Interpretation*, Fourth edition, John Wiley & Sons, Inc., 724 p.

Optech, Inc., 2011. URL: http://www.optech.ca/ (last date accessed: 06 February 2011).

Renslow, M., and B. Mercer, 2003. Workshop 7 – Performance, Accuracy, and Economics of Imaging Sensors, Lidar and IFSAR for DEM Generation, *Technology: Converging at the Top of the World*, ASPRS 2003 Annual Conference, 6 May 2003, Anchorage, Alaska.

Riegl, Inc., 2011. URL: http://www.riegl.com/nc/products/airborne-scanning/ (last date accessed: 06 February 2011).

Romano, M.E., 2007. Lidar processing and software, *Digital Elevation Model Technologies and Applications: The DEM Users Manual*, Second edition (D.F. Maune, editor), American Society for Photogrammetry and Remote Sensing, Bethesda, Maryland, pp. 475-498.

Shan, J., and C.K. Toth, 2009. *Topographic Laser Ranging and Scanning: Principles and Processing*, CRC Press, Taylor and Francis Group, Boca Raton, Florida, 590 p.

Smith, M.J., F.F.F. Asal, and G. Priestnall, 2004. The use of photogrammetry and lidar for landscape roughness estimation in hydrodynamic studies, *Proceedings of the XXXV ISPRS Congress, Istanbul, Turkey.*

8.3 LIDAR/IMAGE COMBINATIONS

8.3.1 Introduction

Lidar by itself is a powerful technology. However, when one adds imagery to it, the combination provides an even more powerful capability to visualize and interpret objects in a way the either technology alone cannot. Successful use of this combination relies on: (a) collecting lidar and image data of the same area, either during separate missions, or at the same time, (b) co-registering and combining the two datasets together, and (c) visualizing and exploiting the combined datasets.

This section discusses the various types of lidars and cameras, the way they can work together, and the various algorithms that can be used to co-register data. The various unique ways that the combined data can be used will also be presented.

8.3.2 Lidar/Imager Systems

The first goal is to accurately determine which lidar points match pixel locations on the images. The simplest case is when the two datasets are acquired during separate missions. In this case the two datasets are combined later through the processes of global co-registration. When the two sensors are mounted together and the two datasets are acquired during the same mission, more streamlined methods can be used to combine these datasets. However, it is more complicated than the registration of two photographs because:

1. Biaxial co-mounted systems, i.e., systems that are mounted side by side with separate optics, introduce perspective-related relative scene displacements even when perfectly synchronized. This so-called parallax is used to advantage in photogrammetry but must be dealt with in matching lidar points with image pixels. Figure 8.8a illustrates this concept.
2. Asynchronous collection introduces absolute scene displacements and distortions of lidar relative to images. Most laser scanners are relatively slow and collect points over a much longer period than cameras do. Co-mounted sensors typically have small data acquisition time offsets associated with the frame rate of each sensor. Of course when the two sensors are deployed separately, they have large data acquisition time offsets. Figure 8.8b illustrates this concept.

Figure 8.8 summarizes the effect of asynchronous and biaxial scanning on what a cube looks like from the perspective of a lidar frame. The cube, because it is moving left to right, is skewed increasingly to the right as the scan progresses from top to bottom. Only when a lidar is instantaneous and coaxially combined with a camera will the two sensors see exactly the same scene at the same time.

The slowest lidar systems use flying spot scanners that direct one laser shot at a time along one axis while the lidar head is either rotated or translated along a second axis so as to cover a scene with shots, e.g., the Riegl LMS Q-560 airborne lidar system(Riegl, 2012). Some lidar systems can handle both axes of scanning, e.g., the electro-magnetic fine steering mirror in Utah State University's targeting lidar built for the US Navy (Pack et al., 2009). In either case, because only a single spot is sensed at a time, it can take several seconds to acquire the entire scene, even at laser shot rates of 500,000 shots per second or higher. These systems require more work in co-registering their data with imagery.

Flash lidars have faster frame rates and consist of either linear detector arrays (*line-flash* lidar) or area detector arrays (*area-flash* lidar). Line-flash lidars spread a single laser pulse across an entire line that is captured simultaneously on a line sensor array, e.g., the line array in the US Navy's CMRTR sensor (Palm et al., 2007). The second dimension of the scene still needs to be covered by either rotating or translating the entire sensor head but the time to collect a frame can be reduced to a fraction of a second. Targeting systems like the CMRTR system have a relatively narrow field-of-view and the lidar data are typically displayed as a range image. Motion compensation algorithms are used to "correct" the range image distortion so that is has the same geometry as a camera image.

The fastest frame rates are achieved by area-flash lidars, e.g., Stettner et al. (2004). In this case, a single high-power laser pulse is spread over the entire scene (just like a camera flash bulb) and the reflected radiation is collected simultaneously on a two-dimensional detector array. These systems are the most like an imager in that a whole scene is collected at the "push of a button." If the frame rate of the lidar is matched with the imager, they are then synchronous and the two frames can be matched without any knowledge of scene motion or geometry.

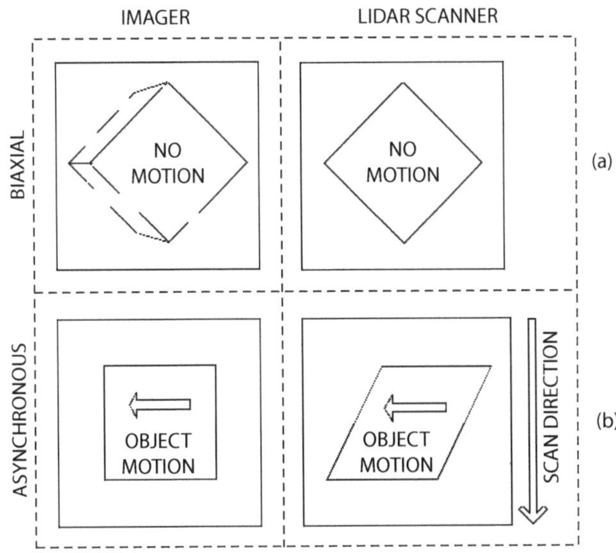

Figure 8.8 Illustration of (a) the effects of parallax on image displacement with a biaxial configuration, and (b) the effects of object motion on image distortion due to asynchronous operation.

Following is a summary of the four specific lidar/imager combinations that pose unique challenges to data co-registration and examples of each that have been developed at Utah State University.

1. Synchronous and coaxial combinations that do not require compensation for motion or parallax. Data is processed in the same way that data from multiple cameras are treated. This type of combination includes lidar systems that have an internal color channel that colorizes each lidar shot, e.g., the Riegl LMS-Q280i sensor that, for each return, provides 8-bit red, green, and blue intensities (Huang et al., 2009). Another type is coaxial flash systems that use a hot or cold mirror to align their optical paths. The downside of combining optical paths is that both signals suffer energy loss where the optical paths are combined. An example of this type of combination is a small hand-held lidar designed for indoor use (Boldt et al., 2007). Figure 8.9a shows a photo and a diagram illustrating the configuration of this unit.
2. Synchronous and biaxial combinations that need compensation for parallax. Area-flash lidars are typically mounted biaxially next to a camera and are an example of this combination. Another less common approach involves using a computer vision camera that reads out a small sub-frame that corresponds with the predicted location of a given laser spot in the field of view. The sub-frame exposure time is synchronized with the laser shot time. The camera is fixed to a single-shot lidar transceiver and the whole unit is scanned along the two axes of a pan-tilt unit mounted to a tripod (Pack et al., 2005). The sub-frame patches can be merged to create effective 3D imagery. Figure 8.9b shows a photo and diagram illustrating the configuration of this unit.
3. Asynchronous and coaxial systems that rely on the use of special optics that can combine two optical paths into one. In a targeting lidar system, the coaxial configuration provides the advantage of being able to use the camera image to point at objects that one would like the lidar to scan. Because the systems are asynchronous, motion compensation is still needed. This approach was used for the "Eye-safe Ladar Testbed" built for the U.S. Navy (Pack et al., 2009), Figure 8.9c shows a photo and diagram illustrating the configuration of this unit.

4. Asynchronous and biaxial systems consist of a lidar and camera that work independently but that are usually time-tagged together. These are the most common in the commercial world. In the broadest sense, the configuration applies to cases where the two systems are flown separately. In all cases, compensation for both parallax and motion is required. This approach was used in the airborne lidar/multispectral camera system used by USU for natural resource mapping (Pack et al., 2011)) Figure 8.9d shows a diagram an example of this type of combination.

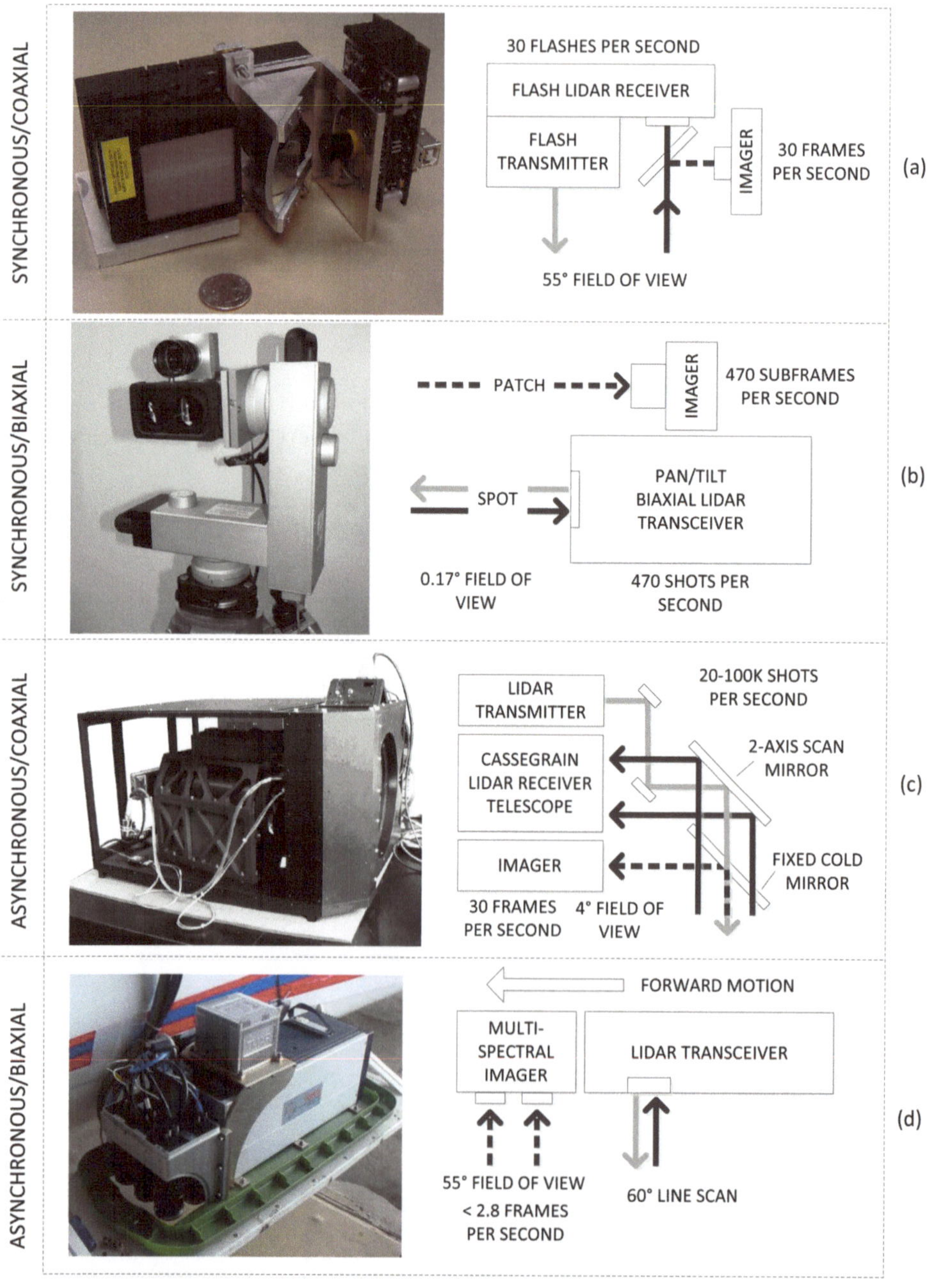

Figure 8.9 The four classes of lidar/imager combinations as illustrated by systems developed at Utah State University.

8.3.3 Co-Registration Algorithms

8.3.3.1 Introduction

Co-registration allows camera pixels and lidar shots to be mapped into a common reference frame. If image pixels are mapped into the lidar-frame, these pixels can be used to create draped 3D surfaces, colored 3D points, and other GIS products. If lidar pixels are mapped into the image-frame, a lidar range image can be created that matches a corresponding camera image. In both cases, a rich combination of lidar and image data is made available for exploitation.

All lidar scanners rely on some kind of position and orientation (POS) system to keep track of where in space the laser is pointed at a given time. The orientation component of the POS system consists of either a pan-tilt mount in the case of a tripod, or an inertial measurement unit (IMU) in the case of a mobile or airborne system. The position component consists of either a kinematic GNSS system or, in the case of a tripod, a surveyed point. As the lidar transmits laser energy into the scene, its position and orientation is accurately tracked such that lidar-derived XYZ points can be computed. This requires the lidar be accurately aligned with the POS using a calibration procedure that relies on an analysis of overlapping point clouds (Skaloud and Lichti, 2006).

The mounting of a camera next to the lidar requires that they be aligned by either by (a) computing independent calibrations to a common POS and using direct georeferencing (DG) techniques, or (b) matching camera points with lidar points in order to compute a relative orientation. Both approaches are explained below.

8.3.3.2 Co-Registration Using a Common Position and Orientation System

This approach aligns the camera with the lidar scanner by mounting them to a common POS. The problem then becomes one of computing two alignments with the POS, one for each sensor. This is illustrated in Figure 8.10a.

As already mentioned, lidar scanners (other than flash lidar) depend on a POS, so this calibration is already available. The alignment of a camera to the POS is calibrated by comparing the POS-estimated camera pose with that obtained by an independent and more accurate method. Originally, ground control points (GCPs) and tie points between overlapping images have be used to aerotriangulate a camera pose for each image. These results could then be compared with each POS estimate and a correction computed. The down-side to this approach is that the establishment of GCPs is expensive as is a test field with preferably signalized control points. A more efficient calibration procedure, called integrated sensor orientation, uses tie point observations to correct the camera to POS alignment (Jacobsen, 2004; TerraPhoto, 2009; Khoshelham, 2009). This is achieved by a simultaneous adjustment of a number of tie points within a bundle adjustment model with additional constraints for refining the alignment. Since every tie point appears in at least two images, the estimation model results in a redundancy of observations that allows for the correction of the alignment.

Once the lidar and camera are both aligned with the POS, no point matching or ground control registration is needed. This approach is popular in airborne mapping applications where the POS consists of a GNSS/IMU combination.

8.3.3.3 Co-Registration Using Matched Features

Where a POS system is unavailable, lidar/image data combinations are co-registered using feature matching techniques. Using matches in overlapping areas to identify tie-points, a relative orientation (RO) can be computed using aerotriangulation (AT) and block adjustment techniques. Four steps are commonly followed: feature detection, feature matching to obtain tie points, transform model estimation, and transformation (Ronnholm et al., 2007). How this basic strategy is implemented varies. Sometimes two iterations are used, one for initial RO and another for accurate RO. Once oriented, each pixel in an image can be mapped into the 3D coordinate system of the lidar. Figure 8.10b shows a diagram of the geometry for this type of co-registration.

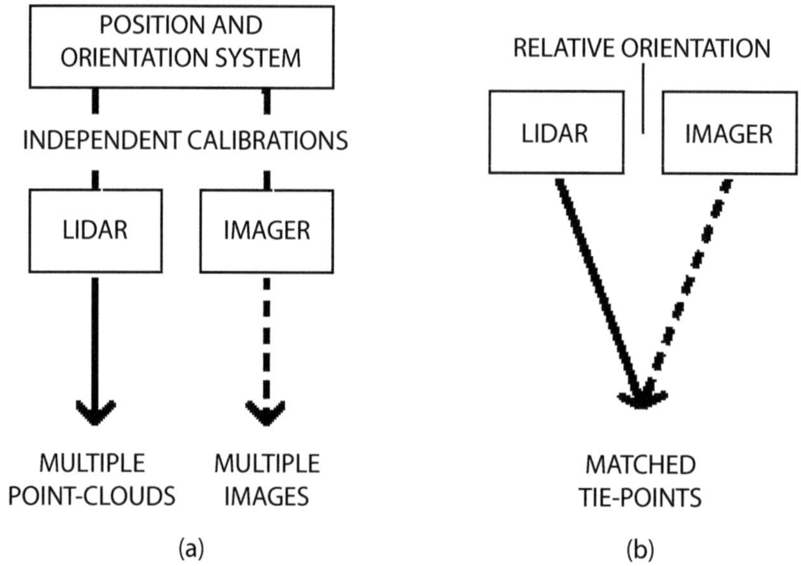

Figure 8.10 Illustration of the two fundamental methods for co-registering lidar/image combinations: (a) through independent calibration with the position and orientation system, or (b) through relative orientation using tie-points.

Common features must be identifiable and accurately measured in both datasets. Depending on the nature of the scene, this can vary in difficulty. If natural features are lacking, synthetic targets may be needed to improve matching fidelity. Extracting tie-features typically follows one of three strategies: (a) 3D lidar to 3D image feature matching; (b) 3D lidar to 2D image feature matching; or (c) 2D to 2D feature matching.

The first strategy, in which 3D lidar features are matched with 3D image features, first requires stereocompilation techniques to obtain the 3D image points. For example, Habib et al. (2004) match photogrammetrically derived 3D lines representing the intersection of roof planes with intersecting planar patches derived from lidar. A 3D similarity transform was then used to compute the RO given these matches.

The second strategy is where 3D lidar points are matched with 2D image points. Wang and Neumann (2009) demonstrate a robust algorithm of this type that matches connected line segment features identified in both aerial images and lidar data. Their paper also includes an excellent review of a variety other feature types and methods that can be used with this strategy.

The third strategy first transforms the lidar data into a 2D range image such that well-know image-to-image matching techniques can be used. Once matched, the 2D image to 3D lidar co-registration is maintained and an RO transformation can be achieved. Visualization of features in the range image is primarily through (a) colorization of range, (b) colorization of elevation, or (c) gray-scale colorization of lidar return intensity. For example, Lee et al. (2011) demonstrate the use this strategy to automatically register lidar intensity data with Ikonos-2 and QuickBird satellite images.

Sometimes a combination of the above strategies can be used to advantage. For example, Swart et al. (2011) co-register panoramic images with lidar by (a) compiling a dense point cloud photogrammetrically from an image pair, and (b) matching this point cloud with the lidar point cloud using the popular iterative closest point (ICP) matching algorithm. This 3D to 3D matching strategy is then followed up with 2D to 2D feature matching between images and lidar return intensity data. It was demonstrated that the first strategy improves the success of the second.

8.3.3.4 Exploiting the Combined Data

Once combined, imagery can be aided by lidar, the lidar can be aided by the imagery, or the lidar/image combinations can be tightly integrated. An example of using lidar to aid imagery is mono-plotting where a user can read out a 3D coordinate by running a mouse over an image (Ressl et al. 2006). Also, the lidar data can be used to warp and mosaic images into orthoimages and image maps (Pritt et al., 2010). Common examples of lidar being aided by imagery are colorizing of lidar points, creating breaklines from stereo images that improve triangulated surface fidelity, or improving registration of laser scans.

Tightly integrated lidar/image combinations enable automated or semi-automated feature classification and extraction. Feature classification can be pixel-based (Brook et al., 2011; Guo et al., 2011) or object based (Tan and Zhan, 2011).

Samples of the types of features that can be more successfully extracted with a combined dataset than with either dataset alone include buildings (Hermosilla et al., 2011; Rottensteiner et al., 2004 and 2007; Salah et al., 2011; Samadzadegan and Mahmoudi, 2011; Sohn and Dowman, 2007; Xin et al., 2011), shorelines (Liu et al., 2011); roads and railways (Wang et al., 2011; Beger et al., 2011; Clode et al., 2004), and vegetation (Straub and Koch, 2011; Koukoulas and Blackburn, 2007; Maltamo et al., 2006).

For example, Brook et al. (2011) used VIS, NIR, and SWIR cameras along with lidar to discriminate roads and building roofs. A quality assurance (QA) value of 96.2 percent was achieved for roads, and 98 percent for building roofs. It was also found that the difference in surveyed building heights from the lidar/MS derived heights had minimum and maximum values of 0.4 m and 2.1 m, respectively. Combining the datasets resulted in a 3D thematic map.

Guo et al. (2011) used aerial Riegl LMS-Q560 lidar data with a density of 2.5 points/m^2 and co-mounted Applanix DSS 22M multispectral image data to study dense urban scenes. The *Random Forests* classification algorithm was used to classify buildings, vegetation, artificial ground, and natural ground. It was found that the most relevant features for all training classes were the height difference in lidar, the red and blue channels in imagery, the lidar return intensity, and the echo cross-section of the lidar waveform. The classification accuracy per class worked out to 82 percent using just RGB, 92 percent with just multi-echo lidar, 93 percent with just full-waveform lidar, and 96 percent with all data sources. Buildings and artificial ground were classified most successfully.

Hermosilla et al. (2011) focus on automatic building detection by combining 0.5 m IRGB (color + infrared) multispectral imagery with lidar data acquired at a density of 0.5 to 2 points/m^2. Two building detection approaches are used: threshold-based detection and object-based classification. For the various industrial, urban, and suburban sites investigated, QA values for the thresholding- and object-based approaches were 66 percent to 92 percent, and 54 percent and 90 percent, respectively, depending on the type of area.

These are but a few examples of studies of automatic feature extraction using combined lidar/image datasets. As shown in the building roof extraction work cited above, success rates can approach 98 percent. However, success rates in the 50 percent range are also reported in the other study. This range in performance is reported in the literature for the other types and classes of features. However, this is an active area of research and therefore continuous improvement in the performance of automatic feature classifiers and extractors is expected.

REFERENCES

Beger, R., C. Gedrange, R. Hecht, and M. Neubert, 2011. Data fusion of extremely high resolution aerial imagery and lidar data for automated railroad centre line reconstruction, *ISPRS Journal of Photogrammetry & Remote Sensing* (in press).

Boldt, B.M., S.E. Budge, R.T. Pack, and P.D. Israelsen, 2007. A handheld texel camera for acquiring near-instantaneous 3D images, *Conference Record of the 41st Asilomar Conference on Signals, Systems and Computers,* Pacific Grove, California.

Brook, A., E. Ben-Dor, and R. Richter, 2011. Modelling and monitoring urban built environment via multi-source intergrated and fused remote sensing data, *International Journal of Image and Data Fusion*, pp.1–31.

Clode, S., P.J. Kootsookos, and F. Rottensteiner, 2004. The automatic extraction of roads with lidar data, *Proceedings of the ISPRS 2004 Congress*, Istanbul, Turkey.

Guo, L., N. Chehata, C. Mallet, and S. Boukir, 2011. Relevance of airborne lidar and multispectral image data for urbane scene classification using Random Forests, *ISPRS Journal of Photogrammetry and Remote Sensing*, 66(1):56–66.

Habib, A., M. Ghanma, M. Morgan, and E. Mitishita, 2004. Integration of laser and photogrammetric data for calibration purposes, *International Archives of Photogrammetry, Remote Sensing and Spatial Information Sciences*, 35(Part B2):170–175.

Hermosilla, T., L. Ruiz, J. Recio, and J. Estornell, 2011. Evaluation of automatic building detection approaches combining high-resolution images and lidar data, *Remote Sensing*, 3(6):1188–1210.

Huang, H., P. Gong, X. Cheng, N. Clinton, and Z. Li, 2009. Improving measurement of forest structural parameters by co-registering of high resolution aerial imagery and low density lidar data, *Sensors*, 9:1541–1558.

Jacobsen, K., 2004. Direct/Integrated Sensor Orientation - Pros and Cons, *International Archives of Photogrammetry and Remote Sensing*, XXXV(B3):829–835.

Khoshelham, K., 2009. Role of tie points in integrated sensor orientation for photogrammetric map compilation, *Photogrammetric Engineering & Remote Sensing*, 75(3):305–311.

Koukoulas, S., and G.A. Blackburn, 2007. Mapping individual tree location, height and species in broadleaved deciduous forest using airborne lidar and multi-spectral remotely sensed data, *International Journal of Remote Sensing*, 26(3):431–455.

Lee, J., C. Lee, and K. Yu, 2011. Autoregistration of high-resolution satellite imagery using lidar intensity data, *KSCE Journal of Civil Engineering*, 15(2):375–384.

Liu, H., L. Wang, D. Sherman, Q. Wu, and H. Su, 2011. Algorithmic foundation and software tools for extracting shoreline features from remote sensing imagery and lidar data, *Journal of Geographic Information System*, 3:99–119.

Maltamo, M., J. Malinen, P. Packalen, A. Suvanto, and J. Kangas, 2006. Nonparametric estimation of stem volume using airborne laser scanning, aerial photography, and stand-register data, *Canadian Journal of Forest Research*, 36:426–436.

Pack, R.T., P. Israelsen, and K. Sealy, 2005. A co-boresighted synchronized ladar/EO imager for creating 3D images of dynamic scenes, *Proceedings of the SPIE Defense and Security Symposium*, SPIE, Orlando, Florida.

Pack, R.T., B.D. Petersen, and K.F. Blonquist, 2011. Lidar scanning the equatorial jungle of central Africa, *Proceedings of the European Lidar Mapping Forum*, Salzberg, Austria.

Pack, R.T., J. Swasey, R. Fullmer, S. Budge, P. Israelsen, B. Petersen, and D. Cook, 2009. Eye-safe ladar test-bed with coaxial color imager, *Proceedings of the SPIE Defense and Security Symposium*, 7323(03):1–9

Palm, H. C., T.V. Haavardsholm, and H.Ajer, 2007. *Detection of Military Objects in LADAR Images*, Norwegian Defense Research Establishment (FFI).

Pritt, M.D., M. Gribbons, and K. LaTourette, 2010. Automated cross-sensor registration, orthorectification, and geopositioning using lidar digital elevation models, *Proceedings of the IEEE 39th Applied Imagery Pattern Recognition Workshop* (*AIPR*), IEEE, Washington, D.C.

Ressl, C., A. Haring, C. Briese, and F. Rottensteiner, 2006. A concept for adaptive monoplotting using images and laserscanner data, *International Archives of Photogrammetry and Remote Sensing* 36(3):98–104.

Riegl Laser Measurement Systems, 2012, URL: http://www.riegl.com/ (last date accessed: 01 May 2012).

Ronnholm, P., E. Honkavaara, P. Litkey, H. Hyyppä, and J. Hyyppa, 2007. Integration of laser scanning and photogrammetry, *International Archives of Photogrammetry and Remote Sensing and Spatial Information Sciences,* 36(Part 3/W52):355–362.

Rottensteiner, F., J. Trinder, S. Clode, and K. Kubik, 2004. Using the Dempster-Shafer method for the fusion of lidar data and multi-spectral images for building detection, *Information Fusion*, 6(4):283–300.

Rottensteiner, F., J. Trinder, S. Clode, and K. Kubik, 2007. Building detection by fusion of airborne laser scanner data and multi-spectral images: Performance evaluation and sensitivity analysis, *ISPRS Journal of Photogrammetry & Remote Sensing*, 62(2):135–149.

Salah, M, J.C. Trinder, and A Shaker, 2011. Performance evaluation of classsification trees for building detection from aerial images and lidar data: A comparison of classification tree models, *International Journal of Remote Sensing,* 32(20):5757–5783.

Samadzadegan, F., and F.T. Mahmoudi, 2011. Information fusion of lidar range and intensity data for automatic building recognition, *International Journal of Image and Data Fusion* 2(1):37–60.

Skaloud, J., and D. Lichti, 2006. Rigorous approach to bore-sight self calibration in airborne laser scanning, *ISPRS Journal of Photogrammetry and Remote Sensing*, 61:47–59.

Sohn, G., and I. Dowman, 2007. Data fusion of high-resolution satellite imagery and lidar data for automatic building extraction, *ISPRS Journal of Photogrammetry and Remote Sensing* 62(1):43–63.

Stettner, R., H. Bailey, and R. Richmond, 2004. Eye safe laser radar focal plane array for three-dimensional imaging, *Proceedings of the SPIE Defense and Security Symposium: Laser Radar Technology and Applications IX.*

Straub, C., and B. Koch, 2011. Enhancement of bioengergy estimations within forests using airborne laser scanning and multispectral line scanner data, *Biomass and Bioenergy,* 35(8):3561–3574.

Swart, A., J. Broere, R. Veltkamp, and R. Tan, 2011. Refined non-rigid registration of a panoramic image sequence to a lidar point cloud, *Photogrammetric Image Analysis: Lecture Notes in Computer Science*, 6952:73–84.

Tan, J., and Q. Zhan, 2011. Object-based feature extraction method using airborne laser scanning data and aerial photographs, P*roceedings of the 2011 International Conference on Remote Sensing, Environment, and Transportation Engineering,* Nanjing, China.

TerraPhoto User's Guide, 2009. TerraSolid Ltd., URL: http://www.terrasolid.fi/en/users_guide/terraphoto_users_guide (last date accessed: 01 May 2012).

Wang, G., Y. Zhang, J. Li, and P. Song, 2011. 3D road information extraction from lidar data fused with aeria-images, *Presentation at 2011 IEEE International Conference on Spatial Data Mining and Geographical Knowledge Services,* Fuzhou, China.

Wang, L., and U. Neumann, 2009. A robust approach for automatic registration of aerial images with untextured aerial lidar data, *IEEE Conference on Computer Vision and Pattern Recognition:* IEEE.

Xin, W., L. Yi-ping, J. Ting, G. Hui, L. Sheng, and Z. Xiao-wei, 2011. A new classification method for lidar data based on unbalanced support vector, *Proceedings at the 2011 International Symposium onf Image and Data Fusion*, Tengchong, Yunnan.

8.4 QUALITY CONTROL

The following sections represent different user communities' requirements and the recommended data specifications:

8.4.1 Terrain Modeling Applications

Terrain modeling is a requirement of nearly all lidar projects, spanning a wide range of uses and specifications. Below are the most common terrain modeling applications requested by users.

8.4.1.1 Contour Generation

The dwindling use of paper (hardcopy) maps combined with advancements in 3D terrain modeling software capabilities have driven down the need for traditional contour generation. The demand for contours and contour specifications in RFPs involving lidar data collection continues, however, despite availability of new terrain modeling and viewing methods, such as 3D rendering and shaded relief maps. To create lidar-based contours that meet cartographic and geometric qualities, lidar data with modest post-spacing of around 2 to 4 meters can be augmented with breaklines derived from image-based photogrammetry. If imagery is not available for breakline production, then a "lidargrammetry" approach is possible. In this method, very dense lidar datasets with post-spacing of around 1-meter are used to create detailed stereomates by draping the lidar intensity image over the lidar DEM; these stereomates are then used to generate breaklines using any stereo-softcopy system. Once the breaklines are collected, either through photogrammetry or lidargrammetry, the lidar points can be thinned to a great degree, depending on the complexity of the terrain. The thinned dataset is then merged with the breaklines to create a digital terrain model (DTM), required for modeling the contours. In addition, all lidar points within a buffered distance around the breaklines should be removed to achieve acceptable contours without sacrificing accuracy.

This process makes sense, as that is how we have always modeled contours from a DTM. The issue of utilizing breaklines in modeling contours from lidar data often gets confused, however, as service providers attempt to mix very dense and accurate lidar data with manually-collected and possibly less accurate breaklines. Without buffering or thinning the lidar points close to the breaklines, the contours will appear problematic whenever a lidar point appears right next to a breakline, as it is supposed to even with a photogrammetrically-collected DTM. In constructing lidar-derived DTMs, we should consider all the best practices previously developed and utilized for modeling photogrammetric DTMs in the past. A good quality DTM is achieved by having accurately modeled breaklines, with minimum mass points outside the breaklines when necessary. Lidar indiscriminately collects dense mass points throughout the project area, including on and around the later collected breaklines. Unless lidar data is thinned and breaklines are buffered and cleared from the points around it, it will be very difficult, if not impossible, to achieve cartographically acceptable contours.

8.4.1.2 3D Terrain Modeling

Most modern lidar data users and providers are equipped with 3D modeling software that allows them to model lidar data for different applications, such as flood and environmental hazard analysis, watershed management, etc. Depending on the required vertical accuracy of the model, many applications can utilize lidar data without the need for breaklines. However, for hydro-enforced terrain modeling where the user expects a downhill uniform flow of water, breaklines or manual 3D modeling is required around such water features to assure the effect. For most applications, lidar data with post-spacing of 1 to 2 meters is adequate. Hydro enforcement of a lidar-derived terrain model is still cumbersome and costly, and logical automation is strongly needed in this field.

8.4.1.3 3D Urban Modeling Applications

High density lidar point clouds are widely accepted in user communities who need high definition elevation datasets for applications other than terrain modeling. Those applications include but are not limited to line-of-sight, 3D city modeling, 3D urban fly-through and simulation, and security and emergency planning. Home land security agencies showed strong interest in dense lidar datasets to arrange inter-cities combat planning and high profile security development. In addition, the emerging capabilities of oblique imaging and modeling added more emphases on high quality and high definition elevation data that would be cost prohibitive without the use of lidar technology. In most urban modeling applications, users are more concerned about the definitions and details of the lidar dataset than the centimeter level accuracy. Most 3D city modeling can be achieved with a lidar point density of five to ten points per square meter. We are witnessing an emerging new market for dense to ultra-dense lidar data and many lidar providers are enhancing their operations with sensors to meet this demand. Figure 8.11 illustrates the quality of lidar intensity data with a post-spacing of about 20 points per square meter. It is amazing how the level of detail such data provides.

Figure 8.11 Sample of lidar intensity with 20 points per square meter.

8.4.1.4 Biomass and Forest Modeling

Lidar point clouds are proven to be very effective in studying and modeling forest floor and canopy. Lidar-derived spatial data provides solutions for the following resource management goals:

- Accurate inventory and composition of forested land
- Harvest planning
- Habitat monitoring
- Watershed protection
- Fuel management (for fire management).

The newly introduced technology of full waveform digitization will greatly benefit foresters and other specialists working in the field of biomass monitoring and assessment. It provides the ability for precision analysis that even denser lidar datasets from current systems would fail to achieve.

AUTHORS

Section 8.1 was written by Martin Flood, General Manager (GeoCue Canada), GeoCue Corporation; Section 8.2 was written by Michael S. Renslow, Renslow Mapping Services; Section 8.3 was written by Robert T. Pack, Associate Professor at the Utah State University Department of Civil & Environmental & Engineering; and; Section 8.4 was written by Qassim Abdullah, Chief Scientist at Fugro Earthdata, Inc.

9
ACCURACY STANDARDS AND GUIDELINES

QASSIM ABDULLAH and HANS KARL HEIDEMANN

9.1 FACTORS EFFECTING LIDAR DATA ACCURACY

The quality of lidar products can be assessed in several ways depending on the practices and available means of the data provider. Having an independent set of ground control points is essential to any technically sound approach. Production procedures play a major role in achieving and guaranteeing the high quality every client expects from a spatial data provider. Assuring high quality production practices followed by thorough quality controls will not only result in the desired accuracy, but it will help the bottom line as it saves long rework hours. The following are some of the main technical reviews and practices that must be followed if to achieve accurate products.

9.1.1 Project Planning Stage

Make sure that you follow the manufacturer's recommendation on the limiting acquisition parameters. Give attention to the optimum performance environment for any of the sensors or auxiliary components that make up the acquisition system. Flight line length or flying time on the line, flying altitude, ground control number and configuration, quality of the GPS receiver, grade of the inertial measurement unit (IMU) used for sensor position and orientation determination, and processing software, if any, are all to be considered carefully in deciding on the resulting accuracy.

9.1.2 Sensor Orientation Determination Stage

The exterior orientation parameter of the aerial sensor directly affects the accuracy of the ground mapping. The lidar boresighting process, for example, should be conducted with surgical precision and no compromise in these efforts should be allowed in order to cut corners or reduce costs. The sigma of the differences between flight lines should be within the expected horizontal and vertical accuracies of the dataset. The residuals of the ground control should also fall well within these expected accuracies. Finally, budgeting extra control points to be utilized as blind checkpoints for the boresighting block adjustment is an excellent tool for a non-bias estimate of the final data accuracy. For the lidar sensor orientation, a number of factors contribute to the final accuracy of the ground height determination, including accurate boresighting of the IMU, flying at the correct altitude and flying speed, distance from the GPS base station, calibration of the internal mechanical parts, such as galvanometer, and encoder, and mirror motion.

9.1.3 Height Bias Determination

Modern technologies associated with the GPS, IMU, laser, and microwave scanning provide direct measurement of ground positioning without the rigorous photogrammetric model that provides a means to discover and correct biases. Therefore, all geospatial data collected by non-conventional photogrammetric means, such as lidar or IFSAR, should be subject to height or planimetric bias determination after the processing of data and prior to its delivery to the client. Ground check or control points are the most reliable means to determine and then remove such bias.

Since its introduction to the mapping industry during the late 1990's, lidar faced two major obstacles that slowed its acceptance in every day mapping practices.

9.1.3.1 Lack of Feature Details

Constructing two-dimensional and three-dimensional building models and other land features requires accurate delineation of sharp building edges and this traditionally has been accomplished using photogrammetric stereo compilation. All past attempts to automate the extraction of ground features from dense lidar datasets with a post-spacing of 1 to 3 meters for the purpose of planimetric mapping have failed to meet user's expectations. The main reason for this is the rounded and jagged edges of delineated buildings and other manmade features. Despite some software efforts to employ smart algorithms that correct the geometrical shape of objects, this type of modeling remains less appealing to service providers and customers alike, as it does not meet horizontal map accuracy requirements for large-scale mapping. The ASPRS standard requires buildings be placed within one foot of their true ground position when compiled from a map with a scale of 1" to 100'. Recent advancements in lidar systems enable the collection of ultra-dense lidar datasets with point density of five to eight points per square meter (ppsm), which makes the data more suitable for use in the aforementioned modeling algorithms. The downside of such demanding software requirements is the high cost associated with the aerial acquisition due to the additional flight lines required to collect the ultra-dense lidar data. Traditionally, a lidar dataset used for terrain modeling is collected with a nominal post-spacing ranging between 1-meter and 3-meters or a point density ranging between 1 ppsm and 0.11 ppsm, respectively. The ratio between the data densities of the normally collected dataset and the ultra-dense dataset ranges from 20:1 to 32:1. This is true if we assume the normal density dataset is collected with 2-meter post-spacing or 0.25 ppsm. This does not necessarily translate to the same ratio in cost increase, but it could come very close. The high cost of acquisition coupled with the massive amount of resulting lidar data prohibits, in many cases, the use of ultra-dense lidar data for accurate building modeling, and may encourage providers and customers to consider other means, such as traditional photogrammetric modeling for this purpose. Finally, while ultra-dense lidar datasets may not currently be cost-effective for large-scale modeling, the technology is impressive in the sense of delineating details. A recent acquisition of an ultra-dense lidar dataset with about 6 ppsm over the City of Baltimore, Maryland, reveals great details of the baseball game underway in the Oriole Park Stadium, as shown in Figure 9.1a, b, and c. Baseball fans can easily observe which base was occupied at that moment!

9.1.3.2 Lack of User's Confidence in Lidar Accuracy

During the past several years, a few data providers have delivered one-foot or better contours derived from lidar datasets. Most of these contracts were characterized by the following:

- Contour accuracy and quality were assessed based on the limited availability of ground control points, which makes it impossible to verify the acceptance criteria. Ground control points are essential to the discrete verification of the terrain surface accuracy. Only a few projects involve detailed field-surveyed profiles to verify the accuracy and quality of lidar-derived one-foot contours. This is understandable as tight project budgets in most cases prohibit such field investigation. This lack of verification has helped fuel the debate about whether lidar can support one-foot contour generation. There is, in fact, strong speculation among users as to whether lidar data will ever meet such criteria.

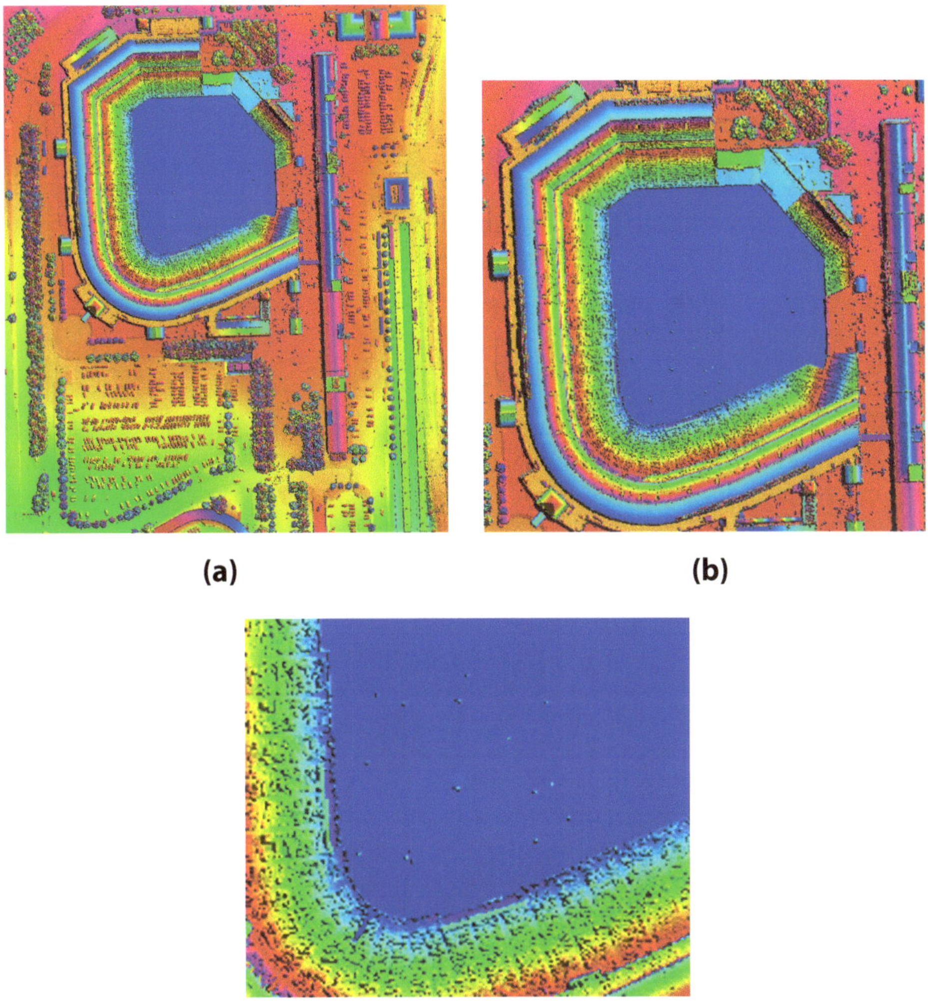

(c)

Figure 9.1 TIN models of Oriole Park Stadium colored by elevation:(a) overview, (b) enlarged to show detail, and (c) baseball players visible on the infield.

- Lidar data providers often are trapped by the conventional process for modeling contours based on photogrammetric practices. Photogrammetric datasets combine highly accurate, manually-collected breaklines with sparsely distributed, as-needed mass points. These breaklines and mass points never contradict each other as they are both collected from the same stereo pairs, by the same operator. This is not the case with lidar data, as it provides very dense mass points regardless of the terrain roughness or slope. What's more, lidar is often combined with photogrammetric breaklines that do not match in accuracy or distribution.

As a result, the lidar-derived contours in many cases do not meet certain cartographic quality standards. Accuracy can also be affected by this practice.

Users and data providers recently engaged in a debate on the following topics:

1. Will the error budget in the different components of the lidar system allow the high accuracy requirement of the one-foot contours?
2. If it does, then will the role of the breaklines be compromised considering the high density of the lidar points? In another words, do we need breaklines when we have a dense lidar surface.
3. If breaklines are needed, then what type and density of breaklines are required to support one-foot contours generation from lidar data?

Later investigations on the subject proved that lidar data can be used for the generation of one-foot contours or better. This ended the debate on the one-foot contours capability, but never laid to rest the issue of breaklines needs and density.

9.1.4 Sources of Error and the Effect on Lidar Data Accuracy

The error budgets in a lidar system involve more than just errors caused by the position of the sensor and the laser ranging. It includes other factors such as mirror shaft encoder sensitivity and stability, atmospheric refraction models, sensor attitudes (rotation angles), thermal stresses and torsion, to name a few. However, most of the later error sources can be modeled accurately and minimized or removed, except for the errors affecting the sensor rotation angles, such as IMU accuracy, reliability, and encoder biases.

Lidar ranging is extremely accurate and most systems can provide ranges with precision of millimeters to a few centimeters from a distance of hundreds or even thousands of miles. Satellite Laser Ranging has been used over the last few decades. The International Earth Rotation and Reference Systems Services (IERS) use laser ranging, with a precision of less than 1 cm, to measure the distance to the geodetic satellite Lageos, which orbits the earth from an altitude of 5,900 km. Such high precision laser ranging was used by the IERS to accurately determine the Earth's orientation parameters. As for the errors caused by inaccurate boresight determination or inadequate Inertial Measurement Unit (IMU) performance, the effect is linearly proportional to the flying altitude and is significant when compared to errors caused by laser ranging. To determine the maximum error due to IMU or boresight performance, let us evaluate the error at the edge of a swath flown with a lidar system with a total field of view of 40 degrees from two different flying altitudes: 3,000 m and 6,000 m.

The horizontal error can be estimated using Equation 9.1:

$$Horizontal-error = \sqrt{(R)^3\left[\Delta\theta_{heading}\right]^2 + 2H^2\left[\Delta\theta_{roll}\right]^2} \qquad (9.1)$$

The vertical error can be estimated using Equation 9.2:

$$Vertical-error = (R)(\Delta\theta_{roll}) = (R)(\Delta\theta_{pitch}) \qquad (9.2)$$

where R = Distance on the ground from nadir in meters, H = Flying altitude in meters, $\Delta\theta_{heading}$ = Heading accuracy in radians, and $\Delta\theta_{roll} = \Delta\theta_{pitch}$ = Roll and pitch accuracy in radians.

Assuming the use of an IMU with an accuracy of 0.005-degree in the roll and pitch and 0.01 degree in the heading, the following errors, resulting from poor IMU and/or boresight performance, can be expected.

Table 9.1 Predicted errors, resulting from poor IMU and/or boresight performance

Flying Altitude (m)	Max. Horizontal Error (m)	Max. Vertical Error (m)
3,000	0.40	0.095
6,000	0.80	0.19

These figures represent the maximum error at the edge of the swath, and therefore one-sigma error should be smaller than those shown in the table above. It is obvious from the error figures in the table that the effect of the attitude errors doubles when we double the flying altitude. This is not the case with the positioning error that is caused by inaccurate airborne GPS. Although poor quality GPS affects the Kalman filter performance of the IMU-measured angles, the effect of the residual errors in the GPS translates to the ground through the sensor position with a ratio of 1:1. In other words, if we have an error of 0.10 m in the vertical component of the lidar system GPS position, we should expect an equal error of 0.10 m in the elevation of the calculated ground position. This is a direct result of the GPS error regardless of the flying altitude. Finally, the correlation between the degradation in accuracy with the increased flying height is mainly due to errors coming from the sensor orientation and not the laser ranging. The laser ranging errors are insignificant when compared to errors caused by inaccurate altitude determination.

9.2 METADATA REQUIREMENTS AND TEMPLATES

The Federal Geographic Data Committee (FGDC) published their Content Standards for Digital Geospatial Metadata (CSDGM) in 1994, and revised it once in June 1998. This standard was established prior to the commercial recognition and acceptance of lidar and does not address the special metadata requirements for lidar data. In October 2011, the FGDC held a Metadata Summit "to bring NSDI Stakeholders up to date on International Organization for Standardization (ISO) metadata, identify challenges, and set a course for implementation." (FGDC, 2011) indicating that the CSDGM would be abandoned in favor of ISO 19115 metadata standards. Although some Federal agencies have begun providing metadata in the ISO format, the adoption of ISO metadata standards is far from widespread, and there are numerous open issues with regard to lidar point cloud data. It is therefore premature to attempt to address the question of ISO 19115 metadata for lidar in detail at this time. This section will focus on how to use the existing CSDGM metadata most effectively for lidar point cloud data in the interim.

Most lidar point cloud users have experienced the frustration of attempting to answer even simple questions regarding lidar data by reading the FGDC metadata. Because the CSDGM does not provide specific entries for basic lidar-specific information such as Nominal Post Spacing (NPS), each vendor or data publisher must place this information in a free text field of their choosing, or perhaps, not include it. This process makes locating the information difficult for a user; worse, it is impossible to automatically locate using software, which greatly defeats the entire purpose of XML-format metadata standards with regard to lidar data.

The USGS-NGP Lidar Base Specification includes both a metadata template and example file which attempt to address this issue. Both are reproduced in the following sections. The template was developed through careful review of the CSDGM, with special consideration for additional information unique to airborne topographic lidar data. A special block of information has been added to store the more commonly needed lidar-specific attributes. This block is currently in review by the National Digital Elevation Program (NDEP) and is expected to undergo some modification to ensure it meets the needs of all Federal agencies. Until it is finalized and approved by the NDEP, it is an *Optional but Recommended* metadata block. After approval, it will be incorporated into the USGS-NGP Lidar Base Specification as a required metadata block. It will also be submitted to the FGDC and the appropriate ISO committees for inclusion in broader metadata standards for lidar point cloud data, as appropriate.

9.2.1 Suggested Lidar-Specific Metadata Tags

```
<lidar>         This is the top-level block. It contains three sub-blocks.
 <ldrinfo>      This block contains information on the lidar data and collection
  <ldrspec>     Published specification(s) the lidar was collected under
  <ldrsens>     Instrument make and model used for the collection
  <ldrmaxnr>    Maximum number of returns that can be detected and recorded
 <ldrnps>       Single-swath Nominal Pulse Spacing of the data (empirical), in METERS
  <ldrdens>     Single-swath Nominal Density of the data, in POINTS PER SQUARE METER
   <ldrfltht>   Flight height, Above Mean Terrain, in METERS
  <ldrfltsp>    Flight speed, in KNOTS
   <ldrscana>   Total scan angle, in DEGREES
   <ldrscanr>   Scan rate, in HZ
   <ldrpulsr>   Pulse rate, in KHZ
  <ldrpulsd>    Laser Pulse Duration, in NANOSECONDS
  <ldrpulsw>    Laser Pulse Width, in METERS
  <ldrwavel>    Laser Wavelength, in NANOMETERS
  <ldrmpia>     Multiple Pulse in the Air 0=No; 1=Yes
  <ldrbmdiv>    Laser Beam Divergence, in MILLIRADIANS
   <ldrswatw>   Swath width, in METERS
   <ldrswato>   Swath overlap, in PERCENT
   <ldrcrshz>   Horizontal Coordinate Reference System, always with units
   <ldrgeoid>   Geoid model
 <ldraccur>     This block contains information on the accuracy of the lidar point cloud data
  <ldrchacc>    Raw lidar horizontal accuracy, in METERS
  <rawfva>      Raw lidar FVA, in METERS
  <clsfva>      Classified lidar FVA, in METERS
  <clscva>      Classified lidar CVA, in METERS
  <clssvas>     Blocks for each SVA landcover type in the project
   <svalctyp>   SVA landcover type
   <svavalue>   Classified lidar SVA, in METERS
 <lasinfo>      This block contains information on the LAS files
   <lasver>     LAS version of the lidar data files
   <lasprf>     Point Record Format used in the LAS files
  <laswheld>    How Withheld (ignore) points are identified within the LAS files
  <lasolap>     How Swath Overage points are identified within the LAS files
   <lasintrz>   The native radiometric resolution of the stored intensity values, in BITS
   <lasclass>   Blocks for each Classification code used in the LAS files
   <clascode>   The classification code (integer)
   <clasitem>   The classification description
```

9.2.2 Suggested Lidar-Specific Metadata Example

```
<lidar>
 <ldrinfo>
 <ldrspec>USGS NGP Base Specification v1</ldrspec>
   <ldrsens>Optech Gemini Airborne Laser Terrain Mappers (ALTM)</ldrsens>
   <ldrmaxnr>4</ldrmaxnr>
   <ldrnps>0.7071</ldrnps>
   <ldrdens>2</ldrdens>
   <ldrfltht>3000</ldrfltht>
   <ldrfltsp>115</ldrfltsp>
   <ldrscana>26</ldrscana>
   <ldrscanr>40</ldrscanr>
   <ldrpulsr>120</ldrpulsr>
   <ldrpulsd>10</ldrpulsd>
   <ldrpulsw>3</ldrpulsw>
   <ldrwavel>1064</ldrwavel>
   <ldrmpia>0</ldrmpia>
   <ldrbmdiv>4.5</ldrbmdiv>
   <ldrswatw>1200</ldrswatw>
   <ldrswato>15</ldrswato>
  <ldrcrs> NAD_1983_UTM_Zone_15N_Meters</ldrcrs>
  <ldrgeoid>National Geodetic Survey (NGS) Geoid09</ldrgeoid>
  </ldrinfo>
  <ldraccur>
```

```
   <ldrchacc>0.5</ldrchacc>
   <rawfva>0.11</rawfva>
   <clsfva>0.09</clsfva>
   <clscva>0.188</clscva>
   <clssvas>
    <svalctyp>Forest</svalctyp>
    <svavalue>0.38</svavalue>
   </clssvas>
   <clssvas>
    <svalctyp>Agriculture</svalctyp>
    <svavalue>0.21</svavalue>
     </clssvas>
   <clssvas>
    <svalctyp>Pasture/Meadow</svalctyp>
    <svavalue>0.15</svavalue>
   </clssvas>
   <clssvas>
    <svalctyp>Scrub/Shrub</svalctyp>
    <svavalue>0.11</svavalue>
   </clssvas>
  </ldraccur>
  <lasinfo>
  <lasver>1.3</lasver>
  <lasprf>6</lasprf>
  <laswheld>Withheld (ignore) points were identified in these files using the standard
           LAS Withheld bit
  </laswheld>
  <lasolap>Swath "overage" points were identified in these files by adding 16 to the
          standard classification values.
  </lasolap>
  <lasintrz>11-bit</lasintrz>
  <lasclass>
   <clascode>0</clascode>
   <clasitem>Never Processed</clasitem>
  </lasclass>
  <lasclass>
   <clascode>1</clascode>
   <clasitem>Undetermined/Unclassified</clasitem>
  </lasclass>
  <lasclass>
   <clascode>2</clascode>
   <clasitem>Bare-earth</clasitem>
  </lasclass>
  <lasclass>
   <clascode>4</clascode>
   <clasitem>All vegetation</clasitem>
  </lasclass>
  <lasclass>
   <clascode>6</clascode>
   <clasitem>All structures except bridges</clasitem>
  </lasclass>
  <lasclass>
   <clascode>7</clascode>
   <clasitem>All noise</clasitem>
  </lasclass>
  <lasclass>
   <clascode>8</clascode>
   <clasitem>Bare-earth Key Points</clasitem>
  </lasclass>
  <lasclass>
   <clascode>9</clascode>
   <clasitem>Water</clasitem>
  </lasclass>
  <lasclass>
   <clascode>10</clascode>
   <clasitem>Bridges</clasitem>
  </lasclass>
  </lasinfo>
</lidar>
```

The importance of properly formatted, factually correct metadata cannot be overemphasized. As the number and rate of lidar collections continue to grow, the need for consistent, reliable metadata that is readable by both humans and computer systems becomes increasingly critical to manage the overall lidar resources for the country. While local and regional collections may feel their needs can be met without such high levels of detail and precision, the ability for users to integrate disparate collections for scientific study and mapping across the boundaries of individual collections is severely handicapped by the lack of consistent metadata content and formatting. Arduous as the task of developing complete, accurate, and compliant metadata may be, the level of effort required to create and maintain metadata is significantly less than that of researching, and re-researching each project by every subsequent user even within the original customer's organization.

Appendix C is an example lidar metadata file and full metadata template, excerpted from the USGS-NGP Lidar Base Specification.

REFERENCES

FGDC, 2011. Metadata Summit, 27 October, URL: http://www.fgdc.gov/metadata/events/metadata-summit/index_html (last date accessed: 30 April 2012).

9.3 EXISTING STANDARDS AND GUIDELINES

Note: Section 9.3 is an excerpt from the US Geological Survey's Lidar Base Specification Version 1.0. *Appendix 5 of this document* (Metadata Requirements and Templates) *is reproduced in section 9.2 of this book, and Appendix 4 of this document* (Lidar Metadata Example) *is reproduced in Appendix C of this book.*

Conversion Factors

SI to Inch/Pound

Multiply	By	To obtain
	LENGTH	
centimeter (cm)	0.3937	inch (in.)
meter (m)	3.281	foot (ft)
kilometer (km)	0.6214	mile (mi)
meter (m)	1.094	yard (yd)
	AREA	
square meter (m^2)	0.0002471	acre
square meter (m^2)	10.76	square foot (ft^2)
hectare (ha)	2.471	acre
hectare (ha)	0.003861	square mile (mi^2) = 640 acres = 1 section
square kilometer (km^2)	247.1	acre
square kilometer (km^2)	0.3861022	square mile (mi^2)

Altitude and Elevation, as used in this report, refers to the distance above the geoid, unless specifically referenced to the ellipsoid.

Height, as used in this report, refers to the height above ground.

Acronyms and Abbreviations

ARRA	American Reinvestment and Recovery Act
ASPRS	American Society for Photogrammetry and Remote Sensing
CLICK	Center for Lidar Information, Coordination, and Knowledge
CONUS	Conterminous United States
CORS	Continuously Operating Reference Stations
CVA	Consolidated Vertical Accuracy
DEM	Digital Elevation Model
DSM	Digital Surface Models
DTED	digital terrain elevation data
DTM	digital terrain model
EDNA	Elevation Derivatives for National Applications
EPSG	European Petroleum Survey Group
FGDC	Federal Geographic Data Committee
FOV	field of view
FVA	Fundamental Vertical Accuracy
GB	gigabyte
GPS	Global Positioning System
GSD	ground sample distance
H&H	hydraulic and hydrologic
IFSAR	Interferometric Synthetic Aperature Radar
IMU	Inertial Measurement Unit
lidar	light detection and ranging
NAD83	North American Datum of 1983
NAVD88	North American Vertical Datum of 1988
NDEP	National Digital Elevation Program
NED	National Elevation Dataset
NGP	National Geospatial Program
NGS	National Geodetic Survey
NIR	near infra-red
NIST	National Institute of Standards and Technology
NPS	Nominal Pulse Spacing
NSRS	National Spatial Reference System
NSSDA	National Standards for Spatial Data Accuracy
OCONUS	Outside the Conterminous United States
QA/QC	Quality Assurance/Quality Control
RMSE	Root Mean Square Error
SVA	Supplemental Vertical Accuracy
TIN	Triangulated Irregular Network
USGS	U.S. Geological Survey
UTM	Universal Transverse Mercator
XML	eXtensible Markup Language

Introductory Material

Purpose and Scope

The U.S. Geological Survey (USGS) intends to use this specification to acquire and procure lidar data, and to create consistency across all USGS National Geospatial Program (NGP) and partner funded lidar collections, in particular those undertaken in support of the National

Elevation Dataset (NED) (Gesch, 2007). Unlike most other "lidar data procurement specifications", which are focused on the products derived from lidar point cloud data; such as the bare-earth Digital Elevation Model (DEM), this specification places unprecedented emphasis on the handling of the source lidar point cloud data. The goal of this document is to assure that the complete acquired source dataset, including the data, metadata, descriptive documentation, quality information, and ancillary data—collected in accordance with the minimum parameters specified within—remains intact and viable to support the wide variety of DEM and non-DEM science and mapping applications that benefit from lidar technology. In the absence of other comprehensive specifications or standards, the USGS NGP hopes that this specification will, to the greatest degree practical, be adopted by other USGS programs and disciplines, and by other federal agencies.

Adherence to the following minimum specifications ensures that bare-earth DEMs derived from lidar data are suitable for ingestion into the NED at the 1/9 arc-second resolution, and can be resampled for use in the 1/3 and 1 arc-second NED resolutions. It also ensures that the point cloud source data are handled in a consistent manner by all data providers and delivered to the USGS in clearly defined formats. This allows straight-forward ingestion into lidar data management systems such as the Center for Lidar Information, Coordination, and Knowledge (CLICK) (Stoker and others, 2006) and simplifies subsequent use of the source data by the broader user community, particularly with regard to cross-collection analysis.

It must be emphasized that this is a base specification, defining minimum parameters for acceptance of the acquired lidar data. It is expected that local conditions in any given project area, specialized applications for the data, or the preferences of cooperators, may mandate more stringent requirements. The USGS encourages the collection of more detailed, accurate, or value-added data. A list of common upgrades to the minimum requirements defined in this report is provided in appendix 1.

Although lidar data have been used in the research and commercial mapping environments for more than a decade, it is still a relatively new technology. Advancements and improvements in instrumentation, software, processes, applications, and understanding are constantly refined or developed. It would not be possible to develop a set of guidelines and specifications that address all of these advances. The current document is based on the NGP's understanding of and experience with the lidar technology being used in the industry at the present time (2012). Furthermore, the NGP acknowledges that there is a lack of commonly accepted best practices for numerous processes and technical assessments (for example, measurement of Nominal Pulse Spacing (NPS), point distribution, classification accuracy, common accuracy and quality indicators, and others). The USGS encourages the development of such best practices through its partners in the industry, other Government agencies, and the appropriate professional governance organizations, and considers them for future revisions to this and other similar documents.

Applicability

These specifications and guidelines are applicable to lidar data and deliverables supported in whole or in part, with either financial or in-kind contributions, by the USGS NGP.

Maintenance Authority

The USGS NGP is the maintenance authority for this document.

Background

The USGS NGP has cooperated in the collection of numerous lidar datasets across the Nation for a wide array of applications. These collections have used a variety of specifications and had a diverse set of product deliverables, the end result was incompatible datasets with cross-project analysis extremely difficult. The need for a single base specification, defining minimum collection parameters and a consistent set of deliverables, is apparent.

Beginning in late 2009, an increase in the rate of lidar data collection because of the American Reinvestment and Recovery Act (ARRA) funding for *The National Map* makes it imperative that a single data specification be implemented to ensure consistency and improve data utility. Although the development of this specification was prompted by the ARRA funding, the specification is intended to remain durable beyond ARRA-funded NGP projects.

In addition, it is recognized that the USGS NGP also uses lidar technology for specialized scientific research and other projects whose requirements are incompatible with the provisions of this specification. In such cases, and with properly documented justification supporting the need for the variance, waivers of any part or all of this specification may be granted.

It is conceivable that in some cases, based on specific topography, land cover, intended application, or other factors, the USGS-NGP may require specifications more rigorous than those defined in this document.

Technical alternatives to enhance delivery of data or associated products and processes as described by parameters and language in this document are encouraged and may be submitted with any proposal and will be given due professional consideration.

Collection

Multiple Discrete Returns

Data collection must be capable of at least three returns per pulse. Full waveform collection is acceptable and welcomed; however, waveform data are regarded as supplemental information. Deriving and delivering multiple discrete returns is required in all cases.

Intensity Values

Intensity values are required for each return. The values are to be recorded in the .las files in their native radiometric resolution.

Nominal Pulse Spacing (NPS)

An NPS of 2 meters or less is required. Dependent on the local terrain and land cover conditions in the project area, a greater point density may be required on specific projects. Assessment of the NPS will be made against single swath, first-return only data, located within the geometrically usable center portion (typically 90 percent) of each swath, acceptable data voids excluded. NPS will be calculated as the square root of the average area per point. Average along-track and cross-track point spacing should be comparable (within 10 percent).

In general, the target NPS for a project should not be achieved through swath overlap or multiple passes. Such collection techniques may be permitted with prior approval.

Data Voids

Data voids within a single swath are not acceptable, except in the following circumstances:

- Where caused by water bodies,
- Where caused by areas of low near infra-red (NIR) reflectivity such as asphalt or composition roofing, or
- Where appropriately filled-in by another swath.

Spatial Distribution

The spatial distribution of geometrically usable points is expected to be uniform. Although it is understood that lidar instruments do not produce regularly gridded points, collections should be planned and executed to produce a first-return point cloud that approaches a regular lattice of points, rather than a collection of widely spaced high density profiles of the terrain. The uniformity of the point density throughout the dataset is important and will be assessed using the following steps:

- Generating a density grid from the data with cell sizes equal to the design NPS times 2, using a radius equal to the design NPS.
- Ensuring at least 90 percent of the cells in the grid contain at least one lidar point.
- The assessment is to be made against individual (single) swaths, using only the first-return points located within the geometrically usable center portion (typically 90 percent) of each swath.
- Excluding acceptable data voids previously identified in this specification.

Note: This requirement may be relaxed in areas of substantial relief where it is impractical to maintain a consistent and uniform distribution.

Note: The process described in this section relates only to the uniformity of the point distribution. It in no way relates to, nor can it be used for the assessment of point density or NPS.

Scan Angle

Scan angle will support horizontal and vertical accuracy within the requirements as specified in the next two sections. See Appendix 2 for additional information.

Note: This requirement primarily is applicable to oscillating mirror lidar systems. Other instrument technologies may be exempt from this requirement.

Vertical Accuracy

Vertical accuracy of the lidar data will be assessed and reported in accordance with the guidelines developed by the National Digital Elevation Program (NDEP) and subsequently adopted by the American Society for Photogrammetry and Remote Sensing (ASPRS). Complete definitions for vertical accuracy assessments are in Section 1.5 of the NDEP Elevation Guidelines (NDEP, 2004).

The minimum vertical accuracy requirement for the unclassified lidar point cloud, using the NDEP/ASPRS methodology, is listed below:

- Fundamental Vertical Accuracy (FVA)<= 24.5 centimeters (cm) Accuracyz (ACCz), 95 percent (12.5 cm Root Mean Square Error (RMSE)z).

The minimum vertical accuracy requirements for the derived DEM, using the NDEP/ASPRS methodology are listed below:

- Fundamental Vertical Accuracy (FVA)<= 24.5 cm ACCz, 95 percent (12.5cm RMSEz);
- Consolidated Vertical Accuracy (CVA)<= 36.3cm, 95th percentile, and
- Supplemental Vertical Accuracy (SVA)<= 36.3 cm, 95th percentile.

Point cloud data accuracy is to be tested against a Triangulated Irregular Network (TIN) constructed from lidar points in clear and open areas. A clear and open area can be characterized with respect to topographic and ground cover variation such that a minimum of 5 times the NPS exists with less than 1/3 of the $RMSE_z$ deviation from a low-slope plane. Slopes that exceed 10 percent should be avoided. Ground that has been plowed or otherwise disturbed is not acceptable. All tested locations should be photographed showing the position of the tripod and the surrounding area ground condition.

Each land cover type representing 10 percent or more of the total project area must be tested and reported with an SVA. In areas where a land cover category is something other than forested or dense urban, the tested point should not have any obstructions 45 degrees above the horizon to ensure a sufficient TIN surface. Additionally, tested areas should not be in proximity to low NIR reflective surfaces such as asphalt or composition roofing materials.

The SVA value is provided as a target. It is understood that in areas of dense vegetation, swamps, or extremely difficult terrain, this value may be exceeded.

The CVA value is a requirement that must be met, regardless of any allowed "busts" in the SVA(s) for individual land cover types within the project.

Checkpoints for each assessment (FVA, CVA, and all SVAs) are required to be well-distributed throughout the land cover type, for the entire project area. See Glossary for definition of well-distributed.

Exceptions: These requirements may be relaxed in cases:

- Where there exists a demonstrable and substantial increase in cost to obtain this accuracy.
- Where an alternate specification is needed to conform to previously contracted phases of a single larger overall collection effort, for example, multi-year statewide collections.
- Where the USGS agrees that it is reasonable and in the best interest of all stakeholders to use an alternate specification.

Relative Accuracy

The requirements for relative accuracy are listed below:

- Within individual swaths:<= 7 cm RMSEz
- Within overlap between adjacent swaths:<=10 cm RMSEz

Flightline Overlap

Flightline overlap of 10 percent or greater is required to ensure there are no data gaps between the usable portions of the swaths. Collections in high relief terrain are expected to require greater overlap. Any data with gaps between the geometrically usable portions of the swaths will be rejected.

Collection Area

Data collection for the Defined Project Area, buffered by a minimum of 100 meters, is required. The buffered boundary is the Buffered Project Area.

In order that all products are consistent to the edge of the Defined Project Area, all products must be generated to the limit of the Buffered Project Area. Since these areas are being generated, they shall also be delivered.

Collection Conditions

Atmospheric conditions must be cloud and fog-free between the aircraft and ground during all collection operations.

Ground conditions must be snow free. Very light, undrifted snow may be acceptable in special cases, with prior approval.

Water conditions must be free of any unusual flooding or inundation, except in cases where the goal of the collection is to map the inundation.

Leaf-off vegetation conditions are preferred, however, as numerous factors beyond human control may affect the vegetative condition at the time of any collection, the USGS NGP only requires that penetration to the ground must be adequate to produce an accurate and reliable bare-earth surface suitable for incorporation into the 1/9 (3-meter) NED. Collections for specific scientific research projects may be exempted from this requirement, with prior approval.

Data Processing and Handling

ASPRS LAS File Format

All processing should be carried out with the understanding that all point deliverables are required to be in fully compliant LAS format, either v1.2 or v1.3. The version selected must be used for all LAS deliverables in the project. Data producers are encouraged to review the LAS specification in detail (ASPRS, 2011).

Full Waveform

If full waveform data are collected, delivery of the waveform packets is required. LAS v1.3 deliverables with waveform data are to use external auxiliary files with the extension .wdp for the storage of waveform packet data. See the LAS v1.3 Specification for additional information (ASPRS, 2011).

Global Positioning System (GPS) Times

GPS times are to be recorded as Adjusted GPS Time, at a precision sufficient to allow unique timestamps for each pulse. Adjusted GPS Time is defined to be Standard (or satellite) GPS time minus $1*10^9$. See the LAS v1.3 Specification for more detail (ASPRS, 2011).

Datums

All data collected must be tied to the datums listed below:

- Horizontal datum reference to the North American Datum of 1983/HARN adjustment (NAD83 HARN) is required.
- Vertical datum reference to the North American Vertical Datum of 1988 (NAVD 88) is required.
- The most recent National Geodetic Survey (NGS)-approved geoid model is required to perform conversions from ellipsoidal heights to orthometric heights.

Coordinate Reference System

The USGS preferred Coordinate Reference System for the Conterminous United States (CONUS) is Universal Transverse Mercator UTM, NAD83 HARN, Meters; NAVD88, Meters. Each discrete project is to be processed using the single predominant UTM zone for the overall collection area.

The USGS will also accept data in other Coordinate Reference Systems that meet the conditions below:

- State Plane and State Coordinate Reference Systems that have been accepted by the European Petroleum Survey Group (EPSG) may be used.
- Coordinate Reference Systems for collections in Alaska, Hawaii, and other areas Outside the Conterminous United States (OCONUS) must be approved by the USGS before collection.

Units of Reference

All references to the unit of measure "Feet" and "Foot" must specify "International", "Intl", "U.S. Survey", or "US".

Swath Identification

Each swath will be assigned a unique File Source ID. It is required that the Point Source ID field for each point within each LAS swath file be set equal to the File Source ID before any processing of the data. See the LAS v1.3 Specification (ASPRS, 2011).

Point Families

Point families (multiple return "children" of a single "parent" pulse) shall be maintained intact through all processing before tiling. Multiple returns from a given pulse will be stored in sequential (collected) order.

Swath Size and Segmentation

Swath files will be 2 gigabytes (GB) in size or less. Long swaths (those which result in a LAS file larger than 2 GB) will be split into segments no greater than 2 GB each.

- Each sub-swath will retain the original File Source ID of the original complete swath.
- Points within each sub-swath will retain the Point Source ID of the original complete swath.
- Each sub-swath file will be named identically to the original complete swath, with the addition of an ordered alphabetic suffix to the name ("-a", "-b" … "-n"). The order of the named sub-swaths shall be consistent with the collection order of the points ("-a" will be the chronological beginning of the swath; "-n" will be the chronological end of the swath).
- Point families shall be maintained intact within each sub-swath.
- Sub-swaths should be broken at the edge of the scan line.
- Other swath segmentation approaches may be acceptable, with prior approval.

Scope of Collection

All collected swaths are to be delivered as part of the Raw Data Deliverable. This includes calibration swaths and cross-ties. This in no way requires or implies that calibration swath data are to be included in product generation. All collected points are to be delivered. No points are to be deleted from the swath LAS files. Excepted from this are extraneous data outside of the buffered project area (aircraft turns, transit between the collection area and airport, transit between fill-in areas, an the like). These points may be permanently removed. Busted swaths that are being completely discarded by the vendor and re-flown do not need to be delivered.

Use of the LAS Withheld Flag

Outliers, blunders, noise points, geometrically unreliable points near the extreme edge of the swath, and other points the vendor deems unusable are to be identified using the Withheld flag, as defined in the LAS specification.

This applies primarily to points that are identified during pre-processing or through automated post-processing routines.

If processing software is not capable of populating the Withheld bit, these points may be identified using Class=11.

Noise points subsequently identified during manual Classification and Quality Assurance/Quality Control (QA/QC) may be assigned the standard LAS classification value for Noise (Class=7), regardless of whether the noise is "low" or "high" relative to the ground surface.

Point Classification

- ALL points not identified as Withheld are to be classified.
- No points in the Classified LAS deliverable will be assigned Class=0.
- Use of the ASPRS/LAS Overlap classification (Class=12) is prohibited.

If overlap points are required to be differentiated by the data producer or cooperating partner, they must be identified using a method that does not interfere with their classification:

- Overlap points are tagged using Bit:0 of the User Data byte, as defined in the LAS specification. (SET=Overlap).

- Overlap points are classified using the Standard Class values + 16.
- Other techniques as agreed upon in advance.

The technique used to identify overlap must be clearly described in the project metadata files.

Note: A standard bit flag for identification of overlap points has been included in LAS v1.4, released on November 14, 2011.

Positional Accuracy Validation

Before classification of and development of derivative products from the point cloud, verification of the vertical accuracy of the point cloud, absolute and relative, is required. The Fundamental Vertical Accuracy (absolute) is to be assessed in clear, open areas as described in the section called Vertical Accuracy above. Swath-to-swath and within swath accuracies (relative) are to be documented. A detailed report of this validation process is a required deliverable.

Classification Accuracy

It is required that due diligence in the classification process will produce data that meet the following tests:

- Following classification processing, no non-withheld points should remain in Class 0.
- Within any 1 kilometer (km) × 1 km area, no more than 2 percent of non-withheld points will possess a demonstrably erroneous classification value.
- Points remaining in Class 1 that should be classified in any other required Class are subject to these accuracy requirements and will be counted towards the 2 percent threshold.

Note: These requirements may be relaxed to accommodate collections in areas where the USGS agrees classification to be particularly difficult.

Classification Consistency

Point classification is to be consistent across the entire project. Noticeable variations in the character, texture, or quality of the classification between tiles, swaths, lifts, or other non-natural divisions will be cause for rejection of the entire deliverable.

Tiles

Note: This section assumes a projected coordinate reference system.

A single non-overlapped tiling scheme (the Project Tiling Scheme) will be established and agreed upon by the data producer and the USGS before collection. This scheme will be used for ALL tiled deliverables.

- Tile size is required to be an integer multiple of the cell size of raster deliverables.
- Tiles are required to be sized using the same units as the coordinate system of the data.
- Tiles are required to be indexed in *X* and *Y* to an integer multiple of the tile's *X*-*Y* dimensions.
- All tiled deliverables will conform to the Project Tiling Scheme, without added overlap.
- Tiled deliverables will edge-match seamlessly and without gaps.

Hydro-Flattening

Note: Please refer to appendix 3 for reference information on hydro-flattening.

Hydro-flattening pertains only to the creation of derived DEMs. No manipulation of or changes to originally computed lidar point elevations are to be made. Breaklines may be used to help classify the point data. The goal of the NGP is for the delivered DEMs to represent water bodies in a cartographically and aesthetically pleasing manner. It is not the goal of the NGP to accurately map water surface elevations within the NED. The requirements for hydro-flattening are listed below.

Inland Ponds and Lakes

- 2 acres or greater surface area (approximately equal to a round pond 350 feet in diameter) at the time of collection.
- Flat and level water bodies (single elevation for every bank vertex defining a given water body).
- The entire water surface edge must be at or below the immediately surrounding terrain. The presence of floating water bodies will be cause for rejection of the deliverable.
- Long impoundments such as reservoirs, inlets, and fjords, whose water surface elevations drop when moving downstream, are required to be treated as rivers.

Inland Streams and Rivers

- 100 feet nominal width: This should not unnecessarily break a stream or river into multiple segments. At times it may squeeze slightly below 100 feet for short segments. Data producers should use their best professional cartographic judgment.
- Flat and level bank-to-bank (perpendicular to the apparent flow centerline); gradient to follow the immediately surrounding terrain. In cases of sharp turns of rapidly moving water, where the natural water surface is notably not level bank-to-bank, it is appropriate to represent the water surface as it exists in nature, while maintaining an aesthetic cartographic appearance.
- The entire water surface edge must be at or below the immediately surrounding terrain.
- Stream channels are required to break at road crossings (culvert locations). The roadway over a culvert should be continuous. A culvert, regardless of size, is defined as having earth between the road surface and the top of the structure.
- Bridges are required to be removed from the DEM. Streams and rivers should be continuous at bridge locations. Bridges are defined as having an elevated deck structure that does not rest on earth.
- When the identification of a structure such as a bridge or culvert cannot be made reliably, the feature should be regarded as a culvert.

Non-Tidal Boundary Waters

- Represented only as an edge or edges within the project area; collection does not include the opposing shore.
- Water surface is to be flat and level, as appropriate for the type of water body (level for lakes; gradient for rivers)
- The entire water surface edge must be at or below the immediately surrounding terrain.

Tidal Waters

- Tidal water bodies are defined as water bodies such as oceans, seas, gulfs, bays, inlets, salt marshes, large lakes, and the like. This includes any water body that is affected by tidal variations.
- Tidal variations over the course of a collection or between different collections, will result in lateral and vertical discontinuities along shorelines. This is considered normal and these anomalies should be retained. The final DEM is required to represent as much ground as the collected data permits.
- Water surface is to be flat and level, to the degree allowed by the irregularities noted in 4.4.2.
- Scientific research projects in coastal areas often have specific requirements with regard to how tidal land-water boundaries are to be handled. For such projects, the requirements of the research will take precedence.

Islands

- Permanent islands 1 acre or larger shall be delineated within all water bodies.

Single-Line Streams

Cooperating partners may require collection and integration of single-line streams within their lidar projects. Although the USGS does not require these breaklines be collected or integrated, it does require that if used and incorporated into the DEMs, the following guidelines are met:

- All vertices along single-line stream breaklines are at or below the immediately surrounding terrain.
- Single-line stream breaklines are not to be used to introduce cuts into the DEM at road crossings (culverts), dams, or other such features. This is hydro-enforcement and as discussed in appendix 3 will create a non-topographic DEM that is unsuitable for integration into the NED.
- All breaklines used to modify the surface are to be delivered to the USGS with the DEMs.

Deliverables

The USGS requires unrestricted rights to all delivered data and reports, which will be placed in the public domain. This specification places no restrictions on the data provider's rights to resell data or derivative products as they see fit.

Metadata

The term "metadata" refers to all descriptive information about the project. This includes textual reports, graphics, supporting shapefiles, and Federal Geographic Data Committee (FGDC)-compliant metadata files. Metadata deliverables include the following items:

- Collection report detailing mission planning and flight logs.
- Survey report detailing the collection of control and reference points used for calibration and QA/QC.
- Processing report detailing calibration, classification, and product generation procedures including methodology used for breakline collection and hydro-flattening (see the section called Hydro-flattening and Appendix 3 for more information on hydro-flattening).
- QA/QC Reports (detailing the analysis, accuracy assessment and validation of the following:
  - Point data (absolute, within swath, and between swath)
  - Bare-earth surface (absolute)
  - Other optional deliverables as appropriate
- Control and calibration points: All control and reference points used to calibrate, control, process, and validate the lidar point data or any derivative products that are to be delivered.
- Georeferenced, digital spatial representation of the precise extents of each delivered dataset. This should reflect the extents of the actual lidar source or derived product data, exclusive of TIN artifacts or raster NODATA areas. A union of tile boundaries or minimum bounding rectangles is not acceptable. ESRI Polygon shapefile or geodatabase is preferred.
- Product metadata [FGDC compliant, eXtensible Markup Language (XML) format metadata]. Metadata files for individual files are not required. One XML file is required for the following examples:
  - The Overall Project: Describing the project boundary, the intent of the project, the types of data collected as part of the project, the various deliverables for the project, and other project-wide information.
  - Each Lift: Describing the extents of the lift, the swaths included in the lift, locations of GPS base stations and control for the lift, preprocessing and calibration details for the

lift, adjustment and fitting processes applied to the lift in relation to other lifts, and other lift-specific information.
- Each tiled deliverable product group:
  - Classified point data
  - Bare-earth DEMs
  - Breaklines (if used)
  - Other datasets delivered under the contract (Digital Surface Models (DSM), intensity images, height surfaces, and others)
- FGDC compliant metadata must pass the USGS metadata parser (mp) with no errors.

Note: Please refer to the metadata templates in appendix 4 for additional information.

Raw Point Cloud

Delivery of the raw point cloud is a standard requirement for USGS NGP lidar projects. Raw point cloud deliverables include the following items:
- All swaths, returns, and collected points, fully calibrated and adjusted to ground, by swath.
- Fully compliant LAS v1.2 or v1.3, Point Data Record Format 1, 3, 4, or 5.
- LAS v1.3 deliverables with waveform data are to use external auxiliary files with the extension .wdp for the storage of waveform packet data. See the LAS v1.3 Specification for additional information.
- Correct and properly formatted georeference information must be included in all LAS file headers.
- GPS times are to be recorded as Adjusted GPS Time, at a precision sufficient to allow unique timestamps for each pulse.
- Intensity values (native radiometric resolution).
- One file per swath, one swath per file, file size not to exceed 2 GB, as described under the section called Swath Size and Segmentation above.
- Vertical accuracy of the lidar point data will be assessed and reported in accordance with the guidelines developed by the NDEP and subsequently adopted by the ASPRS. The complete guidelines on vertical accuracy are in Section 1.5 of the NDEP Guidelines (NDEP, 2004).
- Vertical accuracy requirements using the NDEP/ASPRS methodology for the point cloud are FVA<= 24.5 cm ACCz, 95 percent Confidence Level (12.5 cm $RMSE_z$)

Classified Point Cloud

Delivery of a classified point cloud is a standard requirement for USGS NGP lidar projects. Specific scientific research projects may be exempted from this requirement. Classified point cloud deliverables include the following items:
- All project swaths, returns, and collected points, fully calibrated, adjusted to ground, and classified, by tiles. Project swaths exclude calibration swaths, cross-ties, and other swaths not used, or intended to be used, in product generation.
- Fully compliant LAS v1.2 or v1.3, Point Data Record Format 1, 3, 4, or 5
- LAS v1.3 deliverables with waveform data are to use external auxiliary files with the extension .wdp for the storage of waveform packet data. See the LAS v1.3 Specification for additional information.
- Correct and properly formatted georeference information must be included in all LAS file headers.
- GPS times are to be recorded as Adjusted GPS Time, at a precision sufficient to allow unique timestamps for each pulse.
- Intensity values (native radiometric resolution).

- Tiled delivery, without overlap, using Project Tiling Scheme.
- Classification Scheme (minimum) as listed in Table 9.2.

Table 9.2 Minimum Classified Point Cloud Classification Scheme

Code	Description
1	Processed, but unclassified
2	Bare-earth ground
7[a]	Noise (low or high; manually identified; if needed)
9	Water
10[b]	Ignored Ground (Breakline proximity)
11	Withheld (if the Withheld bit is not implemented in processing software)

[a]Class 7, Noise, is included as an adjunct to the Withheld bit. All noise points are to be identified using one of these two methods.

[b]Class 10, Ignored Ground, is for points previously classified as bare-earth but whose proximity to a subsequently added breakline requires that it be excluded during Digital Elevation Model (DEM) generation.

Bare-Earth Surface (Raster DEM)

Delivery of a bare-earth DEM is a standard requirement for USGS NGP lidar projects. Specific scientific research projects may be exempted from this requirement. Bare-earth surface deliverables include the following items:

- Bare-earth DEM, generated to the limits of the Buffered Project Area.
- Cell size no greater than 3 meters or 10 feet, and no less than the design Nominal Pulse Spacing (NPS).
- Delivery in an industry-standard, GIS-compatible, 32-bit floating point raster format (ERDAS .IMG preferred).
- Georeference information shall be included in each raster file.
- Tiled delivery, without overlap.
- DEM tiles will show no edge artifacts or mismatch. A quilted appearance in the overall project DEM surface, whether caused by differences in processing quality or character between tiles, swaths, lifts, or other non-natural divisions, will be cause for rejection of the entire DEM deliverable.
- Void areas (for example, areas outside the Buffered Project Area but within the tiling scheme) shall be coded using a unique NODATA value. This value shall be identified in the appropriate location within the raster file header or external support files (for example, .aux).
- Vertical accuracy of the bare-earth surface will be assessed and reported in accordance with the guidelines developed by the NDEP and subsequently adopted by the ASPRS. The complete guidelines are in Section 1.5 of the NDEP Guidelines (NDEP, 2004).
- The following threshholds represent the minimum vertical accuracy requirements using the NDEP/ASPRS methodology:
- FVA<= 24.5 cm ACCz, 95 percent Confidence Level (12.5 cm RMSEz)
- CVA<= 36.3 cm, 95th Percentile
- SVA<= 36.3 cm, 95th Percentile
- All QA/QC analysis materials and results are to be delivered to the USGS.
- Depressions (sinks), natural or man-made, are not to be filled (as in hydro-conditioning and hydro-enforcement).
- Water bodies (ponds and lakes), wide streams and rivers (double-line), and other non-tidal water bodies as defined in the section called Hydro-flattening are to be hydro-flat-

tened within the DEM. Hydro-flattening shall be applied to all water impoundments, natural or man-made, that are larger than 2 acres in area (approximately equal to a round pond 350 feet in diameter), to all streams that are nominally wider than 100 feet, and to all non-tidal boundary waters bordering the project area regardless of size. The methodology used for hydro-flattening is at the discretion of the data producer.

Note: Please refer to the section called Hydro-Flattening and Appendix 3 for detailed discussions of hydro-flattening.

Breaklines

Delivery of the breaklines used in hydro-flattening is a standard requirement for USGS NGP lidar projects. Specific scientific research projects may be exempted from this requirement. If hydro-flattening is achieved through other means, this section may not apply. Breakline deliverables include the following items:

- Breaklines shall be developed to the limit of the Buffered Project Area.
- All breaklines developed for use in hydro-flattening shall be delivered as an ESRI feature class (PolylineZ or PolygonZ format, as appropriate to the type of feature represented and the methodology used by the data producer). Shapefile or geodatabase is required.
- Each feature class or shapefile will include properly formatted and accurate georeference information in the standard location. All shapefiles must include a correct and properly formatted *.prj file.
- Breaklines must use the same coordinate reference system (horizontal and vertical) and units as the lidar point delivery.
- Breakline delivery may be as a continuous layer or in tiles, at the discretion of the data producer. In the case of tiled deliveries, all features must edge-match exactly across tile boundaries in both the horizontal (*X*-*Y*) and vertical (*Z*) spatial locations.

References Cited

American Society for Photogrammetry & Remote Sensing (ASPRS), 2011, LAS specification (Version 1.4–R12): Bethesda, Md., ASPRS, 27 p. Available online at *http://www.asprs.org/Committee-General/LASer-LAS-File-Format-Exchange-Activities.html.*

American Society for Photogrammetry & Remote Sensing (ASPRS), 2004. Vertical accuracy reporting for lidar–Version 1.0, 20 p. (Also available at *http://www.asprs.org/a/society/committees/lidar/Downloads/Vertical_Accuracy_Reporting_for_Lidar_Data.pdf*.)

Gesch, D.B., 2007, The National Elevation Dataset, chap. 4 of Maune, D., ed., Digital elevation model technologies and applications—the DEM users manual, (2nd ed.): Bethesda, Md., American Society for Photogrammetry and Remote Sensing, p. 99–118. (Also available at *http://topotools.cr.usgs.gov/pdfs/Gesch_Chp_4_Nat_Elev_Data_2007.pdf.*)

Maune, D.F., 2007, Definitions, *in* Digital Elevation Model Technologies and Applications—The DEM Users' Manual, (2nd ed.), American Society for Photogrammetry and Remote Sensing, Bethesda, Md., p. 550—551

National Digital Elevation Program (NDEP), 2004, Guidelines for Digital Elevation Data—Version 1: 93 p.(Also available at *http://www.ndep.gov/NDEP_Elevation_Guidelines_Ver1_10May2004.pdf*.)

Stoker, J.M., Greenlee, S.K., Gesch, D.B., and Menig, J.C., 2006, CLICK—the new USGS center for lidar information coordination and knowledge: Photogrammetric Engineering and Remote Sensing, v. 72, no. 6, p. 613–616. (Also available at *http://www.asprs.org/a/publications/pers/2006journal/june/highlight.pdf.*)

USGS Federal Geographic Data Committee, 1998, Geospatial Positioning Accuracy Standards Part 3: National Standard for Spatial Data Accuracy, 20 p. (Also available at *http://www.fgdc.gov/standards/projects/FGDC-standards-projects/accuracy/part3/chapter3.*)

Supplemental Information

USGS National Elevation Dataset (NED) Web site:
http://www.ned.usgs.gov
USGS Center for Lidar Information Coordination and Knowledge (CLICK) Web site:
http://www.lidar.cr.usgs.gov
MP-Metadata Parser:
http://geology.usgs.gov/tools/metadata
National Institute of Standards and Technology (NIST) Percentile Information:
http://itl.nist.gov/div898/handbook/prc/section2/prc252.htm

Appendices

Appendix 1 – Common Data Upgrades

- Independent 3rd-Party QA/QC by another Architecture & Engineering (AE) Contractor (encouraged)
- Lower NPS (greater point density)
- Increased Vertical Accuracy
- Full Waveform collection and delivery
- Additional Environmental Constraints
- Tidal coordination, flood stages, crop/plant growth cycles, and the like
- Shorelines corrected for tidal variations within a collection
- Top-of Canopy (First-Return) Raster Surface (tiled).
- Raster representing the highest return within each cell is preferred.
- Intensity Images (8-bit gray scale, tiled)
- Interpolation based on 1st-Returns.
- Interpolation based on All-Returns, summed.
- Detailed Classification (additional classes):

Table 9.3 Additional Point Cloud Classifications

Code	Description
3	Low vegetation
4	Medium vegetation (use for single vegetation class)
5	High vegetation
6	Buildings, bridges, other man-made structures
n	Additional class(es) as agreed upon in advance

- Hydro-Enforced DEMs as an additional deliverable
- Hydro-Conditioned DEMs as an additional deliverable
- Breaklines (PolylineZ and PolygonZ) for single-line hydrographic features
- Narrow streams not collected as double-line, culverts, and other similar features, including appropriate integration into delivered DEMs
- Breaklines (PolylineZ and PolygonZ) for other features (to be determined)
- Including appropriate integration into delivered DEMs

- Extracted Buildings (PolygonZ)
- Footprints with maximum elevation or height above ground as an attribute.
- Other products as defined by requirements and agreed upon in advance of funding commitment.

Appendix 2 – Guidelines

Scan Angle

For oscillating mirror lidar systems, the total field of view should not exceed 40 degrees (within 20 degrees of nadir). U.S. Geological Survey (USGS) quality assurance on collections performed using scan angles wider than 34 degrees will be particularly rigorous in the edge-of-swath areas. Lidar systems that use rotating mirrors/prisms may be exempted from this guideline.

Swath Size

Renaming schemes for split swaths are at the discretion of the data producer. The processing report shall include detailed information on swath segmentation sufficient to allow reconstruction of the original swaths if needed.

Non-Tidal Boundary Waters

The elevation along the edge or edges should behave consistently throughout the project. May be a single elevation (for example, a lake) or gradient (for example, a river), as appropriate. If unusual changes in water surface elevation occur during the course of the collection, then the surface may be treated as a tidal boundary, as described in the next section. The reason for the changes must be documented in the project metadata.

Tidal Waters

Variations in water-surface elevation resulting in tidal variations during a collection should NOT be removed or adjusted, as this would require either the removal of valid, measured ground points or the introduction of unmeasured ground into the digital elevation model (DEM). The USGS National Geospatial Progam (NGP) priority is on the ground surface, and accepts occasional, unavoidable irregularities in tidal water surfaces.

Breaklines and Hydro-Flattening

The USGS does not require any particular process or methodology be used for hydro-flattening or for breakline collection, extraction, or integration. However, if breaklines are developed, the following general guidelines must be adhered to:

Bare-earth lidar points that are in close proximity breaklines should be excluded from the DEM generation process. This is analogous to the removal of masspoints for the same reason in a traditional photogrammetrically compiled digital terrain model (DTM). The proximity threshold for reclassification as Ignored Ground is at the discretion of the data producer, but in general should not exceed the nominal pulse spacing (NPS). These points are to be retained in the delivered lidar point dataset and shall be reclassified as Ignored Ground (class value = 10) so that they may be subsequently identified.

Delivered data must be sufficient for the USGS to effectively recreate the delivered DEMs using the lidar points and breaklines without substantial editing.

The goal of hydro-flattening is to produce a topographic DEM that, with respect to water surfaces, resembles a DEM derived from traditional photogrammetric methods. Best professional judgment should be used to achieve this end.

Appendix 3 – Hydro-Flattening Reference

The subject of modifications to lidar-based digital elevation models (DEM) is somewhat new and there is substantial variation in the understanding of the topic across the industry. The following material was developed to provide a definitive reference on the subject only as it relates to the creation of DEMs intended to be integrated into the U.S. Geological Survey (USGS) NED. The information presented here is not meant to supplant other reference materials and it should not be considered authoritative beyond its intended scope.

The term hydro-flattening also is new, coined for this document and to convey the USGS National Geospatial Program's (NGP) need for DEMs having a specific type of functional surface . It is not, at this time, a known or accepted term across the industry. It is our hope that its use and acceptance will expand beyond the USGS with the assistance of other industry leaders.

Hydro-flattening of DEMs is accomplished predominantly through the use of breaklines, and this method is considered standard. Although other techniques may exist to achieve similar results, this section assumes the use of breaklines. The USGS does not require the use of any specific technique.

The Digital Elevation Model Technologies and Applications: The DEM User's Manual, 2nd Edition (Maune, 2007) provides the following definitions related to the adjustment of DEM surfaces for hydrologic analyses.

Hydrologically Conditioned (*Hydro-Conditioned*) Processing of a DEM or TIN so that the flow of water is continuous across the entire terrain surface, including the removal of all spurious sinks or pits. The only sinks that are retained are the real ones on the landscape. Whereas hydrologically-enforced is relevant to drainage features that are generally mapped, hydrologically-conditioned is relevant to the entire land surface and is done so that water flow is continuous across the surface, whether that flow is in a stream channel or not. The purpose for continuous flow is so that relations/links among basins/catchments can be known for large areas. This term is specifically used when describing Elevation Derivatives for National Applications (EDNA), the dataset of NED derivatives made specifically for hydrologic modeling purposes.

Hydrologically-Enforced (*Hydro-Enforced*) Processing of mapped water bodies so that lakes and reservoirs are level and so that streams flow downhill. For example, a DEM, TIN or topographic contour dataset with elevations removed from the tops of selected drainage structures (bridges and culverts) so as to depict the terrain under those structures. Hydro-enforcement enables hydrologic and hydraulic models to depict water flowing under these structures, rather than appearing in the computer model to be dammed by them because of road deck elevations higher than the water levels. Hydro-enforced TINs also use breaklines along shorelines and stream centerlines, for example, where these breaklines form the edges of TIN triangles along the alignment of drainage features. Shore breaklines for streams would be 3-D breaklines with elevations that decrease as the stream flows downstream; however, shore breaklines for lakes or reservoirs would have the same elevation for the entire shoreline if the water surface is known or assumed to be level throughout. See also the definition for hydrologically conditioned that has a slightly different meaning.

Whereas these are important and useful modifications, they result in surfaces that differ substantially from a traditional DEM. A hydro-conditioned surface has had its sinks filled and may have had its water-bodies flattened. This is necessary for correct flow modeling within and across large drainage basins. Hydro-enforcement extends this conditioning by requiring water bodies be leveled and streams flattened with the appropriate downhill gradient, and also by cutting through road crossings over streams (culvert locations) to allow a continuous flow path for water within the drainage. These treatments result in a surface on which water behaves as it physically does in the real world, and they are invaluable for specific types of hydraulic and hydrologic (H&H) modeling activities. Neither of these treatments is typical of a traditional DEM surface.

A traditional DEM such as the NED, on the other hand, attempts to represent the ground surface more the way a bird, or person in an airplane, sees it. On this surface, natural depressions exist, and roadways create apparent sinks because the roadway is depicted continuously without regard to the culvert beneath, making it an apparent dam. Bridges, it should be noted, are removed in most all types of DEMs because they are man-made, above-ground structures that have been added to the landscape.

Note: DEMs developed solely for orthophoto production may include bridges, as their presence can prevent the smearing of structures and reduce the amount of post-production correction of the final orthophoto. These are special use DEMs and are not relevant to this discussion.

For years, raster DEMs have been created from a digital terrain model (DTM) of masspoints and breaklines, which in turn were created through photogrammetric compilation from stereo imagery. Photogrammetric DTMs inherently contain breaklines defining the edges of water bodies, coastlines, single-line streams, and double-line streams and rivers, as well as numerous other surface features.

Lidar technology, however, does not inherently collect the breaklines necessary to produce traditional DEMs. Breaklines have to be developed separately through a variety of techniques, and either used with the lidar points in the generation of the DEM, or applied as a correction to DEMs generated without breaklines.

To maintain the consistent character of the NED as a traditional DEM, the USGS NGP requires that all DEMs delivered have their inland water bodies flattened. This does not imply that a complete network of topologically correct hydrologic breaklines be developed for every dataset; only those breaklines necessary to ensure that the conditions defined in the section called Hydro-flattening exist in the final DEM.

Glossary

Note: Many of the following definitions are taken from Maune (2007) and are used with permission.

accuracy The closeness of an estimated value (for example, measured or computed) to a standard or accepted (true) value of a particular quantity. Note: With the exception of Continuously Operating Reference Stations (CORS), assumed to be known with zero errors relative to established datums, the true locations of 3-D spatial coordinates of other points are not truly known, but only estimated; therefore, the accuracy of other coordinate information is unknown and can only be estimated.

- **absolute accuracy** A measure that accounts for all systematic and random errors in a dataset. Absolute accuracy is stated with respect to a defined datum or reference system.
- **accuracy$_r$ (ACC$_r$)** The National Standards for Spatial Data Accuracy (NSSDA) reporting standard in the horizontal component that equals the radius of a circle of uncertainty, such that the true or theoretical horizontal location of the point falls within that circle 95 percent of the time. $ACC_r = 1.7308 \times$ Root Mean Square Error $(RMSE)_r$.
- **accuracy$_z$ (ACC$_z$)** The NSSDA reporting standard in the vertical component that equals the linear uncertainty value, such that the true or theoretical vertical location of the point falls within that linear uncertainty value 95 percent of the time. $ACC_z = 1.9600 \times RMSE_z$.
- **horizontal accuracy** The positional accuracy of a dataset with respect to a horizontal datum. The horizontal accuracy reporting standard (ACCr) is defined above.

- **local accuracy** A value that represents the uncertainty in the coordinates of a control point relative to the coordinates of other directly connected, adjacent control points at the 95 percent confidence level. The reported local accuracy is an approximate average of the individual local accuracy values between this control point and other observed control points used to establish the coordinates of the control point.
- **network accuracy** A value that represents the uncertainty in the coordinates of a control point with respect to the geodetic datum at the 95 percent confidence level. For National Spatial Reference System (NSRS) network accuracy classification in the United States, the datum is considered to be best expressed by the geodetic values at the CORS supported by the National Geodetic Survey (NGS). By this definition, the local and network accuracy values at CORS sites are considered to be infinitesimal, that is, to approach zero.
- **positional accuracy** The accuracy of the position of features, including horizontal and vertical positions.
- **relative accuracy** A measure that accounts for random errors in a dataset. Relative accuracy may also be referred to as point-to-point accuracy. The general measure of relative accuracy is an evaluation of the random errors (systematic errors and blunders removed) in determining the positional orientation (for example, distance, azimuth) of one point or feature with respect to another. In lidar, this also may specifically mean the accuracy between adjacent swaths within a lift, adjacent lifts within a project, or between adjacent projects.
- **vertical accuracy** The measure of the positional accuracy of a dataset with respect to a specified vertical datum. The vertical accuracy reporting standard (ACCz) is defined above.

ambiguity resolution Combining the phase data from two or more Global Positioning System (GPS) receivers so that, after eliminating all other significant errors, the unknown number of integer wavelengths can be determined for signals coming from GPS satellites. Redundant L1 and L2 phase observations from two or more receivers, each tracking five or more satellites, provide the information for rapid unambiguous resolution. Once the ambiguities are resolved, the corrected phases for each observed satellite become precise ranges that allow the computation of the baseline vector(s) between the receivers with a typical accuracy of 2–10 centimeters.

artifacts An inaccurate observation, effect, or result, especially one resulting from the technology used in scientific investigation or from experimental error. In bare-earth elevation models, detectable surface remnants of buildings, trees, towers, telephone poles or other elevated features; also, detectable artificial anomalies that are introduced to a surface model by way of system specific collection or processing techniques, for example, corn-row effects of profile collection, star and ramp effects from multidirectional contour interpolation, or detectable triangular facets caused when vegetation canopies are removed from lidar data.

arc-second In angular measurements, 1/3600 of a degree. Commonly used in descriptions of geographic coordinate reference systems, for example, the horizontal resolution of the National Elevation Dataset (NED) (1 arc-second, 1/3 arc-second. 1/9 arc-second). For most of the Contiguous United States (CONUS), 1 arc-second is approximately 30 meters, linear.

aspect The direction which a building or topographic surface faces with respect to points of a compass. The slope direction or steepest downslope across a surface, normally measured clockwise in degrees from due north. The value of each location in an aspect dataset indicates the direction the surface slope faces.

attitude The position of a body defined by the angles between the axes of the coordinate system of the body and the axes of an external coordinate system. In photogrammetry, the attitude is the angular orientation of a camera (roll, pitch, yaw), or of the photograph taken with that camera, with respect to some external reference system. With lidar and Interferometric Synthetic Aperature Radar (IFSAR), the attitude is normally defined as the roll, pitch and heading of the instrument at the instant an active pulse is emitted from the sensor.

bald earth Non-preferred term. *See* bare earth.

bare earth (bare-earth) Digital elevation data of the terrain, free from vegetation, buildings and other man-made structures. Elevations of the ground.

boresight Calibration of a lidar sensor system equipped with an Inertial Measurement Unit (IMU) and Global Positioning System (GPS) to determine or establish the accurate:

- Position of the instrument (*x, y, z*) with respect to the GPS antenna
- Orientation (roll, pitch, heading) of the lidar instrument with respect to straight and level flight.

breakline A linear feature that describes a change in the smoothness or continuity of a surface. The two most common forms of breaklines are as follows:

- A ***soft breakline*** ensures that known *z*-values along a linear feature are maintained (for example, elevations along a pipeline, road centerline or drainage ditch), and ensures that linear features and polygon edges are maintained in a Triangulated Irregular Network (TIN) surface model, by enforcing the breaklines as TIN edges. They are generally synonymous with 3-D breaklines because they are depicted with series of x/y/z coordinates. Somewhat rounded ridges or the trough of a drain may be collected using soft breaklines.
- A ***hard breakline*** defines interruptions in surface smoothness, for example, to define streams, shorelines, dams, ridges, building footprints, and other locations with abrupt surface changes. Although some hard breaklines are 3-D breaklines, they are often depicted as 2-D breaklines because features such as shorelines and building footprints are normally depicted with series of *X*/*Y* coordinates only, often digitized from digital orthophotos that include no elevation data.

bridge A structure carrying a road, path, railroad, or canal across a river, ravine, road, railroad, or other obstacle.

In mapping, this is distinguished from a roadway over a culvert, in that a bridge is a man-made elevated deck for which there is no earth or soil. *See* culvert.

calibration The process of identifying and correcting for systematic errors in hardware, software, or procedures. Determining the systematic errors in a measuring device by comparing its measurements with the markings or measurements of a device that is considered correct. Airborne sensors can be calibrated geometrically and radiometrically.

- **lidar system** calibration Falls into two main categories:
- **instrument calibration** Factory calibration includes radiometric and geometric calibration unique to each manufacturer's hardware, and tuned to meet the performance specifications for the model being calibrated. Instrument calibration can only be assessed and corrected by the factory.
- **data calibration** The lever-arm calibration determines the sensor-to-GPS-antenna offset vector (lever arm) components relative to the antenna phase center; the offset vector components are re-determined each time the sensor or aircraft GPS antenna is moved or repositioned in any way. Because normal aircraft operations can induce slight variations in component mounting, field calibration is normally performed for each project, or even daily, to determine corrections to the roll, pitch, yaw, and scale calibration parameters.

check point (checkpoint) One of the surveyed points used to estimate the positional accuracy of a geospatial dataset against an independent source of greater accuracy.

classification (of lidar) The classification of lidar point cloud returns in accordance with a classification scheme to identify the type of target from which each lidar return is reflected. The process allows future differentiation between bare-earth terrain points, water, noise, vegetation, buildings, other manmade features and objects of interest.

cleanness A subjective term to describe the degree to which artifacts have been removed from a Digital Elevation Model (DEM) or Triangulated Irregular Network (TIN).

confidence level The probability that errors are within a range of given values.

consolidated vertical accuracy (CVA) The result of a test of the accuracy of a lidar digital terrain model using 3-D Quality Assurance/Quality Control (QA/QC) checkpoints in multiple land cover categories combined, normally including open terrain and vegetated terrain representative of the land cover for the area being tested. Computed using a nonparametric testing method (95th percentile), a consolidated vertical accuracy is always accompanied by a fundamental vertical accuracy. *See* percentile.

CONUS Continental United States; the contiguous 48 states.

culvert A tunnel carrying a stream or open drain under a road or railroad. Typically constructed of formed concrete or corrugated metal and surrounded on all sides, top, and bottom by earth or soil.

data void In lidar, a gap in the point cloud coverage, caused by surface non-reflectance of the lidar pulse, instrument or processing anomalies or failure, obstruction of the lidar pulse, or improper collection flight planning. Any area greater than or equal to (4 times the Nominal Pulse Spacing (NPS)) squared, measured using first-returns only, is considered to be a data void. *See* holiday.

datum A set of reference points on the Earth's surface against which position measurements are made, and (often) an associated model of the shape of the earth (reference ellipsoid) to define a geographic coordinate system. Horizontal datums (for example, the North American Datum of 1983 (NAD83)) are used for describing a point on the earth's surface, in latitude and longitude or another coordinate system. Vertical datums (for example, the North American Vertical Datum of 1988 (NAVD88)) measure elevations or depths. In engineering and drafting, a datum is a reference point, surface, or axis on an object against which measurements are made.

digital elevation model (DEM) See four different definitions below:

- A popular acronym used as a generic term for digital topographic and bathymetric data in all its various forms. Unless specifically referenced as a Digital Surface Model (DSM), the generic DEM normally implies x/y coordinates and z-values of the bare-earth terrain, void of vegetation and manmade features.
- As used by the U.S. Geological Survey (USGS), a DEM is the digital cartographic representation of the elevation of the land at regularly spaced intervals in x and y directions, using z-values referenced to a common vertical datum.
- As typically used in the United States and elsewhere, a DEM has bare-earth z-values at regularly spaced intervals in x and y directions; however, grid spacing, datum, coordinate systems, data formats, and other characteristics may vary widely.
- A "D-E-M" is a specific raster data format once widely used by the USGS. DEMs are a sampled array of elevations for a number of ground positions at regularly spaced intervals.

digital surface model (DSM) Similar to Digital Elevation Models (DEMs) or digital terrain models (DTMs), except that they may depict the elevations of the top surfaces of buildings, trees, towers, and other features elevated above the bare earth. DSMs are especially relevant for telecommunications management, air safety, forest management, and 3-D modeling and simulation.

digital terrain elevation data (DTED) A uniform matrix of terrain elevation values produced by the National Geospatial-Intelligence Agency (NGA). It provides basic quantitative data for military systems that require terrain elevation, slope, and gross surface roughness information. Data density depends on the level produced.

- DTED0 post spacing is 30-arc seconds (approximately 1,000 meters), corresponding to small-scale hardcopy products.
- DTED1 post spacing is 3-arc seconds (approximately 100 meters), corresponding to medium-scale hardcopy products.
- DTED2 post spacing is 1-arc second (approximately 30 meters), corresponding to large-scale hardcopy products.
- Specifications for high-resolution DTED levels 3 (10 m), 4 (3 m), and 5 (1 m) also may been developed.

digital terrain model (DTM) See two different definitions below:

- In some countries, DTMs are synonymous with Digital Elevation Models (DEMs), representing the bare-earth terrain with uniformly-spaced z-values.
- As used in the United States, DTMs may be similar to DEMs, but they may also incorporate the elevation of important topographic features on the land. DTMs are comprised of mass points and breaklines that are irregularly spaced to better characterize the true shape of the bare-earth terrain. The net result of DTMs is that the distinctive terrain features are more clearly defined and precisely located, and contours generated from DTMs more closely approximate the real shape of the terrain. Such DTMs are normally more expensive and time consuming to produce than uniformly spaced DEMs because breaklines are less suitable for automated collection; but the DTM may be technically superior to standard DEMs for many applications. DTMs are most commonly associated with ground surface models derived using stereo photogrammetry, as opposed to lidar.

discrete return lidar Lidar system or data in which the significant peaks in the waveform are captured and stored. Each peak represents a return from a different target, discernible in either vertical or horizontal domains. Most modern lidar systems are capable of capturing multiple discrete returns from each emitted laser pulse. *See* full waveform lidar.

drape The superposition of 2-D features over a 3-D surface, normally for viewing of all features in 3-D perspective, for 3-D fly-throughs or walk-throughs in virtual reality. Also a method for conflating values, typically elevation, from a surface to 2D features to create 3D features.

elevation The distance measured upward along a plumb line between a point and the geoid. The elevation of a point is normally the same as its orthometric height, defined as H in the equation:

$$H = h - N, \text{ where } h = \text{ellipsoid height and } N = \text{geoid height.}$$

elevation post The vertical component of a Digital Elevation Model (DEM) lattice mesh point, having height above the vertical datum equal to the z-value of its mesh point.

elevation post spacing The constant sampling interval in x and y directions of a Digital Elevation Model (DEM) lattice. The horizontal resolution of a DEM.

field of view (FOV), lidar The angular extent of the portion of object space surveyed by a lidar sensor, measured in degrees. To avoid confusion, a typical airborne lidar sensor with a field of view of 30 degrees is commonly depicted as ±15 degrees on either side of nadir.

first return (first-return) The first significant measurable portion of a return lidar pulse.

flightline A single pass of the collection aircraft over the target area. Often used incorrectly to refer to the data resulting from a flightline of collection. *See* swath.

full waveform lidar Lidar system or data in which the entire reflection of the laser pulse is fully digitized, captured, and stored. Discrete return point clouds can be extracted from the waveform data during post processing. *See* discrete return lidar.

fundamental vertical accuracy (FVA) The value by which vertical accuracy of lidar can be equitably assessed and compared among datasets. The fundamental vertical accuracy of a dataset must be determined with well-distributed checkpoints located only in open terrain, free of vegetation, where there is a high probability that the sensor will have detected the ground surface. It is obtained using standard tests for Root Mean Square Error (RMSE), where $FVA = ACC_z = RMSE_z \times 1.9600$.

geographic information system (GIS) A system of spatially referenced information, including computer programs that acquire, store, manipulate, analyze, and display spatial data.

geospatial data Information that identifies the geographic location and characteristics of natural or constructed features and boundaries of earth. This information may be derived from, among other things, remote-sensing, mapping, and surveying technologies. Geospatial data generally are considered to be synonymous with spatial data; however, the former always is associated with geographic or cartesian coordinates linked to a horizontal or vertical datum, whereas the latter (for example, generic architectural house plans) may include dimensions and other spatial data not linked to any physical location.

grid post *See* elevation post.

ground sample distance (GSD) The size of a pixel projected to the ground surface, reported as linear units/pixel, such as 1 meter/pixel or 1 foot/pixel. Actual or nominal distance in ground measurements between ground elevation samples. Different from horizontal post spacing in that horizontal post spacing describes the ground-distance interval of a uniform elevation grid, whereas ground sample distance describes the spacing on the ground of the source data. For example, the ground distance between data points collected by a lidar system, although variable, will yield some nominal ground sample distance. A gridded data model produced by resampling the lidar data may be incremented on a different, user-selected horizontal post spacing.

ground truth Verification of a situation, without errors introduced by sensors or human perception and judgment.

hillshade A function used to create an illuminated representation of the surface, using a hypothetical light source, to enhance terrain visualization effects.

holiday An area of missing coverage, caused by missing or unresolvable data, data edits, or incorrectly positioned flightlines, normally identified for further investigation or re-flying. An unintentionally unsurveyed part of a region that was to have been completely surveyed. *See* data void.

horizontal accuracy Positional accuracy of a dataset with respect to a horizontal datum. According to the National Standards for Spatial Data Accuracy (NSSDA), horizontal (radial) accuracy at the 95 percent confidence level is defined as $Accuracy_r$ (ACC_r).

horizontal post spacing The ground distance interval between grid posts in a uniformly gridded data model. It is important to note that features of a size equal to, or even greater than the post spacing, may not be detected or explicitly represented in a gridded model. For gridded elevation data, the horizontal post spacing may be referenced as the cell size, the grid spacing, the posting interval, or the ground sample distance.

hydraulic modeling The use of digital elevation data, rainfall runoff data from hydrologic models, surface roughness data, and information on hydraulic structures (for example, bridges, culverts, dams, weirs, sewers) to predict flood levels and manage water resources. Hydraulic models are based on computations involving liquids under pressure, and there are many other definitions of hydraulic modeling that have nothing to do with terrain elevations, for example, modeling of hydraulic lines in aircraft and automobiles.

hydrologic modeling The computer modeling of rainfall and the effects of land cover, soil conditions, and terrain slope to estimate rainfall runoff into streams and lakes. Digital elevation data are used as part of hydrologic modeling.

hydrologically-conditioned (hydro-conditioned) Processing of a Digital Elevation Model (DEM) or Triangulated Irregular Network (TIN) so that the flow of water is continuous across the entire terrain surface, including the removal of all spurious sinks or pits. The only sinks that are retained are the real ones on the landscape. Whereas hydrologically-enforced is relevant to drainage features that generally are mapped, hydrologically-conditioned is relevant to the entire land surface and is done so that water flow is continuous across the surface, whether that flow is in a stream channel or not. The purpose for continuous flow is so that relations/links among basins/catchments can be known for large areas.

hydrologically-flattened (hydro-flattened) Processing of a lidar-derived surface (Digital Elevation Model (DEM) or Triangulated Irregular Network (TIN)) so that mapped water bodies, rivers, reservoirs, and other cartographically polygonal water surfaces are flat and, where appropriate, level from bank-to-bank. Additionally, surfaces of rivers and long reservoirs demonstrate a gradient change in elevation along their length, consistent with their natural behavior and the surrounding topography. In traditional photogrammetrically-compiled mapping, this process is accomplished automatically through the inclusion of measured breaklines in the digital terrain model (DTM). However, because lidar does not inherently include breaklines, a DEM or TIN derived solely from lidar points will depict water surfaces with unsightly and unnatural artifacts of triangulation. The process of hydro-flattening typically involves the addition of breaklines along the banks of specified water bodies, rivers, ponds, and streams. These breaklines establish elevations for the water surfaces that are consistent with the surrounding topography, and produce aesthetically acceptable water surfaces in the final derived DEM or TIN. Unlike hydro-conditioning and hydro-enforcement, hydro-flattening is not driven by any hydrologic and hydraulic (H&H) modeling requirements, but solely by cartographic mapping needs.

hydrologically-enforced (hydro-enforced) Processing of mapped water bodies so that lakes and reservoirs are level and so that streams flow downhill. For example, a Digital Elevation Model (DEM), Triangulated Irregular Network (TIN) or topographic contour dataset with elevations removed from the tops of selected drainage structures (bridges and culverts) so as to depict the terrain under those structures. Hydro-enforcement enables hydrologic and hydraulic models to depict water flowing under these structures, rather than appearing in the computer model to be dammed by them because of road deck elevations higher than the water levels. Hydro-enforced TINs also use breaklines along shorelines and stream centerlines, for example, where these breaklines form the edges of TIN triangles along the alignment of drainage features. Shore breaklines for streams would be 3-D breaklines with elevations that decrease as the stream flows downstream; however, shore breaklines for lakes or reservoirs would have the same elevation for the entire shoreline if the water surface is known or assumed to be level throughout.

intensity, lidar For discrete-return lidar instruments, intensity is the recorded amplitude of the reflected lidar pulse at the moment the reflection is captured as a return by the lidar instrument. Lidar intensity values can be affected by numerous factors such as the instantaneous setting of the instrument's Automatic Gain Control, angle of incidence, and so forth, and cannot be equated to a true measure of energy. In full-waveform systems, the entire reflection is sampled and recorded, and true energy measurements can be made for each return or overall reflection. Intensity values for discrete returns derived from a full-waveform system may or may not be calibrated to represent true energy.

Lidar intensity data make it possible to map variable textures in the form of a gray-scale image. Intensity return data enable automatic identification and extraction of objects such as buildings and impervious surfaces, and can aid in lidar point classification. In spite of their similar appearance, lidar intensity images differ from traditional panchromatic images in several important ways:

- Lidar intensity is a measure of the reflection of an active laser energy source, not natural solar energy.
- Lidar intensity images are aggregations of values at point samples. The value of a pixel does not represent the composite value for the area of that pixel.
- Lidar intensity images depict the surface reflectivity within an extremely narrow band of the infra-red spectrum, not the entire visible spectrum as in panchromatic images.
- Lidar intensity images are strongly affected by the angle of incidence of the laser to the target, and are subject to unnatural shadowing artifacts.
- The values on which lidar intensity images are based may or may not be calibrated to any standard reference. Intensity images usually contain wide variation of values within swaths, between swaths, and between lifts.

For these reasons, interpretation and analysis of lidar intensity images must be performed with unusually high care and skill.

Kalman filter A recursive minimum variance estimation technique that uses statistical models to weight each new measurement relative to past information and estimate quantities of interest having statistical characteristics. The technique can be considered a recursive least squares adjustment.

last return The last significant measurable portion of a return lidar pulse.

lattice A 3-D surface representation method created by a rectangular array of points spaced at a constant sampling interval in x and y directions relative to a common origin. A lattice differs from a grid in that it represents the value of the surface only at the lattice mesh points or elevation posts of the lattice, rather than the elevation of the cell area surrounding the centroid of a grid cell.

lever arm A relative position vector of one sensor with respect to another in a direct georeferencing system. For example, with aerial mapping cameras, there are lever arms between the inertial center of the Inertial Measurement Unit (IMU) and the phase center of the Global Positioning System (GPS) antenna, each with respect to the camera perspective center within the lens of the camera.

lidar An instrument that measures distance to a reflecting object by emitting timed pulses of light and measuring the time difference between the emission of a laser pulse and the reception of the pulse's reflection(s). The measured time interval for each reflection is converted to distance, which when combined with position and attitude information from Global Positioning System (GPS), Inertial Measurement Unit (IMU), and the instrument itself, allows the derivation of the 3-dimensional point location of the reflecting target's location.

lift A lift is a single take-off/landing cycle of a collection platform (fixed or rotary wing) within a lidar data collection project.

local accuracy *See* accuracy.

metadata Any information that is descriptive or supportive of a geospatial dataset. This would include formally structured and formatted metadata files (for example, eXtensible Markup Language (XML)-formatted Federal Geographic Data Committee (FGDC) metadata), reports (collection, processing, Quality Assurance/Quality Control (QA/QC)), and other supporting data (for example, survey points, shapefiles).

nadir The point or line directly beneath the collection platform , corrected for attitude variations. In lidar, this would correspond to the centerline of a collected swath.

nominal pulse spacing (NPS) A common measure of the density of a lidar dataset, it is the typical or average lateral distance between points in a lidar dataset, most often expressed in meters. It is often and most simply calculated as the square root of the average area per point. This value is predicted in mission planning and empirically calculated from the collected data. In high-density collections (<1 meter NPS), this may be directly expressed as Points Per Square Meter (ppsm). $PPSM = 1/NPS^2$.

OCONUS Outside of CONUS; not within the contiguous United States. *See* CONUS.

overage Those parts of a swath that are not necessary to form a complete single, non-overlapped, gap-free coverage with respect to the adjacent swaths. The non-tenderloin portions of a swath. *See* overlap, tenderloin.

overlap Any potion of a swath that also is covered by any portion of any other swath. *See* overage, tenderloin.

percentile Although there is no accepted single definition of percentile, the following general definition can be used: A percentile is a measure that tells us what percent of the total frequency scored at or below that measure. A percentile rank is the percentage of scores that fall at or below a given score. The National Institute of Standards and Technology (NIST) recommends the following as the preferred method for calculating percentile:

$$n=(P / 100) * (N + 1).$$

NIST notes the following method, used by Microsoft Excel and other software packages, as an alternative:

$$n=((P / 100) * (N - 1) + 1)$$

As both methods yield similar results, particularly with large samples, either may be used.

point classification The assignment of a target identity classification to a particular lidar point or group of points.

point family The complete set of multiple returns reflected from a single lidar pulse.

point cloud A large set of three dimensional points, typically from a lidar collection. As a basic geographic information system (GIS) data type, a point cloud is differentiated from a typical point dataset in several key ways:

- Point clouds are almost always 3D,
- Point clouds have an order of magnitude more features than point datasets, and
- Individual point features in point clouds do not typically possess individually meaningful attributes; the informational value in a point cloud is derived from the relations among large numbers of features

preprocessing In lidar, the preprocessing of data most commonly refers to those steps employed in converting the collected Global Positioning System (GPS), Inertial Measurement Unit (IMU), instrument, and ranging information into an interpretable *X-Y-Z* point cloud. This includes generation of trajectory information, calibration of the dataset, and controlling the dataset to known ground references.

postprocessing In lidar, postprocessing refers to the processing steps applied to lidar data point clouds. This includes point classification, feature extraction (for example, building footprints, hydrographic features, and others), tiling, and generation of derivative products (Digital Elevation Models (DEMs), Digital Surface Model (DSMs), intensity images, and others).

scan angle *See* field of view.

supplemental vertical accuracy (SVA) The result of a test of the accuracy of a lidar digital terrain model using 3-D Quality Assurance/Quality Control (QA/QC) checkpoints in a single land cover category. Computed using a nonparametric testing method (95th percentile), a supplemental vertical accuracy is always accompanied by a fundamental vertical accuracy. *See* percentile.

spatial distribution In lidar, the regularity or consistency of the point density within the collection. The theoretical ideal spatial distribution for a lidar collection is a perfect regular lattice of points with equal spacing on *X* and *Y* axes. Various factors prevent this ideal from being achieved, including the following factors:

- Instrument design (oscillating mirrors),
- Mission planning (difference between along-track and cross-track point spacing), and
- In-flight attitude variations (roll, pitch, and yaw).

swath The data resulting from a single flightline of collection. *See* flightline.

tenderloin The central portion of the swath that, when combined with adjacent swath tenderloins, forms a complete, single, non-overlapped, gap-free coverage. *See* overage, overlap.

triangulated irregular network (TIN) A vector data structure that partitions geographic space into contiguous, non-overlapping triangles. In lidar, the vertices of each triangle lidar points with x-, y-, and z-values. In most geographic applications, TINs are based on Delaunay triangulation algorithms in which no point in any given triangle lies within the circumcircle of any other triangle.

well-distributed For a dataset covering a rectangular area that has uniform positional accuracy, checkpoints should be distributed so that points are spaced at intervals of at least 10 percent of the diagonal distance across the dataset and at least 20 percent of the points are located in each quadrant of the dataset (adapted from the National Standards for Spatial Data

Accuracy (NSSDA)). As related to this specification, these guidelines are applicable to each land cover class for which checkpoints are being collected.

withheld Within the LAS file specification, a single bit flag indicating that the associated lidar point is geometrically anomalous or unreliable and should be ignored for all normal processes. These points are retained because of their value in specialized analysis. Withheld points typically are identified and tagged during preprocessing or through the use of automatic classification routines. Examples of points typically tagged as withheld are listed below:

- Spatial outliers in either the horizontal or vertical domains
- Geometrically unreliable points near the edge of a swath

9.4 QUALITY ASSURANCE/QUALITY CONTROL REPORTING

As for the best method of determining the accuracy of lidar data products, correct practices by the data providers, as previously discussed, are the safest ways to assure the desired accuracy is achieved. Often times data providers limit their accuracy verification to their office activities and seldom send the map to the field for verification. This makes sense as the vendor knows what went into the production of the map, backed by a check list of acceptable practices mentioned earlier. Additional measures can be considered by the data providers for added assurance and increased accuracy of the data. Existing reliable geospatial data is an excellent and inexpensive means to check a map. In addition, a data provider can use automated and manual routines to verify tile to tile edge matching.

In the United States, three different map standards or guidelines are used to determine the accuracy of the geospatial products. Those are: the National Map Accuracy Standard (NMAS), the American Society for Photogrammetry and Remote Sensing (ASPRS) standard, and finally, the National Standard for Spatial Data Accuracy (NSSDA). The three different standards employ the same philosophy of utilizing an independent and more accurate survey technique to assess the measurement made from the map product. However, they differ in their approach of limiting the accuracy figure and the confidence level. In general, at least 20 well-distributed points throughout the project area and visible on the map are to be accurately surveyed, and the map-derived positions are compared to the field survey values and statistically analyzed to come up with the correct accuracy figure.

Use of the latest guidelines for the map accuracy standard or the NSSDA is recommended, as most government agencies both endorsed it and use it. However, the NSSDA is not a standard in the same manner that NMAS and ASPRS are considered. The NSSDA is more a set of guidelines to safeguard the statistical map accuracy checking practices. The NSSDA is based on 95 percent confidence level using the Root Mean Square Errors (RMSE) as the variable. The NSSDA did not specify the values for the RMSE that need to be used for each map scale, but left it to be agreed upon between the data provider and the client.

To follow a scientifically sound approach, I recommend the use of either the standard deviation (sigma) or the RMSE values that are given by the ASPRS standard for use with the NSSDA. In this case you satisfy both groups of map standards, the ASPRS which is widely used among large scale clients, and the new NSSDA which is strongly recommended by the Federal Geographic Data Committee (FGDC). To provide a numerical example for the above argument, let us assume that you are providing orthophoto products with a map scale of 1" to 100' (or 1:1200) and a lidar-derived digital terrain model meeting the accuracy of 2' contour intervals. Assume that your client is endorsing the NSSDA standard for both horizontal and vertical accuracy. According to the ASPRS standard, the two products should meet the following horizontal and vertical accuracies:

- Horizontal accuracy: RMSE = 1 ft or 30 cm,
- Vertical accuracy: RMSE = 1/3 * 2.0 = 0.67 ft or about 20 cm.

According to the NSSDA standard, the two products shall satisfy the following criteria:

- Horizontal Accuracy$r$ = 2.4477 * RMSEx = 2.4477 * RMSEy
  = 2.4477 * RMSEr /1.4142
  = 1.7308 * RMSEr
- Vertical accuracy: 95 Percent of the checked elevation should be within 1.96 $RMSE_v$

If we adopt the RMSE that is given by ASPRS, the two accuracy figures according to NSSDA will be:

- Horizontal Accuracy$r$ = 2.45 ft
- Vertical accuracy of lidar DEM = 1.96 * 0.67 = 1.3 ft or 40 cm

Here you may noticed that we totally ignored the vertical accuracy requirements for the spot height called for by the ASPRS standards, as it is not applicable with lidar data as a lidar dataset has unified accuracy across the data. Table 9.4, provide a sample computation or estimation of the accuracy for a project with 23 check points scattered around the project area. The calculated quantities are Residuals, Mean (which may provide idea about biases in the measurements), standard deviation, and the Root Mean Square Error (RMSE). The residuals is the difference between the measured value from the map or the dataset and the field surveyed value for the same point. Therefore,

$$Residual\ (v_i) = x_{\text{measured, i}} - x_{\text{survey, i}} \quad \text{or} \quad y_{\text{measured, i}} - y_{\text{survey, i}} \quad \text{or} \quad H_{\text{measured, i}} - H_{\text{survey, i}}$$

Also,

$$\text{Mean } (\mu) = \frac{\sum_{i-1}^{n} vi}{n}$$

$$\text{Standard Deviation (StDEV)} = \frac{\sum_{i=1}^{n} (vi-\mu)}{n}$$

$$\text{Root Mean Square Error (RMSE)} = \frac{\sum_{i-1}^{n} vi.vi}{n}$$

where:

$x_{\text{measured, i}}$, $y_{\text{measured, i}}$ are the coordinates of the ith check point in the dataset

$x_{\text{survey, i}}$, $y_{\text{survey, i}}$ are the coordinates of the ith check point in the independent source of higher accuracy

n is the number of check points used for the accuracy verification

i is an integer refer to the points sequence

The troubling issue with former standards comes from relying on only 20 points to evaluate a lidar dataset, while ignoring other quality aspects of the surface model such as point density, noise and misalignments between flying lines and along seam lines, among others.

Lack of understanding of statistical terms often leads to user disappointment in the data they receive. Whether dealing with lidar data, GIS vector database, or ortho-rectified maps, many end-users judge the final products with a standard that may differ from the specification they agreed upon with the data providers. For example, if a user requested a lidar dataset with an accuracy that meets class 1 ASPRS standards for 0.3 m (or 1 ft) contours, the data provider has to guarantee that the delivered product shall have a vertical accuracy of 10 cm as RMSE. The user may reject the data if he or she experienced a localized occurrence where the data did not meet the 10 cm criteria. This is a legitimate concern and one that can be addressed by examining the end-user's accuracy requirement more carefully. When the user awards a contract to the data provider under the accuracy requirements we mentioned earlier, it needs to be understood that by requesting an RMSE of 10 cm, the end-user accepts the statistical reality of the specified vertical accuracy. When we say an RMSE of 10 cm, it refers to an acceptance criterion that requires only about 68% of the data to meet the 10 cm and the remaining 32% of the data could have an error of up to 2.5 to 3 times the RMSE (or 25 to 30 cm in this instance). On the other hand, if the user wants to guarantee that most of the data meets 10 cm, he or she should request a different statistical criteria, such as 95%. The detriment with requesting 95% is reflected in the cost of achieving such tight accuracy. For the dataset to meet 10 cm vertical accuracy at 95%, the RMSE should be around 5.1 cm.

I would like to emphasize that the criteria of RMSE is a sound statistical approach and is widely employed by surveyors and engineers. The issue is not in the statistical approach we use but in the proper understanding of these methods. The vertical accuracy statement of 10 cm as RMSE is another way of saying that about 68% of the data should be within 10 cm and it is the same as saying 95% of the data should be within 19.6 cm. If the user understood the previous interpretation of the RMSE measure, they would not be concerned when a small portion (less than 32%) of the data deviated by more than 10 cm. Education is crucial and it is the responsibility of the guardian of the geospatial data accuracy standard(s) to educate the public on the intent of the standard and the statistical concepts behind it. However, it is also the responsibility of the user and data provider to educate themselves on the standard and how it applies to their data. Without proper knowledge of the standards, the cost of geospatial data will be inflated as data providers strive to achieve perfection in the final product in order to meet exaggerated expectations. Similar problems are also common among users of orthophoto maps because most users expect their dataset to be perfect, to the degree that they will not accept a few shifted pixels between tiles in the dataset, despite the fact that it meets all the routine accuracy testing as set by the National Standard for Spatial Data Accuracy (NSSDA).

The lack of clear and detailed guidelines on how to test and accept different geospatial data types often causes confusion. None of the popular map accuracy standards or guidelines, for example, addresses the quality of seamlines and the shift that may occur between data from adjacent flight lines, be it imagery or lidar. Many users expect the dataset to be seamless despite the fact that it meets all geometrical testing criteria called for by the different map standards.

The common understanding among the geospatial data community for the word "seamless" is a zero tolerance or no tolerance for any shift between tiles or flight lines even if it is only a pixel or two for an ortho map and few centimeters if it is elevation data. Such high quality requirements force the data providers to Photoshop™ the data so adjacent ortho tiles fit together perfectly or to apply some kind of polynomial fitting if it is elevation data. While this enforcement results in aesthetically appealing data, it can cause unwanted warp in the data, rendering it less accurate.

Another example on the weakness of the testing procedure of the NSSDA standard is the check points requirement. The NSSDA standard calls for 20 well-defined, high-fidelity points in order to perform accuracy verification. This requirement often causes confusion among data users and providers alike. For large areas, such as an entire county or state, the project is usually divided into multiple production blocks and users are divided on whether the 20 points is for each block or the entire state. Even if we assume that the points are for the entire state or county and the dataset meets the accuracy criteria on these 20 points, none of the other geometric anomalies that may occur in areas outside the 20 locations of the ground control check points are identified or addressed in any standard. The seamline issue mentioned above falls in this category, especially if it is located far from the location of the 20 check points.

With most lidar datasets, a minor stepping effect of 5 cm or less in a stretch of road that is only a few hundred meters in length does not affect the overall quality of the data. However,

Table 9.4 Sample accuracy computation.

Point ID	Measured Values			Surveyed Values			Residuals		
	Easting (E) *meter*	Northing (N) *meter*	Elevation (H) *meter*	Easting (E) *meter*	Northing (N) *meter*	Elevation (H) *meter*	ΔE (Easting) *meter*	ΔN (Northing) *meter*	ΔH (Elevation) *meter*
207	359584.354	5142449.934	477.027	359584.314	5142450.004	477.098	0.040	−0.070	−0.071
207A	359872.330	5147939.180	412.406	359872.290	5147939.250	412.384	0.040	−0.070	0.022
208	359893.029	5136979.824	487.222	359892.989	5136979.894	487.190	0.040	−0.070	0.032
201	359927.144	5151084.129	393.571	359927.264	5151083.979	393.561	−0.120	0.150	0.010
203	372737.074	5151675.999	451.305	372736.844	5151675.879	451.318	0.230	0.120	−0.013
206	373282.550	5137216.753	466.579	373282.320	5137216.633	466.567	0.230	0.120	0.012
204	383714.817	5135495.256	452.954	383714.587	5135495.136	452.933	0.230	0.120	0.021
202	383908.812	5148006.957	492.029	383908.932	5148006.807	492.067	−0.120	0.150	−0.038
205	384026.003	5153418.148	488.902	384025.773	5153418.028	488.890	0.230	0.120	0.012
110	397996.280	5171300.737	440.296	397996.380	5171300.782	440.322	−0.100	−0.045	−0.026
113	400459.859	5179949.119	361.217	400459.979	5179948.969	361.268	−0.120	0.150	−0.051
104	400758.695	5167097.114	543.830	400759.007	5167097.107	543.816	−0.312	0.007	0.014
101	400794.774	5188864.498	263.666	400795.086	5188864.491	263.719	−0.312	0.007	−0.053
105	411418.398	5177915.793	525.953	411418.458	5177915.713	525.920	−0.060	0.080	0.033
109	413983.652	5170924.872	546.530	413983.752	5170924.917	546.574	−0.100	−0.045	−0.044
102	416177.444	5185743.870	320.442	416177.756	5185743.863	320.467	−0.312	0.007	−0.025
107	435497.773	5180055.008	345.719	435497.833	5180054.928	345.664	−0.060	0.080	0.055
111	435725.456	5165270.316	468.886	435725.556	5165270.361	468.892	−0.100	−0.045	−0.006
108	435979.436	5175221.390	443.009	435979.496	5175221.310	443.026	−0.060	0.080	−0.017
103	439669.309	5188155.815	190.855	439669.621	5188155.808	190.813	−0.312	0.007	0.042
114	448111.544	5184558.142	190.447	448111.664	5184557.992	190.458	−0.120	0.150	−0.011
112	450709.272	5164362.790	433.864	450709.372	5164362.835	433.851	−0.100	−0.045	0.013
106	452302.471	5175490.022	226.278	452302.531	5175489.942	226.230	−0.060	0.080	0.048
						Number of check points	23	23	23
						Mean	**−0.06**	**0.05**	**0.00**
						StDEV	0.17	0.08	0.03
						RMSE	**0.18**	**0.09**	**0.03**

it may causethe user to reject the delivery despite the fact that the dataset meets the accuracy requirements of RMSE = 10 cm when checked against ground check points. Similar concerns arose when users evaluate a lidar-based digital terrain model that contains photogrammetric break lines and lidar points. None of the map standards address the issue or provided guidelines on how to deal with situations where the end user finds out that 6% of the breaklines in one tile are in disagreement with the lidar elevation by more than the RMSE value. For acceptable non-seamless data that meet the NSSDA accuracy check, a standard needs to be developed to clearly provide guidelines on details of testing and to clearly define what is and is not acceptable outside the ground control locations. Such guidelines, once published, will become the industry standard.

The ASPRS is leading a national efforts to develop new geospatial data testing and evaluation. Until such time it is completed, I suggest a different approach to the lidar data accuracy from the three existing national standards mentioned above.

The following guidelines, if adopted, provide different classes for the accuracy and quality measures. Users will have the ability to select different accuracy and quality classes to match their specific needs and applications.

a) Classification According to Lidar Point Accuracy:

1. **Engineering class-I grade lidar data accuracy,** for products with:
   Horizontal accuracy of $RMSE_X = RMSE_Y$ = 20 cm or better
   Vertical accuracy of $RMSE_v$ = 5 cm or better
2. **Engineering class-II grade lidar data accuracy,** for products with:
   Horizontal accuracy of $RMSE_X = RMSE_Y$ = 30 cm or better
   Vertical accuracy of $RMSE_v$ = 10 cm or better
3. **Planning class-I grade lidar data accuracy,** for products with:
   Horizontal accuracy of $RMSE_X = RMSE_Y$ = 0.60 m or better
   Vertical accuracy of $RMSE_v$ = 20 cm or better
4. **Planning class-II grade lidar data accuracy,** for products with:
   Horizontal accuracy of $RMSE_X = RMSE_Y$ = 0.75 m or better
   Vertical accuracy of $RMSE_v$ = 30 cm or better
5. **General purpose grade lidar data accuracy,** for products with:
   Horizontal accuracy of $RMSE_X = RMSE_Y$ = 1.2 m or better
   Vertical accuracy of $RMSE_v$ = 0.50 m or better
6. **User defined Accuracy,** for products that do not fit into any of the previous five categories.

The above accuracy figures should only be guaranteed in open flat and rolling terrains.

b) Classification According to lidar Surface Definitions (quality):

1. **Engineering class-I grade lidar data quality,** for a lidar surface with:
   a) Nominal post spacing of 0.30 m or less
2. **Engineering class-II grade lidar data quality,** for a lidar surface with:
   a) Nominal post spacing of 0.70 m or less
   b) To have optional break lines
3. **Planning class-I grade lidar data quality,** for a lidar surface with:
   a) Nominal post spacing of 1.0 m or less
   b) To have optional break lines
4. **Planning class-II grade lidar data quality,** for a lidar surface with:
   a) Nominal post spacing of 1.5 m or less
   b) To have optional break lines
5. **General purpose grade lidar data quality,** for a lidar surface with:
   a) Nominal post spacing of 2.0 m or larger

b) To have optional break lines

6. **User defined quality,** for products that do not fit into any of the previous five categories.

The above concept clearly describes the overall accuracy and quality of lidar data in a way that can be easily understood by both users and data providers. For example, if the user requests lidar data to meet **engineering class-I grade accuracy and engineering class-II grade quality**, the data provider must commit to the delivery of lidar surface according to the following specifications:

Horizontal Accuracy: $RMSE_X = RMSE_Y = 20$ cm or better

Vertical accuracy of: $RMSE_v = 5$ cm or better

Nominal post spacing: 0.70 m with optional breaklines, if requested by the user.

Additional details can be added to the standard, such as specifying accuracy and quality figures according to land use/land cover classes, specifying quality of seam lines, definition of bare-earth and feature removal, and perhaps different standards for a bare-earth digital elevation model versus reflective surface.

Finally if you are involved in evaluating geometrical accuracy for a lidar dataset, the following guidelines may help you to succeed in your task:

1) Make sure the 20 check points are well distributed around the project areas. Check points in proximity from each others (clusters) should be considered as one check points in computing the statistics
2) Evaluate different ground cover categories separately. In other words, do not mix results of check points situated in flat and open ground with results of check points situated in densely vegetated areas
3) Although it is difficult to prove, try to evaluate the horizontal accuracy using intensity images against existing GIS data such as road centerlines and edges
4) For the vertical accuracy verification, compute each check point in the set and aggregate the residuals of all points to calculate mean, standard deviation, root mean squares error (RMSE), and the error at 95% confidence level. The mean is useful in revealing and estimating vertical biases in the data if it is evaluated against the standard deviation.
5) Examine the quality of the seamlines where two flight lines come together.

There can often be a problem along these lines

6) Do not evaluate gridded or re-sampled data. The original point cloud is the most authentic set to evaluate the delivery. Even if you ask for gridded data, always ask for the cleaned point cloud to run your accuracy check on it.
7) Unwanted or unexpected re-sampling or re-gridding of the data can indicate that the data provider hid noisy data.

AUTHORS

Sections 9.1 and 9.4 were written by Qassim Abdullah, Ph.D., PLS, CP, Chief Scientist at Fugro Earthdata, Inc.; Sections 9.2 and 9.3 were written by Hans Karl Heidemann, Physical Scientist at USGS/EROS.

10
APPLICATIONS

HANS KARL HEIDEMANN, JASON STOKER, DAVID BROWN, MICHAEL J. OLSEN, RON SINGH, KEITH WILLIAMS, ABBY CHIN, ALVAN KARLIN, GORDON MCCLUNG, JAMISON JANKE, JIE SHAN, KYOHYOUK KIM, APARAJITHAN SAMPATH, SERKAN URAL, CHRISTOPHER E. PARRISH, KIRK WATERS, JENNIFER WOZENCRAFT, CHRISTOPHER L. MACON, JOHN BROCK, C. WAYNE WRIGHT, CHRIS HOPKINSON, ALAIN PIETRONIRO, IAN MADIN, and JEREMY CONNER

10.1 DIGITAL ELEVATION MODELS

10.1.1 Introduction

Previous sections have defined that the word "lidar" refers only to the collected point cloud data. This section will explore six commonly delivered products derived from that point cloud. In addition to the point cloud itself, Bare-Earth Digital Elevation Models (DEMs), First-Return Digital Surface Models (DSMs), Lidar Intensity Imagery, and Lidar-Derived Breaklines form the core deliverables of most commercial lidar data collections. Digital Contours continue to be requested, although by fewer users than in years past. Triangulated Irregular Networks (TINs) and the Esri Terrain Feature Class implementation of the TIN are data structures, typically used for bare-earth surfaces not a derivative per se. Because Terrain Models are continuously growing in popularity as a bare-earth surface deliverable, they will be also reviewed in detail here.

The exact methods used to generate derivative products from lidar point data varies among vendors, where each has developed techniques and processes that best serve their existing business models, internal organization and resources, and their client's needs. As with other lidar processing steps there is no single "correct" process. In the following discussions some of the more widely used and accepted techniques will be showcased. This should not be interpreted as a dismissal of lesser-known methods: the mainstream techniques used today were once considered advanced, and customized techniques are often used to develop products with unusual requirements, or from atypical datasets.

Effective production of derivatives from lidar point data requires, in most cases, individual points to possess some level of attribution. Each product relies on specific attributes. Many point attributes are inherent to the system and are populated when the original LAS files are generated. Examples of these include the X-Y-Z values, Return Number, Number of Returns, Intensity, and Scan Angle. Other attributes may be added or populated through analysis during postprocessing: classification, location within the swath or relative to adjacent swaths (overage), height above ground, etc. It is important to bear in mind that the points, as collected, do not inherently possess any information about the target from which they were reflected. Points alone cannot identify the target as a road, tree, or building. This information is largely inferred by skilled analysts using intelligent interpretation of the spatial relationships between points, and by referencing ancillary support data. These processes may be automated or manual. However the attribute values, specifically classification, are achieved, the quality and usability of derived products depends on the quality of all previous processing.

10.1.2 Bare-Earth Digital Elevation Models (DEMs)

Lidar technology was initially developed within the commercial arena primarily as a fast and relatively inexpensive means to produce bare-earth surface models. Its advantages, when compared to traditional photogrammetry, were a significantly higher density of points, the ability to take ground measurements within wooded or vegetated areas, the option of day or night-time collections, and fast turn-around from collection to delivery. Although lidar cannot produce the vertical accuracies of very large-scale (i.e., 1" = 40') stereo photogrammetry, its advantages for municipal and county scale terrain mapping are without question.

The disadvantages of lidar in terrain mapping are more complex. A lidar point will be created from any object that sufficiently reflects the laser pulse: the ground, buildings, trees, birds, other man-made structures, and sometimes water. Unlike photogrammetry, lidar points are located randomly (although there may be a non-random pattern to the points, the operator has no control over the placement of individual points), and it is impossible to take measurements of specific objects or locations. Unlike photogrammetry, linear features within the terrain surface cannot be measured and represented with breaklines. The increased density of points mitigates this deficiency to a degree, but does not remove it. And although lidar does effectively collect measurements in areas of natural shadow and under trees and vegetation, there is no guarantee that the measurements are of the ground, rather than of other objects hidden within the shadows (more vegetation or other structures).

The determination of which points actually represent the ground is highly subjective. Performing point classification solely from the points' elevations can lead to significant errors in the final DEM surface. A lidar collection does not require collecting concurrent imagery; it is often preferable to collect lidar at night to avoid air traffic and turbulence. Collecting concurrent imagery can therefore be an expensive addition to a lidar project. Many lidar projects are processed without the benefit of readily interpretable imagery to assist in point classification. Lidar intensity imagery, available from most modern lidar instruments, can provide some relief in this task, but requires more highly skilled interpretation. The use of other existing orthophotography must be done with careful consideration to the currency of the imagery and any differences in the environmental conditions (water levels, crop cycles, etc.) that exist between the imagery and lidar.

These fundamental points are being reiterated because it is critical for users of lidar-derived DEMs to understand the limitations of the data on which those surfaces are based. In its earliest days, lidar was often touted as a solution to all terrain surface shortcomings. While this tag certainly helped sell the technology and assure its ultimate market success, it simply is not true: Lidar is a merely another tool we can use for terrain mapping; it has its strengths and weaknesses, and there are situations where it is not the best solution.

10.1.2.1 Nominal Pulse Spacing (NPS) versus Ground Sample Distance (GSD)

One common misconception about lidar is that a given NPS will result in the same density of ground points; moreover, that the GSD will be uniform across a project. Equating GSD with NPS may be valid in clear and open areas, but it is not the case if there are buildings, other structures, or vegetation present. For most projects, final density of ground points will be lower (perhaps substantially) than the project's NPS depending on the environmental conditions within the project AOI. Heavily wooded and vegetated areas will be particularly sparse, and the ground surface in those areas will typically be of lower accuracy, detail, and reliability. This should not be any cause for alarm: it is no different from a traditional photogrammetric DEM, in which "obscured areas" are delineated and the contours are rendered as dashed lines to indicate they are in fact, best guesses. Although not currently a standard practice, delivery of some representation of these lower confidence areas (density raster or vector polygon layer) is becoming a more frequently requested deliverable. If a user has a minimum GSD requirement for some areas or the entire project, this requirement should be made clear to the data provider during contract development, and an understanding that even the best vendor's best effort cannot absolutely guarantee a final GSD in all areas.

10.1.2.2 Topographic DEMs versus Hydrologic DEMs

DEM is used to describe a number of different products, and it is necessary to make some distinctions between some of the more common meanings of the term. First, we need to be careful to distinguish between a DEM, which means a raster bare-earth elevation surface of some type; and a (*.dem), the US Geological Survey's data format for their DEMs. *.dems have largely been supplanted by more common industry formats such as the Esri GRID, GeoTIFF, the ERDAS .IMG, and numerous others. This section refers to DEMs in any data format, and there is no implicit association to the USGS *.dem format.

Over the past decade there has been much confusion over what a DEM is, largely due to the generalized use of the term and a lack of clarity and specificity on what the purpose of the DEM is. Generally speaking there are two types of DEMs: Topographic and Hydrologic, though within each type there are further variations. As the name implies, a topographic DEM is a representation of how the *ground looks* from above: "topo-graphic." This is the DEM that most users are familiar with. Topographic DEMs are the basis for most traditional mapping, USGS Topographic Quad Maps, and modern digital contour generation. Historically, they have been produced using a variety of methods, each with minor variations in the character of the final products, but all with some consistent common characteristics:

- Elevated Bridge Decks are removed.
- Surfaces over Culverts are retained (this created apparent "dams" in the water course, making the DEM unusable for many hydrologic applications).
- Water Surfaces, where visible from above, are smooth and represent a natural, logical elevation (uniform or gradient).
- Water Surface Elevations (WSELs) are typically not calibrated or referenced to any metric.

A Hydrologic DEM, as its name implies, represents how *water works* on the surface: "hydro-logic." Once confined to use by Hydrologic and Hydraulic (H&H) engineers, this type of DEM has recently become more common, as interest in and concerns about our water resources have grown. As with topographic DEMs, there are a number of variations of the hydrologic DEM, but most possess these characteristics:

- Elevated Bridge Decks are removed.
- Surfaces over Culverts are *removed* (this leaves cuts through road and other transportation surfaces making the DEM unusable for contour mapping and numerous other topographic applications).
- Water Courses, whether actually visible from above or not, maintain continuity of flow.
- Water Surfaces are smooth and represent a natural, logical elevation (uniform or gradient).
- WSELs are often set to some known stage.

In 2001, the State of North Carolina undertook the first statewide, lidar-based, comprehensive update to its FEMA Flood Insurance Rate Map (FIRM) catalog. Because the lidar was being collected specifically for flood modeling, the H&H engineers required specialized hydrologic DEMs to be developed by the lidar data provider. Although primarily intended as a working product for the engineers, the DEMs also became available to North Carolina's 100 Counties, and to the public. The term "hydro-enforced DEM" went into common usage across the lidar industry and was widely (and incorrectly) perceived to be an improvement over the traditional topographic DEM. More than one county customer specified a hydro-enforced lidar DEM only to find the surface inadequate for the contours they needed (though their

Engineering Departments were usually delighted). Today, we better understand that topographic and hydrologic DEMs are simply two *different* surfaces, each with their own properties and applications. Neither is any better or worse than the other, and both can be generated from the same point cloud and breakline data with planning.

10.1.2.3 Lidar-derived DEM Resolution and Reliability

The highest resolution DEM that can be realistically produced from a given lidar data set will be of no finer resolution than the NPS of the source lidar data. Hence, lidar collected at an NPS of 1.4 meters can be used to produce a DEM at ≥1.4 meter cell size (although typically rounded up to 2 meters), but not at 1 meter. In cases where the NPS is very close to a smaller integer, say ≤1.1 meters, it is not unreasonable to round this down to 1 meter, though there are no explicitly defined rules for when it is acceptable to round down. Users should be aware that arbitrarily rounding down to "improve" the resolution of the DEM are in fact not improving the data; rather, they create a surface that has less fidelity than its resolution implies and risk that the data may be used in applications for which it is unsuitable.

There is some sentiment that the DEM resolution should not be any finer than the density of the bare-earth lidar points. Technically speaking, this is a defensible assertion, though it is not particularly practical. Except in the broadest terms, it is difficult to predict the density of bare-earth returns that will result from a given NPS in vegetated areas. So, if a customer requires a 1 meter DEM, the vendor would have to collect at a much higher estimated NPS to achieve this, e.g., in some environments this may be 20 times greater. But making such an estimate is inexact: What if, after processing, the bare-earth point density in vegetated areas is 1.2 meters? Can a 1 meter DEM still be generated? Or does the vendor have to re-collect the data? And should the value of the higher density data in clear and open areas be disregarded in the DEM surface?

Whether a bare-earth surface is created from lidar, stereo photogrammetry, radar, or ground survey, the simple fact is that the accuracy and reliability of the surface will be lower in vegetated areas than it will be in clear and open areas. This is not a new deficiency unique to lidar. Reflect on our historic use of traditional photogrammetric elevation data as contours: obscured areas, where masspoints and breaklines cannot be compiled, are delineated as polygons, the contours within are tagged as obscured, and they are depicted with dashed lines. Data providers are not penalized for this; it is a simple fact of mapping technology and representation, and as users we learn to work with the data. Lidar is simply a different measurement tool, with its own strengths and weaknesses: some of which are shared with other technologies. A number of methods for identifying areas of lowered confidence in the accuracy or reliability of the DEM surface do exist, and some clients are requiring such disclaimers for their deliverables. At the time of this writing however, there is no standard practice for the development and delivery of any such product.

10.1.2.4 Surface Interpolation Methods

There are numerous mathematical methods for creating a raster surface from an irregular set of X-Y-Z points. All involve the interpolation of Z values at regular X-Z intervals from the source data; most will lose the original X-Y-Z measurements (Mark, 1976). Therefore, all methods result in a surface that is, to varying degrees, less accurate than the source point data. There are appropriate and inappropriate circumstances for each method's application, depending on the type of analysis and the density and distribution of the source points. It is beyond the scope of this subchapter to discuss even the most common methods in full detail, though a brief overview is provided. For additional information on interpolation, refer to (Maune et al., 2007).

- **Point-to-Raster:** The simplest means of converting points to a raster surface, this method identifies the point or points within each output raster cell and assigns a Z value to the cell based on the Z value of those points. Different implementations of the method may allow the user to choose the Min, Max, Mean, or other simple statistical values as the output Z value. If there are no points within the output cell, a NULL value is assigned. This can produce a raster with a large number of void cells and regions. Point-to-Raster is very computationally efficient.

- **Point-TIN-Raster:** Although transparent to the user, many software packages internally use this relatively simple interpolation method, primarily due to its processing efficiency. The irregular points are first triangulated using a Delaunay algorithm (Peuker et al., 1978). The value of each output raster cell is then interpolated from the TIN surface, usually at the center of each cell. Some regard this method as the most "accurate" because it has no adjustable parameters and introduces no user-tunable variances. It does not, however, always produce the most natural looking DEM: In areas where the point density is lower than the output cell size, the faceted character of the intermediate TIN will be transferred to the output DEM. Unless "Erase Polygons" are specifically included as breaklines by the user, it produces a DEM without voids, regardless of variations in point density. In some implementations, this process must be done in two distinct user steps: (a) The creation of the TIN (which may allow more flexibility in including breaklines), and (b) the conversion of the TIN into a Raster.
- **Inverse Distance Weighted (IDW):** IDW is the most basic and common true interpolation method. For each output cell, it calculates a weighted average of the points found within a user-defined search radius. Numerous parameters can be defined by the user: the minimum or maximum numbers of points and/or search distance, as well as the weighting factor for the average. In some implementation, linear barriers can also be included as a simple form of a breakline. Different densities of points often dictate different parameters for best results; this can be a challenge with bare-earth lidar where the point densities can vary widely. IDW can also be very computationally intensive and may require significant computing power on large lidar datasets.
- **Natural Neighbors:** This method is also known as the Sibson or "area-stealing" method. Voronoi polygons are created for the input points, and then for both the input and output points (as raster cell centers) together (Sibson, 1981). Each output point's Voronoi polygon is then intersected with the input Voronoi polygons, and the proportional areas of each input polygon within the output polygon is calculated. These proportions are used as the weighting values for the calculation of the output Z value from the Z values of the points whose Voronoi polygons intersect the output point's Voronoi polygon. This method produces a surface based on localized measurements and relieves the user of determining the most appropriate values for multiple parameters. It produces a fairly natural surface when the point density is close to the output DEM resolution, but creates unnatural artifacts when the points are sparse. Like Point-TIN-Raster, it will not produce unwanted void areas within the output surface. The Esri implementation allows for the inclusion of all standard breaklines as well. This method is more computationally intensive than Point-TIN-Raster.
- **Trend:** This method fits a single polynominal equation to the all of the input points. It produces a surface that retains only the most general trends in the data. These generalized surfaces can be useful in developing ellipsoid and geoid models, globally or locally, but are not appropriate for topographic mapping from lidar data.
- **Spline:** The Spline interpolation produces a surface that passes through all of the input points, while minimizing surface curvature. Because of this, it may be thought to be ideal for converting lidar to raster, however, it uses mathematical functions that may introduce extreme pits and peaks into the surface, particularly in voids or other uncharacteristically sparse areas, and that do not handle sharp changes in the terrain very well. Spline inter-

polations do produce smooth surfaces from which cartographic-quality contours can be generated, if proper care is taken in their configuration (Maune, 2007). It is extremely computationally intensive and is intended to generate a surface from sample points that are spatially sparse compared to the output resolution. Although spline interpolators have advantages and have been favored for topographic mapping in the past, their use with dense, over-sampled lidar data is questionable.

- **ANUDEM (Esri Topo to Raster):** An advanced variant of the spline interpolation, this method was developed to convert contour and hydrographic data into a hydrologically correct raster (Hutchinson, 1989). Although it can be used to convert point data, its primary purpose is to create a surface suitable for H&H modeling, not topographic mapping. It is extremely computationally intensive and requires highly skilled configuration. ANUDEM was originally developed before high density lidar data was available, and was optimized for much sparser source data. Research into refining ANUDEM to better handle dense data such as lidar is ongoing and showing promise (Hutchinson, 2011).
- **Kriging:** Kriging is a highly specialized advanced geostatistical process, intended for use with sparse data points. It is extremely computationally intensive. In most cases, it is not appropriate for topographic mapping from lidar data. In such cases where the user believes the specialized nature of kriging to be of value with their data, a variant known as ordinary kriging is recommended (Issaks and Srivastava, 1989).

10.1.2.5 Topographic DEMS

In the following section, various types of DEMs will be compared. This orthophoto depicts the area of interest and is provided for reference.

10.1.2.6 Photogrammetric DEM (Reference)

This DEM is derived from traditional stereo photogrammetry. Masspoints and breaklines are compiled to create a Digital Terrain Model (DTM); the DTM is then used to generate a DEM, usually using an intermediate TIN (Figure 10.2). This DEM is included in the discussion as a point of reference for the lidar DEMs.

Note in this example that:

- the masspoints are sparse,
- the pond surface is flat,
- the river surface is level from bank to bank but follows the adjacent terrain on a downhill gradient,
- natural depressions (sinks) remain in the surface,
- the bridges have been removed,
- the road surface over the culverts remains,
- the buildings are absent, and
- the ridges are clearly defined.

10.1.2.7 Pure Lidar DEM

As the name implies, a Pure Lidar DEM is generated solely from those lidar points classified as bare-earth (Figure 10.3).

When compared to the Photogrammetric DEM, note that:

- the "masspoints" are very dense and the overall surface is far more detailed,
- stream and pond surfaces, and the building footprint areas, have an unnatural, TIN'd appearance,
- road surfaces are textured,
- stream channels (single-line) are ambiguous, and
- ridges are not sharply defined.

These differences are due to the much greater density of bare-earth lidar over the stereo-compiled masspoints, and the lack of breaklines.

10.1.2.8 Hydro-Flattened Lidar DEM

In this DEM, the water surfaces have been corrected through the addition of hydrographic breaklines. Although the water surface elevations are somewhat arbitrary and of little use for any H&H modeling (and may still contain a few minor "steps") the overall appearance of the DEM for topographic mapping purposes is much improved (Figure 10.4).

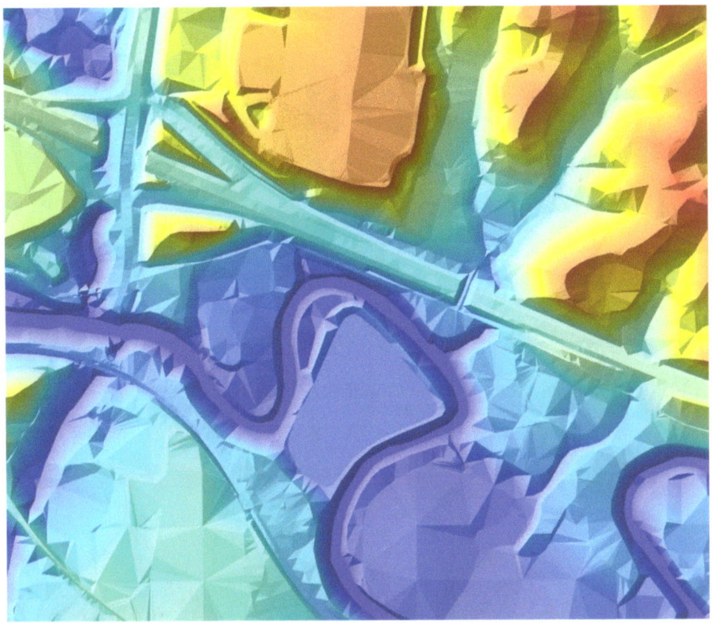

Figure 10.2 DTM TIN created from photogrammetric masspoints and breaklines.

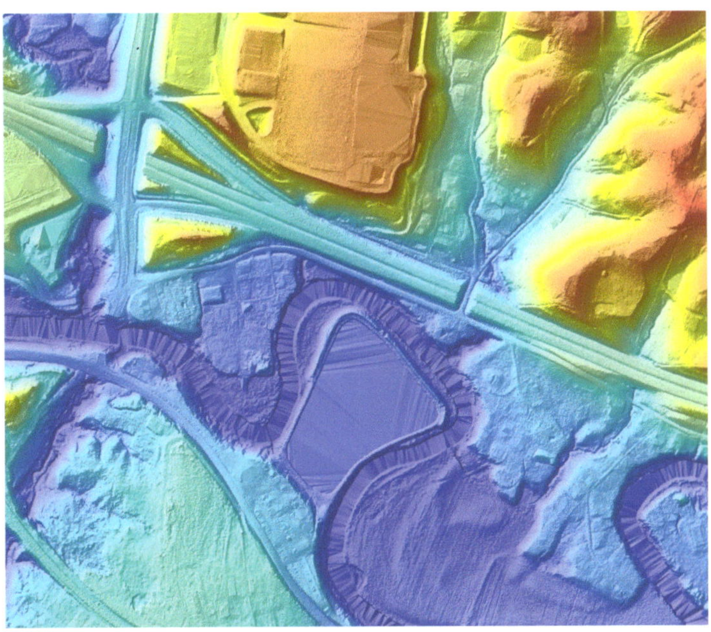

Figure 10.3 TIN from Pure Lidar DEM.

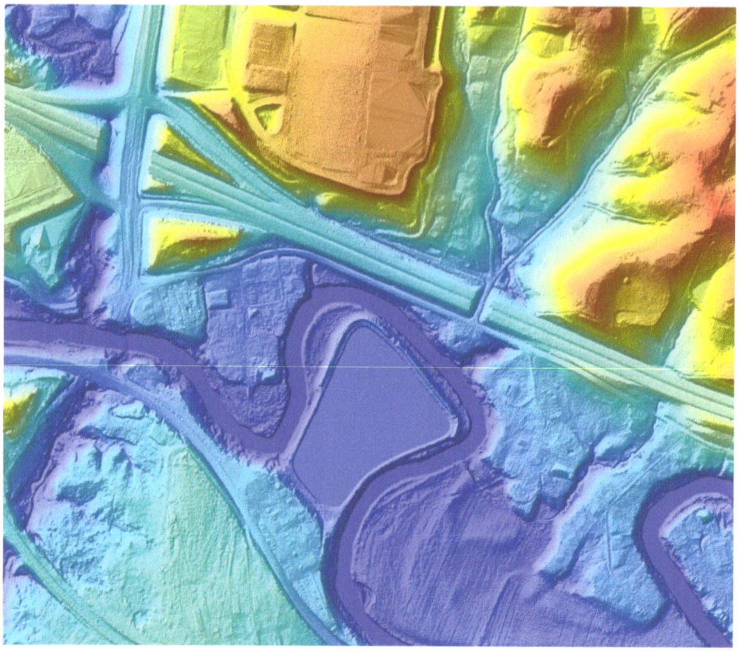

Figure 10.4 TIN from Lidar DEM where the water surfaces have been corrected through the addition of hydrographic breaklines.

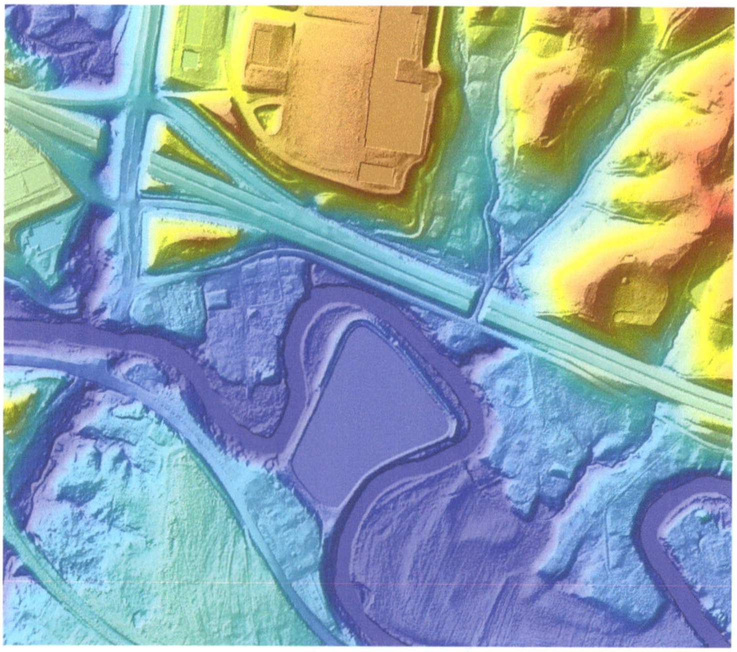

Figure 10.5 TIN created from fully-breaklined Lidar DEM.

Since 2010, this type of DEM has been a required deliverable for all USGS-NGP funded lidar projects destined for inclusion in the National Elevation Dataset (NED). The primary rationale for this deliverable is to maintain a surface character that is consistent not only across all new data, but also with the majority of existing data in the NED. The most commonly used method for producing a hydro-flattened DEM is to develop breaklines (see the *Breaklines Sub-*

section) for the streams and ponds, although other methods can also be used. The breaklines used here were digitized against the lidar DEM, referencing the lidar intensity imagery and other orthophotos, and conflated with elevations from the bare-earth lidar points.

10.1.2.9 Fully-Breaklined Lidar DEM

The technique used for hydro-flattening can be extended further to include other features such as building footprints, ridges, road edges and crowns, and other linear features in the landscape.

In Figure 10.5, the ridgelines are more distinct, selected building footprints are clearly defined, the roads have been smoothed and crowned, and narrow (single-line) streams are decisively indicated in the surface model. The generation of these breaklines is extremely expensive and time-consuming, however, and can increase a project's cost and turn-around time compared to that of traditional photogrammetry. Production of this type of DEM from lidar is uncommon. The breaklines used here are the same as in the previous example, with additional feature types (See the *Breaklines Subsection 10.1.3*).

10.1.2.10 Hydrologic DEMS

10.1.2.10.1 Hydro-Enforced DEM

The basic goal of the hydro-enforced DEM is for the surface to allow water to flow across it as it would on the actual landscape. The challenge is that while in the real world different elevations exist coincidentally (as with a road over a culvert), a digital raster surface can only depict a single elevation at a given location. Hence, some real topographic features must be mitigated to favor depicting the underlying water courses.

In Figure 10.6, the most significant change is the cuts through the highway. These allow the completion of the drainage network on the surface; although the DEM is clearly no longer useful for standard topographic mapping. Often, the elevations assigned to the water surfaces and stream channels in the hydro-enforced DEM are derived from definitive sources (survey, stream gages, etc.) so that the modeling processes are more accurate and can be related to real-world conditions.

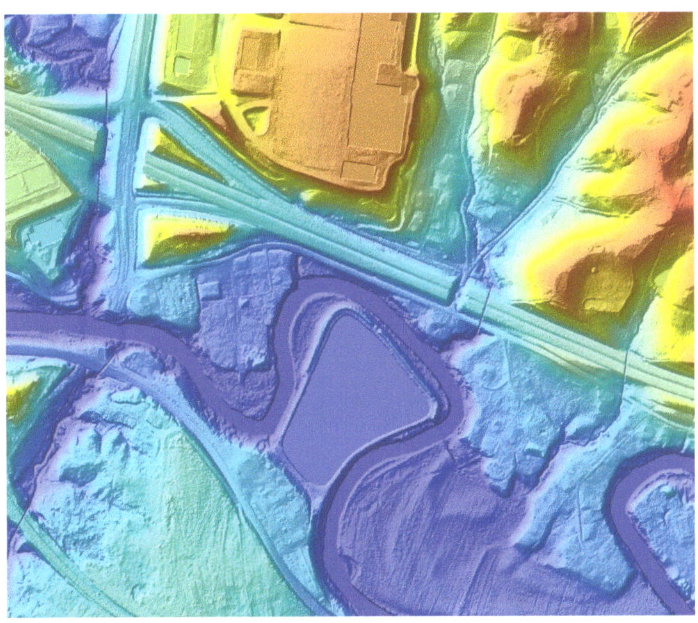

Figure 10.6 TIN from Hydro-enforced Lidar DEM.

10.1.2.10.2 Hydro-Conditioned DEM

Hydro-conditioned DEMs are similar to hydro-enforced DEMs, but extend the modification of the surface another significant step. Naturally-occurring depressions ("sinks") are filled so that modeled water does not accumulate in the sinks. This allows each cell of the raster surface to have a definite route to the lowest point on the surface, i.e., its natural outfall. This type of surface allows additional types of H&H modeling to be performed.

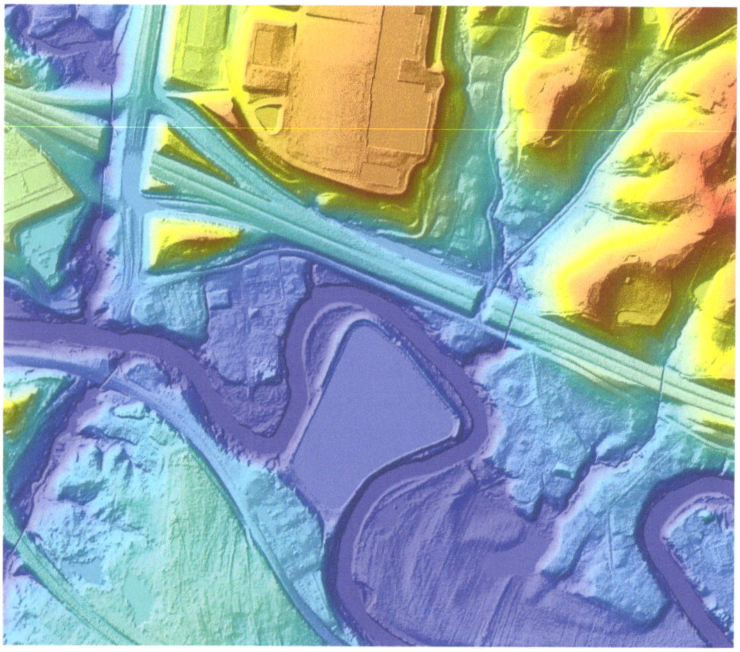

Figure 10.7 TIN from hydro-conditioned Lidar DEM.

10.1.3 Breaklines

The previous section demonstrated the significance of breaklines in the generation of DEMs, whether derived from traditional photogrammetric methods or as an adjunct to lidar point clouds. This section will discuss how breaklines can be used in, and generated from, lidar point cloud data.

In the early days of commercial lidar, pure lidar DEMs were the norm. The increased resolution of the surface was almost overwhelming unto itself, and the need for breaklines was not widely recognized. But, the user community soon recognized that in spite of lidar's vastly increased point density, lidar could not (and cannot) resolve linear features as *true* linear features, and therefore could not produce DEMs that met with the user's traditional expectation of what a DEM should look like. This was most apparent after contours were generated from pure lidar DEMs: contour lines did not cross roads and rivers at right angles; contour-indicated stream channels did not match the channels shown in imagery; contours around ponds and lakes were erratic and crossed over the water surfaces; and the contour lines are jagged and erratic (see Figure 10.15b).

Early efforts to remedy these issues used the breaklines from existing DTMs and other mapping sources (i.e., the NHD). Occasionally successful, this method could make the contouring problems worse and require extensive manual editing of the breaklines.

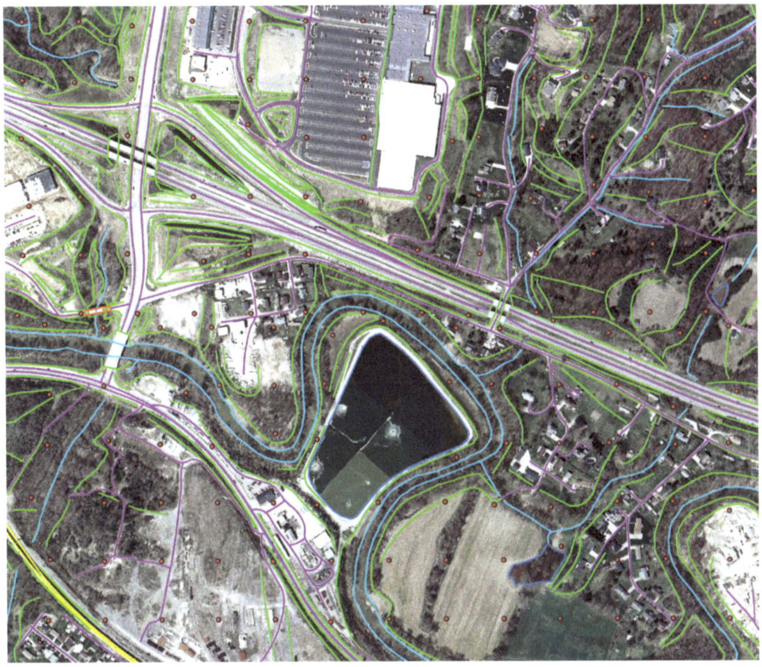

Figure 10.8 Photgrammetrically compiled breaklines.

Combined lidar/orthophotography projects appeared to offer an advantage in that virtually concurrent imagery was available, and the best possible breaklines could be compiled and used with the lidar. This produced improved, but often disappointing results. Slight but noticeable differences in the elevation and position of features would exist between the lidar measurements and the stereo models, creating irregularities in the resulting DEMs (see Figure 10.15c). It became clear that the most suitable breaklines would have to come from the lidar data itself, as those would be the only ones that adequately matched the lidar (see Figure 10.15d).

As noted earlier, the 2001 NC Floodplain Mapping Project was the first large-scale lidar project to require hydro-treated DEMs as a deliverable (note that the deliverables were to the H&H engineering firm performing the modeling, not to the State). And as such, it was also the first large-scale project to successfully incorporate lidar-derived hydrologic breaklines into the DEMs. The method used on this project was imagery-based and lidar-refined, as follows (generalized process):

1. Hydrographic features (single-line streams, double-line streams and stream centerlines, ponds and lakes, shorelines, culverts, etc.) were roughly digitized in 2D from the best available imagery. This was done in advance to prevent having to wait for completion of the lidar classification. Different types of features were tagged with differentiating attributes.
2. This 2D linework was visually compared to the bare-earth lidar surface (hillshaded DEMs and TINs); omissions, positions, and attributes were corrected.
3. Elevations from the lidar points were then conflated to the vertices of the linework creating 3D breaklines.
4. Monotonicity (downhill flow) was then enforced onto the 3D breaklines.
5. Lidar points within a close proximity to the breaklines were removed from the bare-earth classification (as would be in stereo photogrammetry).
6. DEMs were created from the remaining bare-earth lidar points and the new breaklines, though culverts were excluded.

This workflow was initially implemented using Workstation Arc/INFO© software, and all processing was done against Esri coverages through specialized AML scripts, developed in-house by the data producer. For Phase-2 of the project, the tools were converted to ArcView© 3.2 and Avenue scripts, and updated to take advantage of the shapefile as the native data format. In spite of the high level of automation that was incorporated, the process was immensely complex and required an enormous amount of human intervention, checking, and manual editing of vertex elevations.

The resulting DEMs were not, by current definitions, "hydro-enforced" as the streams did not breech the roads at culverts. They were hydro-flattened, with the addition of having monotonic single-line streams enforced into the surface. Flow through culverts was introduced by the H&H engineers during the modeling processes.

In the mid-2000s, the GeoCue Corporation developed a technique called "Lidargrammetry" (See Chapter 8.1). This process generates artificial, pseudo-scopic stereo pair images from the lidar point data, relying on the presence of elevations and intensity values. Lidargrammetry allows breaklines to be compiled, in stereo, directly from lidar data, and puts the existing suite of highly matured compilation tools and software at the disposal of the data producers. Highly detailed, very accurate breaklines could now be generated directly, significantly reducing production steps and costs.

Like other photogrammetric processes, lidargrammetry is highly specialized, requiring dedicated software and hardware tools, as well as specific skills and training. While useful and appropriate for data producers, it is not generally within the reach of most end users who may have more limited needs for breaklines. Several lidar software producers have implemented tools that allow creation of breaklines using methods internally similar to those used on the North Carolina Floodplain Mapping project, though much refined and simplified from the user perspective. These solutions may not meet all possible requirements for breakline development, and the results may not be as refined as those from lidargrammetry, but for many applications they are both adequate and affordable.

The USGS-NGP requirement for hydro-flattened topographic DEMs, the general adoption of the USGS-NGP Lidar Base Specification across the industry, and the tendency to develop such DEMs using breaklines, have all combined to make the generation of at least limited breaklines an almost routine component of many lidar projects. Coincidentally, the demand for hydro-enforced DEMs for use in H&H modeling has led to more frequent delivery requirements of both DEM products. Fortunately, both types of DEMs can be generated from the same source data with relative ease. Although there is often initial disagreement regarding how to conduct the lidar project and what deliverables are necessary, careful planning for classification of the lidar points, and compilation of breaklines can ensure that all stakeholders can receive the products they need without unnecessary expense. Following is a suggested basic scheme for achieving this end.

10.1.3.1 Lidar Point Classifications for Digital Elevation Models

The following lidar classifications should be considered for inclusion:

- Noise (low or high)
- Water (all points contained within any water body or between the banks of a river)
- Bare-Earth
- Road (optional)
- Bridge (optional)
- Building (if desired for urban or line-of-sight modeling)
- Vegetation (if desired for ecological or line-of-sight modeling)
- Other Structures (such as towers, power lines, catenaries, etc. Optional)

Each of these classes may further be tagged as "Swath Overage" (which many users prefer to exclude from certain applications) and/or as "Breakline Withheld" (for points that are too close to a breakline to be useful in constructing the DEM). How these tags are applied and maintained in the dataset is at the discretion of the user and their software. LAS version 1.4 includes a dedicated bit flag for Swath Overage; Breakline Withheld points are often simply assigned to separate classification value.

10.1.3.2 Breakline Types

The following breakline types should be considered for inclusion:

- Shorelines (lines or polygons, as preferred by the workflow)
- Single-line Streams (lines)
- Culverts (lines)
- Stream Banklines (lines or polygons, as dictated by the workflow)
- Ponds and Lakes (lines or polygons, as dictated by the workflow)
- Stream Centerlines (double-line streams; estimated; to complete the hydro network)
- Pond and Lake Centerlines (estimated; to complete the hydro network)
- Spillways (pond to stream connector, overland)
- Other overflow structures (pond to stream connector, buried)
- Buildings (optional)
- Road Edges (optional)
- Road Centerlines (optional)
- Bridges (optional)

Naturally, the extent to which breaklines are compiled depends entirely on the requirements and budget of the project.

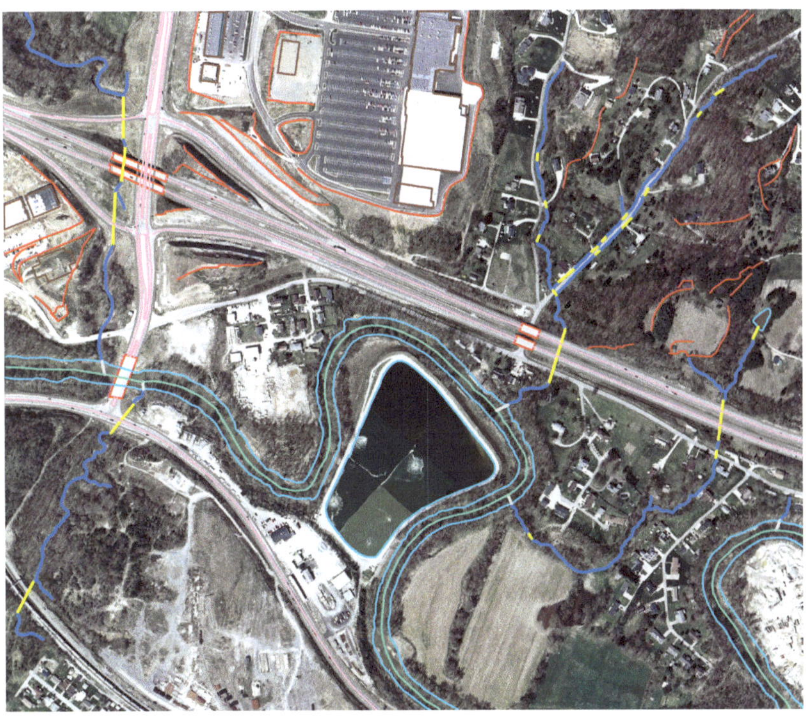

Figure 10.9 Breaklines developed solely from lidar data and reference imagery.

It is usually necessary for the breaklines to depict not only the water bodies, but also the complete interconnected drainage network. This means that not only are the banks of the water bodies necessary, but the approximate centerlines of these features are as well. Practically speaking, these functional centerlines can be of great value to both the H&H engineer and the GIS analyst, as they are most useful in conflating proper elevations to the bank lines. It would be neither necessary nor practical for a DEM to represent a complex underground storm drainage system, however, there are a few such structures that are needed to form a functionally complete drainage network.

Water bodies can be represented by lines, however, the use of polygons makes it easier to reclassify the points within the areas as "Water," and also makes display and map production easier as well. This is also true of road edges, though care must be taken to not create polygons so complex that they cannot be managed by the GIS software.

Each type of breakline should include a buffer value; that is, the distance from the feature within which no lidar points should be included in the surface. This distance may vary between breakline types, depending on the application. For example, a single-line stream might only need to be buffered by 1 to 2 times the NPS of the lidar to make the DEM properly follow the stream; a culvert may need to "cut" a wider path (4 to 5 times the NPS) through the road and fill above it for the modeling software to recognize it.

Depending on how the user wishes to construct their various DEM products, it may be necessary to ensure that breaklines passing under other breaklines be split. For example, consider a wide, double-line stream passing under a road through a culvert: if it is desired to create a hydro-enforced DEM that also represents smooth roads using breaklines, the road linework would have to break to the stream. Similarly, the stream linework will also have to break so that a correct topographic DEM can be generated from the same data. The "split" sections of the breaklines at these intersections must be tagged with an attribute that allows them to be differentiated for exclusion in the creation of various DEM products.

The level of detail described here may seem daunting and likely costly, however, collection of these features is largely a standard process for many lidar data producers who have their roots in traditional stereo photogrammetry. As with all data collection projects (regardless of data type) the key element to success is clearly understanding the applications the data will be used for, the requirements of the data for those applications, and properly pre-planning for the collection to meet those requirements. It is far more cost-effective to develop these strategies before a single collection than to rework a collection after the fact, or to have different users undertake separate collections.

10.1.4 TINs and Terrains (Bare-Earth Surfaces)

A Triangulated Irregular Network (TIN) is a data structure that establishes connectivity between X-Y data points, forming triangular facets. Using the Z values of the three corner points, Z values can then be interpolated (using different algorithms, although linear is most common) for any location within the triangle. Collectively, the facets comprise a continuous surface. So while DEMs and TINs are both continuous surfaces, DEMs are raster-based and TINs are vector-based. TINs have the distinct advantage over DEMs in that they retain the original geometry of the source data points. It is also worth noting that TINs define a connectivity network of only X and Y coordinates; the network is not influenced by the Z-value of the data points. And just as with raster surfaces, the Z-value does not have to be elevation, it can be any numeric attribute of the source points.

For virtually all topographic applications, TINs are generated using the Delaunay method, in which no point in any given triangle lies within the circumcircle of any other triangle. It is important to understand that there are many different algorithms available to create TIN that meets the Delaunay criteria, and that for a given set of points there are multiple solutions to the triangulation. Similarly, for a given algorithm, a minor change in the composition (or even file order) of the source points can induce differences in how the points are connected, and therefore the surface values at locations within the triangles. Users of TINs need to be aware that having a TIN does not equate to having a singularly correct surface.

Because TINs are inherently vectors, they can further offer the advantage of incorporating line and polygon breaklines. As the examples in *Breaklines Subsection* show, breaklines can substantially enhance the appearance and functionality of the TIN surface.

In spite of the many advantages TINs can offer, they pose some disadvantages as well. As vector data, they are often significantly larger in file size than raster DEMs. Until very recently, many operations on and between TINs (i.e., differencing two TINs) was impractical, due to the complex data structure. And unlike rasters, there are no well-established, open, and published TIN data formats in widespread use, largely limiting data exchange to those using the same COTS software.

The Esri Terrain Feature Class overcomes some of the problems with TINs and allows the practical use of lidar bare-earth points as a vector surface. In a Terrain Model, all of the points and breaklines are loaded into the database. To conserve space, lidar point attributes are usually discarded, retaining only the X-Y-Z geometry. Various zoom levels are defined for the terrain, each with specified accuracy tolerances for that scale. The input features are then analyzed and tagged as to which zoom level(s) they are needed for. Large scale zoom levels use more (or all) of the features and are the most accurate. Small scale levels use an intelligently decimated subset of the features. In use, TINs are created on-the-fly from only those features predetermined to be appropriate for the current display or analysis scale.

Terrains allow for the retention of the full resolution of the lidar dataset, without the overhead of managing a complete TIN with hundreds of millions of triangles all of the time. They further allow the incorporation of breaklines and other features that may be valuable to the correct modeling of the surface. More and more analysis routines that work directly against TIN and Terrain surfaces are being implemented, bringing the higher resolution and accuracy of the vector surface model to the fore, and resolving some of the problems inherent with an interpolated raster surface.

As datasets, TINs, and Terrains remain significantly larger than their raster counterparts; this can be a significant consideration for many users. Terrains are an ESRI-proprietary format, and common open formats for TINs have not been broadly adopted. The use of these surfaces can be somewhat limited by the software available to the user.

Finally, the lidar user should be aware that TINs and Terrains, like DEMs, are representations of a single thematic aspect of the lidar point cloud, most commonly the bare-earth elevation surface. None of these data structures allow for the exploitation of the full point cloud.

10.1.5 First-Return Digital Surface Models (DSMs)

Because lidar returns points are from (almost) whatever target it hits, it has proven itself to be an excellent tool to measure the tops of features, such as buildings, vegetation, crops, towers, transmissions lines, etc. Applications have been developed to extract building footprints and heights, power lines, and other items of specific interest from the point cloud. A more generalized use of this characteristic of lidar is the Digital Surface Model (DSM). A DSM most commonly represents the first-return or top-of-canopy surface, although a DSM could be of any single thematic aspect of a point cloud. It provides the ability to perform analyses of line-of-sight, view-sheds, communications coverage, airspace and flight-path safety, and many other similar applications. DSMs can also be used to estimate vegetation density and height, making it extremely useful for studies of biomass, carbon stores, forestry, and agriculture.

Figure 10.10 TIN wireframe of the photogrammetric DTM (stereo compiled masspoints and breaklines).

Figure 10.11 TIN wireframe of the lidar points and breaklines (developed from lidar).

Technically, DSMs could be created using the same general techniques as for DEMs, but there are some unique considerations. In vegetated areas, the 1st-return point cloud will contain points at or near tops of the trees *and* nearby points on the ground. A TIN surface created directly from these points will have extreme vertical variability and show a forest of steep, pointed spires. When converted to a raster DSM, cell values will most often be interpolated down along the sides of these spires and cause a general underestimation of the surface elevation. The Point-TIN-Raster method is not preferred for most general DSM applications.

Similarly, other interpolation methods will also produce DSMs that underestimate the elevation of the canopy, due to the presence of the low ground points in vegetated areas. Populating the elevation of each raster cell with the elevation of the highest lidar point within each cell has been used to avoid the underestimation issue; however, if the raster resolution is not sufficiently larger than the NPS, each cell will become associated with a single point and the low ground points (and void pixels) will again be introduced into the surface. The result can be a more accurate and detailed DSM; this may be desirable in some situations but not in others.

If a less textured, more generalized surface is needed, there are numerous ways to achieve this. It is common to simply use a raster cell size at least 2 to 3 times greater than the NPS, and select the single highest lidar elevation within that cell. Alternatively, the more detailed surface described in the preceding paragraph can be passed through a 3 × 3 neighborhood filter, selecting the highest value in the neighborhood (rather than the more typical average value). A more robust approach compares the detailed DSM surface to the bare-earth DEM; those DSM cells close in elevation to the DEM are replaced with values from neighboring cells.

The estimation and modeling the top-of-canopy is exactly that: an estimation. Therefore, there is no single correct method of producing a DSM from lidar. The methods chosen in any given situation should be based on which produces a surface that is most suitable for the application.

10.1.6 Lidar Intensity Imagery

In addition to recording the timing of laser reflections and deriving point positions, most modern lidar instruments also record the amplitude of the reflection. These values are commonly referred to as the intensity, and by using the positions of the points and these intensity values, images can be generated from the lidar point cloud. These images are single-band, grayscale, usually normalized to 8-bit values (0 to 255), and at first glance resemble panchromatic ("black and white") orthophotos.

The most basic and common lidar intensity image is derived from the first return points; logically, this is the image that best represents what we would see, if we could see in the near infra-red portion of the electromagnetic spectrum (assuming the lidar instrument uses a near infra-red laser). They can be generated using the same techniques implemented for DEMs and DSMs, substituting the intensity values for the elevation values in the interpolation process. Note that in the example image below (Figure 10.8), the void areas (mainly water) have been triangulated over.

Intensity images can also be generated from other collections of points (all returns, last returns, etc.) depending on the requirements of the application. In some cases, it may be preferable to generate the image by taking the sum of all intensity values within each pixel, from all returns, and normalizing the result back to 8-bit. This process would *theoretically* produce an image that is the best estimate of the total energy returned within each pixel, and may be of more value to the user than a generalized image. Note that in Figure 10.12, the void areas (in this case, standing water, which absorbs most of the infrared radiation) remain "NODATA" in the image, and are depicted in blue for reference.

Figure 10.12 Sample lidar intensity image.

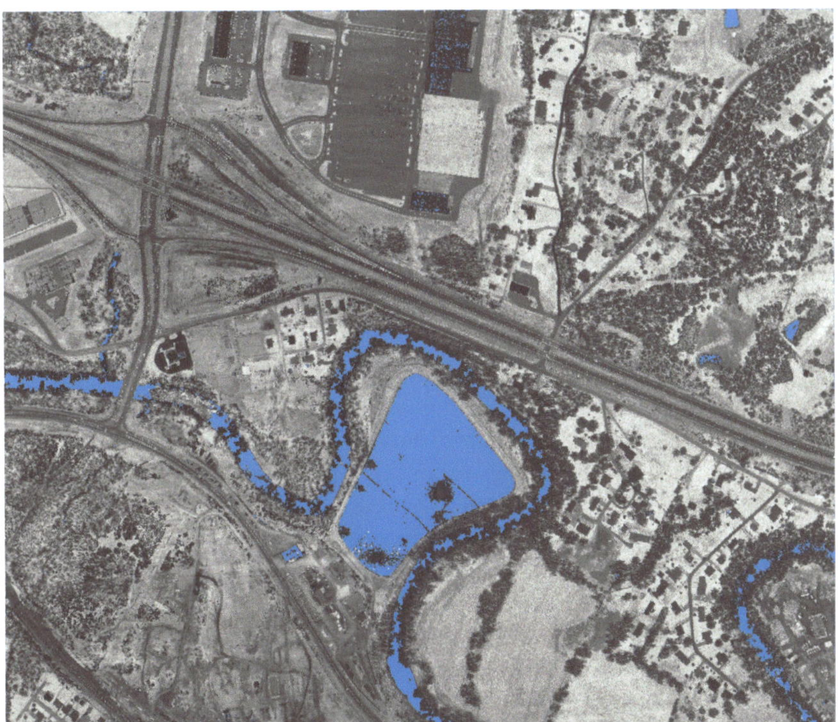

Figure 10.13 Lidar intensity image; areas with "NODATA" are shown in blue.

The earliest lidar instruments did not record intensity values. When this capability was implemented, many vendors were eager to use the new intensity images as a marketing and sales tool. In some cases, the intensity images were touted as a replacement for traditional orthophotography from imagery. In spite of the initial similarity in appearance to panchromatic orthophotos, lidar intensity images are very different, and must be viewed and interpreted with great care and skill.

Intensity imagery is not without its value, however. Proper analysis of lidar points for classification and processing requires some ancillary reference data, usually imagery. Many lidar projects cannot afford concurrent imagery collection, and other recent orthophotos may not be available. Particularly in these cases, the intensity imagery is invaluable as an aid to data processing and analysis. Intensity imagery can be used by lidar analysts to troubleshoot a problem with an instrument. And if expertly interpreted, it can provide a wealth of additional information about the project area. The following are brief discussions of some nuances of lidar intensity imagery.

10.1.6.1 Spectrum

Most topographic lidar systems operate in the near infrared portion of the electromagnetic spectrum, around 1064 nm, which humans cannot see. Viewing a lidar intensity image is similar to viewing only the NIR band of satellite imagery, airborne multi-spectral imagery, or the NIR band from CIR (false-color) photography. The reflective properties of various materials may be very different from what we are accustomed to seeing, and in some cases, deceptive. Users who have inadequate training in the interpretation of NIR imagery may have difficulty with lidar intensity images.

Unlike the sensors in satellites, multispectral cameras, and the emulsion of CIR film, the bandwidth of the laser is extremely narrow, perhaps ±20 nm. The recorded intensity is indicative of only this extremely narrow range of wavelengths. The lidar intensity image is more analogous to a single NIR band from a hyperspectral sensor, and is even more foreign and difficult to interpret.

10.1.6.2 Sampling

By their design, film and digital sensors record an intensity value for the entire area covered by each pixel (or emulsion crystal) within the field-of-view. This brightness of each pixel in the final image represents the total brightness of the entire area of that pixel. With lidar, however, the brightness recorded is derived solely from the reflectivity of the target which caused that return's reflection. It may be a very small, but very bright spot within what will ultimately be a pixel representing a larger ground area, and may be quite unrepresentative of the pixel's total area.

10.1.6.3 Active versus Passive System

Cameras, satellites, and other imaging systems are passive detection systems; that is, they simply record what natural energy is being reflected by the Earth. The natural energy comes from the Sun, and it contains an extremely wide range of spectral values. Because of the great distance from the Sun to the Earth and the Earth's relatively small size when compared to the Sun, the rays of the Sun are effectively parallel when they reach Earth. At any given moment, all of the energy striking a target area and reflecting into the lens of a camera or sensor has the same angle of incidence. Because this view is how we are accustomed to seeing the world, passively-acquired images look quite natural to us.

Lidar, however, is an active sensor. It carries its own illumination system, its own energy source, and it is neither effected by nor does it record energy reflected from the Sun. Moreover, most commercial lidar systems use some sort of scanning mechanism, where the laser is swept back and forth over a field of view of up to 75 degrees. This collection process introduces significant variations in the angles of incidence and reflection, relative to the sensor, which are quite alien to human vision and interpretation. This is not to suggest that as humans we are incapable of interpreting imagery from such a system; doing so requires more thoughtful consideration of what we *think* we see in the imagery.

10.1.6.4 Automatic Gain Control (AGC)

Our eyes are a true wonder of human vision capability. We can see in color or black/white, and the ability changes which it uses depending on ambient conditions. It has a wide focal range, and it can handle an enormous dynamic range of brightness within a given view. Unfortunately, we have not yet developed a detector that can do the same. The range of intensity being reflected back from the Earth is far too great for any laser detector to handle. Either the low reflectance targets will be too dim to be detected; or the bright targets will overload the system. In order to ensure that all reflections can be detected and recorded, an Automatic Gain Control (AGC) is incorporated within the lidar instrument. The AGC continuously monitors the brightness of the incoming reflections, and adjusts the system to maintain optimum sensitivity at the detector.

Unfortunately, what this means to intensity is that the values being recorded have already been modified from their true reflective brightness. The broad dynamic range of the reflections has been compressed in the recorded values. Some lidar systems may correct this during preprocessing, but this is difficult to determine. In any case, the use of lidar intensity values, either directly from the points or as imagery, for quantitative scientific study, is severely handicapped.

10.1.6.5 Radiometric Resolution and Value Stretching

Different lidar instruments have different native radiometric resolutions, depending on the detector and digitizer subsystems used. Most systems are between 8-bits (recording values from 0 to 255) and 12-bits (recording values from 0 to 4,095), though some may be up to 16-bit (recording values from 0 to 65,535). The actual display of intensity images is almost universally limited to 8-bit by the computer systems we commonly use. In order for a computer system (software and hardware) to display the intensity of points or images, it needs to know what the actual range of values is so it can correctly remap the true values to an 8-bit range. LAS versions before version 1.4 simply provided a 16-bit field for the storage of intensity values; no provision was made for defining the data's native radiometric resolution, nor did they require the intensity be pre-stretched to any given resolution. The result has been that most software can only presume the data is16-bit and handle it accordingly. For many datasets, this caused the point intensity display (and the images produced from the points) to appear almost completely black.

In many datasets, users may find that virtually all of the intensity values are fairly normally distributed in the 0 to 4,095 range, but that there are a handful of extremely high outliers near or equal to 65,535. These outliers may be processing artifacts. Building histograms of the intensity values allows the software to properly stretch, display, and process the intensity information, but this capability is not widespread. As lidar software continues to advance, more robust intensity functionalities can be expected, allowing users to apply Min-Max, Standard Deviation, and custom stretches to the display of intensity. In LAS ver.1.4, all intensity values are required to be Min-Max stretched to 16-bits.

10.1.6.6 Underestimation and Calibration

When recording lidar returns, most instruments are triggered by a rapid rise in the magnitude of the pulse's reflection. This causes the detection circuitry to record the time and magnitude and record the values to the storage system. However, the "moment-in-time" that is recorded is not necessarily the peak of the reflection; it often is recorded prior to the maximum returned reflection. Although the timing (and therefore positional) error this induces is modeled out of the data during preprocessing; the error in the recorded intensity value usually is not. The magnitude of this type of error is difficult to assess, and it may vary depending on numerous other factors (which return number a reflection is; whether it followed an ignored return, AGC settings, etc.). Although this error is of little consequence to the casual user, it should be considered if lidar intensity values or images are used for analytical purposes.

The wide variability in recorded intensity values, even within a single swath, makes it impossible to truly calibrate intensity values to any known reference. Values across homogenous surfaces can vary significantly within a single scan line or swath, and values in overlap between swaths and lifts do not match. Changes in instrument settings (e.g., for a different AGL flight altitude within a single project) can greatly affect the recorded intensity values. The following example (Figure 10.14) reveals some of these issues. Note that this is a typical example, not a particularly bad one.

Figure 10.14 Example of intensity data image showing the tonal variation between flight lines.

10.1.6.7 Full-Waveform Lidar

Although full waveform lidar is maturing rapidly, it has not quite yet become commercially viable. This may be due to a combination of the extremely large files required to store fully digitized lidar reflections, a lack of affordable and available software with which to exploit the data, and a lack of practitioners experienced with the waveform data. With regards to intensity imagery, the value of the full-waveform is beginning to be realized. Having full-waveforms allows far more precise calculation of the peak of each return with a single reflection, as well as allowing calculation of the total energy returned (as a function of the areas under the curve(s) of the reflection). These advances will, in time, allow science to improve on the quality and usability of lidar intensity images (see Chapter 2.4).

10.1.7 Contours

Contours continue to be a widely requested deliverable on lidar projects. As do isoline maps of other phenomena, they remain one of the most effective means of visualizing a continuous surface, certainly when a computer display is not available (field work, recreation, etc.). Over the decades, the map user community has developed a somewhat natural skill at interpreting elevation contours, and we have evolved contours to make that task easy for us. Like all isolines, contours follow rules: they do not touch, they do not cross.

As computer systems developed, the photogrammetry industry adopted Computer Aided Drafting and Design (CADD) as the data backbone of stereo compilation. When mapping adopted CADD, digital contours naturally developed uses beyond the printed map. Engineers could perform measurements against digital files and develop their models and designs using the same techniques they had always used, only in an electronic environment. This step was a great improvement in accuracy and precision over measuring features on paper or mylar, and the practice flourished. Geographic Information Systems (GIS) brought far more sophisticated spatial analysis tools into use, building on the foundations being used in the CADD world. Our familiarity, comfort, and habit of using contours ensured their dominance as our means to map the Earth's surface.

Other data models and systems existed for elevation surface modeling: photogrammetrists used the DTMs (masspoints and breaklines) to construct their models of the bare-earth surface. Compared to contours, DTMs were compact and accurate; most importantly, they allowed the compiler to place measurements where they were needed, rather than along arbitrary lines of a single elevation. But as a vector data structure, they did not form a continuous surface, and contours could not be created directly from a DTM. TINs and raster DEM surfaces were created from the DTMs, and contours could then be created from these. Once regarded as intermediate working data, GIS analysts found the raster data model to be remarkably useful and efficient for a host of applications beyond elevation. Computer and software systems continued to advance, and today we can easily work with large DEMs, DTMs, TINs, and 3D surface models on even a modest laptop computer. Affordable and accessible software now exists to allow extensive modeling, analysis, and visualization of elevation without relying on contours as the base data. But many long-standing, entrenched processes remain that require contours, in spite of their shortcomings. Contours are neither bad nor obsolete, but their limitations need to be kept in perspective.

Humans tend to look at contours and think of them as "the truth," but they are not. Numerous processing variations and subjective judgments are applied to a set of contours as the set is developed; notably how the DEM surface is interpolated from the DTM or lidar, and how the contours are smoothed, generalized, and edited to make them readable. Contours are a cartographic construct, a third-generation derivative from the original source measurements. They are designed and created to adhere to a particular set of visual, cartographic, aesthetic standards, even at the expense of absolute accuracy. Their use as primary digital storage of base elevation data or as source data for elevation surfaces, however ingrained or convenient, is inappropriate except when no better digital elevation data exists.

We have come to expect that contours have a certain appearance: they cleanly peak at stream crossings and point uphill; they cross wide rivers and roads at right angles; they curve smoothly; they are not littered with tiny insignificant isolations, etc. These characteristics stem from how DTMs have been created, and from the end goal of making easily readable printed maps. Breaklines and a relatively sparse masspoints allow for this in photogrammetrically-derived contours. Lidar has no breaklines but compensates with extremely dense masspoints. Because the nature of the source data is different, so is the character of the derived contours.

(a)

Figure 10.15 (a),this page: This example shows contours derived from the Stereo DTM. Although not smoothed and edited as finished cartographic contours, the general character can be seen; (b), page 25: This example shows contours derived from the Pure Lidar DEM. The dramatically increased detail and lack of breaklines is plainly evident; (c), page 26: These contours are derived from the lidar points and a full set of lidar-derived breaklines. Although they still contain too much detail to be considered of cartographic quality, their character and behavior is much closer to the traditional ideal; and (d), page 27: Here, stereo DTM breaklines have been used to constrain the lidar points. While an improvement over (b), they continue to display erratic behavior due to discrepancies between the lidar and the stereo geometries.

(b)

(c)

(d)

DISCLAIMER

Any use of trade, product, or firm names is for descriptive purposes only and does not imply endorsement by the U.S. Government.

REFERENCES

Hutchinson, M., Xu, T., and Stein, J., 2011. *Recent Progress in the ANUDEM Elevation Gridding Procedure*, Geomophology 2011, geomorphometry.org , Online, pp. 19–22.

Issaks, E.H. and R.M. Srivastava, 1989. *An Introduction to Applied Geostatistics*, Oxford University Press, New York.

Mark, D.M., 1978. Concept of "Data Structure" For Digital Terrain Models, *Proceedings of the Digital Terrain Models (DTM) Symposium*, 9–11 May 1978, St. Louis, Missouri (American Society for Photogrammetry in cooperation with American Congress on Surveying and Mapping), pp. 24–31.

Maune, D.F., Kopp, S.M., Crawford, C.A., Zervas, C.E., 2007. Introduction. In: Maune, D.F. (Ed.), *Digital Elevation Model Technologies and Applications: The DEM Users Manual*, 2nd Edition. American Society for Photogrammetry and Remote Sensing, Bethesda, pp. 9–16.

Peuker, T.K., R.J. Fowler, J.J. Little, and D.M. Mark, 1978. The triangulated irregular network, *Proceedings of the Digital Terrain Models (DTM) Symposium*, 9–11 May 1978, St. Louis, Missouri (American Society for Photogrammetry in cooperation with American Congress on Surveying and Mapping), pp. 516–540.

Sibson, R., 1981. A brief description of natural neighbor interpolation, *Interpreting Multivariate Data* (V. Barnett, editor), John Wiley & Sons: Chichester, West Sussex, pp. 21–36.

10.2 FORESTRY

10.2.1 Introduction

10.2.1.1 Relevance

While the instrumentation, physics, data and methodologies have been discussed in detail in previous chapters, this chapter will build on earlier chapters and how this information is especially useful for scientists and managers to use lidar for forestry and vegetation-related studies. The capability of lidar to penetrate gaps in forest canopies to reach the ground below not only makes it especially attractive for generating bare earth models, but allows for the derivation of the vegetative characteristics as well. Depending on the instrumentation and collection parameters, a lidar system can record at or near the top of the vegetative canopy and the ground (almost) simultaneously, generate discrete multiple returns per laser pulse as the pulse penetrates through gaps in the canopy, or even digitize the entire energy returned per pulse in a continuous waveform. While the primary use of lidar remote sensing is still the generation of bare earth topographic data and the vegetation information collected in a lidar signal has traditionally been considered "noise", the use of lidar instrumentation in its various forms for forestry applications is one of the more developed non-traditional capabilities of the technology. Forestry applications have precise inventory and mapping needs for understanding the various functions of the forest, including forest management, planning, biodiversity studies and carbon stock estimations. Traditional forest inventory methods were based either on field measurements or photographic and/or photogrammetric interpretations.

10.2.1.2 History

The potential of lidar to map forests was recognized in the published literature about 25 years ago when Arp et al. (1982) used lidar to map forests in Central America. The Canadian Forest

Service also conducted studies of a laser altimeter profiler to characterize height and canopy closure in forest stands (Aldred and Bonnor, 1985). Nelson et al. (1984) used laser profiling to determine forest characteristics. In the early 1990s, NASA's Goddard Space Flight Center began using medium- to large-footprint scanning continuous waveform systems for measuring vertical vegetation structure in forests (Dubayah et al., 2000). The results obtained by these systems were later validated by field measurements (Means et al., 1999; Lefsky et al., 1999a), and showed that these systems were successful at measuring vegetation structure in these areas.

A Scandinavian summary of laser scanning in forestry can be found in Naesset et al. (2004) and a summary of the methods used for forest inventories can also be read in Hyyppa et al. (2008). An excellent chapter called "Forest Inventory Using Small-Footprint Airborne Lidar" by Hyyppa et al. can be found in Topographic Laser Ranging and Scanning: Principles and Processing by Shan and Toth. Dubayah and Drake (2000) describe three main attributes of lidar systems that can differentiate lidar systems for forestry. These include return signal detection (discrete returns or full waveforms), the size of the illumination footprint, the laser pulse rate and scanning pattern created by the scanning mirror.

10.2.2 Background

10.2.2.1 Laser Interactions with Tree Canopies

A laser pulse in a lidar system is in essence a collimated volume of photons that is directionally propagated through the atmosphere to the target. These photons interact with the atmosphere, ideally minimally, continue through to the vegetation canopy where some photons are reflected, some are scattered, and some continue on their path to the ground below and then are reflected back to the sensor, with some scattering occurring on the way back as well (Figure 10.16).

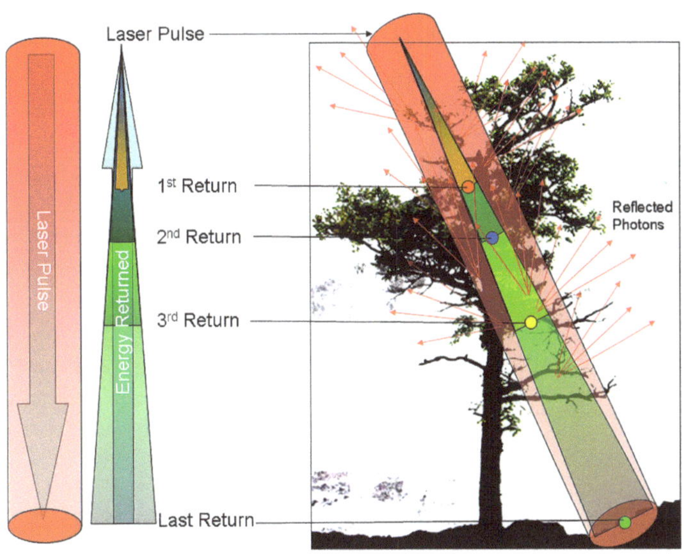

Figure 10.16 An example lidar pulse, interaction with a tree, and multiple returns.

The relationship between the signal recorded by a lidar system and the canopy structure of a forest can be accurately characterized by physically based radiative transfer models that take into account these photon/vegetation interactions (Koetz et al., 2006). The measurement principle of lidar relies on laser pulses propagating through the canopy, with photons scattered within the vegetation are recorded as function of the time it takes them to return to the sensor. The volume of photons backscattered from a canopy is a function of outgoing laser pulse energy, the vertical distribution of gaps in the canopy, the spectral properties of the canopy elements, their angular orientations, and the spectral properties of the ground (Ni-Meister et al., 2001).

The probability of a beam of light passing through a canopy is given by the gap probability function, based upon Beer's law (Sun and Ranson, 2000):

$$p = e-(krS)/cos\theta$$

where

p is the probability that the beam is not reflected,
k is a measure of foliage area projected onto a plane normal to the light beam,
r is the foliage area density, and
S is the distance that the beam travels through the canopy
θ is the off-nadir angle of the beam

Models like this can be used to determine the sampling distribution for lidar measurements $p(yt \mid \underset{\sim}{x})$. As laser pulse energy is reflected by canopy surfaces, such as branches, twigs and leaves, the volume of photons available to return decrease with depth through vegetation. In order to detect the ground in a dense forest, one needs either: 1) a large amount of energy transmitted; 2) a shorter distance for the laser pulse to travel; 3) a detector capable of needing fewer photons returned to record ranges, and/or 4) many opportunities to attempt to penetrate the canopy (high pulse densities, larger footprints or scanning angles).

10.2.2.2 Forestry Measurements Obtainable with Lidar

Obtaining reliable forest inventory data is a labor-intensive and expensive process, and as a result, forest inventories are usually based on plot or stand sampling in the field. Aerial photography has played an important role in the development of forest inventories, and more recently, lidar instruments have emerged as a useful capability for collecting vertical forest structural information (McRoberts and Tomppo 2007). As the availability of lidar data has increased, forest managers are seeing more opportunities to use lidar in order to meet a wide range of forest inventory information needs (Nelson et al. 2003). Lidar data are used to produce accurate depictions of tree height (St-Onge et al. 2008) and structure (Lim et al. 2003) and, when collected with an adequate density of hits on the ground (Lovell et al. 2005), may be used to generate accurate models of individual tree height (Næsset and Økland 2002). Calculating individual tree characteristics in the field is a time consuming and costly process. Attributes such as height can be measured directly in the field using a clinometer or a laser-surveying instrument. Diameter at Breast Height (dbh) is defined as the diameter of a tree 1.37 meters above the ground, and can be measured directly using a dbh tape around a tree. Estimates such as volume can be computed from equations derived from previous model studies (Figure 10.17). The number of trees in a stand must be counted and stem mapped using surveying techniques. Complete ground surveys of all the trees in the stand are very expensive, and it is very difficult to accurately and precisely map the locations of trees in relation to one another.

Estimating forest stand properties in Norway (Næsset, 2002; Næsset and Bjerknes, 2001) are based on a two-step procedure. First, sample plots measured in field are used to train lidar data to relate laser-derived measures of canopy height and density to the biophysical properties of interest. In the second step, the estimated relationships in the form of regression equations are used to predict biophysical properties from the lidar for every stand in the landscape. A major challenge in estimating biophysical variables such as basal area and volume from lidar is the various stem forms of trees in stands within a canopy, creating entirely different relationships between the lidar and the biophysical variables of interest. Unless proper criteria can be found, properties for certain forest types will tend to be over-estimated while others will be under-estimated.

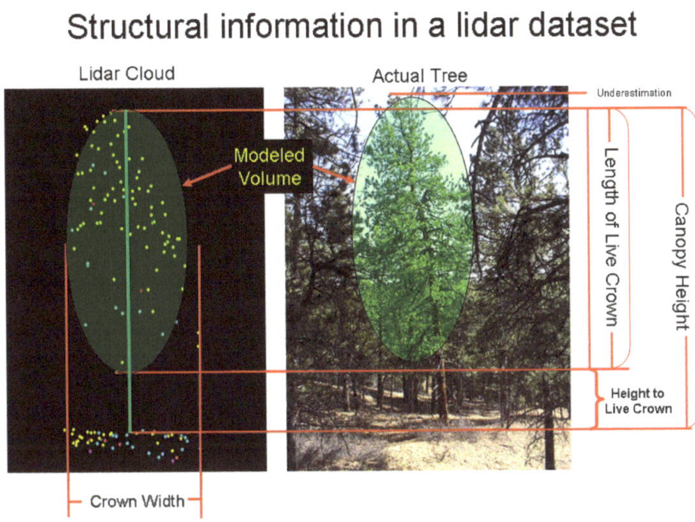

Figure 10.17 Comparison of lidar point cloud and forestry information that can be extracted.

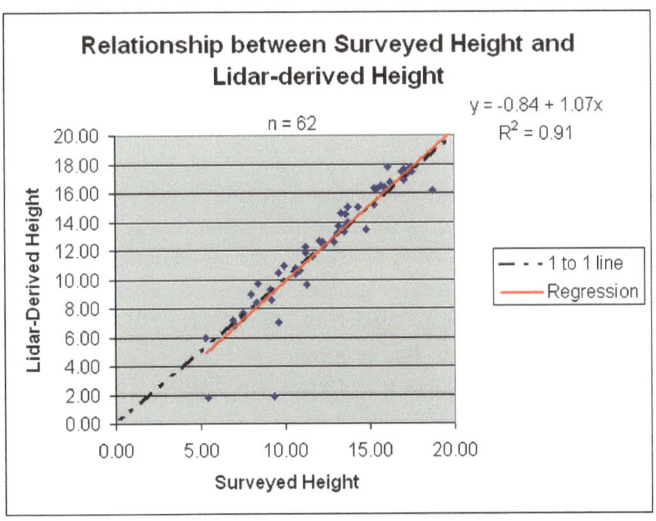

Figure 10.18 Relationship between surveyed and lidar-derived heights for individual trees (From Stoker, 2002).

10.2.2.3 Forest Inventory Parameters

10.2.2.3.1 Canopy Height

Height is a common biophysical characteristic that is important to foresters. Height is a key variable in wood volume equations, determination of site quality (Magnussen and Boudewyn, 1998), biomass, future yields, potential stand treatments (Wulder et al. 2001), and fire behavior (Rothermel, 1991). Both small-footprint and large-footprint lidar systems have had success in estimating tree heights. Mean tree heights have been estimated using small-footprint systems (Magnussen and Boudewyn, 1998; Magnussen et al. 1999; Means et al. 2000; Naesset, 1997a; Young et al. 2000), as well as large-footprint (Lefsky et al. 2002; Means et al. 1999; Lefsky et al. 1999b).

While a canopy height or mean stand height is the vertical distance from an arbitrary point on the forest floor to the topmost part of the canopy at a given location, tree height is the vertical distance from the base of the tree to its topmost terminal bud (Avery and Burkhart 1983; Magnussen et al. 1999). The measurement of tree height for an individual tree is estimated from lidar by subtracting the highest elevation return from the lowest elevation return for a particular tree (Wulder and Seemann, 2003), and has been shown to relate very closely to field surveys (Figure 10.18).

The error associated with lidar measurements of tree height are typically between 0.5 and 1.0 m (Persson et al. 2002; Næsset 1997, 2002; Magnussen and Boudewyn 1998; Magnussen et al. 1999; Næsset and Okland 2002), and lidar is considered more accurate to measure canopy height than even common field-based measurements (Næsset and Økland 2002), due to the difficulties in measuring heights in closed forests. Several studies comparing small-footprint lidar stand height estimates to field surveys found that lidar underestimates stand height (Nillson, 1996; Magnussen and Boudewyn, 1998; Young et al. 2000). This has typically been attributed to the fact that small-footprint systems do not adequately sample treetops. The underestimates of height by small-footprint lidar may be due to failure to obtain reflections from the ground as well as sample the tops of relatively broad trees (Harding et al. 1994). These results, however, were found using small-footprint systems with a relatively low sampling rate (5–8 kHz). Current systems operating today can produce sampling rates from 150 to over 300 kHz, and as a result can more easily collect high-density data.

A requirement for calculating canopy height in both discrete return and full waveform lidar data is the need to penetrate below the canopy and reach the bare ground. In the case of discrete return lidar data, canopy height estimates are usually calculated by taking the difference between a surface representation of the highest non-ground return and a surface representative of the terrain. The surface representation of the terrain is usually interpolated from classified returns after the point cloud has been filtered using a vegetation removal algorithm.

Creating an accurate representation of the ground surface in a discrete return system consists of removing all non-ground returns (e.g., vegetation and buildings) from the three-dimensional point cloud. Axelsson (1999) refers to this process as filtering. Different filtering techniques exist and are either based on statistical criteria (Kraus and Pfeiffer, 1998) and/or morphological criteria (Kilian et al., 1996). An assumption made in most filtering algorithms applied to discrete return lidar data is that the lowest returns in the three-dimensional point cloud are ground returns (Lim et al. 2003). To date, nearly all commercial lidar collections are performed using a discrete return scanning lidar instrument. Full waveform systems are currently used by researchers for science applications and are in the early process of being commercialized.

For full waveform lidar data, canopy heights can be calculated by converting the elapsed time difference between the peaks of the two most prominent modes in the amplitude waveform into range. However, the highest peak in a waveform does not always correspond to the highest point in the canopy, but is instead a function of the canopy structure (Roberts, 1998). Time differences can also be calculated between the leading and trailing edges of the two most prominent peaks in the amplitude waveform as demonstrated by Nilsson (1996). Harding et al. (2001) presented methods of developing representative canopy height profiles of the relative vertical distribution of canopy surface areas from full waveform SLICER data. The true advantage of full waveform data is the fact that there is enough precision in the vertical recording for a better estimation of the exact Z level of the apex of the tree, as the height profiles are not discrete returns caused by the leading edge recording of return responses.

10.2.2.3.2 Length of Live Crown

A similar attribute to tree height is length of the live crown. The length of the live crown in conifers is described as the length of the trunk of the tree that supports live needles. To calculate this metric, one subtracts the highest elevation point recorded from the lowest point recorded in the crown. Naesset and Okland (2002) hypothesized that crown length may be estimated directly from laser data, particularly when crowns of adjacent trees are separated from each other, but also in dense stands where crowns interfere. The computation of this metric is difficult, however, and in discrete-return systems is dependant on the temporal sensitivity of the discrimination of multiple returns. A study by Stoker (2002) shows a fairly strong relationship between surveyed LOLC (Length Of Live Crown) and lidar-derived LOLC in individual free-standing trees in the Colorado Front range (r-square = .71) (Figure 10.19).

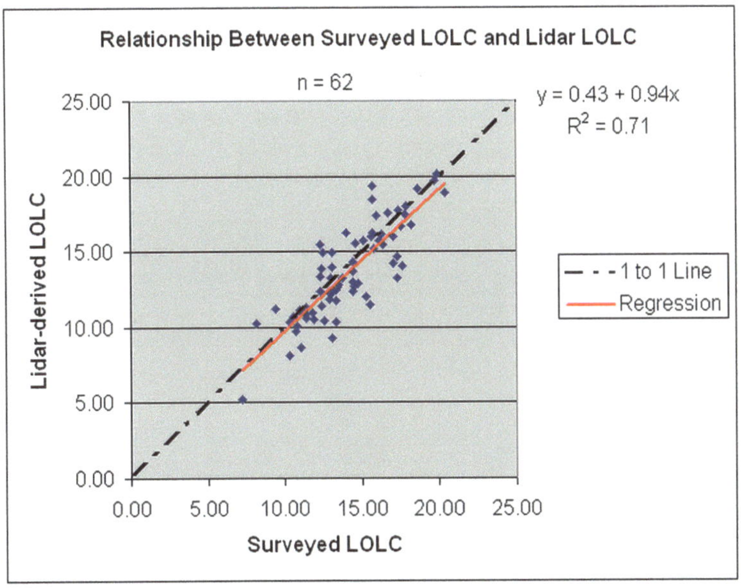

Figure 10.19 Relationship between surveyed and lidar-derived length of live crown for individual trees (From Stoker, 2002).

There is likely much more variability in the length of live crown than in height. The average difference between the two techniques in Stoker (2002) was 1.67 meters. The relationship between the surveyed length of live crown and lidar-derived length of live crown is not as strong as tree height, and is probably due to several factors. First, the probability of having one of the returns in a discrete-return system to hit the very lowest branch is low. Second, there is a good chance that some of the lowest non-ground returns recorded could be from hitting the trunk of the tree or even from shrubs or small trees underneath the tree canopy. Finally, the rule for identifying the lowest branch in the survey was that the branch had to have needles on it to be counted. However, there may have been several dead branches lower than this branch, and these could generate lidar returns.

10.2.2.3.3 Stem Locations

The number of stems in a stand and even the detection of single trees using small-footprint lidar also have been studied. Hyyppa and Inkinen (1999) found that high-pulse rate lidar systems are capable of detecting single trees in the boreal forest zone, and characteristics such as height, location and crown dimensions can be determined. These characteristics can then provide accurate stand estimates for mean height and stem volume. Hyyppa (2000) successfully found the tops of trees, and the number of trees in a stand, but high repetition pulse rates were required in order to detect individual tree crowns. A study by Magnussen et al. (1999) demonstrated how tree count and an estimate of the combined area occupied by the treetops lead to improvements in estimates of tree height.

10.2.2.3.4 Diameter at Breast Height

Indirect estimates involve estimating some fundamental parameter, and inputting these parameters into a predictive model. If these parameters can be calculated from lidar data, they can then be input into simple linear regression models. For example, a simple linear regression model for predicting Diameter at Breast Height (DBH) from tree crown dimensions for southern pines was developed by Minor (1951). Bonnor (1964) investigated the relationship between DBH and the product of crown width and height of lodgepole pine (*Pinus contorta*), and concluded it was unaffected by stand density. Additional studies of the DBH / crown width relationship were made for a number of conifer species (Bonnor 1968). Later work produced a coefficient of determination of 0.8 between DBH and crown diameter for 900 ponderosa pine (Pinus ponderosa) trees in Northern Arizona. (Hitchcock 1974). Lefsky et al. (1999b) used linear regression to develop equations relating height indices to base area and biomass.

10.2.2.3.5 Canopy Cover / Closure

Canopy cover is the area of ground covered by a vertical projection of the canopy, while canopy closure is the proportion of the sky hemisphere obscured by vegetation when viewed from a single point (Jennings et al., 1999). Calculating canopy closure from lidar is defined as the number of returns over a specified height threshold divided by the total number of returns within each cell (McGaughey, 2009):

(# of first returns > height break) / (total number of first returns)

There are two variables to this operation: height break and cell size. To produce meaningful cover estimates, the cell size should be larger than individual tree crowns. With small cell sizes (less than 5 meters) the distribution of cover values of a large area with values near 0 or 100, because each cell serves to test for the presence or absence of a tree instead of providing a reasonable sample area for assessing vegetation cover. For most forest types, cell sizes of 15-meter or larger produce good results (McGaughey, 2009).

10.2.2.4 Biomass / Carbon Parameters

Forests exchange large quantities of CO_2 with the atmosphere and make an important contribution to the global biogeochemical cycling of Carbon (Hamilton et al. 2002). Estimations of biomass are required for calculating the sources (and sinks) of carbon that result from converting a forest to cleared land and vice versa. Carbon budget models often rely on forest inventories for assessment of stand type, forest growth, species composition, and other ecological criteria (Landsberg 2003). Accurate estimations of biomass are essential for better understanding the terrestrial carbon cycles over biomes where forests serve as a primary source of biomass and terrestrial carbon (Andersen et al. 2005). Lidar has the capability of directly measuring vegetation structure, and provides a better estimation of forest aboveground biomass compared to optical sensors that suffer from saturation in spectral response, especially from dense canopies with high biomass (Dubayah and Drake 2000). Forest attributes such as canopy height, bare earth topography, and vertical distributions of canopies can be directly retrieved from lidar data. Attributes that can be predicted using empirical models from lidar data, include above-ground biomass, basal area, mean stem diameter, vertical foliar profiles and canopy volume. The radiative transfer within a canopy depends on the complex 3D canopy structure defined by the geometry, position and density of canopy elements as well as the spectral properties of each canopy element. Lidar remote sensing is capable of providing both horizontal and vertical information, with the horizontal and vertical precision dependent on the type of lidar system used (Figure 10.20).

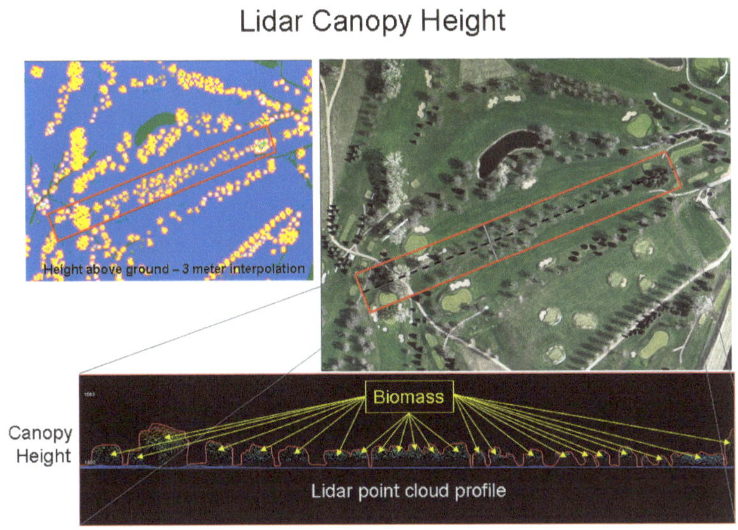

Figure 10.20 Example showing the relationship between lidar point clouds and biomass.

10.2.3 Issues Using Lidar for Forestry Applications

10.2.3.1 Point Density

Determination and characterization of forest structure using the kinds of metrics described above are usually performed at either the individual tree or plot/stand level. A literature review by Säynäjoki et al. (2008) of 34 Nordic lidar studies for forestry showed these studies had an approximate average of 1.6 points / square meter, with a maximum study having a density of 10 points per square meter and one study with a minimum of 1 point per 3 meters. The identification of individual trees requires at least a return density of approximately four returns per square meter (Gatziolis and Andersen, 2008; Reutebuch et al, 2005). This density is usually achieved with discrete-return systems that record multiple returns per pulse. Depending on the type of system, its scanning frequency, its sensitivity in pulse discrimination (e.g., number of returns per pulse), and the type of forest or area being collected, the term return density can be highly variable. When identifying the point density needed in a dataset one should use minimum pulse density and not return density as the specified variable. By influencing the pulse densities and therefore the ability to resolve individual trees and digital elevation model (DEM) accuracies, aircraft altitude and scan angle (measured off nadir) may affect the quantification of forest attributes. For example, Næsset and Økland (2002) found an average point spacing of 1 m was insufficient to accurately estimate individual crown attributes (height to green canopy and crown length) in a Norway spruce (Picea abies) and Scots pine (Pinus sylvestris) forest whilst Takahashi et al. (2005) found greater than 8.8 points/m^2 was successful in estimating tree heights to within 1 m of field estimates in sugi (Cryptomeria japonica) plantation forests, even with slopes up to 38°.

10.2.3.2 Slopes

Slope and vegetation density have been identified as variables which influence the accuracy of Digital Elevation Models (DEMs) developed from lidar data (Clark et al., 2004; Kraus and Pfeifer, 1998; Witte et al., 2000). Since the estimation of vegetation heights with lidar is calculated relative to the associated bare earth surface, any errors incorporated into the DEM will be propagated into height estimates, thus affecting the accuracy of derived forest height metrics (Figure 10.21).

A typical commercial lidar collection can accurately claim vertical errors around 15–18 cm, but these types of errors are typically reported on homogenous flat surfaces, and are not representative of the vertical errors across entire datasets, especially when any slopes exist in the terrain. Depending on the slope of the bare earth, the scan angle of the laser pulse, and the footprint size, which is dependent on flying height and laser beam divergence, the actual vertical error associated with the interaction of the laser pulse and terrain can vary.

For example, at nadir, the vertical error that is actually just the amount of vertical change in the interaction with the laser footprint can be defined as: tan (slope) * Footprint size (Figure 10.22).

So, for a 0.5 m footprint at nadir on a 10-degree slope, the actual amount of vertical change would be approximately 8 cm. On a 60-degree slope, the actual amount of vertical change inside that same footprint would be approximately 86 cm. Combine this with the variability in along-track and across-track slope, scan angle, and flying height, and the variability in vertical error can become very complicated (Figure 10.23).

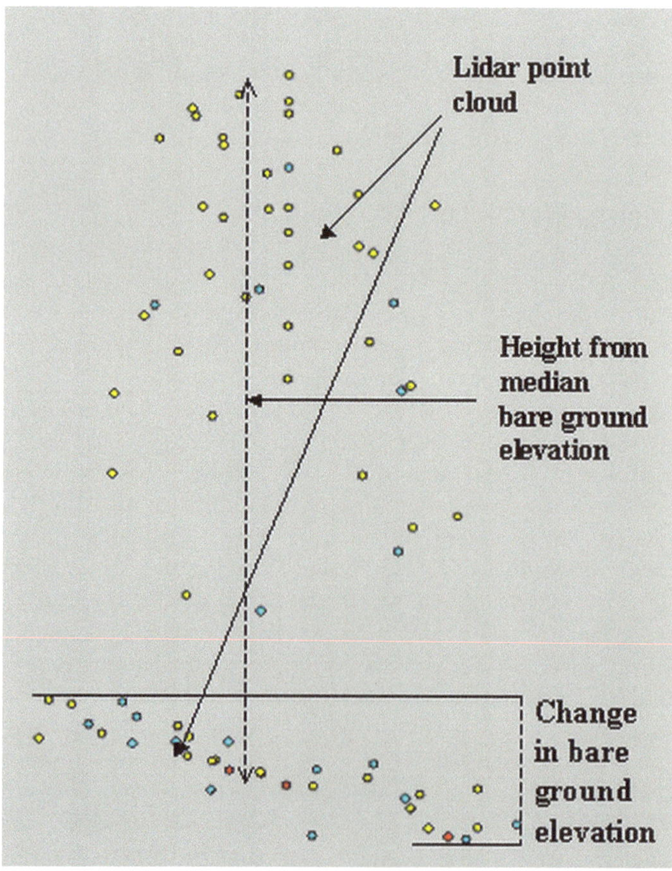

Figure 10.21 Lidar point cloud and the influence of slope on the accuracy of individual tree heights.

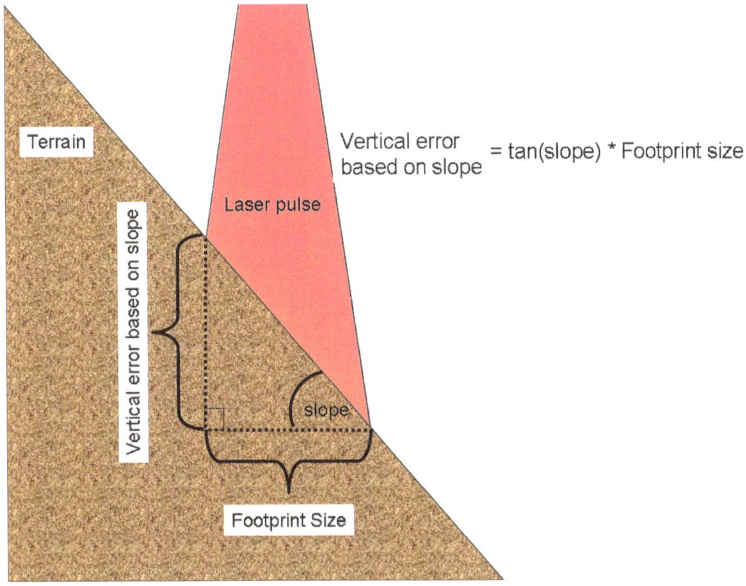

Figure 10.22 Effect of beam width and slope on vertical error.

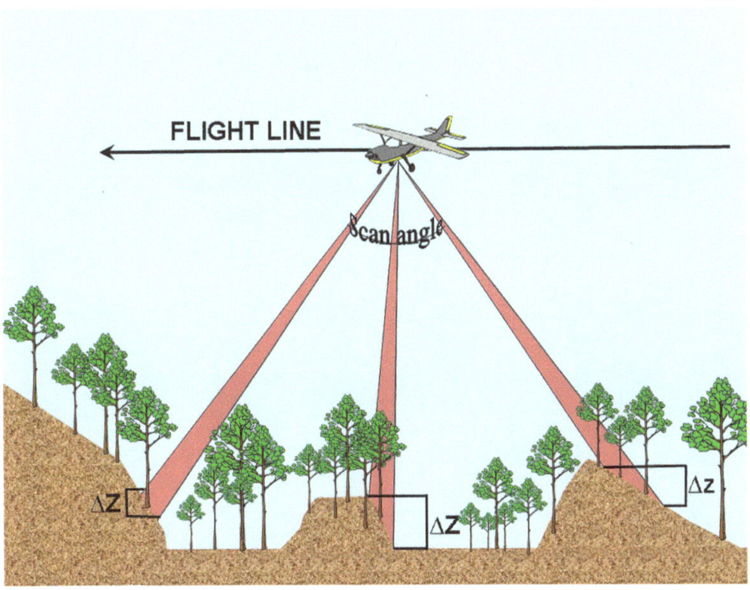

Figure 10.23 Effect of flight direction, scan angle, beam width, and slope on vertical error.

10.2.3.3 Species Discrimination

Tree species identification is important for many natural resource management and monitoring activities, including riparian characterization, wildfire risk assessment, biodiversity monitoring, and wildlife habitat assessment (Kim et al. 2009). The large variations that naturally exist between individual trees, even within trees from the same species, can provide difficulties in species extraction. Brandtberg et al. (2003) used indices derived from laser intensity data as well as height of branches to classify three deciduous species. Holmgren and Persson

(2004) used crown shape-based metrics and intensity-based metrics to differentiate Norway spruce and Scots pine. Hasegawa (2006) used lidar intensity for land cover classification and indicated that it is difficult to extract tree species without supplementary information because trees have a wide intensity range, and the single-band range of lidar overlaps with the range of other objects. Moffiet et al. (2005) did show that certain tree species may influence the return signal characteristics, and intensity information could be used to improve species discrimination. However, the lack of standardization in the calibration of the intensity signal makes using this parameter difficult (Korpela, 2008). Hyperspectral data, with hundreds of spectral bands, allow for a finer discrimination of spectral signatures and have been applied to identifying tree species (Gong et al., 1997). Because spectral reflectance changes depending on the time of a year, especially for deciduous species, acquiring lidar datasets in leaf-on and leaf-off conditions could provide additional information useful for species identification

10.2.3.4 Understory

Optical reflectance measurements from northern, barren pine-dominated forests suggest that spectral signatures of the species in the understory vegetation overlap are strongly dependent on the imaging geometry (Peltoniemi et al., 2005). Under a tree canopy, the understory can be in direct light or in shadow, and in optical remote sensing the reflected signal from the understory is mixed with that from the upper canopy, which complicates the interpretation of images (Eriksson et al., 2006). The same can be said for active lidar remote sensing data (Korpela 2008). There is also the possibility of vertical inter-mixing of the lower branches of overstory trees and the upper branches of understory trees, shrubs or forbs, which makes the identification of understory tree stems or shrubs very difficult (Figure 10.24).

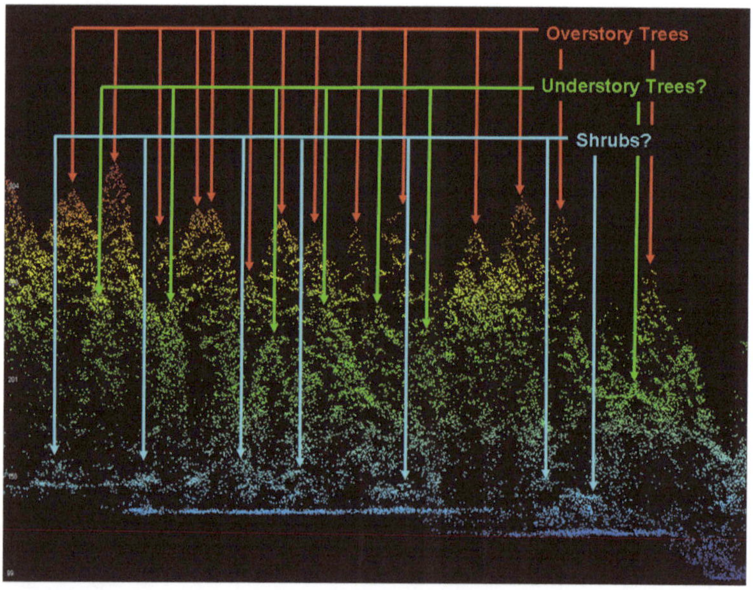

Figure 10.24 Difficulties in isolating overstory versus understory vegetation.

However, Zimble et al. (2003) demonstrated that tree height variance could be used as a measure to distinguish between two vertical structural classes within the different forest types. For applications such as fire behavior modeling, discriminating between overstory and understory trees may not be as important, as the main issue is to identify ladder fuels, which could provide a transition from surface fire to crown fire.

10.2.3.5 Leaf-on versus Leaf-off

Because the primary product from a lidar flight is a bare earth DEM, the majority of lidar collections are flown leaf off; that is at times of the year when the leafs are off deciduous trees and shrubs. This is because with the leaves off of trees, the chances for the laser pulse to make it to the bare ground are greatly increased. This leaf-off collection time could affect the accuracies for forestry-related estimations, because it is lowering the chances of the laser pulse hitting the uppermost part of the canopy or tree, especially in deciduous forests. Naesset (2005) found that in mixed forests the variability of the laser-measured canopy height is significantly higher under leaf-off conditions; however, when using an upgraded airborne lidar where the instrument configuration was changed between the two acquisitions demonstrated that both first and last return measurements can be significantly affected by system configuration. Brandtberg et al. (2003) detected individual leaf-off tree crowns and did not find any significant tendency of underestimation of the true tree heights by comparison with the maximum laser-derived heights. Therefore, it is possible that a significant portion of the laser pulses are reflected from the upper canopy of the deciduous trees, even under leaf-off conditions. The denser leaf-on canopies will tend to give a sample of reflected pulses that are more concentrated in the upper part of the canopy, where leaf-off canopies will tend to give reflected pulses more evenly scattered through the entire canopy. The combination of a leaf-off and leaf-on acquisitions could provide a more thorough accounting of forest properties, and could also help identify and locate deciduous versus coniferous trees.

10.2.4 Change Detection for Forestry Applications

As lidar has matured, the chances of multi-temporal lidar collections over the same area have increased. As a result, using lidar to detect change has become more possible. Change detection methods are traditionally pixel-based, but it is possible to carry out an object-oriented change detection analysis by using object-oriented image analysis and post classification comparison with lidar as well (Yu et al. 2004). The common method for detecting harvested or removed trees are based on difference imaging, where each pixel value in the image corresponding to the canopy height model of time frame A is subtracted from its corresponding pixel value in the image of the canopy height model of time frame B (Figure 10.25).

The resulting difference image represented the pixel-wise changes between the two dates. Negative differences are from removed trees from logging or fire, while positive differences are from tree growth (Figure 10.26).

Conversely, differences in the other direction can be attributed to tree growth. Using stem-locating methods plus computing canopy heights for each stem and applying those attributes to three-dimensional models of trees, virtual forests can be rendered. Replicating these procedures on datasets from multiple time series allows for extremely realistic visualizations of forest change, such as Figures 10.27a and 10.27b, showing 3D representations of Cheesman Lake, Colorado before and after the Hayman fire.

Conversely, differences in the other direction can be attributed to tree growth. Using stem-locating methods plus computing canopy heights for each stem and applying those attributes to three-dimensional models of trees, virtual forests can be rendered. Replicating these procedures on datasets from multiple time series allows for extremely realistic visualizations of forest change, such as Figures 10.27a and 10.27b, showing 3D representations of Cheesman Lake, Colorado before and after the Hayman fire.

10.2.5 Conclusion

While the majority of lidar work has been conducted with the sole purpose of generating a bare earth DEM to date, forestry applications utilizing lidar have been well researched and documented over the past ten years. Many countries are using lidar operationally to help them have a better accounting of their forest stocks. The improvements in technology over the last decade have translated to a more thorough understanding of forests in three dimensions. Although the data being collected is inherently representing three-dimensional information, currently most software and models still only work in two dimensions, and as a result true three-dimensional processes, such as carbon exchange, fire behavior and water flow modeling may not be taking full advantage of the data it is using. Future studies, new modeling procedures and new novel data representations, such as voxels (Stoker, 2009) may help realistically represent processes in three dimensions.

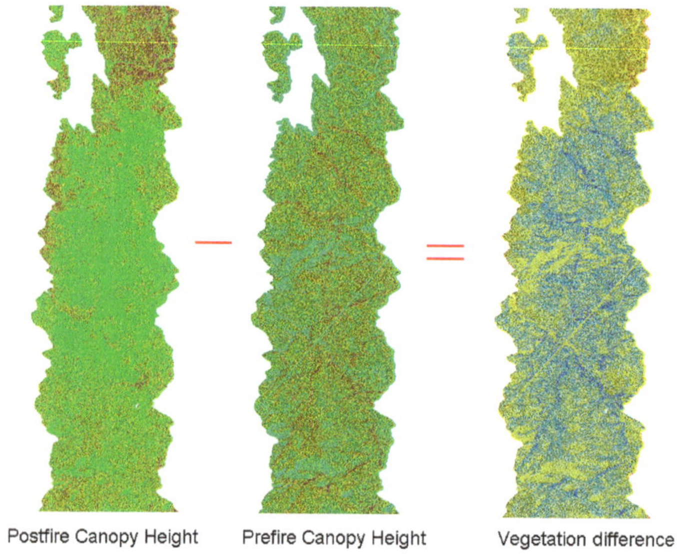

Figure 10.25 Differencing canopy heights from two time periods.

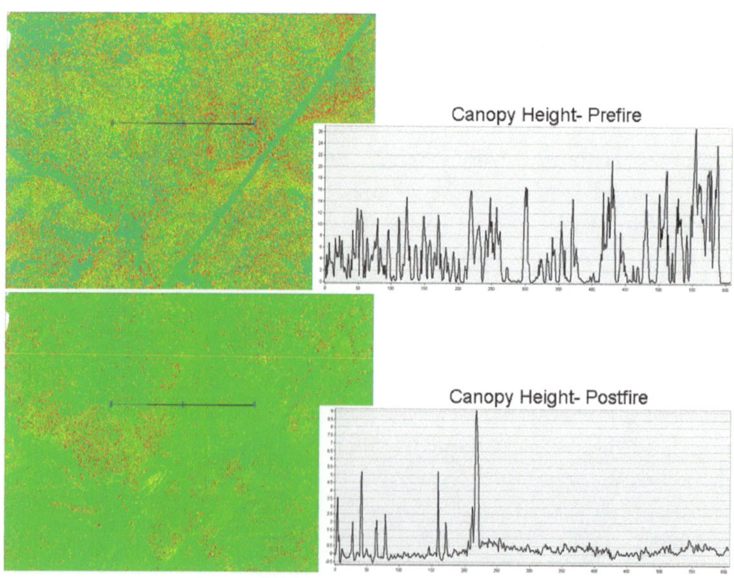

Figure 10.26 Profile of canopy heights before and after fire.

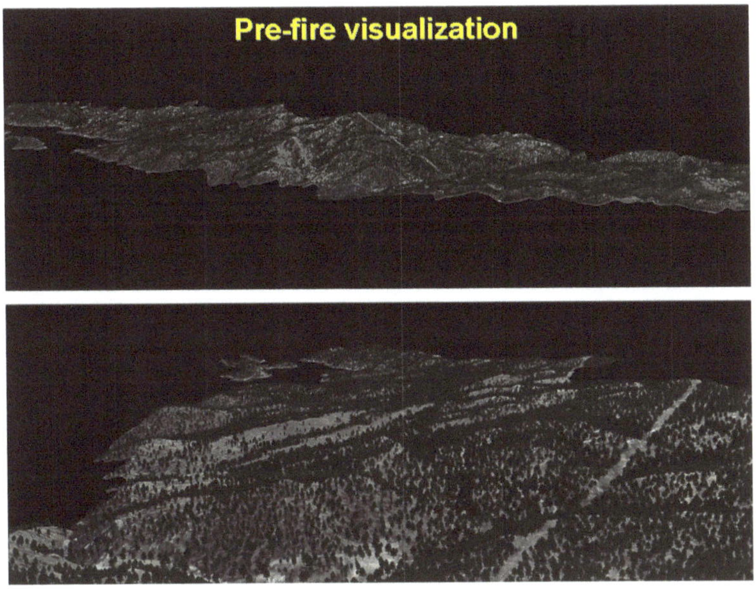

Figure 10.27a. Prefire visualization, Cheesman Lake, Colorado.

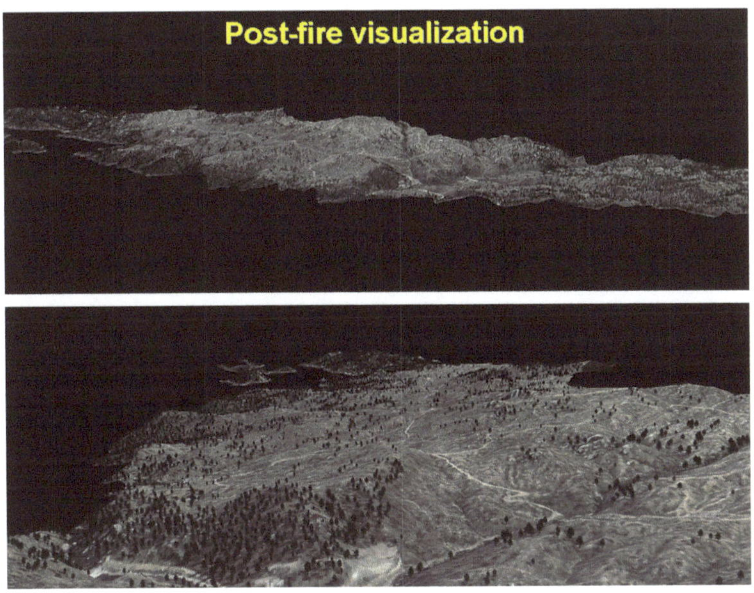

Figure 10.27b. Postfire visualization, Cheesman Lake, Colorado.

REFERENCES

Aldred, A.H., and Bonnor, G.M., 1985. Application of airborne lasers to forest surveys. Petawawa National Forestry Institute, Information Report, No. PI-X-51, pp. 1–62.

Andersen, E.H., J.R. McGaughey and E.S. Reutebuch, 2005. Estimating forest canopy fuel parameters using LIDAR data, *Remote Sensing of Environment,* 94(4):441–449.

Arp, H., J.C. Griesbach, and J.P. Burns, 1982. Mapping in tropical forests: A new approach to using the laser APR. *Photogrammetric Engineering and Remote Sensing,* 48(1):91–100.

Avery, T.E., and Burkhart, H.E., 1983. *Forest Measurements.* 3rd Edition. McGraw-Hill.

Axelsson, P., 1999. Processing of laser scanner data-algorithms and applications. *Remote Sensing of Environment*, 54(2–3):138–147.

Bonnor, G.M., 1968. Stem diameter estimates from crown width and tree height, *Commonwealth Forest Review*, 47(1): 8–13.

Bonnor, G.M., 1964. The influence of stand density on the correlation of stem diameter with crown width and height for lodgepole pine, *Forestry Chronicles*, 40(3):347–349.

Brandtberg, T., T.A. Warner, R.E. Landenberger, and J.B. McGraw, 2003. Detection and analysis of individual leaf-off tree crowns in small footprint, high sampling density lidar data from the eastern deciduous forest in North America, *Remote Sensing of Environment*, 85 (3):290–303

Clark, M.L., D.B. Clark, and D.A. Roberts, 2004. Small-footprint lidar estimation of sub-canopy elevation and tree height in a tropical rain forest landscape, *Remote Sensing of Environment*, 91(1):68–89.

Dubayah, R.O., and J.B. Drake, 2000. Lidar remote sensing for forestry, *Journal of Forestry*, 98(6):44–46.

Dubayah, R., R. Knox, M. Hofton, J.B. Blair, and J. Drake, 2000. Land Surface Characterization Using Lidar Remote Sensing, *Spatial Information for Land Use Management*, M. Hill and R. Aspinall, (eds.), International Publishers Direct, Singapore, pp. 25–38.

Evans, D. L., S.D. Roberts, J.W. McCombs, and R.L. Harrington, 2001. Detection of regularly spaced targets in small-footprint Lidar data: Research issues for consideration, *Photogrammetric Engineering and Remote Sensing*, 67(10):1133–1136.

Eriksson, H., L. Eklundh, A. Kuusk, and T. Nilson, 2006. Impact of understory vegetation on forest canopy reflectance and remotely sensed LAI estimates. *Remote Sensing of Environment*, 103:408–418.

Gatziolis, D., and H. Andersen, 2008. A guide to LIDAR data acquisition and processing for the forests of the pacific northwest, USDA Forest Service – General Technical Report PNW-GTR, (768), 1–32.

Gong, P., P. Ruilianp, and Y. Bin, 1997. Conifer species recognition: An exploratory analysis of in situ hyperspectral data, *Remote Sensing of Environment*, 62(2):189–200.

Hamilton, J. G., E.H. DeLucia, K. George, S.L. Naidu, A.C. Finzi, and W.H. Schlesinger, 2002. Forest carbon balance under elevated CO2, *Oecologia*, 131(2):250–260.

Harding, D. J., J.B. Blair, J.G. Garvin, and W.T. Lawrence, 1994. Lidar altimeter canopy height profiles: Methods and validation for closed canopy, broadleaf forests, *Remote Sensing of the Environment*, 76:283-297.

Harding, D. J., M.A. Lefsky, G.G. Parker, and J.B. Blair, 2001. Laser altimeter canopy height profiles methods and validation for closed-canopy, broadleaf forests, *Remote Sensing of Environment*, 76(3):283–297.

Hasegawa, H., 2006. Evaluations of LIDAR reflectance amplitude sensitivity towards land cover conditions, *Bulletin of GSI*, 53:43–49

Hitchcock, H., 1974. Constructing an aerial volume table from existing tariff tables, *Journal of Forestry*, 72:148–149.

Holmgren, J., and Å Persson, 2004. Identifying species of individual trees using airborne laser scanner, *Remote Sensing of Environment*, 90(4):415–423.

Hyyppä, J., H. Hyyppä, D. Leckie, F. Gougeon, X. Yu, and M. Maltamo, 2008. Review of methods of small-footprint airborne laser scanning for extracting forest inventory data in boreal forests, *International Journal of Remote Sensing*, 29(5):1339–1366.

Hyyppa, J., 2000. Feasibility for estimation of single tree characteristics using laser scanner. *IEEE International Geoscience and Remote Sensing Symposium*, July 24–28, Honolulu Hawaii, unpaginated CD-ROM.

Hyyppa, J., and M. Inkinen, 1999. Detecting and estimating attributes for single trees using laser scanner, *The Photogrammetric Journal of Finland,* 16:27–42.

Jennings, S.B., N.D. Brown, and D. Sheil, 1999. Assessing forest canopies and understory illumination: canopy closure, canopy cover and other measures, *Forestry*, 72: 59–74.

Kilian, J., N. Haala, and M. Englich, 1996. Capture and evaluation of airborne laser scanner data, *International Archives of Photogrammetry and Remote Sensing*, 31(B3):383–388.

Kim, S., R.J. McGaughey, H. Andersen, and G. Schreuder, 2009. Tree species differentiation using intensity data derived from leaf-on and leaf-off airborne laser scanner data, *Remote Sensing of Environment,* 113(8):1575–1586.

Koetz, B., F. Morsdorf, G. Sun, K.J. Ranson, K. Itten, and B. Allgöwer, 2006. Inversion of a lidar waveform model for forest biophysical parameter estimation, *IEEE Geoscience and Remote Sensing Letters,* 3(1):49–53.

Korpela, I.S., 2008. Mapping of understory lichens with airborne discrete-return LiDAR data, *Remote Sensing of Environment*, 112(10, 15):3891–3897.

Kraus, K., and N. Pfeifer, 1998. Determination of terrain models in wooded areas with airborne laser scanner data, *ISPRS Journal of Photogrammetry and Remote Sensing*, 53(4):193–203.

Landsberg, J., 2003. Modelling forest ecosystems: State of the art, challenges, and future directions, *Canadian Journal of Forest Research,* 33(3):385–397.

Lefsky, M.A., W.B. Cohen, G.G. Parker, and D.J. Harding, 2002. Lidar remote sensing for ecosystem studies, *Bioscience*, 52(1):19–30.

Lefsky, M.A., D. Harding, W.B. Cohen, G. Parker, and H.H. Shugart, 1999a. Surface Lidar remote sensing of basal area and biomass in deciduous forests of eastern Maryland, USA, *Remote Sensing of Environment,* 67:83-98.

Lefsky, M. A., W.B. Cohen, S.A. Acker, G.G. Parker, T.A. Spies, and D. Harding, 1999b. Lidar remote sensing of the canopy structure and biophysical properties of Douglas-fir western hemlock forests, *Remote Sensing of Environment,* 70:339–361.

Lim, K., P. Treitz, M. Wulder, B. St-Ongé, and M. Flood, 2003. LiDAR remote sensing of forest structure, *Progress in Physical Geography*, 27(1):88–106.

Lovell, J. L., D.L.B. Jupp, G.J. Newnham, N.C. Coops, and D.S. Culvenor, 2005. Simulation study for finding optimal lidar acquisition parameters for forest height retrieval, *Forest Ecology and Management,* 214(1–3):398–412.

Magnussen, S. and P. Boudewyn, 1998. Derivations of stand heights from airborne laser scanner data with canopy based quantile estimators, *Canadian journal of forest research,* 28(7):1016–1031

Magnussen, S., P. Eggermont, and V.N. LaRiccia, 1999. Recovering tree heights from airborne laser scanner data, *Forest Science*, 45:407–422.

McGaughey, R.J., 2009. FUSION/LDV: Software for LIDAR Data Analysis and Visualization. United States Department of Agriculture Forest Service Pacific Northwest Research Station, 127p.

McRoberts, R.E., and E.O. Tomppo, 2007. Remote sensing support for national forest inventories. Remote Sensing of Environment 110 (4):412–419.

Means, J.E., S.A. Acker, J.F. Brandon, M. Renslow, L. Emerson, and C.J. Hendrix, 2000. Predicting forest stand characteristics with airborne scanning lidar, *Photogrammetric Engineering and Remote Sensing,* 66(11):1367–1371.

Means, J. E., S.A. Acker, D.J. Harding, J.B. Blair, M.A. Lefsky, W.B. Cohen, M.E. Harmon, and W.A. McKee, 1999. Use of large-footprint scanning airborne Lidar to estimate forest stand characteristics in the western cascades of Oregon, *Remote Sensing of Environment,* 67: 298–308.

Minor, C.O., 1951. Stem-crown diameter relations in southern pine, *Journal of Forestry*, 49: 490–493.

Moffiet, T., K. Mengersen, C. Witte, R. King, and R. Denham, 2005. Airborne laser scanning: Exploratory data analysis indicates potential variables for classification of individual trees or forest stands according to species, *ISPRS Journal of Photogrammetry and Remote Sensing*, 59(5): 289–309.

Næsset, E., 2005. Assessing sensor effects and effects of leaf-off and leaf-on canopy conditions on biophysical stand properties derived from small-footprint airborne laser data, *Remote Sensing of Environment*, 98(2–3): 356–370.

Næsset, E., and K. Bjerknes, 2001. Estimating tree heights and number of stems in young forest stands using airborne laser scanner data, *Remote Sensing of Environment*, 78(3): 328–340.

Næsset, E., T. Gobakken, J. Holmgren, H. Hyyppä, J. Hyyppä, M. Maltamo, 2004. Laser scanning of forest resources: The nordic experience, *Scandinavian Journal of Forest Research*, 19(6): 482–499

Naesset, E., 1997a. Determination of mean tree height of forest stands using airborne laser scanner data, *ISPRS Journal of Forest Research*, 11: 76–84.

Naesset, E., and T. Okland, 2002. Estimating tree height and tree crown properties using airborne scanning laser in a boreal nature reserve, *Remote Sensing of Environment*, 79: 105–115.

Næsset, E., 2002. Predicting forest stand characteristics with airborne scanning laser using a practical two-stage procedure and field data, *Remote Sensing of Environment*, 80(1): 88–99.

Nelson, R., W. Krabill, and G. Maclean, 1984. Determining forest canopy characteristics using airborne laser data, *Remote Sensing of Environment*, 15: 201–212.

Nelson, R., G. Parker, and M. Hom, 2003. A portable airborne laser system for forest inventory, *Photogrammetric Engineering and Remote Sensing*, 69(3):267-273

Nilsson, M., 1996. Estimation of tree heights and stand volume using an airborne Lidar system, *Remote Sensing of Environment*, 56:1–7.

Ni-Meister, W., D.L.B. Jupp, R. Dubayah, 2001. Modeling Lidar Waveforms in Heterogeneous and Discrete Canopies, *IEEE Transactions on Geoscience and Remote Sensing*, 39(9):1943–1958.

Peltoniemi, J., Kaasalainen, S., Näränen, J., Rautiainen, M., Stenberg, P., Smolander, H., 2005. BRDF measurement of understory vegetation in pine forests: dwarf shrubs, lichen, and moss, *Remote Sensing of Environment*, 94:343–354.

Persson, Å, J. Holmgren, and U. Söderman, 2002. Detecting and measuring individual trees using an airborne laser scanner, *Photogrammetric Engineering and Remote Sensing*, 68(9):925-932.

Reutebuch, S. E., H. Andersen, and R.J. McGaughey, 2005. Light detection and ranging (LIDAR): An emerging tool for multiple resource inventory, *Journal of Forestry*, 103(6), 286-292.

Rothermel, R. C. (1991). Predicting behavior and size of crown fires in the northern Rocky Mountains, Research Article INT-438, USDA Forest Service, Intermountain Research Station, Fort Collins, CO, 46pp.

Sun, G. R., and K. Jon Ranson, 2000. Modeling lidar returns from forest canopies, *IEEE Transactions on Geoscience and Remote Sensing*, 38(6):2617–2626.

Säynäjoki, R., M. Maltamo, and Kari T. Korhonen, 2008. Forest inventory with sparse resolution Airborne Laser Scanning data – a literature review, URL: http://www.metla.fi/julkaisut/workingpapers/2008/mwp103.htm

St-Onge, B., Y. Hu, and C. Vega, 2008. Mapping the height and above-ground biomass of a mixed forest using lidar and stereo ikonos images, *International Journal of Remote Sensing*, 29(5):1277–1294

Stoker, J., 2009. Volumetric Visualization of Multiple-return Lidar Data: Using Voxels, *Photogrammetric Engineering and Remote Sensing*, 75(2):109–112.

Stoker, J., 2002. Evaluating Small-Footprint Multiple-Return Lidar to Identify Individual Tree Characteristics. Master's Thesis, Department of Forest Sciences, College of Natural Resources, Colorado State University.

Takahashi, T., K. Yamamoto, Y. Senda, and M. Tsuzuku, 2005. Predicting individual stem volumes of sugi (*cryptomeria japonica D. don*) plantations in mountainous areas using small-footprint airborne LiDAR, *Journal of Forest Research*, 10(4):305–312.

Witte, C., Norman, P., Denham, R., Turton, D., Jonas, D., and P. Tickle, 2000. Airborne laser scanning — A tool for monitoring and assessing the forests and woodlands of Australia, *Forest ecosystem research and assessment technical (pp. Paper 00-10)*, Queensland Department of Natural Resources, Brisbane, Australia.

Wulder, M. A., and D. Seemann, 2003. Forest inventory height update through the integration of lidar data with segmented landsat imagery, *Canadian Journal of Remote Sensing*, 29(5):536–543.

Wulder, M., D. Seemann, and Alain Bouchard, 2001. Within polygon grid based sampling for height estimation with Lidar data, *Proceedings of the Twenty-Third Annual Canadian Symposium on Remote Sensing*, Quebec, QC, August 20–24, 2001.

Young, B., D.L. Evans, and R.C. Parker, 2000. Methods for comparison of Lidar and field measurements of Loblolly pine, *Proceedings of the Second International Conference Geospatial Information in Agriculture and Forestry*, Lake Buena Vista, Florida, Vol I, pp. 193–199.

Yu, X., J. Hyyppä, H. Kaartinen, and M. Maltamo, 2004. Automatic detection of harvested trees and determination of forest growth using airborne laser scanning, *Remote Sensing of Environment*, 90(4):451–462.

Zimble, D.A., D.L. Evans, G.C. Carlson, R.C. Parker, S.C. Grado, and P.D. Gerard, 2003. Characterizing vertical forest structure using small-footprint airborne LiDAR, *Remote Sensing of Environment*, 87(2-3), 171-182.

10.3 CORRIDOR MAPPING

10.3.1 Transmission Line Mapping

10.3.1.1 Introduction

Lidar technology is very important to the utility industry. Before lidar, utilities had to rely on field surveying to plan and maintain powerlines. With lidar the corridors and can be very efficiently surveyed from both helicopters and fixed wing aircraft.

10.3.1.2 Plan and Profile

Nowhere in the industry is a plan and profile sheet more useful than in the utility industry. Operations need to have an accurate ground profile of the right of way as well as tower heights and the catenary of the conductor. A slight change in the location of a tower can significantly change the clearance between the ground and the conductor.

Many of the transmission lines that we currently rely on for power were constructed before we had computers and CADD. The right-of ways were surveyed and the lines were constructed. It is unrealistic to think that as-built surveys were done on many of the transmission lines after construction. The mapping projects that we are performing today with lidar are the first true as-built survey for many of the lines.

10.3.1.3 Lidar Systems

The early lidar systems were a single beam laser mounted in an aircraft. The beam was oriented at a nadir angle and provided a very good surface profile below the transmission line. There was no way to guarantee that the profile was directly under the conductor, but it did provide a very accurate elevation profile.

In the late nineties the development of the scanning lidar was a major breakthrough for the utility industry. The early systems used an oscillating mirror to deflect the laser pulse perpendicular to the direction of flight. The result was a sinusoidal distribution of lidar returns covering the width of the flight path. For surveying a transmission line, this is the perfect tool (Figure 10.28). This utility requires a high-resolution elevation model that covers the width of the right of way along the length of the transmission line. This model allows the utility engineer to have accurate side slope data or offsets when designing the line.

Figure 10.28 Example of lidar coverage along a utility corridor.

Scanning lidar systems transmit laser pulses at very high frequencies; many of the reflected returns are from the towers, poles, and conductors. It became evident that the ability to define the above ground utility features in a lidar point cloud is as powerful as mapping the terrain. By classifying the returns from the conductor, the catenary and its relationship to the ground can be modeled. The newer modern lidar systems now operate at such high frequencies that even the narrow fibre optic wire and distribution wires can be detected in the point cloud.

10.3.1.4 Transmission Engineering Design Software

Parallel to the advancement of lidar technology was the development of transmission line engineering software. These are very advanced programs that allow an engineer to thoroughly design a transmission line in a CAD environment. Very specific parameters about the physical properties of the conductor and the supporting towers can be input into these programs.

Transmission lines are designed to operate at a fixed voltage, and they have constant resistance. During the course of a normal day the draw for power changes which results in an increase in the amperage passing though the conductor. As the amperage increases the conductor will heat up and sag.

The conductors used for transmission lines are very well designed and will not fail due to the sag from high loads, however the clearance to the ground or vegetation is potentially a dangerous issue (Figure 10.29). The famous northeast power outage in 2003 was caused by a transmission line shorting out to a tree on a very hot summer day when the AC usage was extremely high.

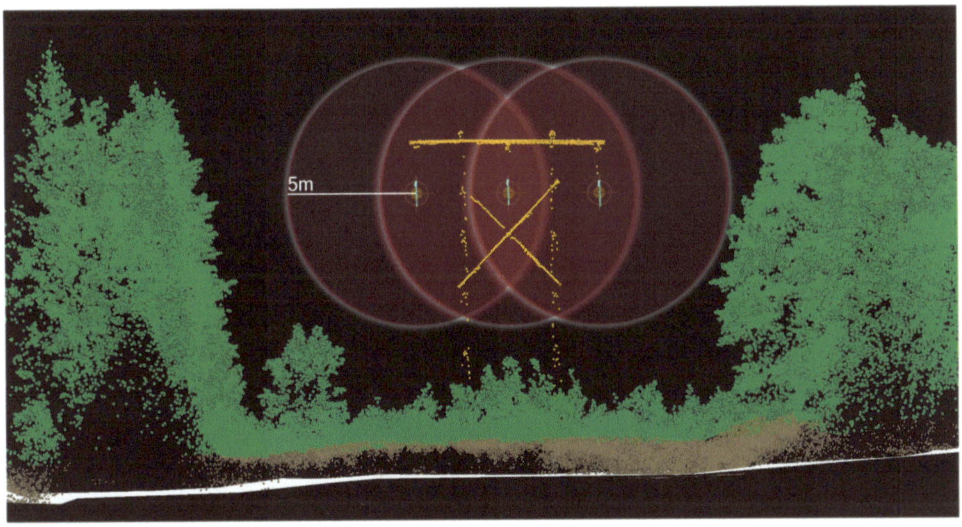

Figure 10.29 Profile view of a transmission line corridor showing the tower (yellow), vegetation (green), and radii of acceptable clearance (red).

Ice build-up is another condition that can cause sag in transmission line. Many of the larger power outages have occurred during ice storms. Wind condition must also be considered for the clearance of a transmission line. The conductor in a 230kV transmission line requires 35 feet of clearance.

With transmission line engineering software, the engineer can accurately determine the maximum amount of sag that a conductor will have under extreme operating conditions. Automated tools are now available that will allow for hazard identification and conducting tree trimming operations and removing manmade obstructions that may exist in the ROW of a transmission line. The operators only need to accurately know where the conductor is, the terrain of the ROW and the height of the trees, distribution lines, telephone poles, and any other potential obstruction that may exist.

10.3.1.5 Vegetation Management

Utility companies spend significant amounts of money on vegetation management programs. Before lidar, this was accomplished by having arborists literally walking the line and trimming the trees or applying herbicides, which is very expensive and difficult to manage. Using lidar provides a much more efficient way to monitor the vegetation clearances.

The above ground returns that are reflected from the tree branches can be classified as a vegetation layer. Branches that are too close to the conductor are then identified in a clearance report and flagged for removal. The violations can be color coded on a map and used in the field to help guide the tree trimmers (Figure 10.30). Many utilities are scheduling regular lidar surveys as part of their vegetation management plan.

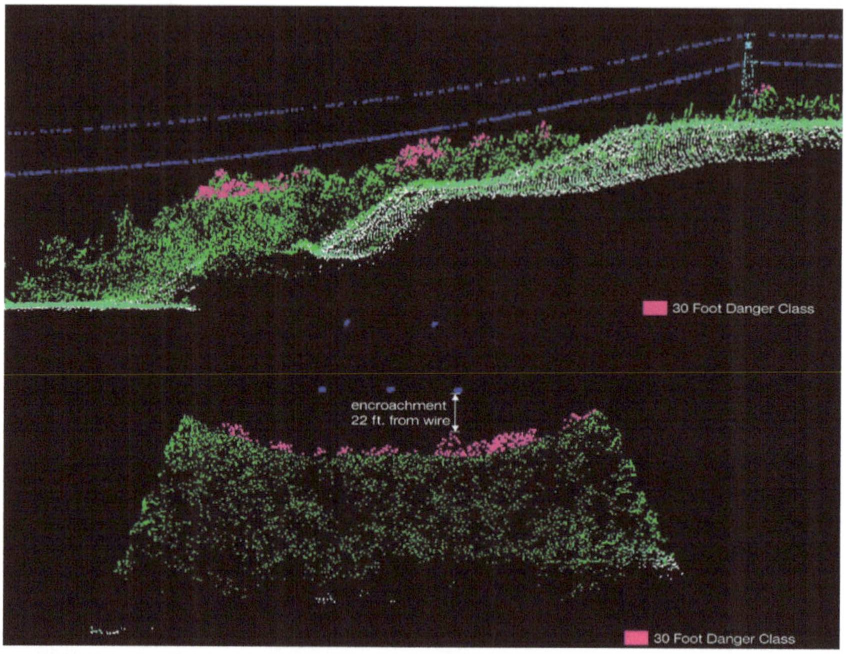

Figure 10.30 High-resolution lidar along a transmission line corridor with vegetation (green) within the 30-foot danger class (pink). Note the location of the tower (light blue) and the conductors shown in dark blue.

10.3.1.6 Line Rating

In addition to managing clearances, utility managers need to control the amount of power that they deliver through a transmission line. With the significant increase in demand for electricity and the high cost of building power lines, the utility needs to operate the line at maximum capacity. Currently, most utilities operate well below the capacity of the conductor. Using engineering design software and a lidar survey, the utility can maximize the current in the conductor to increase the amount of electricity that it can deliver to its customers.

To perform a line rating, the utility needs to know the accurate position of the conductor, the current or load at the time of the survey, and the weather conditions when the lidar mission is flown. Most lidar companies have an on board sensor to measure the wind speed, wind direction, and ambient air temperatures. The wind speed and wind direction help determine the true position of the conductor. The air temperature is used to estimate the temperature of the conductor. The software will then use this information to determine the maximum amount of sag a conductor can have before it exceeds the clearance requirements.

10.3.1.7 Imagery

Most lidar systems used today have digital camera systems incorporated in the sensor. The camera is oriented in a nadir position to collect continuous imagery along with the lidar data. The imagery can be orthorectified using the elevation model and positional data from the lidar collection. The individual orthophotos can be mosaicked into continuous images for plan viewing.

Other types of imagery that are often requested are digital oblique images of towers and poles. Using high-resolution cameras, the structures are photographed at approximately a 30-degree angle. The images are usually deliver in JPEG or TIFF format in are indexed with a CAD file. These images are invaluable for modeling and inspecting the condition of the hardware and insulators on the structures.

Digital high-resolution video is also commonly recorded during a lidar mission. The video is delivered in a MPEG file that is time stamped so it can be synchronized to the engineering design software. Thermal video can also be collected, which will show "hot spots" on the conductor that could indicate a potential problem.

10.3.1.8 NERC Alert

NERC (National Energy Regulatory Corporation) is a not for profit "electrical reliability organization" that assists in managing reliability standards for the bulk power or electric system in North America. NERC reports to the Federal Energy Reliability Commission (FERC) which is the governmental regulatory agency that oversees power reliability. NERC was founded after the blackout of 2003 where an estimated 50 million people were left without power when a transmission line was grounded to a tree growing in the right of way.

NERC has established reliability standards for transmission lines. FAC-003-1 is the standard for Transmission Vegetation Management Program. This standard requires the utility to monitor and control vegetation clearances within the right of way to prevent outages. FAC-009-1 is the standard for Facility Ratings (http://www.isone.com/rules_proceds/nerc_npcc/wkshp/corroborating_evidence_interp_no_7_rev_0.pdf). The utility operator must establish and monitor the operation of the generation and transmission of power through their facilities. Violations of these standards are to be reported to NERC and significant fines can be levied for non-compliance.

On 07 October 2010, the National Energy Regulatory Corporation (NERC) issued an alert to all member utilities that they must verify the field conditions of their transmission lines meet the required clearances established by the industry.

The alert, which is a recommendation not a requirement, was that all utilities should be able to show that the as-built condition of a transmission line meets the clearance distances for the voltage of the line. This action was prompted when a utility provider on the US East Coast had a vegetation related outage. Using lidar and transmission design software, the utility surveyed other lines for vegetation management. It was determined that were also a number of clearance-to-ground violations because the lines were not constructed according to the original design.

This information prompted NERC to issue the alert. The alert applies to lines above 100 kV; in the US this is estimated to be 450,000 circuit miles. The alert originally recommended the utilities submit their plan to address the issue and execute the plan within six months which was unrealistic. The time frame is now extended to three years (2013), with critical infrastructure to be addressed in the first and secondary infrastructure in the subsequent years.

The interesting aspect of this action is that the utility providers specifically referenced "lidar technology" as a tool to survey the line and commercial engineering software. Clearly both the regulators and industry have determined that lidar is the best tool for surveying power lines.

10.3.2 Transportation Engineering

10.3.2.1 Geospatial Technology in Transportation

Departments of Transportation (DOTs) are responsible for maintaining a substantial amount of public infrastructure. Singh (2009) discusses how geospatial technology can provide substantial benefits to the transportation sector in this role. Lidar is one of these technologies critical to providing accurate, dense geospatial data that can be used for a variety of purposes. It is also an important element to transition from traditional 2D, paper based workflows to full 3D, digital workflows. Lidar data from multiple sensors has been used for transportation applications, including airborne (ALS), helicopter (HLS), mobile (MLS, Figure 10.31), and terrestrial (TLS) laser scanning systems.

Figure 10.31 Mobile laser scanner operated by the Oregon Department of Transportation.

10.3.2.2 Considerations

Several challenges exist when performing laser scanning for transportation applications, including:

- The road generally cannot be shut down and is open to traffic during scanning. This can lead to several obstructions and noise in the point cloud. For mobile scanning, a rolling slowdown can be implemented.
- As with any static survey instrument, TLS can be difficult to perform due to vibration and wind from passing vehicles disturbing the instrument set-up.
- TLS data on flat roadways will degrade significantly with distance due to poor incidence angles (Figure 10.32). This also leads to variable point sampling, with dense measurements close to the scanner origin and sparse measurements farther from the scanner.
- Airborne systems will generally miss details underneath bridges and overpass due to the look angle. Also, data may not be of sufficient resolution and accuracy for some applications. Hence, some applications may require data from multiple sources to be fused together.
- Because lidar technology is so new, few DOTs have developed specifications, procedures, or guidelines available for their use, which can lead to difficulties in contracting. The CALTRANS Survey Manual (CALTRANS, 2011) now contains specifications for both static and mobile laser scan data.
- Various applications in transportation require different levels of accuracy and resolution, which should be considered during data acquisition and processing.
- Mobile scan systems are limited to obtaining data along navigable paths and generally obtain data within 100 m of the roadway.

Despite these limitations, lidar has many important benefits to transportation. Several studies have been implemented to examine the role of lidar in transportation. For example, Missouri DOT (Vincent and Ecker, 2010) evaluated the accuracy, cost and feasibility of airborne, mobile, and static terrestrial lidar for typical projects. They determined that all systems met their accuracy requirements. The report also highlights current hurdles including software and computing challenges. The authors also conclude that traditional surveying and/or static scanning may still be required to fill in gaps from mobile scanning. Yen et al. (2011) show that mobile lidar technology presents multiple benefits, including safety, efficiency, accuracy, resolution, technical, and cost savings. Yen et al. (2010) provides a comparison of many available mobile scan systems.

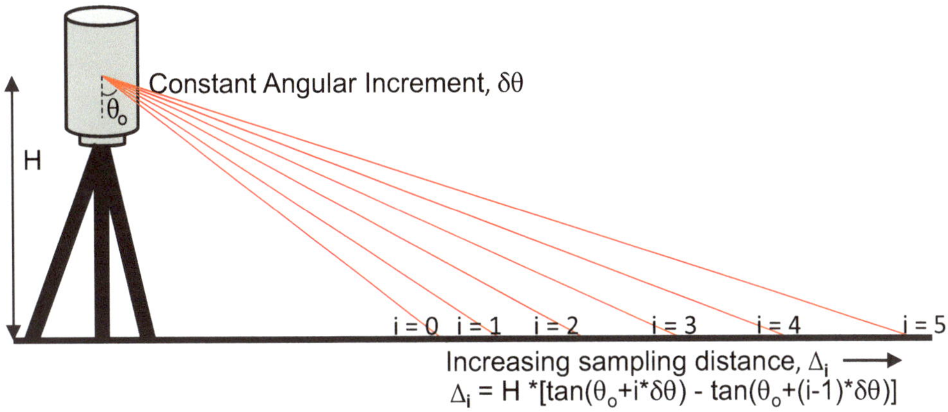

Figure 10.32 Degraded angle of incidence and sampling spacing with distance on flat, horizontal surfaces with TLS.

10.3.2.3 Sample Applications

Due to the evolving nature of laser scan technology, it is not possible to discuss all potential ways that the data can be used in transportation applications. The following discussion will focus on the most common applications.

10.3.2.3.1 Measurements

Lidar data provides a representation of real world conditions that transportation professionals can virtually explore for measurements of interest. Many measurements such as lane widths, shoulder widths, length of striping, and clearances are important to safety and maintenance evaluations of a highway. Data collected for roadways can also be used for several geometric analyses including stopping sight distances, adequate curve layouts, slope, super-elevation, drainage properties, lane width, and pavement wear. Traditionally, many of these measurements are obtained manually with a tape measure or wheel, which poses substantial safety concerns. Grafe (2008) provides examples of a roadway digital surface model, cross sections, and a highway interchange that have all been surveyed using MLS. Lidar systems provide a permanent record of site conditions that can be queried at any time after the initial collection of point data, enabling users to remotely measure length, elevation, deflection, smoothness, camber, curvature, volume, and other parameters (Jaselskis et al., 2005). Figure 10.33 demonstrates how linear measurements can be made quickly within a point cloud to determine lane widths, sidewalk widths, and building dimensions through virtual surveying. Laser scan data can also be used to measure bridge clearances. These measurements can then be input into traffic safety studies, ensuring appropriate visibility and stopping distances.

10.3.2.3.2 Bridge and Sign Clearances

The Oregon Department of Transportation (Oregon DOT) has actively pursued use of lidar for its operations. As permitting demands for taller, wider and longer loads on the state highway system increase, in conjunction with changing conditions such as periodic overlays to existing pavement, the need for accurate and reliable vertical and horizontal clearance information also increases. High accuracy is especially important in cases where existing clearance values are close to the permitted or statutory values. Lidar data enables:

1. Unlimited clearance measurements to be made from the point cloud in any plane or direction: vertical, horizontal, or perpendicular to the roadway surface (Figure 10.34).
2. The creation of a 3D Solid Object Model of the structure opening and underlying surface.
3. Visualization of the structure and surroundings.

In extraordinary situations when trying to determine whether a very large and complex shaped load can transit through the structure opening, the load itself can also be scanned, modeled and virtually transported through the modeled structure opening to wnsure the real load can pass through. In addition to bridges, lidar can also be used to determine clearance of other overhead structures and obstructions such as cantilevered signs, sign bridges, or power lines (Figure 10.35).

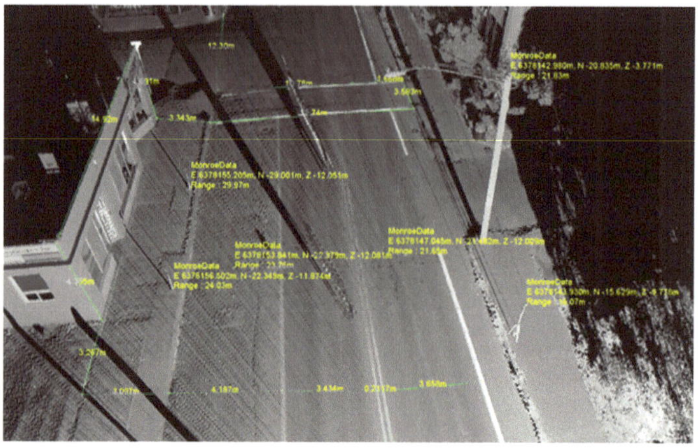

Figure 10.33 Linear measurements in a dataset.

Figure 10.34 Clearance Values measured from Point Cloud perpendicular to roadway surface.

Figure 10.35 Colorized point cloud of a steel pedestrian structure.

Bridges over navigable waterways (Figure 10.36) have similar issues and requirements with the added complication of a variable water surface and the possibility of being on a different vertical datum. Lidar data of the structure used in conjunction with bathymetric data of the river bottom and water level at the time of the expected transit helps determine the maximum size load that can pass under the structure.

10.3.2.3.3 Topographic Mapping

Topographic mapping, including earth work computation, is an important application of lidar in transportation. Jaselskis et al. (2003) performed a comparative study of total station and lidar based volume calculations, showing only a 1.2 percent difference between volumes calculated from each method. Slattery et al. (In Press) also evaluated the use of static scanning for construction applications such as determining earthwork quantities. Yen et al. (2010) evaluate the quality of pavement DTMs created from laser scan data and determined that although the technology does not yet meet CALTRAN's specification, additional refinement of the technology will probably make it suitable in the near future.

Precise topography is important to evaluate and analyze drainage conditions on highways, which can have significant safety implications. Laser scan data has been shown to be effective to determine cross-slope measurements to ensure safe driving conditions.

10.3.2.3.4 Tunnels

In addition to providing vertical and horizontal clearance measurements, high accuracy scans using TLS systems provide a baseline dataset for deformation studies of tunnels and other complex features. The Oregon DOT now scans all the tunnels in the State on an annual basis to monitor any movement so they can detect deformations before they become problematic.

10.3.2.3.5 Asset Management and Inspection

Duffell and Rudrum (2005) show inventory mapping as a benefit of a point cloud, including extraction of any structure, pavement, signage, traffic signaling devices, etc.. Becker and Haala (2007) emphasize the need for detailed 3D modeling of urban landscapes for city planning. They demonstrate an automated façade grammar building tool that can model building facades beyond the line-of-sight of the scanner by hypothesizing further facades based on the adjoining style. Jochem et al. (2011) also proposes using MLS to model building facades; however, the focus is to select the facades with the highest solar potential by extracting individual structures from the point cloud and assigning solar potential ratings to each façade.

Kingston et al. (2006) focus on both manual and automated feature extraction from lidar data and also demonstrate the ability of software to automatically detect road signs and classify them by shape as defined by the Manual on Uniform Traffic Control Devices. Semi-automatic or fully automatic extraction of signs is necessary to efficiently locate signs in a large point cloud. Brenner (2009) presents a method of pole (any vertical narrow structure) extraction by use of cylindrical stacks containing a core with data surrounded by a torus of no data. McQuat (2011) discusses methods for automatic detection ofseveral different structures (signs, facades, bays, automobiles, curbs, etc.). They also provide insight on how these structures can then be converted to useful shapes for storage in a GIS.

Figure 10.39 and Figure 10.40 show examples of how intensity measurements can be used to identify and extract features such as striping, manhole locations, and signs.

(a)

(b)

Figure 10.36 (a) Boat loads passing under the Longview Bridge, and (b) Lidar was used to model the bridge to determine clearances for a given tide level.

Figure 10.37 Colorized point cloud of tunnel portal, showing smoke damage from vehicle fire.

Figure 10.38 Close examination of the intensity shaded point cloud shows additional, minor damage to concrete.

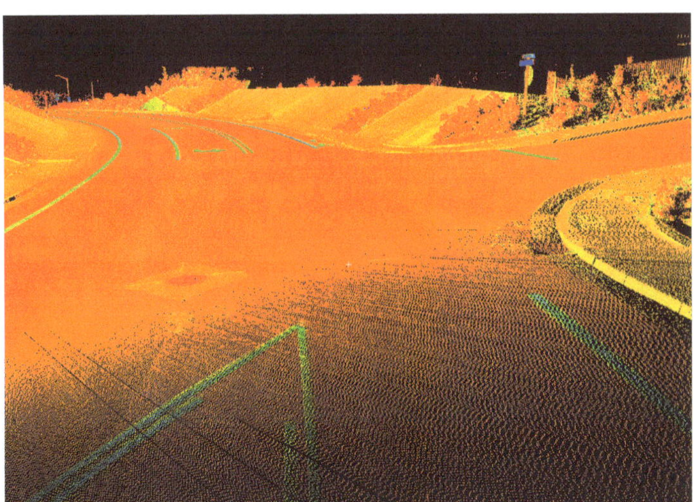

Figure 10.39 The painted street markings, manhole cover, curbs, and other objects can be distinguished in the intensity return image.

Figure 10.40 Example of sign extraction from TLS data. Because the signs are highly reflective, they can be easily extracted from the scan data based on intensity values.

10.3.2.4 Project Development Surveys

Airborne or helicopter lidar data captured across very large areas surrounding project sites can be utilized to locate ancient landslides, map wetlands, define watersheds, locate stream channels, model floods, help with archeological investigations, study alternate routes, etc.

In addition to capturing data for topographic mapping, point cloud data can be utilized to verify or create accurate as-built plans that either do not exist or are unreadable for structures constructed many decades ago. Figures 10.41 and 10.42 show examples of point cloud data to generate as-built plans. Singh (2009) discussed the vital role of a living survey database through all stages of the infrastructure life cycle: planning, design, construction, and maintenance. Lidar provides critical information for such a database for all phases.

10.3.2.4.1 Construction

Following planning and design, there are many applications for the use of lidar data during and following construction. Examples include:

1. Verifying that pre-fabricated structural components were built to specifications and will fit existing or newly constructed foundations or substructures prior to shipping and disrupting traffic for installation.
2. Measurement for interim payments for earthwork quantities, etc.
3. Documentation and as-built data.

Lidar data can be extremely useful during construction to record as-built information especially for components that get encased or covered up by other material throughout the construction process. Examples include capturing rebar placement prior to a concrete pour, underground utilities, drainage structures prior to backfill, and subgrade, base rock and other structural roadbed layers prior to the installation of a subsequent layer within the roadbed prism.

The 3D data capture of these obscured components coupled with point clouds of the ground surface and above ground roadway elements provides a very detailed and accurate as-built record for a project. Depending on the need, portions of this as-built point cloud could be modeled or utilized for feature extraction. Su et al. (2006) and Singh (2009) emphasize that as-built records provided through lidar can provide significantly more detail compared to traditional methods.

Figure 10.41 Point cloud data utilized to create as-built plans.

Figure 10.42 Point cloud data utilized to create as-built plans of old structure.

Figure 10.43 TLS scan of a culvert prior to construction completion.

10.3.2.4.2 Construction Automation and Machine Control

Singh (2009) discusses the role of lidar in machine automation for transportation applications for improved efficiency. Rybka (2011) demonstrates an entirely digital project from site-planning to construction through a "Design to Dozer" workshop. Periodic scans with a laser scan system permit initial design, estimates of percent completion, project compliance, and as-built records at project completion. The pure digital workflow also eliminated the need for traditional staking improving both safety and efficiency for highway construction. Grafe (2008) demonstrates how a controlled and guided roadway milling machine can be set to automatically cut the road using the digital surface model created from lidar data.

10.3.2.4.3 Pavement Analysis

Pavement structures require continual maintenance and upkeep, particularly given increasing traffic loads. Pavement smoothness evaluation can have significant financial incentives/disincentives for contractors on new highway construction projects. Current methods measure only the wheelpaths, and data collection may require the road to be closed to traffic. Lidar has shown important benefits to provide data for pavement analyses. Herr (2010) presents a paradigm shift from traditional profiling measurements to full 3D scanning of pavement surfaces, which can be used for a variety of purposes. Lidar acquires more data, enabling a more complete analysis of the road surface. Chang et al. (2006) performed tests to compare the use of TLS, Multiple Laser Profiler (MLP), and rod and level surveys. Comparison of the data showed that the use of lidar accurately measured the IRI (International Roughness Index, a measure of pavement roughness) with a correlation coefficient of 0.95 compared to the rod and level survey and 0.99 compared to the MLP data.

Because rougher surfaces will degrade the return pulse, scanner intensity information can also be used to highlight damaged sections of concrete or asphalt pavement. Scan data can also measure rutting depths (Figure 10.44) Lidar can be used in conjunction with GIS to display deviations from a planar surface. These images may be used in roadway applications to show areas of localized roughness on the pavement surface.

Tang et al. (2011) evaluated scanning and algorithms to determine the flatness of concrete, providing permanently documented results of the flatness defects. Kim et al. (2008) verified super-elevation slope values, curb design, and soundproofing wall design by creating cross sections of a roadway at 5 m intervals.

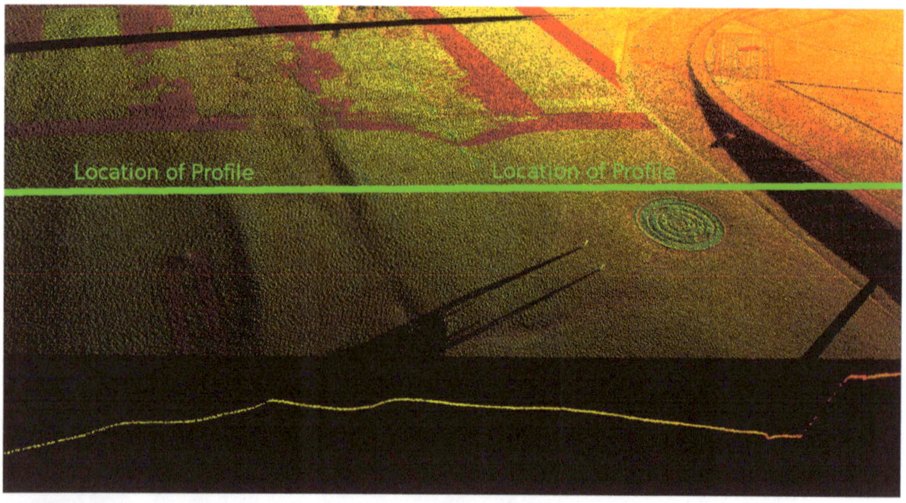

Figure 10.44 Cross-section at an intersection to demonstrate rutting evaluation.

10.3.2.4.4 Rockfall and Slope Stability

The natural hazards subchapter in this manual focuses on use of lidar for landslide studies in general; hence, this section will focus on TLS use for slope assessments along transportation corridors. Su et al. (2006) describes the use of lidar data for geotechnical monitoring of excavations, particularly in urban areas. In these urban excavations, real-time monitoring of the excavation site as well as surrounding infrastructure is critical in maintaining integrity. Miller et al. (2008) also demonstrate the use of TLS in assessing slope instability, and provide two examples along transportation corridors, noting the challenge and safety issues that arise from setting up a stationary TLS instrument along the side of a busy transportation corridor. Kemeny and Turner (2008) evaluated the use of laser scanning for highway rock slope stability analyses and found that ground-based lidar offered several advantages compared to traditional techniques including safety, accuracy, access, and analysis speed. Kemeny et al. (2008) used lidar to evaluate several rockfall sites near highways in Utah and Colorado. Turner et al. (2006) discuss processing procedures to use TLS to evaluate the stability of rocky slopes and how scan data can be integrated into geotechnical and geologic investigations. Lato et al. (2009) demonstrate how rock fall hazards along transportation corridors can be monitored using MLS on both railway and roadway-based systems. In both situations, MLS provided increased efficiency, safety, and the ability to investigate hazards. Olsen et al. (2012a) developed an *in-situ* change detection program to enable immediate geo-referencing and comparison of new scan data to baseline surfaces to determine the distribution and quantity of change. Olsen et al. (2012b) used TLS in conjunction with a geotechnical testing investigation to determine soil strength parameters to evaluate surficial slope failures occurring on fill embankment slopes for the US-20 Pioneer Mountain to Eddyville, Oregon re-alignment project. Additional scans were acquired to determine the dipping plane of larger failures observed on cut slopes (Figure 10.45).

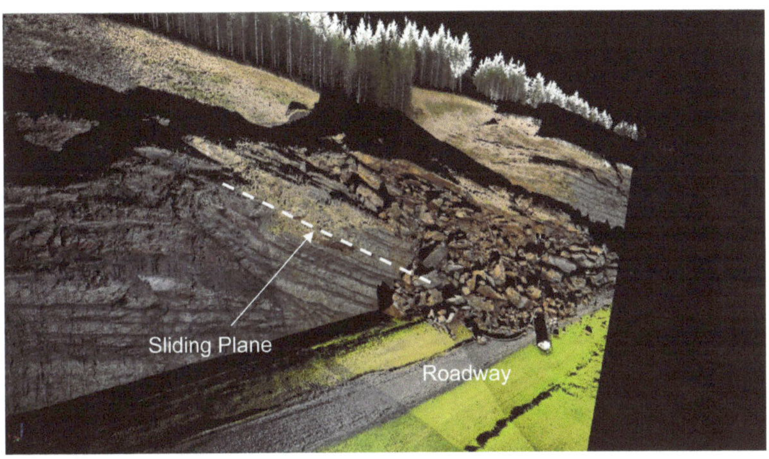

Figure 10.45 Lidar scan of a deep slide on at the US-20: Pioneer Mountain to Eddyville, Oregon re-alignment, which occurred in January, 2011.

10.3.2.4.4.1 Infrastructure Monitoring

Lidar can be used to monitor important infrastructure for transportation networks including bridges, retaining walls, and embankments. This example illustrates the potential application of lidar for Mechanically Stabilized Earth (MSE) wall monitoring. ODOT performed several scans of an older wall in Central Point, Oregon, that was suspected to be moving because of observed cracking, concrete spalling, and other damage on the wall face. Change analysis was performed using the laser scan data provided by ODOT between the December 2010, January 2011, and

February 2011 surveys. No significant differences were observed between the December 2010 and January 2011 surveys. Figures 16 and 17 show the observed change between scans from January and February 2011. Note the "blue splotches" on the face of the wall. These were a result of a maintenance crew repairing sections of spalled concrete that occurred between the surveys as opposed to actual movement of the wall. While it does not appear that the wall is moving during this short time period, it is recommended that scans be completed on a semi-annual or annual basis. Also note that the error and sensitivity of the measurements between the surveys and modeling was estimated to be slightly less than 1 cm. Hence, very small movements (mm-level) could not be detected, unless a scanner was permanently mounted at the site.

10.3.2.4.5 Accident Investigation

TLS systems have been used to document accident scenes, permitting the accidents to be moved off the roadway sooner, and allowing investigators to continue the investigation after all physical evidence has been removed from the scene. 3D Laser Mapping, Inc. (2011) reports that accident scene investigation can be 50 percent faster than total station surveying, resulting in a 1.5 hour reduction in roadway closure. Duffell and Rudrum (2005) and Mettenleiter et al. (2008) show that lidar provides an important resource to not only document pre-accident conditions, but also long accident scenes from high-speed crashes, which are difficult to survey using traditional methods.

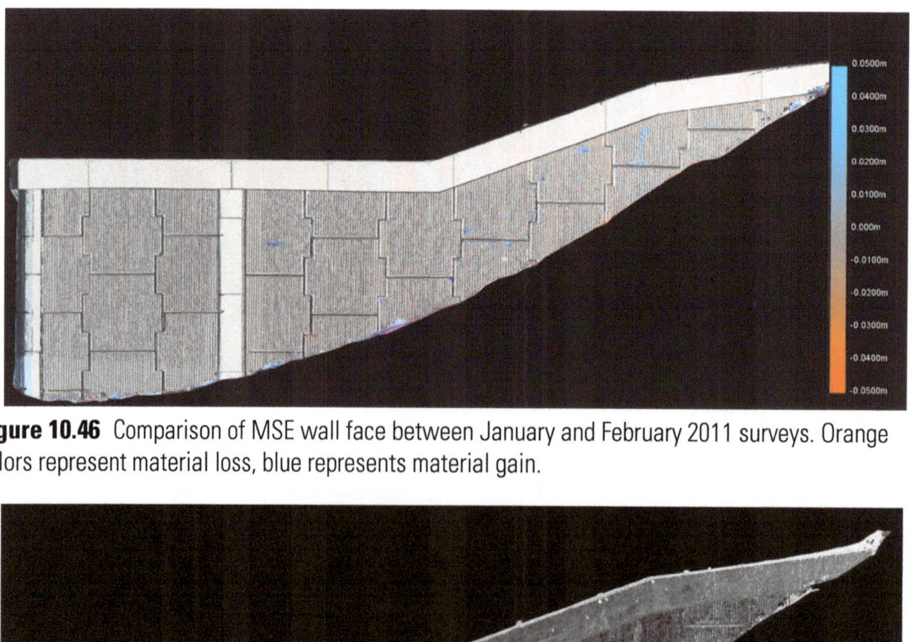

Figure 10.46 Comparison of MSE wall face between January and February 2011 surveys. Orange colors represent material loss, blue represents material gain.

Figure 10.47 Intensity map of MSE wall face between January and February 2011 surveys. Note that the new concrete patches reflect light differently than the worn concrete.

Figure 10.48 Point cloud of Pole Truck utilized to determine turning characteristics.

10.3.2.2.6 Vehicle Analysis

Lidar data can also prove extremely useful when trying to solve unusual engineering problems such as determining the turning radius of unusual vehicles. Figure 10.48 shows an example of a pole truck that was analyzed using measurements from laser scan data to determine its necessary turning radii and other characteristics to ensure that it could safely navigate its course.

ACKNOWLEDGMENTS

The Oregon State University (OSU) geomatics lab and Oregon Department of Transportation collected much of the scan data used in this section. We thank Leica Geosystems and Maptek I-Site for generously providing software used for analyzing and presenting the data shown in this section. Funding for these case studies was provided by the Oregon Department of Transportation, The Oregon Transportation Research Education Consortium (2011-398), and Granite Construction. Opinions, findings, and conclusions do not necessarily reflect those of sponsoring partners.

REFERENCES

3D Laser Mapping, 2011. 3DLM Helps Reduce Motorway Congestion. URL: http://www.3dlasermapping.com/index.php?option=com_content&view=article&id=193:-3d-laser-mapping-helps-reduce-motorway-congestion&catid=7:news&Itemid=39 (last date accessed: 08 February 2012).

Becker, S., and N. Haala, 2007. Grammar supported facade reconstruction from mobile LIDAR Mapping, *International Archives of Photogrammetry, Remote Sensing and Spatial Information Sciences*, 38:229–234.

Brenner, C., 2009. Extraction of features from mobile laser scanning data for future driver assistance systems, *Advances in GIScience* (M. Sester, L. Bernard, and V. Paelkke, editors), *Proceedings of the 12th AGILE Conference*, Hannover, Germany, 2-5 June, pp. 25–42.

CALTRANS, 2011. Interim Mobile Terrestrial Laser Scanning Guidelines, Caltrans District 4, URL: http://www.dot.ca.gov/hq/row/landsurveys/SurveysManual/Caltrans-D4_Interim_MTLS_Guidelines.pdf (last date accessed: 08 February 2012).

Chang, J.R., K.T. Chang, and D.H. Chen, 2006. Application of 3D laser scanning on measuring pavement roughness, *Journal of Testing and Evaluation*, 34(2).

Duffell, C.G., and D.M. Rudrum, 2005. Remote sensing techniques for highway earthworks Assessment, *Proceedings of the Sessions of the Geo-Frontiers 2005 Congress.*

Grafe, G., 2008. Kinematic 3D laser scanning for road or railway construction surveys, *Proceedings of First International Conference on Machine Control and Guidance.*

Herr, W., 2010. Unification of roadway surface dimensional measurements, URL: http://www.phnx-sci.com/PPS/New_Paradigm_files/New%20Paradiam%202010_0121.pdf (last date accessed: 08 February 2012).

Jaselskis, E.J., Z. Gao, and R.C. Walters, 2005. Improving transportation projects using laser Scanning, *Journal of Construction Engineering and Management*, 131(3):377–384.

Jaselskis, E.J, Z. Gao, A. Welch, and D. O'Brien, 2003. Pilot study on laser scanning technology for transportation projects, *Proceedings of the 2003 Mid-Continent Transportation Research Symposium.*

Jochem, A., B. Höfle, and M. Rutzinger, 2011. Extraction of vertical walls from mobile laser scanning data for solar potential assessment, *Remote Sensing*, 3(4):650–667.

Kemeny, J., B. Norton, J. Handy, and J. Donovan, 2008. Three-dimensional digital imaging for the identification, evaluation and management of unstable highway slopes, *Final Report for Highway IDEA project 119*, TRB NAS.

Kemeny, J., and A.K. Turner, 2008. Ground-based lidar rock slope mapping and assessment, *Publication No. FHWA-CFL/TD-08-006.*

Kim, J.S., J.C. Lee, I.J. Kang, S.Y. Cha, H. Choi, and T.G. Lee, 2008. Extraction of geometric information on highway using terrestrial laser scanning technology, *The International Archives of Photogrammetry, Remote Sensing and Spatial Information Sciences*, 37(5):539–544.

Kingston, T., V. Gikas, C. Laflamme, and C. Larouche, 2006. An integrated mobile mapping system for data acquisition and automated asset extraction, GEO-3D, Inc.

Lato, M., J. Hutchinson, M. Diederichs, D. Ball, and R. Harrap. 2009. Engineering monitoring of rockfall hazards along transportation corridors: Using mobile terrestrial lidar, *Natural Hazards and Earth System Sciences*, 9:935–946.

Mcquat, G.J., 2011. Feature Extraction Workflows for Urban Mobile-Terrestrial Lidar Data, MS Thesis, Queen's University, Kingston, Ontario, Canada, URL: http://qspace.library.queensu.ca/bitstream/1974/6530/1/Mcquat_Gregory_J_201105_MSc.pdf (last date accessed: 08 February 2012), 157 p.

Mettenleiter, M., N. Obertreiber, F. Härtl, M. Ehm, J. Baur, and C. Fröhlich, 2008. 3D laser scanner as part of kinematic measurement systems, *Proceedings of the International Conference on Machine Control Guidance*, 1:1–9.

Miller, P., J. Mills, S. Barr, M. Lim, D. Barber, G. Parkin, B. Clarke, S. Glendinning, and J. Hall, 2008. Terrestrial laser scanning for assessing the risk of slope instability along transport Corridors, *International Archives of Photogrammetry, Remote Sensing and Spatial Information Sciences*, 37:495–500.

Olsen, M.J., S. Butcher, and E.P. Silvia, 2012a. Real-time change and damage detection of landslides and other earth movements threatening public infrastructure, *OTREC Final Report 2011-22 and ODOT Final Report RS 500-500*, 80 p.

Olsen, M.J., A.M. Rikli, and D.N. Sillars, 2012b. Investigation of straw wattle influence on surficial slope stability, *Proceedings of the Transportation Research Board (TRB) Annual Meeting*, 2012.

Rybka, R., 2011. Down-to-earth lidar, *Lidar News*. Vol. 1, No 2, URL: http://www.lidarnews.com/newsletter/Vol1No2.htm (last date accessed: 08 February 2012).

Singh, R., 2009. *Engineering Automation*, Key Concepts for a 25 Year Time Horizon, Version 3, ODOT Report.

Slattery, K.T., D.K. Slattery, and J.P. Peterson, 2011. Road construction earthwork volume calculation using 3D laser scanning, *Journal of Surveying Engineering*, URL: http://dx.doi.org/10.1061/(ASCE)SU.1943-5428.0000073 (last date accessed: 08 February 2012).

Su, Y.Y., Y.M.A. Hashash, and L.Y. Liu, 2006. Integration of construction as-built data via laser scanning with geotechnical monitoring of urban excavation, *Journal of Construction Engineering and Management*, 132:1234.

Tang, P., D. Huber, and B. Akinci, 2011. Characterization of laser scanners and algorithms for detecting flatness defects on concrete surfaces, *Journal of Computing in Civil Engineering*, 25:31.

Turner, J.P., 2006. *Rock-socketed Shafts for Highway Structure Foundations*, Vol. 360, Transportation Research Board National Research.

Vincent, R., and M. Ecker, 2010. *Light Detection and Ranging (LiDAR) Technology Evaluation*, Missouri Department of Transportation, 107 p.

Yen, K.S., B. Ravani, and T.A. Lasky, 2011. *Lidar for Data Efficiency*, University of California, Davis, Department of Mechanical & Aerospace Engineering, AHMCT Research Center, 98 p.10.4

10.4 FLOOD-PRONE AREA MAPPING: A CASE STUDY

10.4.1 Introduction

Surface water mapping in flood-prone areas focuses on modeling the volume of water that can be carried over a given portion of the Earth's surface. Factors that influence the amount of water remaining on the surface following a rain event includes, storage volume (ponds, lakes and other depressed area features), connectivity between surface features (ditches and other conveyances), the slope of the surface, and the soil porosity (Chow, Maidment, and Mayes, 1988). Older studies traditionally used photogrammetric or survey derived spot elevations to construct Triangular Irregular Networks (TINs) from which contours were derived to model the surface topography. Today, modern analyses depend on lidar and lidarmetric derived breaklines, or Synthetic Aperture Radar (SAR) for the detailed surface elevation measurements that participate in the flood models. This subchapter will focus on the use of aerial lidar to construct the surface for hydrologic modeling in southwestern Florida.

10.4.2 Flood Prone Areas

Southwestern peninsular Florida is hydrologically isolated from the mainland United States (US) by a hydrological divide that snakes through Florida from Cedar Key on the Gulf of Mexico to New Smyrna Beach on the Atlantic (Patton and DeHan, 1998; and see inset Figure 10.49). The major river systems that drain the southeastern US empty into either the Atlantic Ocean or the Gulf of Mexico without passing through the Southwest Florida Water Management District's (SWFWMD) 16 county area (See Figure 10.49). Although there are four major river systems, the Oklawaha, the Hillsborough, the Peace, and the Withlacoochee Rivers that drain the SWFWMD, the prevalent geomorphology is termed "deranged" because water accumulates in small pools (puddles), that with additional rainfall, connect to each other to form larger pools, etc. (Smithson and Briggs, 2006). This process of pool filling and flowing into each other continues until the water reaches a river (or stream). The combination of minimal topological relief, the deranged terrain, high annual rainfall that can exceed 50 inches/year in combination with an occasional hurricane/tropical storm that can deliver over 10 inches of rain per hour, results in flood prone areas.

Deranged terrains are characterized by having no coherent pattern of rivers and lakes, but rather a series of loosely connected drainage basins, sometimes called sub-basins, contributing areas, or catchments (Smithson and Briggs, 1993; Maidment and Djokic, 2000). Each catchment is characterized by its lowest elevation, a sink, and the irregular weir that forms its ridge. During rain or storm events catchments may begin filling at the elevation of the sink depending on the soil saturation and porosity. Depending on the rain intensity and duration, soil type, land use, and other physical parameters, the water may percolate through the soil or accumulate and fill the catchment. When the water level reaches the rim of the catchment, if the rain event continues as the water may continues to rise, it will spill over to an adjacent catchment at the lowest point, the saddle or spill point, on the ridge. So depending on the surface topography and the water volume, a catchment may spill over into one, or several adjacent catchments (Olivera et al., 2002). In Figure 10.50, Catchment A has a sink node (= lowest point in the catchment) with elevation of 162.8 feet. Its boundary is shared with Catchments B, C, D, and E with boundary nodes (= lowest point along the shared boundary) of 167.3 feet, 169.1 feet, 164.2 feet, and 165.3 feet elevation, respectively (See Figure 10.50). This reticulated network characterizes deranged terrains.

Figure 10.49 West-Central Florida showing the Four Major River Basins draining the Southwest Florida Water Management District. The Deep Creek Gully Watershed in the southwestern portion of the Peace River Basin is shown in red. Inset Map shows the Florida Hydrological Divide.

In addition to the deranged terrain, Florida also contains coastal geomorphology (Kautz et al., 1998). The Florida Suncoast, the area between Cedar Key at the northern end and Punta Gorda on the southern end (see Figure 10.49) stretches over 20 barrier islands, 35 miles of white sand beaches, and a total of over 275 miles of coastline. During the hurricane season, between June and November, this coastline, as well as, several miles inland is subject to Tropical Storm and Hurricane storm surge. High winds pushing water has accounted for over 20 feet of water in low-lying, coastal flood prone areas.

Flood protection is one of the Southwest Florida Water Management District's (SWFWMD or the District) core mission components (SWFWMD Strategic Plan, 2011). Toward this end, the District employs a variety of analytical techniques to assist in predicting the severity and extent of flooding associated with a specific storm event. These analytical methods involve computer simulations and models, all of which have, as their fundamental data input, a model of the surface topography. Prior to 2004, the District typically used photogrammetrically-generated contours as the basis of a Digital Terrain Model (DTM). Since 2004, a lidar-derived DTM has been used as the basis of the hydrological models.

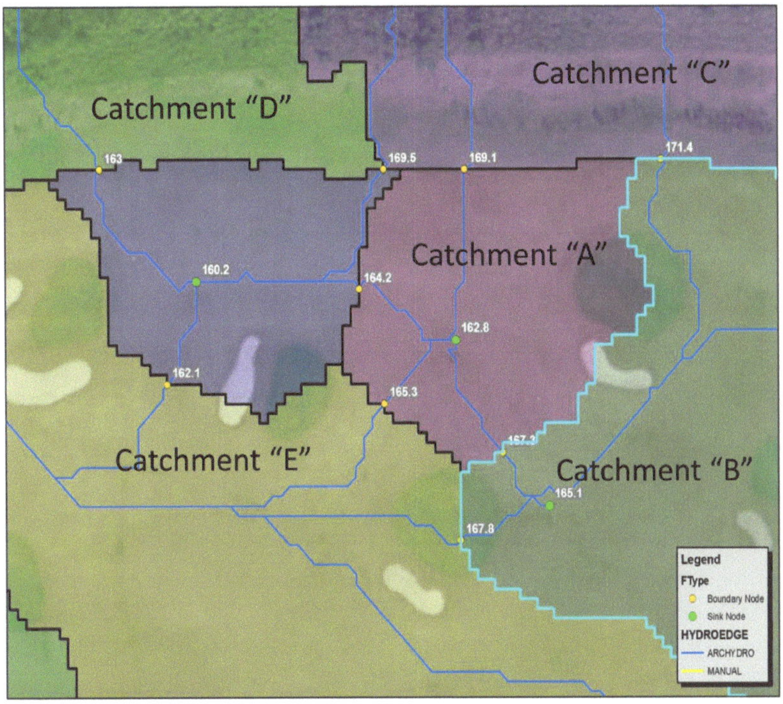

Figure 10.50 ArcHydro®-derived Catchments showing Flow Network (blue lines) between Sinks (green points) and Lowest Points along Catchment Boundaries (yellow points) in typical Deranged Terrain.

10.4.2.1 Using Lidar For FEMA Floodplain Assessment

The Federal Emergency Management Agency (FEMA) has constructed a set of guidelines specifically focused on using lidar data for constructing Flood Risk Zone estimates (FEMA, 2003). Appendix A of the FEMA Guidelines and Specifications for Flood Hazard Mapping Partners, details pertaining to using lidar data are presented. The accuracy specifications are referenced to the National Standards for Spatial Data Accuracy (NSSDA) and differ for flat (i.e., low topographic relief; 1.2 feet at the 95% confidence level) and rolling to hilly (i.e., high topographic relief; 2.4 feet at the 95% confidence level) terrains, but an emphasis remains on point spacing of 5 meters or less to assure that sufficient surface detail is captured. The specifications further state that supplementary information may be needed for lidar-obscured areas, deep or turbid water, and to obtain sub-surface channel and hydraulic structure geometry. The specification permits the use of specific frequency airborne lasers for shallow water and near-shore coastal mapping.

As the goal of lidar mapping is to produce a high accuracy Digital Terrain Model, data voids are particularly troublesome to hydrological modeling. Data voids resulting from instrument malfunction, high winds resulting in errant flight lines, and other mechanical issues are not tolerable and recollection is the only option. Voids resulting from dense vegetation are subject to additional guideline depending on whether or not the void occurs within the floodplain. Vegetative data voids with the floodplain must be addressed with supplemental ground survey while voids outside of the floodplain can be filled through interpolation of surrounding laser points.

Lidar artifacts also influence the quality of the DTM. Artifacts, such as spikes, ridges, roughness, etc. can result from numerous causes, including systematic errors, environmental conditions, and incomplete or post-processing. Because removal of artifacts would lead to data voids, they are approached in much the same manner. Artifacts outside the floodplain, but inside the watershed, have no bearing on the hydraulic modeling are neglected. Those artifacts that would bear on the model are removed and replaced with supplemental ground survey.

As with all lidar products, the accuracy of derived surfaces is reported based on a Fundamental Vertical Accuracy (FVA), Supplemental Vertical Accuracy (SVA), and Consolidated Vertical Accuracy (CVA) as described by the American Society of Photogrammetry and Remote Sensing Guidelines (ASPRS; Blak, 2007). For the purpose of FEMA Flood Zone mapping, the Guidelines specify testing secondary products over the following: (a) Bear-earth and low grass, (b) High grass, weeds and crops, (c) Brush lands and low trees, (d) Forested, fully covered by trees, and (e) Urban areas. As these categories comprise the majority of land-use classes that commonly occur, FEMA suggests that they may be sub-divided, as needed, to provide better analysis of accuracy.

In general, FEMA guidelines focus on using the lidar data for the specific purpose of hydrological modeling. The guidelines are highly focused on correctly mapping the floodplain and not very concerned with data voids, artifacts, etc. in areas outside of the floodplain. Testing and reporting of vertical accuracy is important as is the distribution of any associated error.

10.4.2.2 Using Lidar for Surface Water Modeling

The USGS, NOAA, EPA, USACE, and many private consultants have constructed surface water quantity and storm surge models. In Florida, the Sea, Lake, and Overland Surges from Hurricanes (SLOSH) model from the National Hurricane Center (NHC) has been used to estimate storm surge heights and winds resulting from predicted hurricanes. SLOSH models the storm surge and storm tide based on several parameters, including the topography of the ocean shelf and adjacent inland area for up to several hundred feet. In 2007, the Florida Division of Emergency Management collected lidar data along the entire SunCoast for use with SLOSH modeling. The use of lidar data for SLOSH modeling is a separate topic and will not be further addressed in this chapter.

The District uses one of several computer models to estimate the amount and duration of water in flood prone areas. The preferred model for deranged terrains is a commercially available model, Interconnected Channel and Pond Routing Model (ICPR) produced by Streamline Technologies (Streamline Technologies, Inc. Winter Springs, Florida 32708). There are three primary building blocks to this model: (a) Basins (=Catchments), (b) Nodes (=sinks), and (c) Links. Stormwater runoff hydrographs are generated for basins and then assigned to nodes in the drainage network. Nodes are used to represent ponds and specific locations in the drainage network such as along channels, streams, rivers, and junctions in pipe systems. Stages are calculated at each of the nodes and links such as pipes, channel segments, weirs, control structures, and bridges are used to connect nodes together. Flow rates are calculated for links based on stages at nodes. The Environmental Protection Agency's (EPA's) Storm Water Management Model (SWMM), the United States Geological Survey (USGS) Soil and Water Assessment Tool (SWAT) and others are used on an as-needed basis. The remainder of this chapter will focus on how lidar are used with ICPR.

10.4.2.3 Advantages and Disadvantages to Using Lidar Data

Traditionally, photogrammetric or ground survey data were used to construct a TIN from which contours were derived (Li et al., 2005). The contours represented a "smoothed" continuous surface that presented little impediment to the computationally intensive finite-difference models. The lidar surface, containing much finer detail, presents a new layer of complexity to already complex calculations. So while the surface is more detailed and may be a better representation of the actual watershed, there is a tradeoff between the amount of detail available and the processing that can be performed.

10.4.3 An Example – Deep Creek Gully, DeSoto County Florida

10.4.3.1 Introduction to the Study Area

The Deep Creek Gully watershed in the southern portion of the SWFWMD (see Figure 10.49; red area northeast of Punta Gorda) in the Peace River Basin provides a typical example of mapping flood-prone areas in the deranged terrain of the Gulf Coast of Florida. Repeated flooding between 2000 to the present has occurred in this watershed where elevations vary from +20 feet NAVD88 (North American Vertical Datum of 1988) to only +31 feet NAVD88 over a distance of approximately 35 miles. This very subtle 11-foot elevation gradient is typical of deranged terrains.

During the course of the SWFWMD's annual mapping program, standard one-foot contours were compiled photogrammetrically from 2003 DeSoto County/SWFWMD aerial photography. Following the 2004 hurricane season, when Hurricane Charley passed through southwestern Florida, the District had the opportunity to collect lidar data during the early winter of 2005 (30 January 2005 to 20 February 2005). The lidar point cloud was flown at a density sufficient to support a maximum final post spacing of 6-feet for unobscured areas. These two topographic datasets, one photogrammetric and the other lidar, separated by only two flight seasons, provide the basis for comparison.

Since 2004, the District has been modifying the Esri ArcHydro® data model (Maidment, 2002) for use in deranged terrains. The District's modifications have been formalized into a Geographic Watershed Information System (GWIS), and customized tools have been used for preprocessing data to be used in the Interconnected Channel and Pond Routing (ICPR; Singhofen, 2009) hydrologic and hydraulic surface water model. ICPR uses three building blocks, Basins, Nodes, and Links to construct the drainage network. ArcHydro/GWIS preprocesses the topographic data to construct catchments, which are accumulated into ICPR Model Basins, sinks, and drainage points which are used as nodes, and HydroEdges which are used as links in the ICPR Model.

10.4.3.2 Methodology

The 2003 data were delivered to the District as mylar sheets containing the imagery with the 1-foot contours. The mylars were scanned on a Hawking Contex Hawk-Eye 36G Wide Format scanner at 300 dpi. The resulting scanned image was georeferenced and map-to-map rectified to the 2005 orthoimagery. The 1-foot contours were "heads-up" digitized using Esri ArcGIS 9.3.1. Lines were attributed with their elevations as on the scanned images. Elevations were converted to NAVD88 values using CorpsCon 6.0.

The ArcGIS shapefiles containing the digital contour information were processed using Esri 3D Analyst® into a triangular irregular network (TIN). A 5-foot × 5-foot raster digital elevation model (DEM) was extracted from the TIN and used for ArcHydro® pre-processing.

For the 2005 lidar data, the point cloud was classified by the lidar vendor into the following ASPRS LAS Classes: Class 1 (unclassified), Class 2 (Ground), and Class 9 (water). QCoherent LP360/Esri ArcGIS 9.3.1 software was used to triangulate the lidar point cloud along with vendor provided breaklines (top of bank, centerline of creek, edge of pavement). A 5-foot × 5-foot floating point DEM was extracted and converted to an Esri DEM using the Esri Spatial Analyst®.

10.4.3.3 Results and Discussion

ArcHydro® preprocessing may include steps to fill small depressions (sinks) in the DEM resulting from computational errors (triangulation issues). We chose not to eliminate the minor sinks so that any catchments resulting from the preprocessing would be visualized. The results of the preprocessing is shown in Figure 10.51a for the contour-derived surface and Figure 10.51b for the lidar-derived surface.

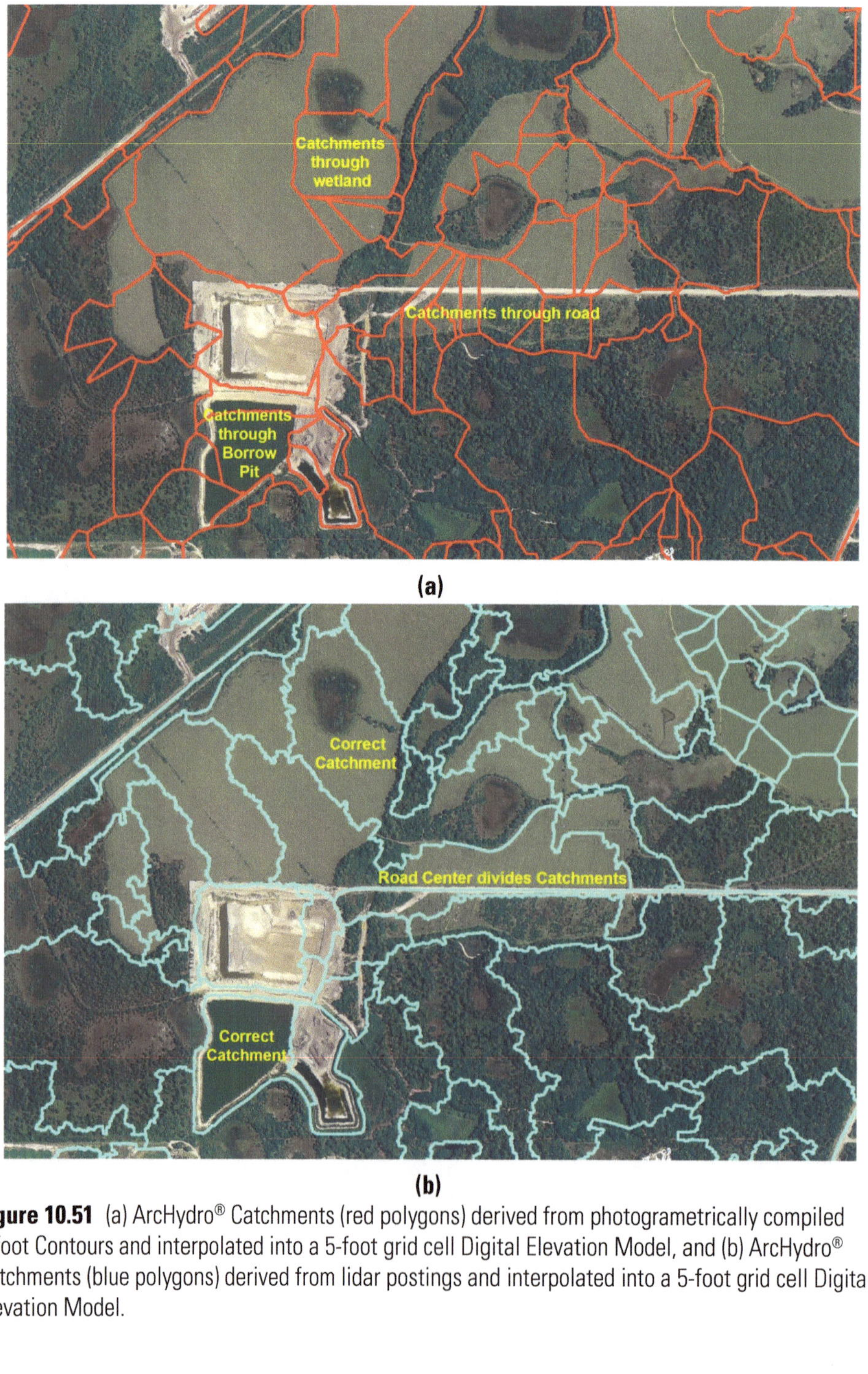

Figure 10.51 (a) ArcHydro® Catchments (red polygons) derived from photogrametrically compiled 1-foot Contours and interpolated into a 5-foot grid cell Digital Elevation Model, and (b) ArcHydro® Catchments (blue polygons) derived from lidar postings and interpolated into a 5-foot grid cell Digital Elevation Model.

As ArcHydro® catchment ridges represent the "top of the bowl" for a contributing area, we expect that wetlands and ponds, which represent the sink points in a catchment, have to be singular. That is, a catchment ridge should not traverse a pond or wetland. Similarly, as roads, either paved or graded, frequently have a crown to facilitate water drainage, we would expect catchments to be divided by that crown.

Figure 10.51a shows catchments constructed on the basis of the contour-derived DEM. In this analysis of the surface, it is obvious that there are several catchments going through the pond in the south and the wetland in the north-central area. Also, we observed several catchments crossing the road running through the center of the study area.

Figure 10.51b shows catchments constructed from the lidar-derived DEM. It is clear that the catchments delineate water features and the wetland feature correctly. Also, the catchments are separated by the crown of the road. The catchments derived from the lidar surface appear to be more hydrologically reasonable with respect to the landscape. The contributing areas flow to sinks and waterbodies contain a single sink. The catchments constructed from the lidar surface could easily be combined into ICPR Model Basins for further analysis, whereas those constructed from the contour surface require significant modification.

In deranged terrains such as this, small differences in elevation can potentially result in differences in catchment delineation. It is not uncommon for the "spillover points" along the catchment ridge to differ from the sink point of the catchment by only 0.3-foot to 0.6-foot. Hence, vertical accuracy of the DEM is critical to correct catchment delineation.

To test the vertical accuracy of the two DEMs, twenty (20) vertical accuracy checkpoints were collected throughout the study area. The difference between the surveyed elevation and the elevation derived from the contour-derived DEM and lidar-derived DEM were used to compute Root Mean Square Errors (RMSE) for the elevation. The results (Table 10.1) show that for the same ground checkpoints, the lidar-derived DEM showed a smaller error range, a smaller RMSe and an overall greater vertical accuracy, National Standards for Spatial Data Accuracy (NSSDA = ±0.823-foot; Table 1).

Table 10.1 Vertical Accuracy Comparison of Contour-derived and Lidar-derived DEM.

	Contour-Derived Surface*	Lidar-Derived Surface**
Error Range (in feet)	−1.31' – +0.82'	−0.66' – +0.43'
RMSe	0.707	0.420
NSSDA Accuracy 95% Confidence Interval	±1.385	±0.823'

10.4.3.4 Conclusions

In the flood prone, deranged terrain typical of the Southwest Florida Water Management District, where elevation gradients are very low, lidar data with a nominal 6-foot posting, provide a DEM that is superior to a contour-derived DEM. The lidar-derived DEM contains more accurate elevations which are critical in determining the water flow direction through the flood prone area. This leads to more accurate catchment and model-basin delineation, providing a more accurate and detailed hydrological model, which, in turn, produces a more accurate floodplain determination.

10.4.3.5 Future Advances Using Lidar in Flood Prone Areas

To date, an emphasis has been placed on the bare earth digital terrain model (DTM) where the laser point-cloud has been classified into ASPRS LAS Class 2 (Ground) and Class 12 (Ground

Overlap). While this approach has been successful, as detailed above, it has been applied primarily in rural to semi-urban areas, where man-made terrain alterations have a minimal effect of surface water flow.

In moderate and heavily urbanized flood prone areas, man-made infrastructure, such as pipes, gutters, drop-basins, and other inlets, serve to route water toward retention/detention ponds and other man-made storage facilities. This "below surface" flow is not visible to lidar, but can be modeled by modern network techniques.

Other man-made features in flood prone urban areas may present obstacles to surface water flow. In a typical flood prone urban area, buildings and houses are built on building pads that are slightly elevated above the surrounding bare earth. The laser points falling onto this slightly elevated surface are usually classified as ground and then all other points representing the building are removed. This may result in shallow depressions where a building resides (a catchment) or the water may appear to flow through the building if the laser points falling onto the building pad are not significantly higher than the bare-earth surrounding them. In either case, the water is not being properly routed.

To accommodate man-made surface features that distort the actual water flow in urbanized areas, the District has been incorporating ASPRS LAS Class 6 (Building) points into the bare earth DTM to construct a partial digital surface model (pDSM). In the high density urban developments of Florida, building pads are graded to route water (a) toward the front of the lot, (b) to the front and rear of the lot, or (c) toward the read of the lot. Hence, catchments should reflect the slope of the lot rather than routing water through (or under) the building pad.

In Figure 10.52a the catchments (blue polygons) were constructed from the lidar bare-earth DTM. Catchments are correctly delineated to either side of the paved road, indicating the presence of the road's crown. However, several catchments, such as those identified (see Figure 10.52a), include both front and rear sides of the house's roof indicating that water would be accumulating under this house.

In Figure 10.52b the catchments (red polygons) were constructed from the lidar pDSM which included the Class 6 laser points. As the road was not affected by the inclusion of Class 6 points, there is no difference in the catchments along the road. However, the catchments in the built-out area now follow the roof ridge line and indicate drainage properly. There is a catchment that accumulates water in the front and another catchment that accumulates water toward the rear of the house. The lidar defines the roofs so well, that where roofs slope toward each other, catchments are delineated between houses. In this medium density residential neighborhood, the inclusion of Class 6 Building points more accurately defines the water accumulation areas than the bare-earth DTM.

Finally, the identification and delineation of directly connected impervious area (DCIA), pipes that route stormwater directly to streams or ponds and non-directly connected impervious areas, sidewalks, driveways, roofs, etc. that route water to pervious surfaces, plays a role in surface water modeling. The District, through one of its contractors, has been working to develop a process using lidar elevations, planar filtering, and the lidar intensity value to extract rooftops, sidewalks, driveways, parking lots and streets (curb to curb) in urban and residential flood prone areas.

REFERENCES

Blak, T.A., 2007. Chapter 12: DEM Quality Assessment, *Digital Elevation Model Technologies and Applications: The DEM Users Manual*, Second edition (D.F. Maune, editor), American Society for Photogrammetry and Remote Sensing Bethesda, Maryland, pp. 425–448.

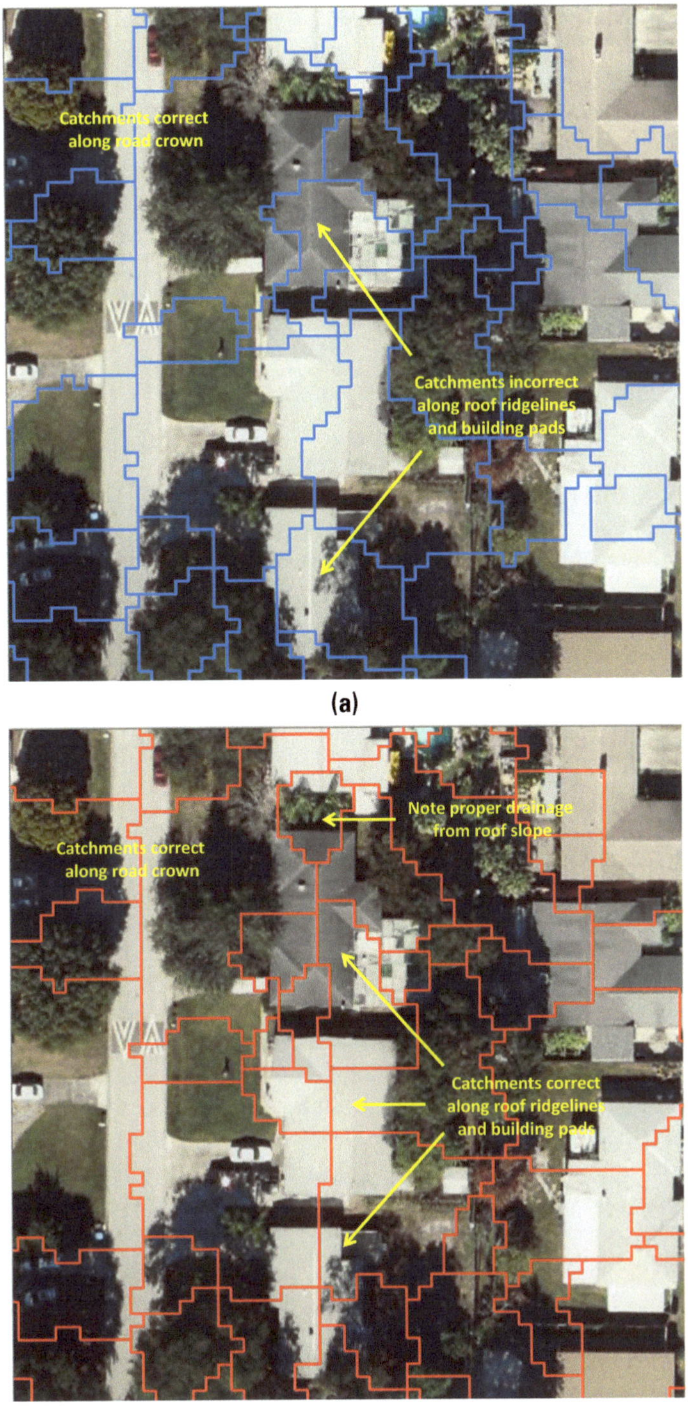

(a)

(b)

Figure 10.52 (a) ArcHydro® catchments (blue polygons) constructed from bare earth DTM, and (b) ArcHydro® catchments (red polygons) constructed from bare earth DTM containing Building (Class 6) laser points.

Chow, V.T., D.R. Maidment, and L.W. Mays. 1988. *Applied Hydrology: McGraw-Hill Series in Water Resources and Environmental Engineering*, New York, McGraw- Hill, Inc.
FEMA. 2003. Guidelines and Standards for Flood Hazard Mapping Partners, Appendix A: Guidance for Aerial Mapping and Surveying, URL: http://www.fema.gov/pdf/fhm/frm_gsaa.pdf (last date accessed: 27 September 2011)
Kautz, R.S., K. Haddad, T.S. Hoehn, T. Rogers, E.D. Estevez, and T. Atkeson, 1998. Natural Systems, *Water Resources Atlas of Florida* (E.A. Fernald and E.D. Purdum, editors), Institute of Science and Public Affairs, Florida State University Press, Tallahassee, Florida, pp. 82–113.
Li, Z., Q. Zhu, and C. Gold. 2005. *Digital Terrain Modeling: Principles and Methodology*, CRC Press, Boca Raton, Florida.
Maidment, D., and D. Djokic, 2000. *Hydrologic and Hydraulic Modeling Support with Geographic Information Systems*, Esri Press, Redlands, California.
Olivera, F., J. Furnans, D. Maidment, D. Djokic, and Z. Ye, 2002. Drainage systems, *Arc-Hydro: GIS for Water Resources* (D.R.Maidment, editor) Esri Press, Redlands, California.
Patton, D.J., and R.S. DeHan, Water issues: Global, national, state, ecosystem, *Water Resources Atlas of Florida* (E.A.Fernald and E.D. Purdum, editors), Institute of Science and Public Affairs, Florida State University Press, Tallahassee, Florida, pp. 1–14.
Singhofen, P.J., 2009. Modeling percolation from multiple ponds in close proximity using ICPR, 25th Annual ASCE Water Resources Seminar, Orlando, Florida, URL: http://www.streamnologies.com (last date accessed: 27 September 2011).
Smithson, P. and D.J. Briggs, 1993. *Fundamentals of Physical Geography*, Routledge Press, New York..
Southwest Florida Water Management District, 2011. *2012-2016 Strategic Plan*, Brooksville, Florida.

10.5 BUILDING EXTRACTION AND RECONSTRUCTION FROM LIDAR POINT CLOUDS

10.5.1 Introduction

In the last two decades, lidar (Light Detection and Ranging) has been increasingly utilized as an emerging technology for remote sensing of the Earth. Many methods and algorithms have been developed and implemented to exploit dense 3D data provided by airborne lidar systems to extract information on Earth's topography as well as the natural and man-made features of the terrain. Building extraction has been among the main interests of researchers who try to acquire useful information from lidar data for urban applications. 3D lidar point locations already provide sampled representations of the terrain and the objects on the terrain without any further processing. However, simpler representations are required for practical purposes for many applications as far as data size, operability, and analysis are concerned. The continuous increase in the use of lidar data acquisition puts automated building extraction at the forefront of an active area of research that has its challenges.

Modeling buildings from lidar data requires classification of the points reflected from buildings so that they are separated from the points that pertain to the terrain, vegetation, and other objects in the area. Building extraction is typically carried out on a subset of lidar points, which are those left once the bare ground points are removed from the data. This process, which results in the classification of points as "ground" and "non-ground," is commonly called filtering. There are different filtering algorithms in the literature. Kraus and Pfeifer (1998), Axelsson (2000), Vosselman (2000), Zhang et al. (2003), Sithole and Vosselman (2004), Shan and Sampath (2005), Pfeifer and Mandlburger (2009), and Mongus and Žalik (2012) provide details on various filtering methods.

Usually, a building detection step is employed for the recognition and localization of the buildings within a dataset. This process is often performed following ground filtering, but it may also be a part of the filtering algorithm utilized, such that the filtering result includes a building class. Some building detection algorithms perform on the raster datasets obtained by resampling the lidar points on a regular grid, which allows the use of various well established image processing algorithms to analyze these range images. Typically, a normalized Digital Surface Model (nDSM) is calculated by removing the resampled filtered terrain (i.e., Digital Terrain Model (DTM)) from the Digital Surface Model (DSM). Then, different methods are applied by analyzing the height, discontinuities, texture, etc. on the nDSM for building detection. Hug and Wehr (1997) used thresholds on the difference images, elevation texture, local variance, and directional distribution of the surface gradients to detect natural and artificial objects. Maas and Vosselman (1999) applied thresholding, binning, morphological filtering, and connected component analysis techniques to acquire the points describing a building. Morgan and Tempfli (2000) similarly employed morphological filtering together with connected component labeling, minimum area thresholding, and change analysis of elevation and slope with Laplacian and Sobel filters for building detection. Nardinocchi et al. (2003) implemented a region growing segmentation method based on slope difference thresholds, followed by a hierarchical rule-based classification that considers the geometric and topological properties of the segments to obtain several different classes including a building class. Alharthy and Bethel (2002) generated a discrepancy map as the difference between the first and last returns and used this map to determine and filter penetrable objects. In addition, they employed a second filter utilizing the root mean square errors (RMSE) of local planar fits as an indicator for the irregularity of the surfaces. Some methods also use 3D lidar points, either in combination with the resampled grids or individually for building detection. Tarsha-Kurdi et al. (2006), as an example of using a combination of grid and point based approaches, used the thresholds on the topological features derived for each grid cell based on the number of points per cell, the maximum vertical distance between the points, and the maximum slope angle of the best fitting plane combined with a region growing algorithm to classify buildings and vegetation from off-ground points. Verma et al. (2006) performed eigenvalue analysis within the local neighborhood of each point in the lidar point cloud to filter out all non-flat points and points that lie on vertical walls. After the non-flat points were removed, they voxelized the flat points and performed a connected component analysis to separate the roofs and the ground.

Ancillary data such as building plans (Haala and Brenner, 1999; Hofmann et al., 2002; Overby, 2004; Schwalbe et al., 2005; Tack et al., 2012; Vosselman and Dijkman, 2001; You and Lin, 2011) or imagery (Awrangjeb et al., 2010; Awrangjeb et al., 2011; Demir and Baltsavias, 2010; Hermosilla et al., 2011; Rottensteiner et al., 2007; Schenk and Csathó, 2002; Sohn and Dowman, 2007) are also used, together with lidar data, to obtain building outlines or breaklines.

Once the building points are obtained, building reconstruction may be achieved either by following a model-based or a data-driven method. Model-based methods are based on the assumption that buildings may be represented by parametric models. For each building in the point cloud, the best fitting predefined model is determined and its parameters are estimated. Maas and Vosselman (1999) used closed solutions for the determination of the parameters of simple building models based on invariant moments with raw point data. They also formulated closed solutions for additional parts (e.g., dorms) to standard gable roofs separately; however, higher order solutions for non-rectangular plans can become complicated. Weidner and Förstner (1995), on the other hand, used two different groups of models. The first group consisted of parametric models for simple buildings with the length, width, height, eave-base, and height of ridge-eave parameters. The second group consisted of prismatic models for complex buildings, which consider building blocks as sets of closed polygons with orthogonality constraints at neighboring edges. They used an approach based on the minimum description length (MDL) to find the optimum building model. Yu et al. (2009) established a model template database for determining roof types based on the angle between the roof planes and the number of samples on two intersecting planes. They clustered the normal vectors of the triangular faces of the triangulated irregular network (TIN) and matched pairs of clustered point sets with the models. The authors indicated that the cases of failure were due to the incompleteness of the building template database and suggested expansion of the template database or extension of the method to consider decomposition of complex roofs into simpler ones. Similar concerns were also raised by Tarsha-Kurdi et al. (2007b), who compared the model- and data-based approaches. They stated that the main disadvantage of model-based approaches is their dependency on the predefined existing building types in the model libraries. Such concerns led to model-based methods, which consider complex buildings as a combination of simpler roof shapes. Verma et al. (2006) modeled the class of buildings that may be constructed with a union of simpler parametric roof types. They established a topology graph as the dual of the roof structure such that each planar surface was represented by a vertex. The buildings were modeled with an algorithm that searches sub-graphs to find and match with a set of simple parametric roof shapes. Lafarge et al. (2008) established a library of 3D blocks of roof forms, including monoplane, multiplane, and curved forms with the variants specific to each form. They employed a Bayesian formulation for the optimal configuration decision using a Reversible Jump Markov Chain Monte Carlo (RJMCMC) sampler. Huang et al. (2011) improved a similar approach by allowing overlapping of roof primitives such that a more flexible reconstruction can be achieved. Model-based approaches are considered to be faster than data-based approaches because they provide models without visual deformations while data-based approaches can represent parts of the buildings closer to the polyhedral model with the downside of visual deformations and slower computation time (Tarsha-Kurdi et al., 2007b).

Contrary to the higher level model assumptions of model-based approaches, data-based approaches usually rely on more generic assumptions such as buildings that consist of planar patches. Thus, extracting geometric primitives such as lines and planes from the lidar dataset is an essential step for building modeling for most algorithms. Segmentation of the lidar point cloud provides groups of points with homogenous properties (e.g., that are on the same roof plane). Such decomposition of the building allows determination of the topological relations between the components of the building for reconstruction. Numerous algorithms exist for the segmentation of lidar point clouds. Some representative examples are listed below.

Vosselman and Dijkman (2001) used Hough transform to extract planar points on building roofs and then obtained planar roof faces by merging and expanding the initial segments. Tarsha-Kurdi et al. (2007a) used an extended RANSAC (RANdom SAmple Consensus) algorithm for segmenting the roof planes. Rottensteiner (2003) used region growing segmentation with an analysis of the DSM normal vector variations to detect roof planes. Alharthy and Bethel (2004) employed least squares moving surface analysis with region growing segmentation for the extraction of building faces. Dorninger and Nothegger (2007) clustered the features to define the local regression planes of the points followed by region growing segmentation. Sampath and Shan (Sampath and Shan, 2007 and 2010) performed eigen analysis in the Voronoi neighborhood of the roof points and clustered the surface normal vectors for roof plane segmentation. Filin and Pfeifer (2006) presented a segmentation algorithm on 3D lidar point clouds using feature clustering with slope adaptive neighborhoods. Kim and Shan (2011) presented a multiphase level set approach to building roof segmentation for modeling.

The final stage of building extraction from a data-based perspective deals with the reconstruction of the building by determining its vertices, breaklines, edges, and segments together with their topologies. Sampath and Shan (2010) formed an adjacency matrix based on a distance measure between planar segments. To regularize building footprints, they developed a least squares framework (Sampath and Shan, 2007). Toshev et al. (2010) represented buildings as a tree of volumetric parts that are covered by planar patches in their parsing architecture. Dorninger and Pfeifer (2008) generated building outlines using α-shapes followed by a generalization algorithm based on the deviations in the angular directions of polygon segments. Regularization is performed by enforcing parallelism or orthogonality. Kim and Shan (2011) determined roof structure points by identifying the intersection of non-parallel roof segments. They also determined the intersection of two line segments by applying an α-shape algorithm for tracing the boundary followed by the Douglas-Peucker algorithm (Douglas and Peucker, 1973) to extract the corner points. They then fit lines to the boundary points between these corner points with RANSAC.

The remainder of this subchapter is organized as follows. The next section provides details on segmentation algorithms for building extraction, followed by a discussion of different approaches for building reconstruction.

10.5.2 Building Extraction

The crucial part of executing 3D building reconstruction is to detect individual roof planes from the lidar point cloud of individual buildings. In the literature, this task is often referred to as extraction, for which segmentation or feature extraction algorithms are applied. Popular algorithms are commonly involved, which often have been revised to be adequate to this task. This section presents four existing algorithms. The first two algorithms are most commonly referenced in the literature and use a parameterized model, while the other two, which are from the recently published papers, aim to find homogeneous properties existing in the data set, for which a feature vector needs to be determined.

10.5.2.1 3D Hough Transform

The 2D Hough transform (Hough, 1962) is a well-known feature extraction algorithm in the field of digital image processing. The principle is to find geometric primitives, such as straight lines, circles, or ellipses, in the parameter space rather than the object space. For instance, a straight line defined in the 2D Cartesian coordinate is represented as:

$$y = ax + b \tag{10.1}$$

where two parameters (a, b) determine the characteristics of the line. In the parameter space, this line is represented by a point with a coordinate (a, b). Similarly, a set of points on the line are represented by lines following the equation, $b = x_i a + y_i$, which pass through the point (a, b) as shown in Figure 10.53a and 10.53b.

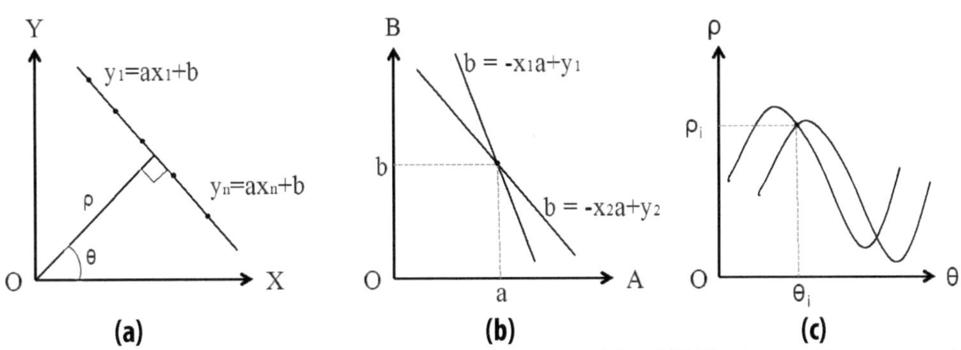

Figure 10.53 Representation of a line in the parameter space: (a) and (b) Line in the object space and parameter space, and (c) Line of normal form in the parameter space of θ, ρ.

To support the case where the slope of a line approaches infinity, normal representation (also termed normal form) is used. In this form, a line is represented as $\{P_i : \text{xi, yi}\}$

$$\cos\theta \cdot x + \sin\theta \cdot y = \rho \tag{10.2}$$

where (θ, ρ) determine the normal vector of the line passing through the coordinate origin (Figure 1a). In this representation, the points on the line are then represented by a set of sinusoidal curves intersecting at a point (θ_i, ρ_i), which corresponds to the parameters of the corresponding line as shown in Figure 1c.

The same principle can be extended to a 3D space to detect roof planes from the lidar point cloud (Tarsha-Kurdi et al., 2007a; Vosselman and Dijkman, 2001). Normal representation of a plane is defined as Equation 3 (Overby, 2004):

$$\cos\theta \cdot \cos\varphi \cdot x + \sin\theta \cdot \cos\varphi \cdot y + \sin\varphi \cdot z = \rho \tag{10.3}$$

where parameters (θ, φ, ρ) determine the normal vector of the plane. The relations between the plane, its normal vector $\vec{n}$, and associated parameters are illustrated in Figure 10.54.

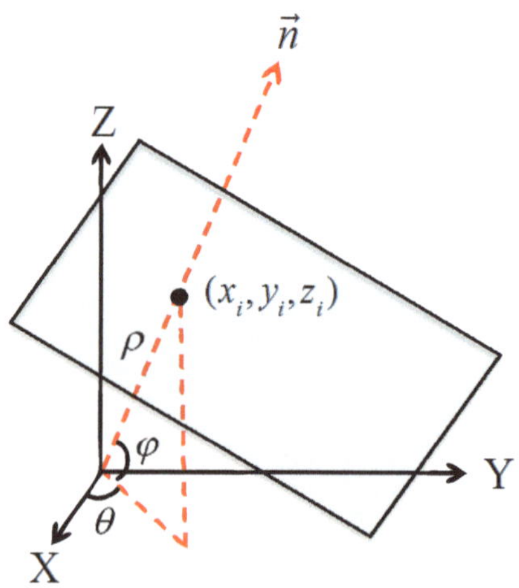

Figure 10.54 A 3D plane and associated parameters (θ, φ and ρ).

A point (x_i, y_i, z_i) on the plane is then represented by a sinusoidal surface in the parameter space. In practice, θ and φ are discretized based on pre-specified intervals (i.e., $d\theta$, $d\varphi$) and ρ is calculated for each pair of (θ, φ) for all points, thereby forming a 3D array called an accumulation matrix. Once all lidar points are evaluated, each cell of the 3D accumulation matrix contains the number of lidar points that are represented as corresponding plane parameters. Therefore, a cell with larger bins is more likely to represent a 3D roof plane. As Vosselman and Dijkman (2001) pointed out, this algorithm does not check whether the points found are in the same roof plane and often finds spurious planes. Therefore, additional criteria, e.g., points forming a connected piece of the TIN of a minimum size (Vosselman and Dijkman, 2001), are necessary for accepting the detected points as an actual roof plane.

10.5.2.2 RANSAC

RANSAC (RANdom SAmple Consensus) (Fischler and Bolles, 1981) is another popular feature extraction algorithm. In 3D building reconstruction, RANSAC has been used for different tasks (Tarsha-Kurdi et al., 2007a), such as roof planes detection (Ameri and Fritsch, 2000; Engels et al., 2008; Tarsha-Kurdi et al., 2007a; Tarsha-Kurdi et al., 2008), roof plane refinement

(Forlani et al., 2006), or 2D building boundary regularization (Kim and Shan, 2011). This section mainly focuses on roof plane detection.

The principle of RANSAC is to find the best geometric primitive by selecting a certain number of data points in order to define the corresponding model (e.g., two and three points are selected for a line and a plane respectively). One of RANSAC's distinctive features, compared to the least squares method, is that RANSAC estimates model parameters only from inliers, which is determined based on a pre-specified threshold tolerance. Therefore, better fitting results are often obtained with RANSAC if the input data contain significant outliers.

To detect a 3D roof plane from the lidar point cloud, RANSAC selects randomly three points, thereby determining the initial plane and associated parameters (e.g., (θ, φ, ρ) in Figure 10.54). Thereafter, the remaining data points are tested with the plane, and the points belonging to this plane are determined based on a certain pre-specified threshold. The vertical distance of the point to the plane, r_i shown in Figure 10.55, is commonly analyzed. Points within the specified threshold are then selected as inliers for the plane.

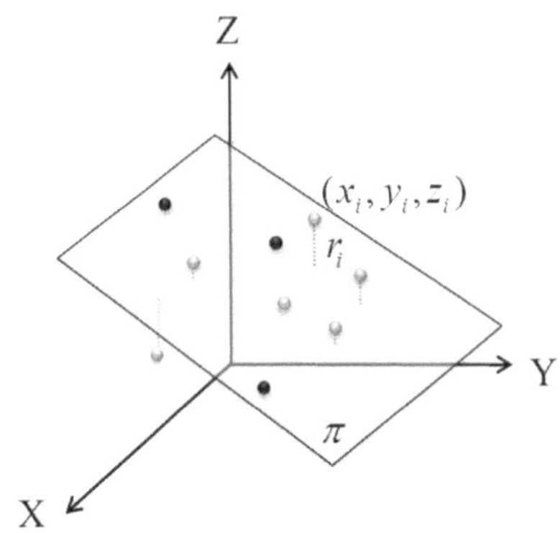

Figure 10.55 Determination of inliers to the plane : (black dots) points used for generating the plane, (gray dots) remaining points to be tested.

The algorithm repeats this procedure N times. In each iteration, the newly obtained result is compared with the last one saved. If a new result outperforms the old one, it replaces the last one saved. After N iterations, the saved plane parameters and corresponding data points are selected as a roof plane.

The original RANSAC algorithm tends to find a pure mathematical plane as noted by Tarsha-Kurdi et al. (2008), which leads to a similar problem identified in the 3D Hough transform (i.e., detected points possibly belong to different roof planes). Therefore, accepting the detected points as a consensus set representing a correct roof plane is another issue in this algorithm. For this purpose, Tarsha-Kurdi et al. (2008) defined a limit on the minimum number of points for a roof plane and the standard deviation in the final fitted plane. Another characteristic of the RANSAC algorithm in practice is that it generates only one set of parameters when the process is finished. Therefore, to extract all of the roof planes in a polyhedral building, the same procedure has to be applied iteratively and in a subtractive manner (Sampath and Shan, 2010).

10.5.2.3 K-means Clustering

In statistics and data mining, K-means clustering is one of the popular algorithms for partitioning input data into K clusters, in which all data points are labeled, equivalently assigned, to one of the K clusters with the nearest mean. Rather than using a parameterized model as in the previous two algorithms, it requires feature vectors that can enhance the dissimilarity among the clusters to be segmented, for which normal vectors are commonly involved.

The original K-means clustering algorithm requires an initial estimate of the number of unique clusters existing in the data set (i.e., K) and an initial assignment of the clusters' means. This information significantly affects the clustering results and performance. Therefore, different initial values may produce different clustering results. Furthermore, wrong clustering results may be derived, if initial values are too different from their actual values. In this context, Sampath and Shan (2010) proposed a potential-based fuzzy K-means clustering algorithm. Figure 10.56 illustrates their work process.

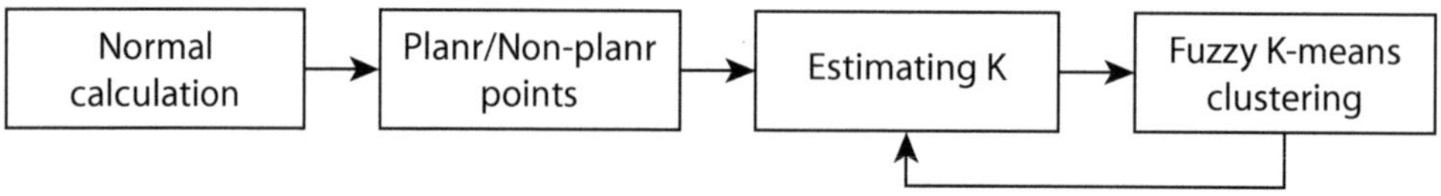

Figure 10.56 Flowchart of fuzzy K-means clustering algorithm (Sampath and Shan, 2010).

In the first step, normal vectors (N_i) are determined for all points P of a building. Sampath and Shan (2010) determined planar points and non-planar points by the dimensionality analysis of the Voronoi neighborhood (V_p) of a point. Let V_p consist of a set of m_p points $\{X_i\ (i = 1, \ldots, m_p)\}$ and $\bar{X}$ be their mean vector, the covariance matrix Σ_{XX} is calculated as:

$$\Sigma_{XX} = \sum_{i=1}^{m_p} (X_i - \bar{X})(X_i - \bar{X})^T \tag{10.4}$$

The eigenvalues λ_1, λ_2 and λ_3 ($\lambda_1 \le \lambda_2 \le \lambda_3$) of the above covariance matrix and corresponding eigen vectors e_1, e_2, e_3 are determined. Ideally, planar points lying on a roof plane should have two non-zero eigenvalues. Considering the noise inherent in the data, the normalized eigenvalue is calculated as follows:

$$\bar{\lambda} = \lambda_1/(\lambda_1+\lambda_2+\lambda_3). \tag{10.5}$$

A pre-specified threshold (e.g., 0.05 in the experiment) is used to determine if a point belongs to planar or non-planar points. Thereafter, non-planar points are excluded for the subsequent segmentation processing. As shown in Figure 10.57, most non-planar points are located along the breaklines, such as roof ridges or surface discontinuities. Therefore, excluding non-planar points significantly reduces noisy components in the feature space as shown in Figure 10.57c.

In the next step, a potential-based fuzzy K-means clustering algorithm is applied to divide the remaining planar points into individual roof segments. The overall procedure is summarized below.

1. Choose initial K and cluster centers.
2. Each point (i.e., normal vector in the feature space) is assigned a weight towards all. clusters depending on the distance of the point to the cluster center.
3. Cluster center (C^j) are determined by:

$$C^j = \sum_{i=1}^{m} (W_i^j N_i) / \sum_{i=1}^{m} W_i^j.$$

4. Each data point is newly assigned to its closest cluster center.

The number of clusters (K) is estimated by a potential-based clustering, which corresponds to the first line of above algorithm. In this context, "potential" denotes the likelihood for the given data point to be a cluster center. Sampath and Shan (2010) used Chiu's potential (Chiu, 1994), in which the potential p_i is determined as

$$p_i = \sum_{j=1}^{m} \exp\left\{-\frac{4}{r_a^2} \| N_j - N_i \|^2\right\} \tag{10.6}$$

where m is the number of all data points, r_a is a (+) constant defining a neighborhood, and $||N_j - N_i||$ denotes the distance between two normal vectors. A data point that has many neighboring data points has a high potential and vice versa. It should be noted that a neighborhood and distance are defined in the feature space. The first cluster center is then chosen as the data point that has the maximum potential value. Thereafter, the potential values for other points are reduced based on Equation 10.7. This procedure is intended to avoid the possibility of the next cluster center being close to the first one.

$$p_{i\_new} = p_{i\_old} - p_1^* \cdot \left\{-\frac{4}{r_a^2} \| N_i - N_1^* \|^2\right\} \tag{10.7}$$

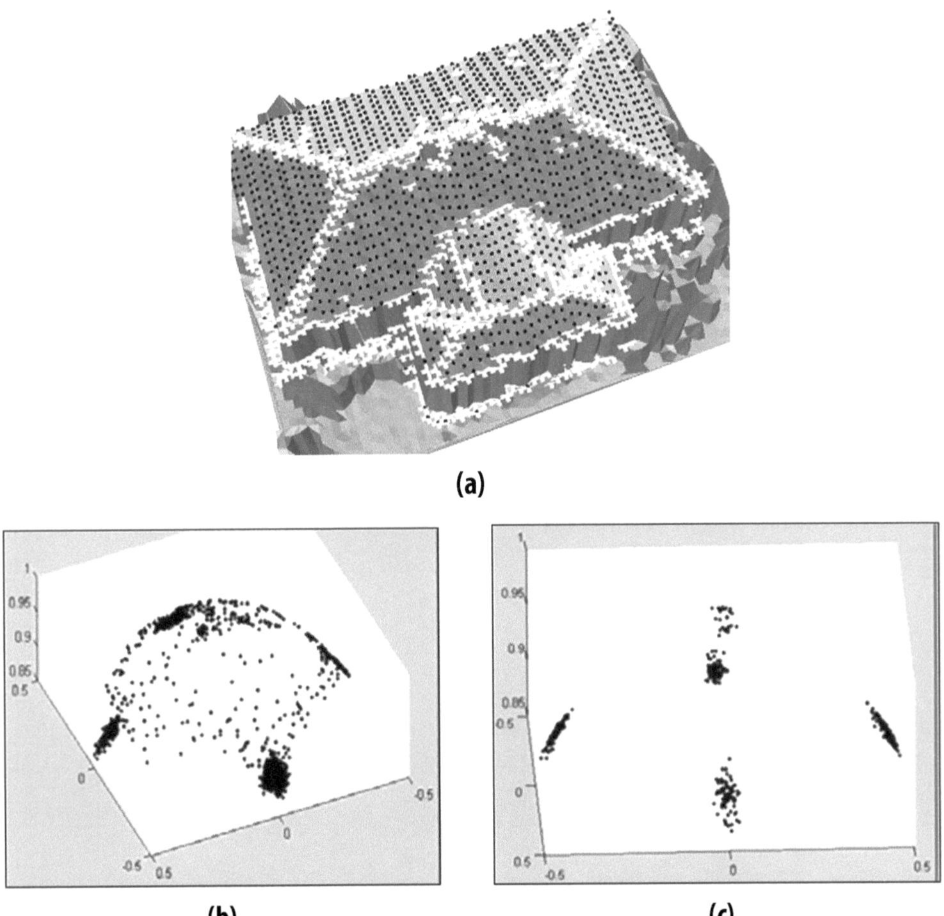

Figure 10.57 Planar versus non-planar points: (a) Non-planar points (white cross) and planar points (black dots), (b) normal vectors of all points, and (c) normal vectors of planar points (Sampath and Shan, 2010)

where p_{i_new} and p_{i_old} are the new and old potentials of point i, N_1^* and p_1^* are respectively the first selected cluster center and its potential. The second cluster center is chosen according to the same principle. This procedure is repeated until a determined potential is significantly smaller than the first potential. Sampath and Shan (2010) pointed out that it is not easy to determine the optimal value for r_a and r_b. To overcome this difficulty, the above procedure is applied iteratively by increasing r_a from its initial value. The resulting K and corresponding clusters' centers are used for the subsequent processes (i.e., 2 to 4 of above algorithm).

10.5.2.4 Level Set-based Segmentation

Kim and Shan (2011) proposed a roof plane segmentation algorithm based on the level set theory. Among numerous variations of the level set, they revised a multiphase level set approach (Chan and Vese, 2001) in order to be adequate for the roof plane segmentation. Similar to the K-means clustering algorithm, the normal vector of each data point is determined first and used as an input feature vector. With two level set functions, it always segments up to four disjoint sub-regions, each of which has the most similar property (i.e., it represents a planar roof plane). For the reader's convenience, the principle of the level set is introduced below.

The level set theory (Osher and Sethian, 1988) has been used in a range of image segmentation and boundary detection problems in computer vision. The fundamental basis is to embed a curve into one higher dimensional space. Let Ω be a fixed rectangular grid in $\mathbb{R}^2$. A 2D image u_0 is then defined as $u_0 : \Omega \rightarrow R$. The evolving curve C in Ω is considered as the boundary of an open subset $\omega \in \Omega$ as shown in Figure 10.58a. In other words, ω corresponds to the foreground objects to be detected (or segmented), and the curve C corresponds to the boundary.

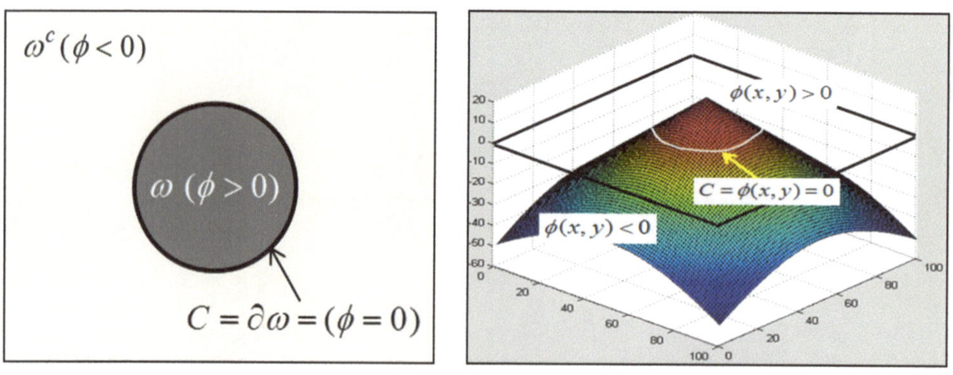

Figure 10.58 Implicit representation of the curve C: (a) $C = (\phi{=}0)$, and (b) implicit representation of C by the zero level contour of ϕ.

In the level set formulation, the boundary C is represented implicitly as the zero level contour of one higher dimensional function ϕ such that Ω is divided into two disjoint sub-regions, i.e.,

$$\begin{aligned} &\phi(x, y)>0 : \text{inside } \omega \\ &\phi(x, y)<0 : \text{outside } \omega \quad . \\ &\phi(x, y) = 0 \text{ on } \partial\omega = C \end{aligned} \tag{10.8}$$

The function ϕ is called the level set function and represented as a signed distance function as shown in Figure 10.54b. The deformation of the curve C is then described by a partial differential equation by introducing a time viable t to the level set function such that:

$$\frac{\partial \phi(x,y,t)}{\partial t} = F\,|\nabla\phi| \text{ with } \phi(x,y,0) = \phi_0(x,y) \tag{10.9}$$

where t denotes the evolving time, $\phi_0(x, y)$ defines the initial level set functions (i.e., $t = 0$), F is the speed function for curve propagation, and ∇ is the gradient operator.

For application to the image segmentation problem, the evolving curve needs to be stopped at the object boundary, for which an edge-stopping function or a certain energy minimization criterion can be used. As one of the methods in the latter category, Chan and Vese (2001) utilized the level set theory to solve the general image segmentation problem, which is called the Mumford and Shah model (Mumford and Shah, 1989), who formulated image segmentation as an energy minimization problem as:

$$E = \mu\cdot Length(C) + v\cdot Area(C_{in}) + F(C_{in}) + F(C_{out}) \tag{10.10}$$

where E is the energy function to be minimized, C is the object boundary, $\mu \geq 0$, $v \geq 0$ are the fixed weight factors, and C_{in} and C_{out} respectively, denote the inside and outside of the curve C. The two fitting terms $F(C_{in})$ and $F(C_{out})$ are defined as Equation 10.11.

$$F(C_{in}) = \lambda_1 \int_{C_{in}} |u_0 - c_1|^2\,dxdy, \quad F(C_{out}) = \lambda_2 \int_{C_{out}} |u_0 - c_2|^2\,dxdy \tag{10.11}$$

where $\lambda_1 > 0$, $\lambda_2 > 0$ are the fixed weight factors, and c_1, c_2 are respectively the average intensity (or other feature) values inside and outside the curve C. The above energy function is minimized when lies on the object boundary because the two fitting terms vanish. Equation 10.10 can be reformulated with the level set function , for which two additional functions (Heaviside and Dirac delta) are used.

$$\delta(x) = H'(x) = \frac{d}{dx}H(x)$$

$$H(x) = \begin{cases} 1 & if\ x \geq 0 \\ 0 & if\ x < 0 \end{cases} \tag{10.12}$$

Provided that the level set function ϕ takes (+) and (-) values inside and outside C, respectively, (see Figure 10.58, its Length (C) and Area (C_{in}) is rewritten as:

$$\begin{aligned} Area(C_{in}) &= \int_\Omega H(\phi)\,dxdy \\ Length(C) &= \int_\Omega |\nabla H(\phi)|\,dxdy = \int_\Omega \delta(\phi)|\nabla\delta|\,dxdy \end{aligned} \tag{10.13}$$

In a similar way, the two fitting terms $F(C_{in})$ and $F(C_{out})$ can be rewritten as:

$$\begin{aligned} F(C_{in}) &= \lambda_1 \int_\Omega |u_0 - c_1|^2 H(\phi)\,dxdy \\ F(C_{out}) &= \lambda_2 \int_\Omega |u_0 - c_2|^2 (1 - H(\phi))\,dxdy \end{aligned} \tag{10.14}$$

Inserting Equations 10.13 and 10.14 into the energy function of Equation 10.10, we have

$$\begin{aligned} E = {} & \mu \int_\Omega \delta(\phi)|\nabla H(\phi)|\,dxdy + v \int_\Omega H(\phi)\,dxdy \\ & + \lambda_1 \int_\Omega |u_0 - c_1|^2 H(\phi)\,dxdy + \lambda_2 \int_\Omega |u_0 - c_2|^2 (1 - H(\phi))\,dxdy \end{aligned} \tag{10.15}$$

To detect more than two segments, a multiphase level set is introduced (Vese and Chan, 2002). In this formulation, N level set functions can segment Ω into 2^N disjoint sub-regions. For instance, two level set functions divide Ω into four sub-regions, allowing for overlap and vacuum as shown in Figure 10.59.

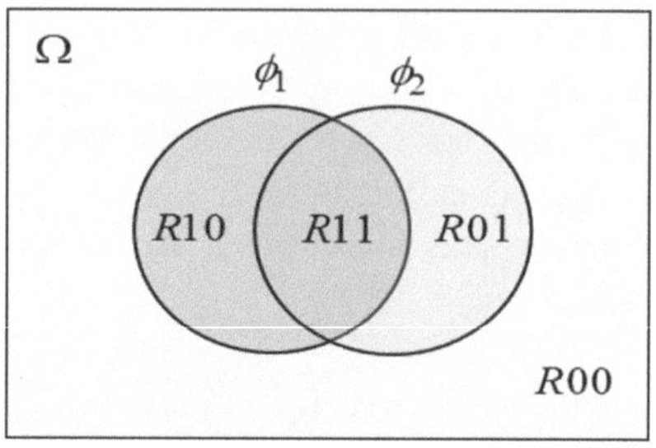

Figure 10.59 Four regions represented by two level set functions.

The different signs of two level functions can determine each sub-region. For instance, region R10 of Figure 10.59 is determined by the condition $\phi_1(x, y)>0$ and $\phi_2(x, y)<0$ while region R00 is determined by $\phi_1(x, y)<0$ and $\phi_2(x, y)<0$. The energy function with two level set functions is then extended as Equation 10.16:

$$\begin{aligned} E = \; & \lambda_1 \int_\Omega |u_0 - c_{11}|^2 H(\phi_1)\, H(\phi_2) dxdy + \lambda_2 \int_\Omega |u_0 - c_{10}|^2 H(\phi_1)(1 - H(\phi_2)) dxdy \\ & + \lambda_3 \int_\Omega |u_0 - c_{01}|^2 (1 - H(\phi_1)) H(\phi_2) dxdy \\ & + \lambda_4 \int_\Omega |u_0 - c_{00}|^2 (1 - H(\phi_1))(1 - H(\phi_2)) dxdy \\ & + v_1 \cdot \int_\Omega H(\phi_1) dxdy + v_2 \cdot \int_\Omega H(\phi_2) dxdy \\ & + \mu_1 \cdot \int_\Omega |\nabla H(\phi_1)| dxdy + \mu_2 \cdot \int_\Omega |\nabla H(\phi_2)| dxdy \end{aligned} \tag{10.16}$$

where c_{11}, c_{10}, c_{01}, c_{00} represent the average feature values of the four regions and λ_i, v_i, μ_i are the fixed weight factors. Finally, the minimization problem of Equation 10.16 is solved by deriving a corresponding Euler-Lagrange equation with respect to ϕ_1 and ϕ_2. For example, the final evolving equation of the first level set function is represented as follows.

$$\begin{aligned} \phi_1^{n+1} = \phi_1^{n} + \Delta t \cdot \delta(\phi_1^n) \{ & \mu_1 \cdot K_1 - v_t \\ & + (\lambda_3 \|u_0 - c_{01}\|^2 - \lambda_1 \|u_0 - c_{11}\|^2) H(\phi_2^n) \\ & - (\lambda_2 \|u_0 - c_{10}\|^2 - \lambda_4 \|u_0 - c_{00}\|^2)(1 - H(\phi_2^n)) \} \end{aligned} \tag{10.17}$$

Once the normal vectors of all points are determined, up to four roof segments are recursively determined. Unlike the previous three algorithms, it produces the segmented roof points as well as a 2D labeled image that represents the spatial extents of the individual roof segments. The algorithm is summarized below.

1. Specify Ω with cell size Δg.
2. Initialize two level set functions, ϕ_1^0 and ϕ_2^0 using the signed distance function.
3. Repeat the following:
   - Compute $H_\varepsilon(\phi_i^n)$ and $\delta_\varepsilon(\phi_i^n)$.
   - Compute mean curvature K_i.
   - Compute mean normal vectors (c_{00}~c_{11}) of four sub-regions.

- Update new ϕ_i^{n+1} using Equation 10.17.
- Check convergence.

The Ω (line 1 of the algorithm) denotes a 2D regular grid that encloses all data points. Two initial level set functions ϕ_1^0 and ϕ_2^0 are then defined as a signed distance function. To improve the performance and convergence, an initial level set function is constructed from a set of regularly spaced zero level curves as shown in Figure 10.60. The segmentation is then carried out iteratively performing Step 3. The intermediate plots and final segmentation results are shown in Figure 10.61.

(a) (b) (c)

Figure 10.60 Initial level set function: (a) $\phi_1^0 = 0$, (b) , $\phi_2^0 = 0$, and (c) 3D representation of ϕ_1^0.

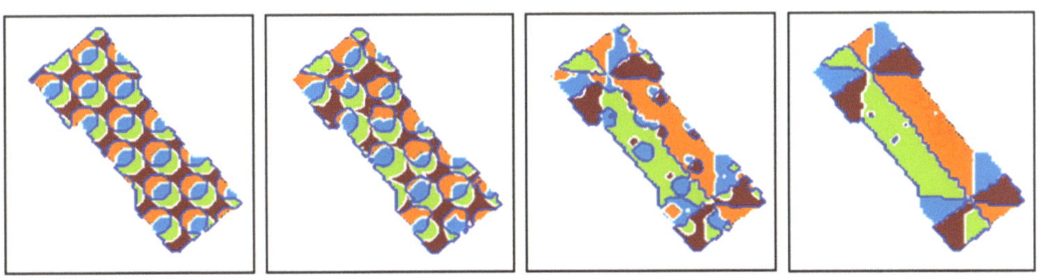

(a) (b) (c) (d)

Figure 10.61 Intermediate plot: (a) $\phi_1^0 = 0$, (b) and (c) $\phi_2^0 = 0$, and (d) 3D representation of ϕ_1^0.

10.5.3 Building Reconstruction

From the previous extraction step, we have groups of points, each of which corresponds to an individual roof plane. The reconstruction step begins with these segmented points. The remaining tasks are to determine the topological primitives (e.g., vertices and edges) and to generate a topologically consistent 3D building model. To achieve this objective, the spatial relations (also termed adjacency) among roof segments need to be determined. Once the spatial relations are determined, topological primitives are extracted. Methods determining the spatial relations vary depending on the results from the extraction step. The first three algorithms produce segmented roof points, while the last one (Kim and Shan, 2011) includes a 2D labeled image as well.

10.5.3.1 Adjacency of Roof Segments

Sampath and Shan (2010) determined the adjacency of roof segments by analyzing the minimum distance of any pair of roof segments in the object space. For instance, the closest distance between two roof segments P and Q can be determined as follows:

$$d(P, Q) = \min(d(p_i, q_i)) \quad \forall p_i \in P_i, \ \forall q_i \in Q \tag{10.18}$$

where $d(P, Q)$ denotes the distance between two roof segments, and $\min(d(p_i, q_i))$ denotes the minimum distance between any two points p_i and q_i, respectively, in the segments P and Q. If the minimum distance is less than a pre-specified threshold, two roof segments are marked "adjacent" (i.e., "X" in Table 10.2). The same procedure is applied to all pairs of roof segments, leading to an adjacency matrix as shown in Table 10.2. The numbers shown in both rows and columns correspond to the roof planes shown in Figure 10.62.

In Sampath and Shan (2010), two times lidar point spacing was used as a threshold, which was therefore affected by the average point spacing of the lidar data used. The vertices and edges were then determined by analyzing the adjacency matrix. For instance, roof plane 1, shown in Figure 10.62, is adjacent to roof planes 2, 3, and 4. The intersection of roof planes 1 and 2 is the edge EF, with E and F denoting the vertices. To determine these two vertices, the planes that are adjacent to both planes 1 and 2 are needed. The authors determined this relationship from the adjacency matrix by looking at the rows of planes 1 and 2 and determining that planes 3 and 4 are adjacent to planes 1 and 2. Under the assumption that the intersecting planes do not form a volume (i.e., we are dealing with a surface), the intersection of the three adjacent roof planes (1, 2, 3) determines the vertex F. Similarly, E is obtained as the intersection of planes (1, 2, 4).

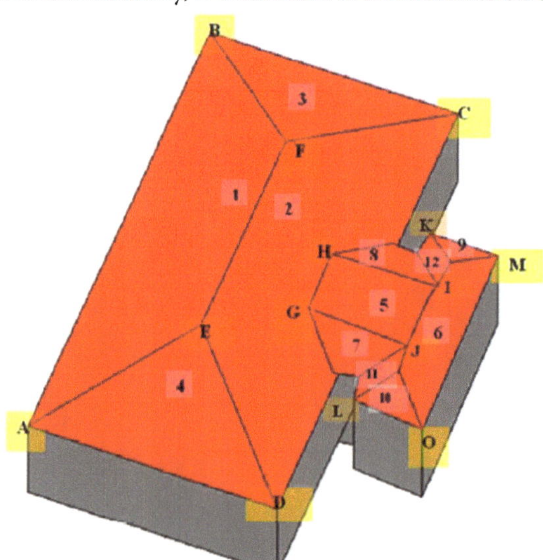

Figure 10.62 Adjacency of roof planes and vertex determination for building reconstruction (Sampath and Shan, 2010).

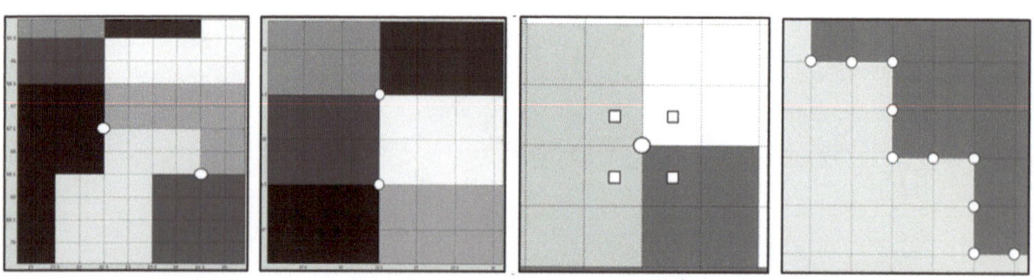

(a) **(b)** **(c)** **(d)**

Figure 10.63 Determination of vertices: (a) Case 1: three roof planes, (b) Case 2: four roof planes, (c) Case 3: two roof planes (light and dark grey) and boundary, and (d) Case 4: points on the roof ridge.

Table 10.2 Adjacency matrix of the building shown in Figure 10.

Planes	1	2	3	4	5	6	7	8	9	10	11	12
1		X	X	X								
2	X		X	X	X		X	X				
3	X	X										
4	X	X										
5		X				X	X	X			X	X
6					X		X	X	X	X	X	X
7		X			X	X					X	
8		X			X	X						X
9						X						X
10						X					X	
11					X	X	X			X		
12						X		X	X			

Unlike other algorithms, the segmentation results from Kim and Shan (2011) include roof segments as well as a 2D labeled image. Based on the level set formulation, all pixels are assigned to one of roof segments. They used this labeled image to determine the vertices directly, rather than determining the spatial relations in the original data space. First, the labeled image is divided with a grid interval of 0.5 pixels. For every grid point, it checks how many unique roof segments exist at its four neighboring cells (i.e., within ±0.25 intervals). In Figure 10.63, four different cases are presented. Circles and squares denote grid points of interest and their four neighboring cells, respectively.

If a grid point has at least three unique roof segments (Cases 1 and 2 of Figure 10.63), it is considered as an intersection of the corresponding roof segments. If two or three unique roof segments and at least one building outside meet at a grid point (Case 3 of Figure 10.63), it is an intersection of the corresponding roof segments and a vertical plane, which is determined from the building boundary. If only two unique roof segments are found at a grid point (Case 4), this point lies on either a roof ridge where two neighboring roof segments meet or a step edge. Therefore, the height difference of the corresponding two roof segments is further analyzed. The identified vertices through this analysis are shown in Figure 10.64a. Based on this process, most of the vertices induced by the intersection of the adjacent roof segments are identified. In order to represent the roof shape correctly, the intersection of two line segments of the building boundary also need to be determined. To determine these points, the authors first applied the α-shape algorithm to trace the building boundary, followed by the Douglas-Peucker algorithm (Douglas and Peucker, 1973) to find the corner points as shown in Figure 10.64b. All boundary points between two consecutive corner points are fitted by RANSAC to a line. Figure 10.64c shows the candidate roof structure points determined by intersecting line segments on the building boundary. When some candidate vertices from these two groups are very close to each other, the points induced by non-parallel planes are used because the vertices by the intersecting three non-parallel planes are considered more accurate. As a result of this procedure, all vertices that constitute a building roof are determined as shown in Figure 12d. The determined vertices are then assigned to a roof plane and reorganized in a correct order for representing each roof polygon as shown in the last image in Figure 10.64e. Finally, building facades are generated from the regularized line segments of building boundary.

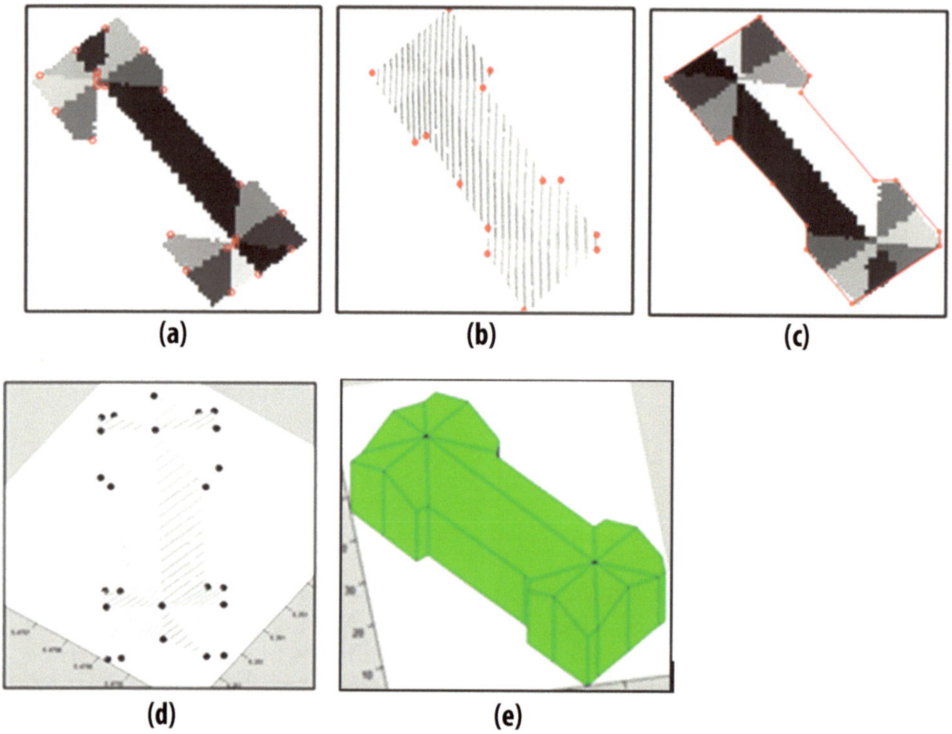

Figure 10.64 Identification of roof vertices: (a) Identified vertices by the intersection of adjacent roof segments, (b) Corner points, (c) Vertices induced by building boundary, (d) All vertices determined, and (e) Reconstructed 3D model.

REFERENCES

Alharthy, A., and J. Bethel, 2002. Heuristic filtering and 3D feature extraction from lidar data, *International Archives of Photogrammetry, Remote Sensing and Spatial Information Sciences*, 34(PART 3/A):29-34.

Alharthy, A., and J. Bethel, 2004. Detailed building reconstruction from airborne laser data using a moving surface method, *International Archives of Photogrammetry, Remote Sensing and Spatial Information Sciences*, 34(PART B3):213–218.

Ameri, B., and D. Fritsch, 2000. Automatic 3D building reconstruction using plane-roof Structures, *Proceedings of the ASPRS Annual Conference*, Washington D.C.

Awrangjeb, M., M. Ravanbakhsh, and C.S. Fraser, 2010. Automatic detection of residential buildings using LIDAR data and multispectral imagery, *ISPRS Journal of Photogrammetry and Remote Sensing*, 65(5):457–467.

Awrangjeb, M., C. Zhang, and C.S. Fraser, 2011. Automatic reconstruction of building roofs using LIDAR and multispectral imagery, *Proceedings of the 2011 International Conference on Digital Image Computing : Techniques and Applications*, pp. 370–375.

Axelsson, P., 2000. DEM generation from laser scanner data using adaptive TIN models, *International Archives of Photogrammetry and Remote Sensing*, 33(B4/1; PART 4):111–118.

Chan, T.F., and L.A. Vese, 2001. Active contours without edges, *IEEE Transactions on Image Processing*, 10(2):266–277.

Chiu, S.L., 1994. Fuzzy model identification based on cluster estimation, *Journal of Intelligent and Fuzzy Systems*, 2(3):267–278.

Demir, N., and E. Baltsavias, 2010. Combination of image and lidar data for building and tree Extraction, *International Archives of Photogrammetry, Remote Sensing and Spatial Information Sciences*, 38:131–136.

Dorninger, P., and C. Nothegger, 2007. 3D segmentation of unstructured point clouds for building modelling, *International Archives of the Photogrammetry, Remote Sensing and Spatial Information Sciences*, 35(3/W49A):191–96.

Dorninger, P., and N. Pfeifer, 2008. A comprehensive automated 3d approach for building extraction, reconstruction, and regularization from airborne laser scanning point clouds, *Sensors*, 8(11):7323–7343.

Douglas, D., and T. Peucker, 1973. Algorithms for the reduction of the number of points required to represent a digitized line or its caricature, *Cartographica: The International Journal for Geographic Information and Geovisualization*, 10(2):112–122.

Engels, J., H. Arefi, and H.G. Maas, 2008. Generation of roof topologies using plane fitting with RANSAC, *International Archives of Photogrammetry, Remote Sensing and Spatial Information Sciences*, 37(B3a):119–126.

Filin, S., and N. Pfeifer, 2006. Segmentation of airborne laser scanning data using a slope adaptive neighborhood, *ISPRS Journal of Photogrammetry and Remote Sensing*, 60(2):71–80.

Fischler, M.A., and R.C. Bolles, 1981. Random sample consensus: A paradigm for model fitting with applications to image analysis and automated cartography, *Communications of the ACM*, 24(6):381–395.

Forlani, G., C. Nardinocchi, M. Scaioni, and P. Zingaretti, 2006. Complete classification of raw LIDAR data and 3D reconstruction of buildings, *Pattern Analysis & Applications*, 8(4):357–374.

Haala, N., and C. Brenner, 1999. Extraction of buildings and trees in urban environments, *ISPRS Journal of Photogrammetry and Remote Sensing*, 54(2-3):130–137.

Hermosilla, T., L.A. Ruiz, J.A. Recio, and J. Estornell, 2011. Evaluation of automatic building detection approaches combining high resolution images and LiDAR data, *Remote Sensing*, 3(6):1188–1210.

Hofmann, A.D., H.G. Maas, and A. Streilein, 2002. Knowledge-based building detection based on laser scanner data and topographic map information, *International Archives of Photogrammetry and Remote Sensing*, 34(3/A):169–174.

Hough, P.V.C., 1962. *Method and Means for Recognizing Complex Patterns*, Google Patents.

Huang, H., c. Brenner, and M. Sester, 2011. 3D building roof reconstruction from point clouds via generative models, *Proceedings of the 19th ACM SIGSPATIAL International Conference on Advances in Geographic Information Systems*, pp. 16–24.

Hug, C., and a. Wehr, 1997. Detecting and identifying topographic objects in imaging laser altimeter data, *International Archives of Photogrammetry and Remote Semsing*, 32(3-4W2):19–26.

Kim, K.H., and j. Shan, 2011. Building roof modeling from airborne laser scanning data based on level set approach, *ISPRS Journal of Photogrammetry and Remote Sensing*, 66(4):484–497.

Kraus, K.,and N.Pfeifer, 1998. Determination of terrain models in wooded areas with airborne laser scanner data, *ISPRS Journal of Photogrammetry and Remote Sensing*, 53(4):193–203.

Lafarge, F., X. Descombes, J. Zerubia, and M. Pierrot-Deseilligny, 2008. Automatic building extraction from DEMs using an object approach and application to the 3D-city modeling, *ISPRS Journal of Photogrammetry and Remote Sensing*, 63(3):365–381.

Maas, H., and G. Vosselman, 1999. Two algorithms for extracting building models from raw laser altimetry data, *ISPRS Journal of Photogrammetry and Remote Sensing*, 54(2–3):153–163.

Mongus, D., and B. Žalik, 2012. Parameter-free ground filtering of LiDAR data for automatic DTM generation., *ISPRS Journal of Photogrammetry and Remote Sensing*, 67:1–12.

Morgan, M., and K. Tempfli, 2000. Automatic building extraction from airborne laser scanning data, *International Archives of Photogrammetry and Remote Sensing*, 33(B3/2; PART 3):616–623.

Mumford, D., and J. Shah, 1989. Optimal approximations by piecewise smooth functions and associated variational problems, *Communications on Pure and Applied Mathematics*, 42(5):577–685.

Nardinocchi, C., G. Forlani, and P. Zingaretti, 2003. Classification and filtering of laser data, *International Archives of Photogrammetry and Remote Sensing*, 34(3/W13).

Osher, S., and J.A. Sethian, 1988. Fronts propagating with curvature dependent speed: Agorithms based on Hamilton-Jacobi formulations, *Journal of Computational Physics*, 79(1)12–49.

Overby, J., 2004. Automatic 3D building reconstruction from airborne laser scanning and cadastral data using Hough Transform, *International Archives of Photogrammetry and Remote Sensing*, ISPRS 2004 Congress Proceedings, Commission III.

Pfeifer, N., and G. Mandlburger, 2009. LiDAR data filtering and DTM generation, *Topographic Laser Ranging and Scanning: Principles and Processing* (C. Toth and J. Shan, editors), pp. 307–333.

Rottensteiner, F., 2003. Automatic generation of high-quality building models from lidar data, *IEEE Computer Graphics and Applications*, 23 (6)6:42–50.

Rottensteiner, F., J. Trinder, S.Clode, and K. Kubik, 2007. Building detection by fusion of airborne laser scanner data and multi-spectral images: Performance evaluation and sensitivity analysis, *ISPRS Journal of Photogrammetry and Remote Sensing*, 62(2):135–149.

Sampath, A., and J. Shan, 2007. Building boundary tracing and regularization from airborne lidar point clouds, *Photogrammetric Engineering & Remote Sensing*, 73(7):805–812.

Sampath, A., and J. Shan, 2010. Segmentation and reconstruction of polyhedral building roofs from aerial Lidar point clouds, *IEEE Transactions on Geoscience and Remote Sensing*, 48(3):1554–1567.

Schenk, T., and B. Csathó, 2002. Fusion of LIDAR data and aerial imagery for a more complete surface description, *International Archives of Photogrammetry and Remote Sensing* and Spatial Information Sciences, 34(3/A):310–317.

Schwalbe, E., H.G. Maas, and F. Seidel, 2005. 3D building model generation from airborne laser scanner data using 2D GIS data and orthogonal point cloud projections. Proceedings of the *International Archives of Photogrammetry and Remote Sensing*, Enschede, The Netherlands, pp. 209–214.

Shan, J., and A. Sampath, 2005. Urban DEM generation from raw lidar data: A labeling algorithm and its performance, *Photogrammetric Engineering & Remote Sensing*, 71(2):217–226.

Sithole, G., and G. Vosselman, 2004. Experimental comparison of filter algorithms for bare-Earth extraction from airborne laser scanning point clouds, *ISPRS Journal of Photogrammetry and Remote Sensing*, 59 (1-2), 85–101.

Sohn, G., and I. Dowman, 2007. Data fusion of high-resolution satellite imagery and LiDAR data for automatic building extraction, *ISPRS Journal of Photogrammetry and Remote Sensing*, 62(1):43–63.

Tack, F., G. Buyuksalih, and R. Goossens, 2012. 3D building reconstruction based on given ground plan information and surface models extracted from spaceborne imagery, *ISPRS Journal of Photogrammetry and Remote Sensing*, 67:52–64.

Tarsha-Kurdi, F., T. Landes, P. Grussenmeyer, and E. Smigiel, 2006. New approach for automatic detection of buildings in airborne laser scanner data using first echo only, *International Archives of Photogrammetry, Remote Sensing, and Spatial Information Sciences*, 36:25–30.

Tarsha-Kurdi, F., T. Landes, and P. Grussenmeyer, 2007a. Hough-transform and extended RANSAC algorithms for automatic detection of 3D building roof planes from Lidar Data, *Proceedings of the ISPRS Workshop on Laser Scanning 2007 and SilviLaser 2007*, Espoo, Finland, pp. 407–412.

Tarsha-Kurdi, F., T. Landes, P. Grussenmeyer, and M. Koehl, 2007b. Model-driven and data-driven approaches using lidar data: analysis and comparison, *International Archives of Photogrammetry, Remote Sensing and Spatial Information Sciences*, 35(3/W49A), 87-92.

Tarsha-Kurdi, F., T. Landes, and P. Grussenmeyer, 2008. Extended RANSAC algorithm for automatic detection of building roof planes from LIDAR data, *Photogrammetric Journal of Finland*, 21(1):97–109.

Toshev, A., P. Mordohai, and B. Taskar, 2010. Detecting and parsing architecture at city scale from range data, *Proceedings of the 2010 IEEE Computer Society Conference on Computer Vision and Pattern Recognition, Proceedings of CVPR'10*, IEEE Computer Society, San Francisco, Califoria., pp. 398–405.

Verma, V., R. Kumar, and S. Hsu, 2006. 3D Building detection and modeling from aerial LIDAR data, *Proceedings of the 2006 IEEE Computer Society Conference on Computer Vision and Pattern Recognition, CVPR'06*, IEEE Computer Society, Washington, D.C., pp. 2213–2220.

Vese, L.A., and T.F. Chan, 2002. A multiphase level set framework for image segmentation using the Mumford and Shah model, *International Journal of Computer Vision*, 50(3):271-293.

Vosselman, G., 2000. Slope based filtering of laser altimetry data, *International Archives of Photogrammetry, Remote Sensing and Spatial Information Sciences*, 33(B3/2; PART 3):935–942.

Vosselman, G., and S. Dijkman, 2001. 3D building model reconstruction from point clouds and ground plans, *International Archives of Photogrammetry, Remote Sensing and Spatial Information Sciences*, 34(PART 3/W4) :37–43.

Weidner, U., and W. Förstner, 1995. Towards automatic building extraction from high-resolution digital elevation models, *ISPRS Journal of Photogrammetry and Remote Sensing*, 50(4):38–49.

You, R.J., and B.C. Lin, 2011. A quality prediction method for building model reconstruction using LiDAR data and topographic maps, *IEEE Transactions on Geoscience and Remote Sensing*, (99), pp. 1–10.

Yu, Y., B. Buckles, and X. Liu, 2009. Residential building reconstruction based on the data fusion of sparse LiDAR data and satellite imagery, *Advances in Visual Computing*, pp. 240–251.

Zhang, K., S.-C. Chen, D. Whitman, M.-L. Shyu, J. Yan, and C. Zhang, 2003. A progressive morphological filter for removing nonground measurements from airborne LIDAR data, *IEEE Transactions on Remote Sensing and Geoscience*, 41(4):872–882.

10.6 AIRPORT SURVEYING

Airport surveying and mapping is an increasingly prevalent application of lidar. Figure 10.65 shows an example of a lidar data set acquired by NOAA's National Geodetic Survey (NGS) at Stafford Regional Airport (RMN) in Stafford, Virginia. Lidar data sets covering airports and the surrounding areas are collected for a wide variety of uses, some proven and others still in various phases of testing and development, including:

- Topography and features for airport geographic information systems (GIS) and airport layout plans (ALPs),
- Orthorectification of aerial imagery of airports,
- Terrain awareness and warning systems (TAWS) and synthetic vision systems (SVS),
- Runway surveys, and
- Airport obstruction surveys

The list above is by no means exhaustive; the airport and aviation communities are rapidly discovering new applications of lidar data. In fact, this "multi-use" aspect of airport lidar data (i.e., the ability to collect one data set, which can then serve several key application needs simultaneously) is a compelling factor in the decision to collect lidar at many airports. To further explain the wide-ranging utility of airport lidar, as well as some of the challenges, we will next consider each of the above-mentioned applications in greater detail.

Figure 10.65 Perspective view of lidar data set for Stafford Regional Airport (RMN), color-coded by elevation and intensity.

10.6.1 Topography for Airport GIS and ALPs

While airport geospatial data have traditionally been portrayed on hardcopy cartographic products, a recent focus has been on collecting, storing, and disseminating these data in digital, GIS-compatible formats. The FAA's Airports GIS program comprises a major initiative to centralize airport survey data in an integrated, web-based GIS (FAA, 2009a). The airport survey requirements for Airports GIS are contained in three Advisory Circulars (ACs) within the AC 150/5300 series: 16A containing guidance and specifications for establishment of geodetic control for aeronautical surveys (FAA, 2007a), 17C describing acceptable uses and standards for different remote sensing technologies for acquiring airport data (FAA, 2011), and 18B containing guidance and specifications for field data collection and GIS standards (FAA, 2009a) (These ACs will be referred to as 16A, 17C, and 18B, without the series number, in this section.). The most recent of these ACs, 17C, which was released in September 2011 and supersedes the previous version, contains new sections on the use of lidar for collecting airport data. Of particular noteworthiness is the statement that lidar mapping "can collect many airport data elements listed in AC 150/5300-18" (FAA, 2011).

Included in the FAA Airports GIS program is an initiative to implement electronic airport layout plans (eALPs) (FAA, 2009b). Briefly, an airport layout plan (ALP) is a graphic depiction of existing and proposed airport facilities and land use necessary for airport operations and development and to show compliance with applicable standards (FAA, 1989; FAA, 2007; Wells and Young, 2004). An approved ALP is a requirement for an airport to receive funding under the Airport Improvement Program (AIP), and airports are responsible for keeping ALPs current (FAA, 2007). By migrating to electronic ALPs, the FAA aims to help keep ALPs up to date, take advantage of new technology, and satisfy International Civil Aviation Organization (ICAO) requirements for electronic terrain and obstacle (eTOD) data.

Elevation contours are an important component of ALPs and are beneficial in planning, design, drainage, and environmental analyses. According to the new AC 17C, lidar data can be used in generating the elevation contours required by 18B, although this requires "interpretation from the data" (FAA, 2011). In particular, appropriate contouring algorithms and procedures must be applied. The lidar data collected in support of the topographic surveying requirements can simultaneously be used to collect other airport feature classes. While rigorous cost-benefit analyses are still needed, the all-digital workflow of lidar acquisition and processing and the ability to output data in a variety of vector and raster formats (e.g., point clouds, TINs, DEMs, contours, all in various file formats) are potential added benefits for the eALP program and possible future Airports GIS initiatives.

On a more local level than the FAA's Airports GIS program, many individual airports (ranging in size from Atlanta's Hartsfield-Jackson International Airport and Boston's Logan International Airport to John C. Tune in Nashville and Martin State Airport in Baltimore) use GIS (e.g., Murphy, 2003; Croft, 2004). These and many other airports are employing GIS for a variety of applications, including lease and property management, facilities maintenance, utilities mapping, asset inventories and condition assessment, and airfield operations reporting. In some cases, airport authorities are using GIS so extensively that they have established dedicated GIS departments to maintain GIS data, provide GIS applications, and prepare specialized maps. Accurate terrain data, such as that generated using lidar, is an important base layer in an airport GIS and can support hydrologic modeling, viewshed analysis, construction planning, and a variety of other activities. Airports also collect lidar as a means of producing a digital elevation model used to orthorectify aerial imagery (see below). As more advanced lidar collection equipment and processing procedures are developed, the use of lidar data for airport applications will continue to grow.

10.6.2 Airport Orthoimagery

One frequent use of lidar data is in orthorectification of aerial images, a process that entails geometrically correcting the imagery such that it has constant scale and can be used as a planimetric map. Airport orthoimagery is used in a wide range of activities, including digitizing aircraft movement areas and other planimetric detail, generating base layers for GIS (see above), and providing reference data for airport planning and development applications. The FAA ACs listed above specify that orthophotos must be delivered for all airport surveys in which aerial imagery is acquired (FAA, 2009a and 2011).

The orthorectification process requires an elevation model to correct for relief displacement, i.e., the radial displacement of imaged features, due to elevation variation. The high spatial resolution and accuracy of lidar data are advantageous for this process. An additional advantage is that, by using lidar-derived digital surface models (DSMs) containing reflective-surface elevations, so-called "true orthos" can be generated (Lohr, 2003; Günay et al., 2007). Unlike orthoimages generated with the more common "bare earth" elevation models, true orthos depict the tops of buildings and other vertical objects in their correct locations. Typically, some manual editing of the lidar data is required to avoid artifacts in the output true-orthos, but the advantages of having elevated features in their correct geographic locations often outweigh the additional time and effort required to produce them. A final benefit of using lidar for this application is that the output orthoimages and lidar data form complementary data sets (see, e.g., Sections 8.2 and 8.3 of this Manual), enabling a wide range of geospatial analysis.

10.6.3 TAWS and SVS

In 2000, the FAA mandated that all U.S.-registered, turbine powered aircraft with six seats or more have a Terrain Awareness and Warning System (TAWS) (Federal Register, 2000). TAWS is the new, enhanced technology replacing Ground Proximity Warning Systems (GPWS). Standard GPWS capabilities include providing warnings for excessive rate of descent, excessive closure rate to terrain, altitude loss after takeoff, negative climb rate, flight into terrain when not in landing configuration, and excessive downward deviation from glideslope. TAWS adds two very important enhancements to GPWS: forward looking terrain avoidance (FLTA), and premature descent alerts (PDA).

The rapid and direct transformation of lidar data to terrain information allows terrain databases to be kept up to date. While TAWS is mandated as an alert-only system, many avionics manufacturers incorporate high-relief terrain data sets "visualized" as surface models on the TAWS display. Lidar data can provide greater fidelity in these visual displays through enhanced spatial resolution and accuracy.

This type of real-time, computer-generated display of terrain, as well as obstacles and other geospatial data to enhance situational awareness in the cockpit are the defining characteristics of a synthetic vision system (SVS) (Theunissen et al., 2005, de Haag et al., 2000). Figure 10.66 depicts an example of an SVS produced by Honeywell International, Inc. A key goal of SVS is to improve safety in reduced-visibility weather conditions and, in particular, to decrease instances of so-called "controlled flight into terrain" (CFIT), in which a pilot unintentionally flies a perfectly-functioning aircraft into terrain. SVS requires even greater resolution data sets than TAWS, and the data must be updated more frequently, as the objective is to provide pilots with the capability of using the SVS data to visualize navigating through a virtual world. This type of visualization is particularly important for helicopter operations in brownout or otherwise occluded conditions. Although SVS is not currently as mature or widely-used as TAWS, the increasing availability of high-resolution, high-accuracy airport lidar data sets could aid the continued development and implementation of this technology.

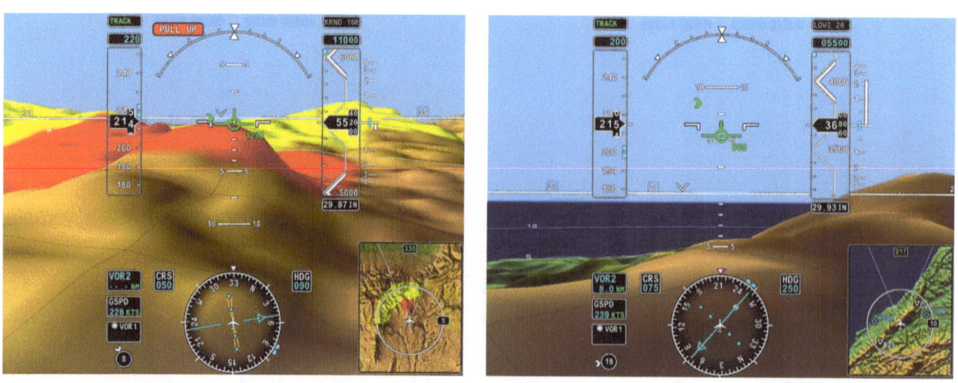

Figure 10.66 Synthetic Vision System (SVS) examples showing terrain visualization (©Honeywell International, Inc.).

10.6.4 Runway Surveys

Runway elevation profiles are a critical part of airport surveys, as the data are used in determining runway gradients, establishing the airport elevation and touchdown zone elevations, and analyzing pavement roughness (FAA, 2009). It is extremely challenging to meet the 0.25 ft. (7.6 cm) absolute vertical accuracy (at the 95 percent confidence level) and 0.5 inch (1.3 cm) relative accuracy requirement for adjacent runway profile points contained in AC 18B (FAA, 2009) with airborne lidar. For many years, frequently-quoted vertical accuracies for airborne lidar were on the order of 15 cm (RMSE), which equates to 29 cm at the 95 percent confidence level. Recently, numerous researchers and practitioners have demonstrated the capability to improve vertical accuracy through rigorous and carefully-conducted system calibration and short GNSS baselines. One promising method for improving vertical accuracy is through the use of lidar-specific ground targets (e.g., Csanyi and Toth, 2007). However, to the best of our knowledge, this technique has not been tested for surveying airport runways with airborne lidar, and the typically-achieved accuracies with airborne systems are not yet at a level at which they can be used to satisfy the stringent specifications for runway surveys.

An alternative, which has been tested (at least on a limited basis) is the use of terrestrial (ground-based) and/or mobile (vehicle-based) lidar for runway surveys. Figure 10.67 shows examples of a runway survey conducted by Terrapoint USA, a division of Ambercore Software, Inc., at Hanscom Airfield (BED) using a truck-mounted mobile lidar system. An accuracy assessment performed in collaboration with NGS on data collected with this mobile lidar system at Suffolk Executive Airport (SFQ) showed that the 0.25 ft. (7.6 cm) vertical accuracy requirement at the 95 percent confidence level was met in this mobile lidar survey, although the relative point-to-point accuracy was not tested. Benefits of mobile lidar runway surveys include: (a) a much more detailed representation of the entire runway surface (not just the centerline or centerline and edges), enabling an analysis of "ponding" (i.e., where puddles will form during a rainstorm), (b) the ability to geolocate runway markings using the lidar intensity data, and (c) the ability to complete the survey quickly, minimizing the time the survey crew and vehicle need to be on the runway.

Figure 10.67 Mobile lidar runway survey at Hanscom Airfield (BED). Clockwise, from upper left: perspective view of lidar point cloud along the Runway 11 centerline; camera image along the same runway; high-speed taxiway off of Runway 5-23; the TITAN mobile lidar system. Images courtesy of Terrapoint USA, a division of Ambercore Software Inc.

10.6.5 Airport Obstruction Surveys

Of all the aeronautical applications of lidar data listed above, airport obstruction surveying arguably holds the greatest potential for time savings, yet has required the most extensive testing to verify the capabilities and limitations. A primary goal in an airport obstruction survey is to accurately geolocate vertical objects (VOs), such as towers, buildings, trees, poles, and antennas that penetrate an airport's obstruction identification surfaces (OIS). These OIS are three-dimensional, mathematically-defined surfaces encompassing the airfield and runway approaches, as shown in red, blue, and gold in Figure 10.68. Any object, including terrain, that penetrates (i.e., sticks up above) these surfaces is, by definition, an obstruction (alternately referred to as an "obstacle"). The data from these obstruction surveys are used for designing approach and departure procedures that ensure safe clearance for aircraft, and other applications, ranging from certifying airports for certain types of operations, to determining maximum takeoff weights for aircraft, to updating aeronautical publications (FAA, 2009a).

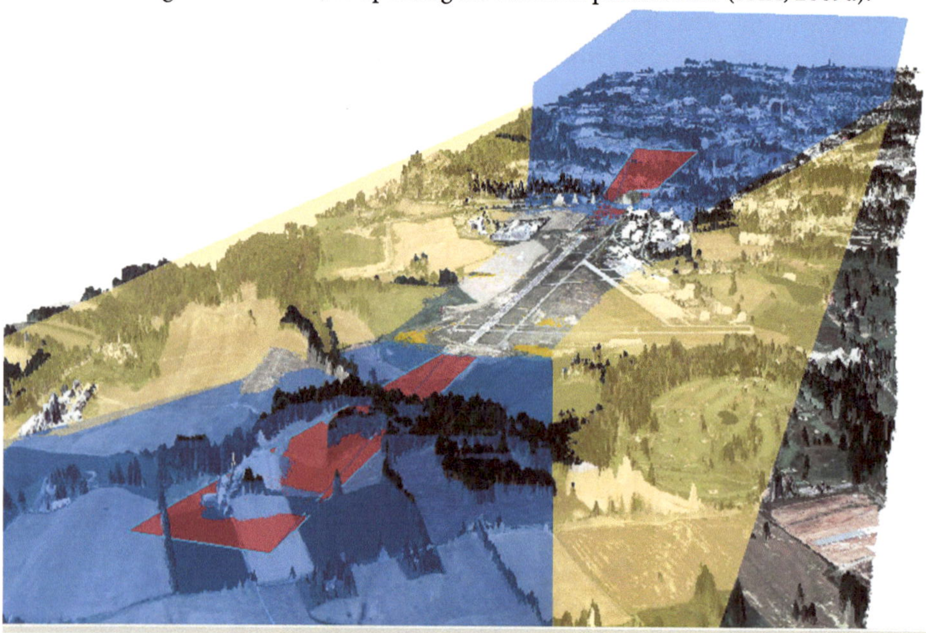

Figure 10.68 AC 150/5300 18-B obstruction identification surfaces (OIS) and penetrations for Frederick Municipal Airport (FDK). Here, the OIS are plotted on a lidar-derived DSM with an orthophoto overlaid.

Airport obstruction surveys have long been conducted through a combination of field and photogrammetric surveying. However, the FAA goal to provide WAAS-based instrument approach procedures to eligible runway ends has greatly increased the need for obstruction surveys, and demand for survey data is rapidly outpacing supply. Due to the quick survey turnaround time afforded by lidar, and the ability to generate extensive data for evaluating penetrations to a wide-variety of OIS, while simultaneously supporting terrain mapping and other airport GIS projects, this technology has come to the attention of many airports and private firms engaged in obstruction surveys. The new AC 17C now permits the use of lidar in *supporting* obstruction analysis, although complimentary imagery acquisition and analysis are required (FAA, 2011).

Because of the criticality of airport obstruction data to flight safety, extensive testing of the technology was required before lidar could even be considered for potential use in obstruction surveying. One of the first empirical tests of the capability to detect airport obstructions with airborne lidar was carried out jointly by NGS, the FAA, and University of Florida in 2001 (Parrish et al., 2004 and 2005). The results of that initial study showed that several small-diameter, low-reflectance VOs, such as poles and antennas, were not detected with the lidar systems. However, two important conclusions came out of this study: (a) the limiting factor in the ability to meet obstruction surveying standards with lidar is not the achievable spatial accuracy of georeferenced lidar points; rather it is the ability to achieve sufficient 3D point density and signal-to-noise ratio (SNR) to detect these small-diameter, low-reflectance VOs, and (b) lidar system configurations and flight parameters used for typical terrain surveys (e.g., floodplain mapping and other bare-earth applications) are poorly suited for airport obstruction surveying. The latter point motivated an effort to determine optimal configurations, flight, and processing parameter settings for airport obstruction surveying.

NGS, in collaboration with other research partners, proceeded to complete three additional phases of research into lidar obstruction surveying, each aimed at building on the previous study and addressing remaining technical and procedural problems. The recommendations for lidar obstruction surveying stemming from these research phases included:

1. Ensure high target irradiance (W/m^2) in order to maximize the probability of receiving a detectable return (i.e., above the receiver's detection threshold) from small-diameter, low-reflectance VOs. This is typically achieved through low flying heights, narrow beam divergence (for systems with adjustable divergence) and system settings that increase transmitted power.
2. Ensure very high point density through large swath overlap and other acquisition parameters.
3. Enable multiple "looks" (i.e., multiple viewing geometries) to each portion of the survey area. This can be achieve using variable sensor tilt angles (e.g., nadir and 20° forward tilt) and/or a flight pattern employing cross lines and large swath overlap.
4. Use sufficient swath overlap to avoid missing the tops of tall obstructions, which may fall in the gaps between swaths, even if the corresponding areas on the ground are covered (Figure 10.69).

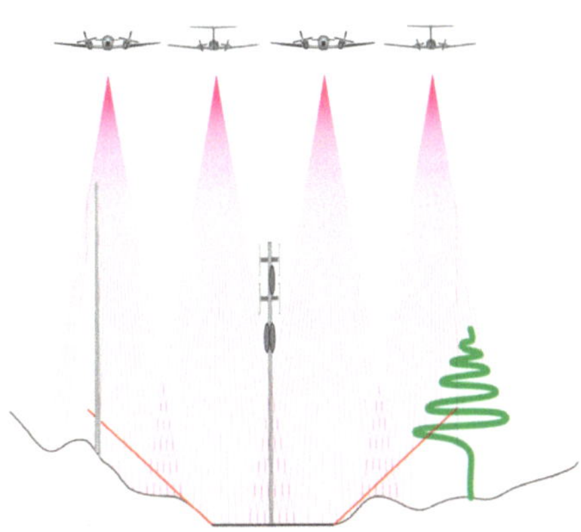

Figure 10.69 Example of the top of a tower being missed (center of image), due insufficient swath overlap, even though the ground areas are completely covered. It is also important to ensure that coverage of the outer edges of the survey areas is sufficient to capture the tops of objects near the OIS boundaries.

The results from these research projects led to: (a) further recommendations for system configurations and parameter settings for lidar obstruction surveying (Parrish et al., 2004 and 2005) (b) a demonstration of the first instrument approach procedure in the US based on a lidar survey (Anderson, F., internal FAA report, 2004), and (c) a recommended lidar/imagery workflow and simultaneous demonstration of the benefits of full-waveform (FW) data for capturing vertical structure of objects, thereby improving VO detection and recognition (Parrish and Nowak, 2009). These outcomes, in turn, contributed to the development of a draft specifications document for use in contracting lidar airport obstruction surveys published online by NGS (National Geodetic Survey, 2009) and to a Transportation Research Board (TRB) study into survey specifications for lidar airport obstruction surveys (Uddin et al., 2011). The NGS specifications document has been utilized by private sector lidar service providers in airport surveys (e.g., Mann and Wagg, 2008), and portions of it have been incorporated into FAA AC 17C (FAA, 2011).

In summary, airport-related applications of airborne lidar data are numerous and rapidly increasing. The ability to support multiple applications through a single lidar mission is one of the key benefits of airport lidar data. As a final point regarding this multi-use concept, it is important to carefully consider the requirements and specifications for all of the potential applications of a particular airport lidar data set and to ensure that all are met. Typically, this means adopting the most stringent specifications for any of the potential uses of the data set. For example, airborne lidar collected to obstruction surveying specifications can typically meet the specifications for terrain mapping and extraction of certain airport infrastructure GIS features, but the reverse is not necessarily true. Through application of these principles, as well as ongoing development and testing, it is anticipated that lidar will continue to become an increasingly valuable tool for the airport and aeronautical communities.

REFERENCES

Croft, J., 2004. Profiting from geography: Airports boost efficiency, safety, savings with GIS, *Airport Magazine*, Vol. 32.

Csanyi, N., and C.K. Toth, 2007. Improvement of lidar data accuracy using lidar-specific ground Targets, *Photogrammetric Engineering & Remote Sensing*, 73(4):385–396.

de Haag, M.U., J. Campbell, and R. Gray, 2000. A terrain database integrity monitor for synthetic vision systems, *Proceedings of the 19th Digital Avionics Systems Conferences (DASC)*, Vol. 1, pp. 2C3/1–2C3/8.

Federal Aviation Administration (FAA), 1989. *Advisory Circular 150/5300-13: Airport Design*, US Department of Transportation, Washington, D.C.

FAA, 1993. *FAA Order 8260.19: Flight Procedures and Airspace*, US Department of Transportation, Washington, D.C.

FAA, 2007a. *Advisory Circular 150/5300-16A: General Guidance and Specifications forvAeronautical Surveys: Establishment of Geodetic Control and Submission to the National Geodetic Survey,* US Department of Transportation, Washington, D.C.

FAA, 2007b. *Advisory Circular 150/5070-6B: Airport Master Plans*, US Department of Transportation, Washington, D.C.

FAA, 2009a. *AC 150/5300-18B: General Guidance and Specifications for Submission of Aeronautical Surveys to NGS: Field Data Collection and Geographic Information System (GIS) Standards*, Federal Aviation Administration, Washington, D.C.

FAA, 2009b. Airport Policy Guidance, *Airports-GIS News*, Vol. 2, No. 1. URL: https://airports-gis.faa.gov/public/data/newsletters/2009_1Q.pdf (last data accessed: 29 December 2011).

FAA, 2011. *AC 150/5300-17C: Standards for Using Remote Sensing Technologies in Airport Surveys*, US Department of Transportation, Washington, D.C. URL: http://www.faa.gov/documentLibrary/media/Advisory_Circular/150_5300_17c.pdf (last date accessed: 29 December 2011).

Federal Register, 2000. *Terrain Awareness and Warning Systems, Federal Register - Rules and Regulations, FR Doc 00-7595*, Vol. 65(61):16735–16756.
Günay, A., H. Arefi, and M. Hahn, 2007. True orthophoto production using LIDAR data, *International Archives of Photogrammetry, Remote Sensing and Spatial Information Sciences*, Vol. XXXVI–4/W45.
Lohr, U., 2003. Precise Lidar DEM and True Ortho Photos, *Proceedings of Photogrametric Week 03* (D. Fritsch, editor), Wichmann, pg. 111–116.
Mann, J., and D. Wagg, 2008. KAOS - Airport obstruction analysis completed using LiDAR, *Proceedings of the 2008 ILMF Conference*, 21–22 February, Denver, Colorado.
Murphy, R., 2003. High quality geographic information helps airport managers make planning decisions. *ICAO Journal*, No. 3, pp. 26-34.
National Geodetic Survey, 2009. *Light Detection and Ranging (LIDAR) Requirements: Scope of Work for Airport Surveying under the NOAA Aeronautical Survey Program*, URL: http://www.ngs.noaa.gov/RSD/AirportSOW.pdf (last date accessed: 29 December 2011).
Parrish, C.E., and R.D. Nowak, 2009. Improved approach to LIDAR airport obstruction surveying using full-waveform data, *Journal of Surveying Engineering*, 135(2):72–82.
Parrish, C.E., G.H. Tuell, W.E. Carter, and R.L. Shrestha, 2005. Configuring an airborne laser scanner for detecting airport obstructions, *Photogrammetric Engineering & Remote Sensing*, 71(1):37–46.
Parrish, C., J. Woolard, B. Kearse, and N. Case, 2004. Airborne LIDAR technology for airspace obstruction mapping, *Earth Observation Magazine (EOM)*, Vol. 13, No. 4.
Uddin, W., B. Gutelius, and C. Parrish, 2011. Airborne laser survey specifications and quality management protocols for airport obstruction surveys, *Transportation Research Record: Journal of the Transportation Research Board*, No. 2214, pp. 117–125.
Wells, A., and S. Young, 2004. *Airport Planning and Management*, Fifth edition, McGraw-Hill, New York.

10.7 COASTAL APPLICATIONS

10.7.1 Introduction

Demand for lidar data is high for a variety of areas, but perhaps none more so than the coastal zone. Primary factors influencing this demand include the high population density in coastal areas, coastal erosion, potential impacts of climate change on sensitive coastal ecosystems, and the economic significance of marine transportation and commerce. Specific uses of coastal lidar data include shoreline mapping, regional sediment management, storm vulnerability analysis, sea level change and inundation modeling, habitat mapping, and a host of related coastal science and management applications. However, lidar acquisition in the coastal zone is challenging, leading to unique considerations in everything from sensor design, to flight planning, to data processing and analysis.

A particular challenge that the coastal zone poses to mapping agencies and technologies is that it straddles the land-water interface. One of the first questions that must be answered for any coastal lidar acquisition is whether the project requires topographic, bathymetric, or seamless topographic-bathymetric lidar (Figure 10.70). Most commercial lidar systems operating today are topographic systems, and their operation principles and use are described extensively in other chapters of this manual. Bathymetric lidar systems operate in a similar manner, but use 532 nm (green) laser light to penetrate the water column and obtain ranges to the seafloor in relatively shallow (~0 to 60m, although perhaps slightly deeper in the clearest water conditions) coastal waters. Where waveform-based processing is a fairly new approach in topographic lidar mapping, it has been the standard in bathymetric lidar depth determination since the first systems were developed in the early 1980s. Many of the bathymetric lidar systems used for hydrographic surveying/nautical charting purposes employ both near-infrared (NIR) and green beams for detection of the sea surface and seabed, respectively. However, nearshore, green-only, topographic-bathymetric systems are becoming more prevalent, especially in environmental applications (see discussion of the EAARL below). Water clarity is a key consideration in bathymetric and topographic-bathymetric lidar, as it limits the maximum depths that can be detected, and high turbidity can preclude acquisition altogether. While an in-depth description of bathymetric lidar technology is beyond the scope of this chapter, Guenther (2007) is a recommended reference for further details related to the history, theory, and operational considerations.

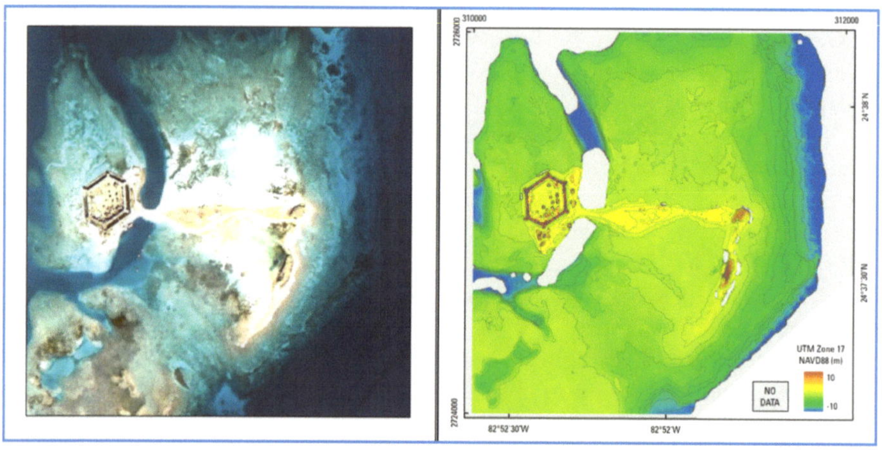

Figure 10.70 (a) Satellite image of Dry Tortugas Fort Jefferson, and (b) corresponding seamless topobathy lidar of same extent.

In this chapter, we describe some of the more prominent applications of coastal lidar and a few of the systems. Many of the activities described here are coordinated through the Joint Airborne Lidar Bathymetry Technical Center of Expertise (JALBTCX) interagency consortium, which is dedicated to operations, research, and development in airborne lidar bathymetry and complementary technologies. JALBTCX member agencies include the US Army Corps of Engineers (USACE), the US Naval Meteorology and Oceanography Command, and the National Oceanic and Atmospheric Administration (NOAA). The US Geological Survey (USGS) is also a frequent partner. The activities of this group include efforts to develop common standards and specifications for lidar coastal mapping, and the annual JALBTCX workshop, which draws attendees from the coastal lidar communities in academia, government, and the private sector.

10.7.2 Coastal Management Applications

One issue of increasing concern is changing sea level and predicting future inundation for planning purposes. Historical records of sea level indicate trends that are spatially variable, with some areas having rising tide levels and some falling. Grand Isle, Louisiana, has one of the highest rates of relative sea level rise in the US (Figure 10.71) while Juneau, Alaska has one of the highest rates of sea level drop. Many coastal areas are relatively flat (particularly in the Southeastern US and Gulf of Mexico) and the flatter an area is, the greater the resulting horizontal error is due to a given vertical error. For this reason, many coastal states with low relief are asking for elevation data that supports one-foot (30 cm) contours. Sea level rise mapping is being done by many groups with varying degrees of complexity and scale. One example covering a large area using lidar data can be seen at www.csc.noaa.gov/slr. The following are some additional examples of coastal management applications of lidar data:

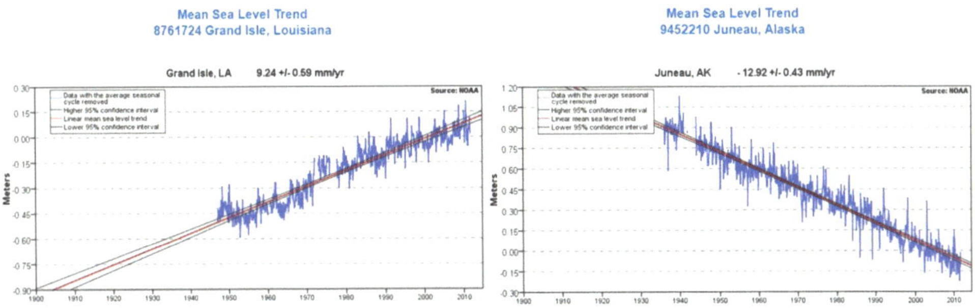

Figure 10.71 Long-term water level data and linear mean sea level trend for (a) Grand Isle, Louisiana, and (b) Juneau, Alaska.

10.7.2.1 Wetlands and Habitat Delineation

Elevation data, particularly lidar, has been found to be very helpful in wetland and habitat delineation in the coastal zone. Six inches (15 cm) of elevation can make a difference in how an area is mapped. This can include identifying depressions that may contain jurisdictional wetlands subject to regulation through the Clean Water Act of 1977.

10.7.2.2 Setback Lines

There are a variety of methods used by communities to set limits on where buildings can be constructed. Along the coast, buildings must be some specified distance from a physical feature related to the ocean or other water body, with this distance establishing the setback line. The feature can often be economically determined using lidar. Examples include the mean high water line and the crest of the primary dune. South Carolina is one state that uses the primary dune and has found it more economical to use lidar to find the crest than to use their previous method of walking the dune with a GPS.

10.7.2.3 Assessment of Severe Storm Damage

The USGS and others have used lidar for many years to look at the change in beach morphology after severe storms (Figure 10.72). This requires baseline pre-storm data followed by post-storm data and allows a study of elevation change, shoreline change, and volumetric changes in beaches, dunes, and nearshore. If the data can be processed quickly enough, they can provide valuable information for emergency responders regarding the amount of sand to move or debris to be cleared. This is simply a coastal example of the more general application of change analysis that can be applied to many fields, such as open pit mining, waste management, and landslides.

10.7.2.4 Tsunami and Storm Surge Modeling

Models of water inundation such as storm surge and tsunami models can benefit from accurate elevation data. In general, they are not able to take advantage of, and do not need, the high spatial resolution that lidar provides. While the ocean floor shape is very important to the models, particularly in the near shore area, most current models have grid or node spacing of 100 meters or greater due to computational limitations. Although lidar can appear to be spatial oversampling on both the bathymetric and topographic side, it is still an important source of elevation information since it can provide the near-shore bathymetry economically for relatively clear water. On the terrestrial side, the high spatial resolution is important for identifying levees and other obstructions to water flow that must be incorporated into the models as vector features even though they are smaller than the model spacing. FEMA uses lidar data extensively in inundation modes that support mapping of flood hazard zones for flood insurance rate maps.

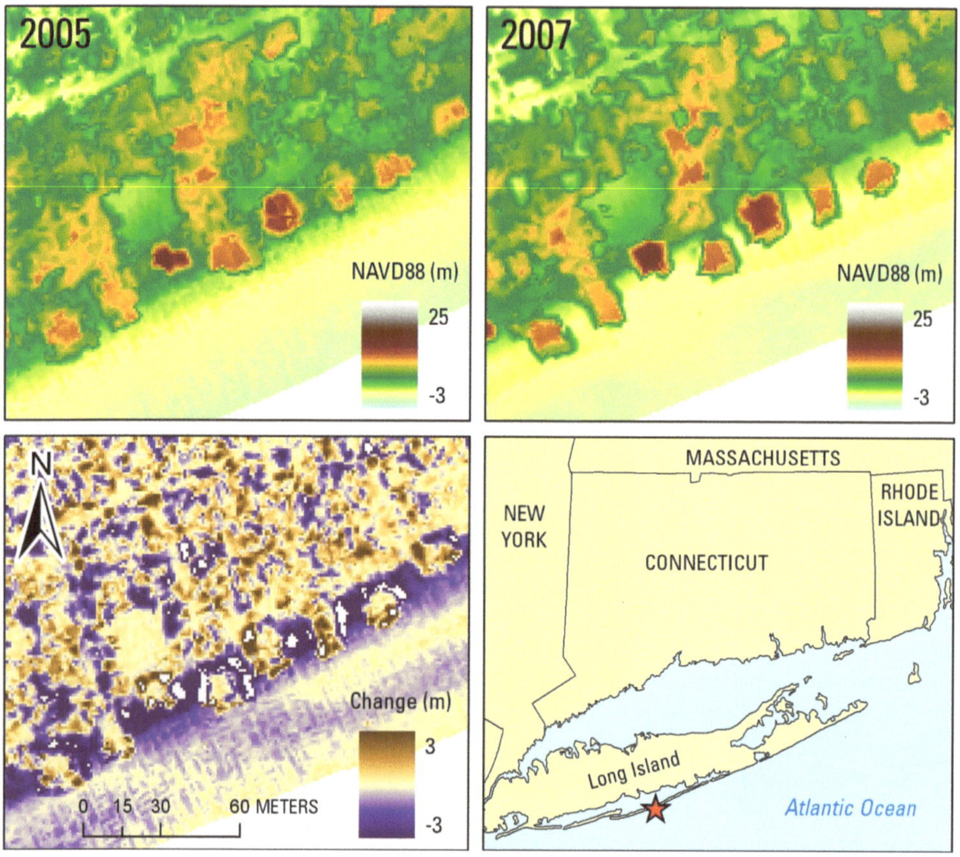

Figure 10.72 Fire Island National Seashore topographic lidar from 2005 and 2007. A nor'easter hit in April 2007 before the 2007 lidar was collected and is largely responsible for the significant erosion (change map included to show erosion).

10.7.2.5 Shoreline Extraction

Shoreline is needed for a number of applications ranging from safe navigation, to property rights issues where the division between public and private land is a tidal elevation line, to shoreline change analysis and modeling. Shoreline mapping is becoming such a widespread use of coastal lidar data that we cover it in a separate section (see below).

10.7.2.6 Sand Resources

A coastal application of lidar that has supplied valuable data for other applications is the USACE mapping for regional sediment management, construction, operations, and regulatory functions in the coastal zone. Using topographic and bathymetric lidar and imagery, they map sand and other resources around the conterminous United States. This often provides baseline data for other applications such as severe storm damage assessment (see USACE National Coastal Mapping Program below).

10.7.2.7 Safe Navigation

Bathymetric lidar is sometimes used by the creators of nautical charts to provide the bathymetry in shallow relatively clear water. In these areas it is more economical than ship surveys because of the wider swaths possible. It is also valuable as an initial survey in uncharted waters where it may be dangerous to do ship surveys, such as near the rocky coasts in Alaska. The US federal government investment in bathymetric lidar was initially driven by hydrographic survey requirements to support dredging activities in the USACE (Lillycrop and Banic, 1993).

To further illustrate some of the specific benefits and uses of coastal lidar data, the following sections expand on a few of the applications listed above. However, it is important to note that this is by no means a compressive treatment of coastal lidar applications; the published scientific literature contains numerous examples of uses of coastal lidar above and beyond those considered here.

10.7.3 Shoreline Mapping

One of the most frequent uses of coastal lidar data is shoreline mapping. Various groups and organizations map shoreline for a range of purposes. Here we restrict our focus to the National Shoreline mapped by NOAA's National Geodetic Survey (NGS) and depicted on NOAA nautical charts. The shoreline produced by NGS is tidally-referenced; both the mean high water (MHW) and mean lower low water (MLLW) lines are mapped for most projects. The current, mathematical definition of these lines is the intersection of two surfaces: (a) that of the tidal datum (MHW or MLLW), and (b) the coastal topographic surface, at the time of data acquisition. Tidal datums are defined for a specific epoch of 18.6 years, and the coastal topographic surface can change on time scales ranging from a single tidal cycle to seasonally to decadal or longer, in response to sediment transport, erosion, accretion, sea level change, coastal engineering, and a range of other drivers. It is for this reason that the definition includes the statement "at the time of data acquisition." Beyond supporting safe marine navigation, this National Shoreline has legal significance, in that it is used in defining the boundaries between federally-owned, state-owned, and privately-owned lands. It is further used in defining the nation's international marine boundaries, including the Exclusive Economic Zone (EEZ) and Territorial Sea (Graham et al., 2003; Parrish, 2012).

Since the 1930s, the primary method of mapping the National Shoreline has been from aerial imagery using photogrammetric techniques. Of course, numerous technological and procedural advancements have occurred over the years, including the transition from analog (i.e., optical-mechanical) to analytical (partly automated) to softcopy (fully digital) stereoplotters, and the ongoing transition from film to digital aerial imagery. The current photogrammetric techniques typically entail extracting the land-water interface from stereoscopic, tide-coordinated imagery on a softcopy photogrammetry workstation. The term "tide coordinated" means that the imagery is acquired when the instantaneous water level in the project site coincides with the tidal datum corresponding to the shoreline to be mapped: MHW or MLLW. Because it is impossible to acquire all of the imagery for a project site instantaneously, the acquisition procedures allow for the image acquisition to be performed within a small "window" around the specified water level, specified in terms of a tide tolerance. While these photogrammetric techniques are well proven and have been tested, refined, and documented over a period of several decades, lidar is becoming an increasing-prevalent, complementary technology for shoreline mapping.

One advantage that lidar offers for shoreline mapping is the ability to alleviate some data acquisition constraints. For example, as an active remote sensing technology, lidar data can be collected at night (unless project instructions call for concurrent imagery). Furthermore, lidar data for shoreline mapping typically only need to be acquired below some specified water level, which is a less restrictive requirement than for tide-coordinated aerial imagery, which must be acquired within a narrow tide window. Another advantage is that the end-to-end workflow for lidar shoreline mapping is highly automated. This has the benefits of both reducing human time and making the shoreline mapping process more repeatable (White, 2007; Parrish, 2012).

On the other hand, there are some operational challenges associated with lidar shoreline mapping. The biggest is simply the fact that the ability to extract shoreline from a lidar data set requires continuous, high-quality data in the intertidal zone. The difficulties posed by the intertidal zone can be summarized, as follows:

- When submerged, water depths in the intertidal zone are typically very shallow (<1m), which can cause challenges for some bottom detection algorithms in bathymetric lidar. Breaking waves, bubbles, and suspended sediment pose further challenges.
- When exposed, the intertidal zone is usually still wet (from waves and/or the previous high tide), which can lead to weaker returns (and, hence, lower signal to noise ratio) for topographic lidar systems operating in the near infrared.

However, the benefits of lidar for shoreline mapping generally outweigh the challenges, as evidenced by the increasing number of shoreline mapping projects that are now incorporating lidar (Stockdon et al., 2002; Robertson et al., 2004; Liu et al., 2007, White, 2007; White et al., 2011).

NGS' production lidar workflow is described in detail in White (2007) and White et al. (2011). A key component of this workflow is the use of NOAA's vertical datum transformation utility, VDatum (Myers et al., 2007). This freely-available software tool enables transformation between 29 different vertical datums, including three-dimensional (realized through space-borne systems), orthometric (defined relative to a form of mean sea level), and tidal datums (based on a tidally-derived surface). Although beyond the scope of this chapter, the extraction technique can be summarized, as follows:

- Coastal lidar point clouds in LAS format are ingested and cleaned in the vicinity of the shoreline, by properly classifying points within the LAS structure (e.g., ground, water surface, etc.)
- The lidar point cloud is processed through VDatum to convert from ellipsoid heights to MHW tidal datum heights (assuming a MHW shoreline is to be mapped).
- The data set is interpolated to a regular grid and contoured. A zero-elevation contour is then extracted, representing the modeled intersection of the tide datum surface and coastal topographic-bathymetric surface (i.e., a shoreline vector).
- The resulting shoreline vector is cleaned, attributed (typically using imagery), and quality controlled, after which it is submitted for application to the nautical chart.

These procedures have recently been incorporated into a production lidar shoreline mapping workflow in NGS. The process outlined above can also be used to map MLLW shoreline. However, topographic-bathymetric lidar data with continuous coverage across the entire intertidal zone is usually required, as it is challenging to acquire topographic lidar data at a sufficiently low water level to enable extraction of MLLW.

10.7.4 USACE National Coastal Mapping Program

The USACE National Coastal Mapping Program (NCMP) was initiated in 2004 to produce the recurring, regional, high-resolution, high-accuracy, data necessary to implement regional sediment management practices at USACE coastal projects (Wozencraft and Millar, 2005). NCMP is executed by the JALBTCX using the Compact Hydrographic Airborne Rapid Total Survey (CHARTS) sensor suite. A bathymetric lidar, topographic lidar, digital aerial camera, and hyperspectral imager deployed on a single aircraft enable concurrent physical and environmental characterization of beaches, wetlands, marshes, estuaries, and barrier islands. The dual-lidar system measures high-accuracy land elevations, and water depths in areas of relatively clear water. The digital camera provides high spatial resolution imagery at 30 cm ground sampling distance, while the hyperspectral imager provides high spectral resolution imagery in the visible and near-infrared range. The program focus area is the active portion of the beach and nearshore (Figure 10.73), and as such, data are collected along sandy shorelines in a swath that extends from 500 m onshore to 1,000 m offshore.

Figure 10.73 Marquette Harbor, Michigan, Lake Superior: federal navigation project and adjacent shoreline.

In the aftermath of the 2004 hurricane season, it became apparent that lidar point clouds and individual image frames presented a processing challenge to engineers seeking to use these data in post-storm analyses. To alleviate the need for USACE engineers and scientists to do basic processing on these data, a suite of GIS information products is generated for immediate ingestion into the USACE Navigation and Coastal Data Bank:

- Bathymetric/topographic digital elevation models (DEMs) in two forms, first return and bare earth. First return includes information about coastal infrastructure and vegetation, while bare earth surface exposes the underlying hydrology to support coastal modeling.
- Laser reflectance images contain information about the reflectivity of the seafloor and nearshore, which can be used to discriminate land-cover and benthic habitats;
- Shoreline vector;
- Aerial photo mosaics;
- Hyperspectral image mosaics;
- Basic land cover classification identifying water, bare ground, vegetation, buildings, and Roads.

The lidar point clouds are made available to the public through NOAA's Digital Coast, and DEMs and imagery are made available to the emergency response community through the USGS Hazard Data Distribution System.

A major focus of the NCMP is fully exploiting the datasets to provide engineering, economic, and environmental information to support management of USACE activities in the coastal zone on a regional scale. New information products currently in development fall into three categories: products that quantify changes to the physical and environmental characteristics of the beach and nearshore; products that exploit the datasets to extract parameters critical to understanding the condition of coastal infrastructure and vulnerability of the shoreline to hazards; and products developed through advanced data fusion techniques to map seafloor and water column characteristics (Reif et al., 2012). Standardized, semi-automated processing ensures consistent products that can be compared from survey to survey to accurately quantify regional changes on scales meaningful to engineers, and timely product delivery.

For example, automated processes have been developed to extract the location of the dune peak, the mean and maximum dune peak heights, and the beach width for a certain stretch of coast (Dunkin and McCormick, 2011). Each of these geomorphic features have been shown to be indicators of vulnerability to inundation and wave impacts (Houser et al., 2008; Saye et al., 2005; Stockdon et al., 2009). Tracking these metrics through time can quantify the cumulative effects of storms, or recovery in times of storm quiescence. Comparison of bare-earth topobathy DEMs quantifies elevation and sediment volume changes on the beach and nearshore and can help identify sand transport pathways through the coastal zone (Wozencraft and Lillycrop, 2006; Wozencraft 2001; Mohr et al., 1999; Irish and Lillycrop, 1997). Other automated processes extract the length and average height of coastal navigation structures (Reif et al., 2012), and assess the availability of navigation channels (Dunkin et al., 2011).

Combining the height information in the lidar data with spectral information available in imagery improves the capability of both data sets to characterize environmental conditions in the coastal zone. New techniques for analyzing coastal land cover (Figure 10.74) incorporate the structural components of lidar elevation data to gain new insights into coastal land cover change, through interrogation of volume change within the land cover types (Reif et al., 2011). On land, hyperspectral imagery and lidar elevation can be combined for improved discrimination of vegetation species like phragmites or other invasives which pose management problems for confined disposal facilities and the wider coastal environment as a whole (Reif et al., 2012). Optimization algorithms use lidar-derived depth, attenuation, and bottom reflectance as constraints in the spectral decomposition of the water column and seafloor (Kim et al., 2010). The resulting images can identify benthic habitats like sand, mud, hard-bottom, and seagrass (Aitken et al., 2010). These techniques have been used to discriminate submerged vegetation species (Reif et al., 2012a) and show promise in estimation of water quality parameters (Reif, 2011).

While the data and information products generated by the NCMP are designed to support regional sediment management, they support a number of other activities in the USACE: regulatory, flood damage reduction, asset management, emergency operations, and environmental stewardship. The data are widely used by the others, specifically for FEMA RiskMap modeling efforts; USGS coastal and marine geology and extreme storm studies; NOAA nautical chart production; and shoreline management, environmental permitting, emergency management, marine spatial planning, and planning for resilient communities by federal, state, and local agencies.

The NCMP also continues evolution of technologies for regional characterization and change detection of engineering, environmental, and economic conditions along the shoreline. The Coastal Zone Mapping and Imaging Lidar (CZMIL) development effort advances the state-of-the-art in data exploitation workflows, algorithms, hardware, and software, and provides a sustained focus and collaboration among academia, industry, and the federal government to constantly review, refine, and expand our capability to produce a wider range of information products over a broader range of operating environments. The CZMIL effort has resulted in new airborne technologies and supporting software that improves operational efficiency, decreases time between data collection and final decision-support product, expands the variety of products derived from the basic datasets in a data fusion approach, improves performance in very shallow and turbid waters, improves navigation hazard detection, and improves data accuracy.

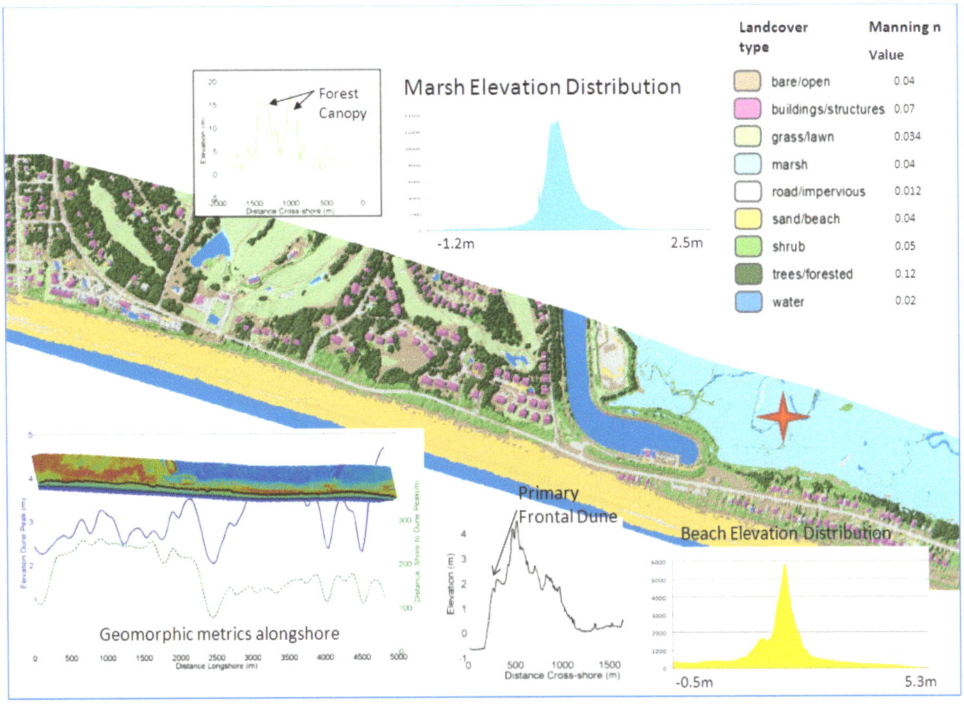

Figure 10.74 Coastal land cover classification.

10.7.5 EAARL

Topographic-bathymetric lidar technology is currently advancing quite rapidly, with innovations being made in industry, government, and academia. Here, we describe advances being made within the USGS Experimental Advanced Airborne Research Lidar (EAARL) program. Originally developed during 1999 to 2002 at NASA Goddard Space Flight Center, Wallops Flight Facility, Virginia, the EAARL (Figure 10.75) is a research instrument specifically designed to measure submerged topography and coastal land elevations seamlessly, and to address some of the challenges that are encountered by traditional bathymetric lidar systems in mapping near-shore coastal environments. The EAARL system is green-only (no NIR lidar beam) and uses a very low-power pulse (60 μJ) compared to traditional bathymetric lidar systems (3,000 μJ). The lower power laser pulse energy develops the same power density at the surface as traditional bathymetric lidar; however, the EAARL illuminated spot is much smaller (0.20 meter diameter versus 2.6 meters). The smaller EAARL surface spot, coupled with the EAARL's much shorter pulse and wider detection bandwidth, enable fine-scale spatial sampling in shallow water and greatly reduced geometric beam spreading due to the laser beam angle of incidence at the water surface. The EAARL system also automatically adjusts its pulse repetition rate to achieve more even, non-overlapping point spacing on the surface. The system has been used for mapping features/areas ranging from coral reefs to coastal vegetation to rivers (Brock et al., 2004; Nayegandhi et al., 2006; McKean et al., 2008).

The EAARL-A sensor, now owned and operated by the USGS St. Petersburg Coastal and Marine Science Center, is currently being upgraded to improve measurement of bathymetry in shallow and turbid water environments. Enhancements currently being incorporated into the EAARL-B instrument include a more powerful laser with a greater pulse repetition frequency and shorter pulse width (700 ps FWHM), and an integrated GPS-INS unit. The upgraded laser transmitter offers up to 400 μJ of laser pulse energy, which is greater than a 5× increase from the original laser transmitter. In order to maintain eye-safe operation, the outgoing beam will be optically segmented into three small foot print "beamlets," thereby eliminating the need to spread the outgoing beam while retaining low geometric spreading. This incorporation of beam splitting into beamlets will also effectively increase the point density by a factor of three for shallow bathymetric mapping. An additional 2× increase will also be accomplished by doubling the number of laser triggers across each scan from 120 to 240. Upgrades to the receiver electronics will include four multi-channel waveform digitizers that will detect laser pulses from the three beamlets, and the fourth detector/digitizer that encompasses the field-of-view of the entire set of three beamlets plus an additional large field-of-view area better suited for detecting the bottom in deeper water, thereby enabling the ability to map in deeper and more turbid waters by capturing beamlet photons that have been scattered out of the "beam-C" region by suspended sediment. The EAARL-B will also include two high-resolution digital cameras with high-precision time stamps enabling greatly improved geo-location and photo registration. One camera provides Red, Green, and Blue bands, while the second camera provides Near Infrared.

This unique design philosophy of the EAARL-B system is expected to produce 6× greater data density, better water penetration, improved ability to discriminate water surface reflections from seabed reflections, better overall accuracy, and better spatial resolution of small topographic features. New software methods will also be implemented to extract bare earth and seabed reflections from the waveforms. Initial test flights of the EAARL-B system are planned to occur during the Spring of 2012.

10.7.6 Coastal Lidar Distribution Systems

The distribution of lidar data in the United States is still evolving, often reflecting the many parties engaged in data collection. Fortunately, lidar data collected by the federal government and many of the states is freely available. The distribution is somewhat fragmented with some redundancy and a range of services offered. Here, we will list just a few of the current systems providing free data and their attributes, recognizing this will likely be out of date soon. The first three systems are also making efforts toward interoperability and showing the holdings of the other systems.

Digital Coast (www.csc.noaa.gov/digitalcoast) is probably the oldest lidar distribution system, as it absorbed the Lidar Data Retrieval Tool (LDART) begun in 1999. The Digital Coast focuses on data in coastal states and provides topographic and bathymetric lidar data in point, raster, and contour formats: all derived dynamically from the points upon request. Users are able to select projections, datums, grid spacing, and interpolation methods to create custom products. Requests for point data are limited to 250 million points currently, though derived products may exceed that.

The Center for Lidar Information Coordination and Knowledge (CLICK: lidar.cr.usgs.gov) provides streamed access to LAS point format files. Projection and datum of the data vary depending on the original data provider. The focus is terrestrial data throughout the United States. CLICK also provides forums discussing various aspects of lidar data.

The OpenTopography (www.opentopography.org) facility is supported by the National Science Foundation and provides access to science-oriented topography data. It provides custom products similar to the Digital Coast. Unregistered users are limited to requests of 50 million points currently, though registered users could request up to 150 million points with a justification.

While the three systems above provide data from a variety of sources, there are also distribution systems with more locally focused data, typically from state-wide collections or a consortium. Some examples of these are state-wide collection in North Carolina (www.ncfloodmaps.com) and Louisiana (atlas.lsu.edu) and consortium collections such as Puget Sound (pugetsoundlidar.org). While these sites have a smaller geographic focus than the more national ones, they may have additional organization and ancillary information particular to their data sets.

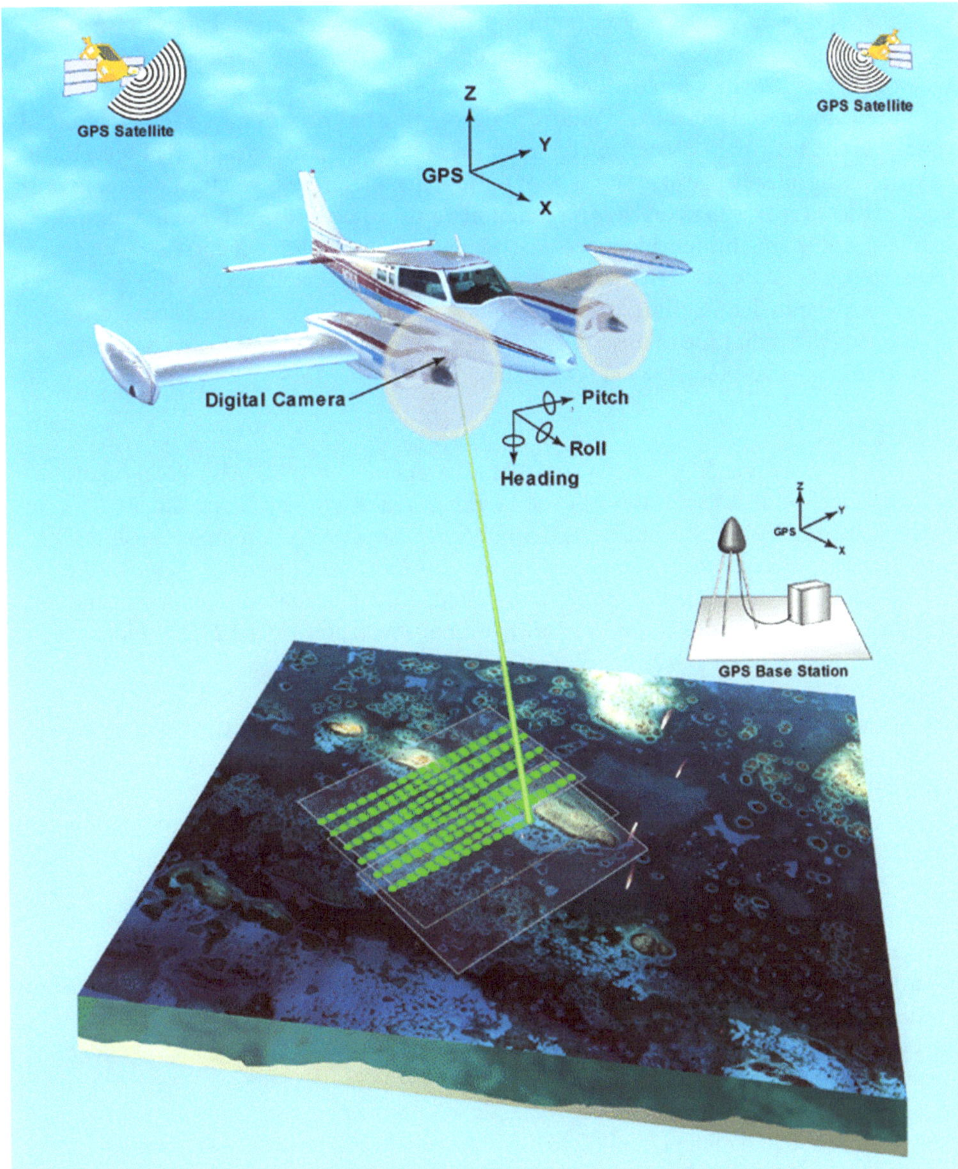

Figure 10.75 The USGS Experimental Advanced Airborne Research Lidar (EAARL).

10.7.6.1 Summary and Future Outlook

In this chapter we have highlighted several applications of coastal lidar data, ranging from shoreline mapping, to regional sediment management, to coral reef mapping. However, we believe that coastal managers and scientists have only begun to scratch the surface of the potential of lidar in the coastal zone. New technological developments, including those discussed here, are enabling new uses of the data. In fact, a great benefit of coastal lidar is that a single data set can typically support multiple projects, programs, and application areas.

Topographic-bathymetric lidar technology is evolving especially rapidly. One recent emphasis has been enhancing use of the systems and data in environmental applications. While earlier generations of commercial bathymetric systems focused on extending operational ranges into deeper water for hydrographic surveying in support of nautical charting, current systems are increasingly pushing into shallower waters and filling the data gap at the land-water interface. In addition to opening up new applications of the data, we believe this enhanced focus on shallow water and environmental information products is leading to a blurring of lines between traditional topographic and bathymetric systems, communities, and software. We are already beginning to see an increasing interchange of ideas between the topographic and bathymetric lidar communities and expect to see further emphasis on common formats, standards, and specifications.

While already high, the demand for coastal lidar is likely to continue to increase with growing awareness of and focus on effects of climate change in the coastal zone, including sea level rise and coastal inundation. Thus, the technological development mentioned above is occurring at exactly the right time to meet the rising need. We anticipate that the next decade will be an extremely active one in the field of coastal lidar.

REFERENCES

Brock, J.C., C.W. Wright, T.D. Clayton, and Am Nayegandhi, 2004. LIDAR optical rugosity of coral reefs in Biscayne National Park, Florida, *Coral Reefs*, Vol. 23, pp. 48–59.

Dunkin, L.M., and J.W. McCormick, 2011. Monitoring shore protection projects using 3D spatial data, *Proceedings of the Coastal Engineering Practices Conference*, August, 12 p.

Dunkin, L.M., J.W. McCormick, and J.M. Wozencraft, 2011. Changes to geomorphic and bathymetric features due to extreme storm events, *Proceedings of the 2011 Solutions to Coastal Disasters Conference*, June, pp. 176–187.

Aitken, J., V. Ramnath, V. Feygels, A. Mathur, K. Kim, J.Y. Park, and G. Tuell, 2010. Prelude to CZMIL: Seafloor imaging and classification results achieved with CHARTS and the rapid environmental assessment (REA) processor, *Proceedings of SPIE 7695*, 76950S (2010); doi:10.1117/12.851915.

Graham, D., M. Sault, and J. Bailey, 2003. National Ocean Service Shoreline-Past, present, and Future, *Journal of Coastal Research*, Special Issue 38, pp. 14–32.

Guenther, G.C., 2007. Airborne lidar bathymetry, *Digital Elevation Model Technologies and Applications: The DEM Users Manual*, Second Edition (D. Maune, editor), American Society for Photogrammetry and Remote Sensing, Bethesda, Maryland, pp. 253–320.

Houser, C., C. Hapke, and S. Hamilton, 2008. Controls on coastal dune morphology, shoreline erosion and barrier island response to extreme storms, *Geomorphology*, 100:223–240.

Irish, J.L., and W.J. Lillycrop, 1997. Monitoring New Pass, Florida, with high density lidar bathymetry, *Journal of Coastal Research*, 13(4):1130–1140.

Kim, M., J.Y. Park, and G. Tuell, 2010. A constrained optimization technique for estimating environmental parameters from CZMIL hyperspectral and lidar data, *Proceedings of SPIE 7695*, 769513 (2010); doi:10.1117/12.851989.

Lillycrop, W.J., and J.R. Banic 1993. Advancements in the U.S. Army Corps of Engineers hydrographic survey capabilities: The SHOALS system, *Marine Geodesy*, 15:177–185.

Liu, H., D. Sherman, and S. Gu, 2007. Automated extraction of shorelines from airborne light detection and ranging data and accuracy assessment based on Monte Carlo simulation, *Journal of Coastal Research*, 23(6):1359–1369.

McKean, J.A., D.J. Isaak, and C.W. Wright, 2008. Geomorphic controls on salmon nesting patterns described by a new, narrow-beam terrestrial-aquatic lidar, *Frontiers in Ecology and the Environment*, 6:125–130.

Mohr, M.C., J. Pope, and J.K. McClung, 1999. Coastal response to a detached breakwater system; Presque Isle, Pennsylvania, USA, *Proceedings of the 4th International Symposium on Coastal Engineering and Science of Coastal Sediment Processes*, American Society of Civil Engineers, 3:2010–2025.

Myers, E., K. Hess, Z. Yang, J. Xu, A. Wong, D. Doyle, J. Woolard, S. White, B. Le, S. Gill, and G, Hovis, 2007. VDatum and strategies for national coverage, *Proceedings of the Marine Technology Society / IEEE OCEANS Conference*, Vancouver, British Columbia.

Nayegandhi, A., J.C. Brock, C.W. Wright, and M.J. O'Connell, 2006. Evaluating a small footprint, waveform-resolving lidar over coastal vegetation communities, *Photogrammetric Engineering & Remote Sensing*, 72(12)1407–1417

Parrish, C.E., 2012. Shoreline mapping, *Advances in Mapping from Aerospace Imagery: Techniques and Applications* (X. Yang and J. Li, editors), CRC Press, Boca Raton, Florida (in press).

Reif, M., 2011. *Remote Sensing for Inland Water Quality Monitoring: A U.S. Army Corps of Engineers Perspective*, WOTS Technical Report Collection, Vicksburg, Mississippi, U.S. Army Engineer Research and Development Center.

Reif, M., C.L. Macon, and J.M. Wozencraft, 2011. Post-Katrina land-cover, elevation, and volume change assessment along the south shore of Lake Pontchartrain, Louisiana, U.S.A., *Journal of Coastal Research: Special Issue 62 - Applied Lidar Techniques*, pp. 30–39.

Reif, M., C. Piercy, J. Jarvis, B. Sabol, C. Macon, R. Loyd, P. Colarusso, H. Dierssen, and J. Aitken, 2012. A case study of ground truth sampling to support sensing research and development: Submersed aquatic vegetation species discrimination using an airborne hyperspectral/lidar system, *DOER Technical Notes Collection*, Vicksburg, Mississippi, U.S. Army Engineer Research and Development Center.

Reif, M.K., J.M. Wozencraft, L.D. Dunkin, C.S. Sylvester, and C.L. Macon, 2012a. U.S. Army Corps of Engineers Airborne Coastal Mapping in the Great Lakes, *Journal of Great Lakes Research* (in press).

Robertson, W., D. Whitman, K. Zhang, and S.P. Leatherman, 2004. Mapping shoreline position using airborne laser altimetry, *Journal of Coastal Research*, 20(3):884–892.

Saye, S.E., D. van der Wal, K. Pye, and S.J. Blott, 2005. Beach-dune morphological relationships and erosion/accretion: An investigation at five sites in England and Wales using LIDAR data, *Geomorphology*, 72:128–155.

Stockdon, H.F., K.S. Doran, and A.H. Sallenger, Jr., 2009. Extraction of lidar-based dune-crest elevations for use in examining the vulnerability of beaches to inundation during hurricanes, *Journal of Coastal Research*, Vol. 53,

Stockdon, H.F., A.H. Sallenger, Jr., J.H. List, and R.A. Holman, 2002. Estimation of shoreline position change using airborne topographic lidar data, *Journal of Coastal Research*, 18(3):502–513.

White, S., 2007. Utilization of LIDAR and NOAA's vertical datum transformation tool (VDatum) for shoreline delineation, *Proceedings of the Marine Technology Society/IEEE OCEANS Conference*, Vancouver, British Columbia, Canada.

White, S.A., C.E. Parrish, B.R. Calder, S. Pe'eri, and Y. Rzhanov, 2011. Lidar-derived national shoreline: Empirical and stochastic uncertainty analyses, *Journal of Coastal Research*, Special Issue, 62:62–74.

Wozencraft, J.M., and D. Millar, 2005. Airborne lidar and integrated technologies for coastal mapping and charting, *Marine Technology Society Journal*, 39(3):27–35.

Wozencraft, J.M., J.L. Irish, and L.S. Lillycrop, 2001. Sand volumes and transport pathways for Gulf of Mexico regional sediment management, *Proceedings of Coastal Dynamics '01*, ASCE Press, pp. 690–702.

Wozencraft, J.M., and W.J. Lillycrop, 2006. JALBTCX coastal mapping for the USACE, *International Hydrographic Review*, 7(2):28–37.

10.8 HYDROLOGICAL APPLICATIONS OF AIRBORNE LASER SCANNING

This subchapter provides a review of recent literature dealing with the application of airborne topographic lidar to the research of water resources and related hydrological problems. A somewhat "geographic" approach has been taken by dividing the discussion into terrestrial land-cover and terrain features that broadly starts in mountainous headwater environments and ends at the coastline.

10.8.1 Introduction

There is substantial overlap in airborne lidar literature dealing with hydrological and water resources topics, with individual papers covering a number of hydrological themes. The themes chosen here are: glaciers, snowpack, terrain morphology, ground surface elevation, vegetation height, canopy structure, wetland environments, fluvial and coastal geomorphology, and flood modeling. To promote an efficient overview of the literature, an attempt has been made to confine discussion of individual studies to the dominant theme of the research being published. This has not always been possible in cases where an individual paper has contributed greatly in more than one area of water resources or hydrological research. Moreover, this approach will undoubtedly lead to a certain amount of omission but it is hoped that the material presented provides a concise while almost complete overview of the current status of the use of airborne lidar technology in the hydrological sciences.

It should be noted that while all of the studies discussed utilize airborne lidar data as an input to the research, in many cases lidar is not the focus of the study, merely a means of acquiring information that would otherwise not be available. As such, the literature discussed is varied as it not only covers a number of broad themes within the hydrological sciences but also ranges from purely "proof of concept," where the objective is evaluation of the technology for a particular hydrological investigation, through to "proven methodology" where there is little explicit mention of its use. Further, it should be noted that some of the papers cited are not even from the hydrological sciences or water resources body of literature. This is particularly the case in the discussion on vegetation. However, it was felt important to include some of the key references from this body of work, as it is becoming more highly relevant as we start to realize new applications for the technology in terms of hydrological model parameterizations.

10.8.2 Glacier Surfaces

Early research based lidar sensors have been successfully utilised in *profiling* mode for various glaciological applications over Greenland (Krabill et al., 1995), the ice caps and mountain glaciers of the Canadian Arctic Islands (Abdalati et al., 2004), and large glacier complexes in Alaska (Echelmeyer et al., 1996). Recently, a satellite laser altimeter (ICESat) was put into space for the purpose of short-term ice surface elevation monitoring over large Arctic and Antarctic ice sheets (e.g., Csatho et al., 2005). An early demonstration of the efficacy of airborne scanning lidar for alpine glacier surface mapping in an alpine mountain environment was provided by Kennet and Eiken (1997). High correspondence between adjacent laser shots was found with absolute errors of approximately 10 cm. Work by Favey et al. (1999) illustrated the sensitivity of lidar measurements over glacierised surfaces to errors in the position and orientation measurement components of the lidar sensor technology.

Modern airborne lidar data has been demonstrated as an effective tool to map all areas of a glacier surface at high accuracy and high resolution (Hopkinson et al., 2001; Favey et al., 2002; Arnold et al., 2006). This capability makes lidar the ideal tool for glacier surface monitoring (Baltsavias et al., 2001), particularly in support of mass balance (Geist and Stotter, 2002) and glacial water resources assessment (Hopkinson and Demuth, 2006). The challenge of obtaining accurate photogrammetric elevation estimates in snow covered areas of minimal surface texture and high reflectance is overcome by the active nature of the lidar sensor (Favey et al., 1999); however limited horizontal accuracy leads to reduced elevation accuracy in areas of steep slopes or crevasses (Favey et al., 2000; Hopkinson and Demuth, 2006). In areas of steep relief, small horizontal errors can easily propagate into vertical elevation errors (Hodgson et al., 2004), and this can pose a challenge for glacier surface change detection. In Hopkinson and Demuth (2006), horizontal uncertainty on the order of 1 to 2 m in one or both lidar DEMs used in the change detection analysis led to apparent surface growth of several meters in areas immediately surrounding steep cliffs.

The active infrared imaging capability of lidar has been postulated as a useful tool for ice, firn, and snow facies discrimination (Favey et al., 2000; Hopkinson et al., 2001; Favey et al., 2002; Arnold et al., 2006) and was implemented by Hopkinson and Demuth (2006) as a means to separate the accumulation and ablation zone so that appropriate water equivalent values could be applied to the respective volumes of down wasting observed over a two year period. For the time period investigated, it was found that there was a reduction in volume totalling 33×10^6 m^3 from the Peyto glacier basin. In addition, it was possible to quantify the contribution from down wasting ice-cored moraines which was estimated at 6 percent of the total glacier basin runoff contribution (Hopkinson and Demuth, 2006). This observation is significant because glacier melt water generation from periglacial moraine environments is rarely monitored and clearly illustrates the potential for lidar-based monitoring techniques in these environments.

The high resolution of lidar DEMs over actively melting surfaces such as glaciers makes the data ideal for the mapping and tracking of hydrological features like crevasses, supraglacial melt streams, and moulins (Hopkinson et al., 2001; Arnold et al., 2006). This type of information monitored through time could provide invaluable data on the surface melt stream generation and the motion of the glacier as expressed at the surface, and thus provide validation data for glacial dynamics models. Moreover, high-resolution morphological data enables radiation loading (e.g., Chasmer and Hopkinson, 2001) and energy balance models (Hopkinson et al., 2007) to be applied over the glacier surface at a scale appropriate to the scales of melt-related hydrological processes.

In Hopkinson et al. (2007) a GIS energy balance model was applied to a lidar DEM of Peyto Glacier to investigate the influence of changing DEM resolution to predicted melt. Initial melt results performed over both a 1,000 m plot and over the entire glacier surface illustrated a noticeable increase in melt of between 2.5 percent to 4 percent across three orders of magnitude of resolution. This difference was even more marked over the ablation stake network, where an 11 percent increase was observed from the 1 m to 1,000 m DEM resolution. However, this systematic increase in melt was effectively removed at the plot and glacier scale when melt estimates were adjusted based on a slope correction to account for the difference between the GIS planar (horizontal) surface area and the actual DEM terrain surface area. Also, as resolution decreases, slope and aspect values are increasingly influenced by surrounding terrain. Therefore, as resolution decreases, apparently random behavior is introduced (i.e., variance increases) into the melt prediction. Such analysis of scaling influences to process representation in hydrological models is made possible only by the availability of high-resolution accurate DEMs. Future work of this type will focus on refining mass balance, melt, and glacier motion models over longer periods of time and using multi-temporal lidar derived DEMs for model validation.

10.8.3 Snowpack Depth

The utility of multitemporal lidar acquisitions for snowpack depth and volume mapping under various canopy conditions was demonstrated by Hopkinson et al. (2004).The study presented faced challenges due to the relatively shallow average snowpack depth of between 25 and 50 cm being little more than two to three times the typically accepted 15 cm accuracy of the lidar sensor. However, it was demonstrated that lidar DEMs of ground and snowpack surface be compared to generate a "difference" surface characteristic of realistic snowpack distribution patterns, with observed variability commensurate with topographic and canopy closure controls. The type of forest canopy and density of understory were found to play an important role in controlling the accuracy of snowpack depth estimates. For example, snow depth estimates were most accurate under conifer plantation, where there was no appreciable understory, while systematic underestimations of snowpack depth were observed in areas of dense understory. However, the depth underestimate was associated with an upwards bias in the ground surface DEM due to a lack of pulse penetration into the understory foliage (Hopkinson et al., 2004). Based on these findings, it was postulated that the utility of lidar technology for snowpack mapping would be greatest in areas prone to deep snowpack conditions, where instrument precision is less important, and in remote regions where ground access is difficult and costly. In mountainous headwater environments, therefore, lidar snowpack surveys may soon provide an economical supplement or even an alternative to traditional snow course monitoring techniques.

The concept of applying lidar to snowpack depth mapping in mountainous headwater environments for water resources assessment was studied by Fassnach and Deems (2006) and Deems et al. (2006). Both studies utilized variogram analysis to investigate the spatial scaling properties of snowpack depth distribution over various headwater environments. Part of the rationale behind these studies was to better understand snowpack depth distribution behavior so that field sampling routines could be optimized. In Deems et al. (2006) it was found that the dominant snowpack depth controlling processes could be separated into two scale ranges with a separation in process dominance between 15 m and 40 m. Based on similarity with snow pack variogram scale peaks, it was concluded that terrain and vegetation cover potentially influenced the scaling behavior of snowpack. Further, variation in the snow depth fractal dimension was qualitatively related to prevailing wind direction and large-scale topographic orientation (Deems et al., 2006). The value of lidar over more traditional remote sensing techniques is that it allows the variation in depth to be directly mapped rather than just spatial extent, and these variations can be directly related to landscape and meteorological driving mechanisms. This is important because from a water resources perspective, once snow cover is complete and providing density, it can be sampled or estimated; the variation in depth is the dominant control on basin snow-water equivalent.

10.8.4 Terrain Morphology

A number of studies have utilized lidar terrain data to investigate hill slope mass movement mechanisms and the associated hydrological interactions (Stock and Dietrich, 2003; Haneberg et al., 2005; Rosso et al. 2006; Corsini et al., 2007). For example, using lidar slope morphology information, Stock and Dietrich (2003) were able to establish that the slope limit for the steady state inverse power law relationship between channel slope and drainage area was rarely greater than 0.03 to 0.10. From this observation and supplemental field data in unglaciated valleys, they proposed that much of the morphological reworking of steep valley sides was due to debris flow events (Stock and Dietrich, 2003). Haneberg et al. (2005) utilized lidar DEM data supported by field investigations to assist with the development of a morphological hazards map near a gold mine in Papua New Guinea. They were able to delineate alluvial and debris fan complexes within the map and even observe that some of the features must be recent due to the lack of a well established drainage network and high number of low order stream channel segments (Haneberg et al., 2003). Another hazards related study by Rosso et al. (2006) explicitly investigated the hydrological control on shallow land slide processes using lidar data to provide the high-resolution slope parameter information. From their high-resolution model simulations, they were able to estimate the return period and magnitude of shallow landslide events based on local hydrological conditions (Rosso et al., 2006).

As with the discussion on snowpack mapping and glacier melt research, some valuable investigations of hillslope hydrological processes in headwater environments utilize multiple lidar datasets for the purpose of surface morphological change detection. Corsini et al. (2007) compared two airborne lidar data sets collected months apart for the purpose of post-failure landslide characteristics research. They observed that dewatering of the landslide mass led to a lowering of the surface by up to 10 m in parts. While the focus of the study was landslide hazards, it was shown that lidar could be used to map changes in surface morphology associated with the hydrological conditions of the slope.

In contrast to the above noted studies where lidar was used to study or illustrate the influence of hydrological processes to surface morphology, a number of studies have taken the opposite approach and used lidar topographic data to investigate morphological controls on hydrology (e.g., Schmidt and Persson, 2003; Lane et al., 2004; Lindsey et al., 2004; Lindsey and Creed, 2005; Tenenbaum et al., 2006; Liu et al., 2005). Three of these studies (Schmidt and Persson, 2003; Lane et al., 2004; Tenenbaum et al., 2006) utilize lidar's capability to provide high-resolution morphological information to study the topographic wetness index (TWI) of Beven and Kirby (1997). Schmidt and Persson (2003) found that while TWI can be used to assess potential soil moisture patterns, topography did not have the main control on water flow in low gradient field environments. Slightly more optimistic results were reported in Tenebaum et al. (2006) where it was shown that while TWI does show some correlation with actual soil moisture conditions, the strength of the correlation varies with the environments and the resolution of the DEM. For example, the accuracy of TWI moisture predictions calculated over an urbanizing environment were increased during wetter conditions and at high DEM resolutions of 0.5 m; while in a forested environment, lidar was unnecessary, as accurate soil moisture predictions were achievable for low DEM resolutions (10 m to 20 m) and the correlation was high under all moisture conditions (Tenebaum et al., 2006). Lidar did not necessarily provide an advantage to lower resolution DEM data sources in part due to the level of extraneous microtopographic details and "noise" at the forest floor level (Tenenbaum et al., 2006).

In general, it is known that DEM resolution impacts predicted flow routing (Quinn et al., 1997), predicted soil salinity due to evaporation (Liu et al., 2005), and the amount of grid node elevation modification necessary to remove hydrological sinks and flow barriers within the DEM drainage network (Bruneau et al., 1995; Lindsey and Creed, 2005). A major benefit of lidar for the derivation of drainage networks is that it can be used to generate DEMs at resolutions approaching and better than 1 m grid spacing and therefore pick out zero and first order drainage features. Further, it can be used to map high-resolution landscape depressions of hydrological importance even beneath forest canopies (Lindsey et al., 2004), where traditional stereo photogrammetric methods tend to be weak due to canopy shadowing. In general, TWI and surface flow pathways predicted from high resolution DEMs tend to be more accurate than low resolution alternatives, but regardless of the DEM resolution, drainage network connectivity can be sensitive to a small number of critically located grid cells. For this reason, the propagation of elevational uncertainty within the DEM should be assessed to mitigate against the disproportionate influence of a few grid cells to the overall drainage network (Lane et al., 2004).

10.8.5 Ground Surface Elevation

As already stated, the merits of lidar for hydrological process research are predominantly due to the ability to map surface morphology at a high resolution and under a wide range of land surface conditions. However, the accuracy of lidar elevation data, and therefore hydrological features or indices associated with surface morphology, is influenced by data acquisition parameters (Hopkinson and Chasmer, 2007), terrain slope (Hodgson and Bresnahan, 2004), and vegetation cover (Bowen and Waltermire, 2002; Töyrä et al., 2003; Hopkinson et al., 2005).

For a well-calibrated lidar sensor operating at typical altitudes (1,000 m to 2,000 m above ground) and scan angles (15 to 20 degrees), it is common to achieve root mean square errors (RMSE) in the elevation data below 10 cm (e.g., Hopkinson et al., 2005; Hopkinson and Demuth, 2006). Hodgson and Bresnahan (2004) observed RMSE values ranging from 17 to 26 cm for paved and forested areas, with the magnitude of error being twice as large over the steeper slopes sampled (e.g., 25 degrees). A number of studies have observed that over vegetated surfaces, there tends to be an upwards bias in lidar elevation data and an increase in the level of noise in the data. Ground height biases up to 0.2 m have been observed for wetland and riparian vegetation covers (Bowen and Waltermire, 2002), and these biases have been found to vary with the species and structure of vegetation (Töyra et al., 2003; Hodgson and Brenahan, 2004; Hopkinson et al., 2005). In Hopkinson et al. (2005) the vertical ground elevation offset computed relative to GPS field validation points varied from no significant difference for grass and herbs to +15 cm for aquatic vegetation due to weak laser backscatter from the saturated soil and open water conditions at the base of the stems. Beneath forested canopies the error in lidar ground elevations typically increases to 30 cm and above due to the combination of a weakened laser signal and forest undergrowth (Reutebuch et al., 2003).

10.8.6 Vegetation Height

Vegetation height is one of the parameters controlling aerodynamic and hydrodynamic resistance to wind and runoff across the landscape, respectively. A valuable feature of lidar data within the hydrological sciences, therefore, is the ability to sample both the ground and the vegetation overlying the ground surface simultaneously through the ability to record multiple reflections from a single emitted pulse. This capability enables the spatial variability of canopy height to be mapped. Many studies have investigated the use of lidar for tree height measurement and found good relationships to field measures with R2 values typically ranging from 0.85 to 0.95 (Maclean and Krabill, 1986; Ritchie, 1995; Naesset, 1997; Magnussen and Boudewyn, 1998; Means et al., 2000; Witte et al., 2001; Naesset, 2002; Naesset and Okland, 2002; Popescu et al., 2002; Lim et al., 2003a). While there is an abundance of literature demonstrating various lidar-based canopy height research for commercial forestry or ecological applications (see Lim et al., 2003b), few studies have put this into a hydrological processes or water resources context.

Roughness length calculations for evaporative loss prediction from profiling (Menenti and Ritchie, 1994) and scanning lidar (Ritchie et al., 2001) estimates of vegetation height have been shown to agree well with field measurements over relatively arid grass and shrub-land areas. In a similar study conducted by Weltz et al. (1994), some underestimation of canopy height was noted: a result common to many scanning lidar studies (e.g., Magnussen et al., 1999; Gaveau and Hill, 2003). Underestimating canopy height is typically attributed to: (a) laser pulse penetration into the foliage (Gaveau and Hill, 2003; Hopkinson and Chasmer, 2007), (b) insufficient representation of canopy apices due to low sample point density (St-Onge et al., 2003), or (c) ground height overestimation due to minimal pulse penetration through dense vegetation (e.g., Weltz et al., 1994; Reutebuch et al., 2003). Hopkinson et al. (2005) found pulse penetration into various foliage types within a boreal wetland environment varied from approximately 10 cm for shrub vegetation up to over 80 cm for marsh vegetation. The reason for the different levels of penetration was believed due to the different foliage orientation and projected surface area of the foliage (Hopkinson et al., 2005).

Few studies have investigated the estimation of short (near ground surface) vegetation height. Hopkinson et al. (2005) found that comparing localized lidar elevation minima and maxima for boreal wetland vegetation shorter than 2 m in height did provide a reasonable correlation with field measured vegetation heights despite the observed penetration into foliage. However, all heights were significantly underestimated. The work of Davenport et al. (2000) and Cobby et al. (2001) demonstrated that crop vegetation up to approximately 1.2 m in height could be predicted from the standard deviation of topographically detrended laser pulse returns, thus negating the need to assume that lidar is accurately capturing the upper vegetation surface. This work was expanded upon by Hopkinson et al. (2006) to develop a robust method of estimating all types of vegetation canopy height regardless of the species, height, or data acquisition configuration. It was found that multiplying by 2.5 the height standard deviation of lidar points sampled from an area of almost any vegetation cover provides a first approximation of the local canopy height (Hopkinson et al., 2006).

10.8.7 Canopy Structure

Perhaps the most commonly referred to component of vegetation structure is leaf area index (LAI), which is defined as one half the total leaf area per unit ground surface area (m2 m-2) (Chen et al., 2006). LAI and canopy transmittance (T) are key input parameters in many ecological and hydrological models as they enable the prediction of energy transmission through the canopy to lower layers of biomass or to ground level (e.g., Pomeroy and Dion, 1996). This information is essential in growth (e.g., photosynthesis) and hydrological (e.g., melt and evaporation) process modeling in forested environments. Accurate and consistent LAI measurements are often labor intensive and may also be difficult to collect in remote or difficult to access areas.

A number of studies have examined the use of lidar for obtaining LAI, gap fraction (P), the fraction of incoming photosynthetically active radiation absorbed by the canopy (FIPAR) and extinction coefficients (k) from lidar (e.g., Magnussen and Boudewyn, 1998; Parker et al. 2001; Todd et al. 2003; Morsdorf et al. 2006; Thomas et al. 2006). However, most lidar estimates of LAI require some modification, as lidar cannot directly differentiate between green foliage and woody material, and thus far it is yet to be demonstrated that lidar can accurately map the spatial variability in canopy clumping.

Laser pulses that are returned from within the canopy have intercepted enough foliage or branch material to be recorded by the receiving optics within the lidar system, while some of the remaining laser pulse energy continues until it intercepts lower canopy vegetation, the low-lying understory, and the ground surface. Laser pulse returns from the ground surface have inevitably passed through canopy gaps both into and out of the canopy. Increasing numbers of gaps within the canopy will result in gap fractions approaching 100 percent, whereas fewer gaps within the canopy will result in a gap fraction closer to zero. The main geometric difference between the canopy interaction of solar and airborne lidar laser pulse radiation is that solar radiation is incident at all zenith angles while laser pulses are typically incident only at overhead ($\theta = 0$ to 30 degrees) angles. Therefore, any direct lidar estimate of P will be for overhead gap fraction only and for a path length close to the height of the canopy. However, using the Beer-Lambert Law of radiative transfer and by assuming randomly dispersed foliage elements and an isotropic canopy radiation environment (i.e., equal transmittance in all directions), it is possible to derive a first approximation of LAI as a function of the overhead gap fraction:

$$LAI = -\frac{Ln(P)}{k} \tag{10.19}$$

Several studies have used this or a similar approach to estimate LAI from lidar data. In particular, Solberg et al. (2006) used this approach and assumed that P could be approximated by the ratio of below canopy returns to total returns. A similar but simpler approach was taken by Barilotti et al. (2006) where the same ratio was found to linearly correlate with LAI.

The assumptions of the two previous studies were corroborated by Riaño et al. (2004) and Morsdorf et al. (2006) where the ratio of lidar canopy returns to all returns was found to be a reasonable indicator of the inverse of gap fraction; i.e., fractional canopy cover (and even LAI as observed by Magnussen and Boudewyn (1998)). Morsdorf et al. (2006) compared canopy lidar fractional cover estimates with field-based digital hemispherical photography (DHP) fractional cover and found the best correlation was returned when using first return data only ($R^2 = 0.73$). A method for estimating LAI that utilized laser profiling techniques was presented by Kusakabe et al. (2000), where field plot measures of L were compared to the cross-sectional area contained within the lidar surface profile across the plots. The rationale underlying this approach was that L would increase with tree height and stem density, and both of these physical attributes would act to increase the cross sectional area of a lidar profile across a plot.

Laser pulse intensity has implicitly been used in estimates of canopy gap fraction in the full waveform lidar literature where the strength of the returned signal from within or below the canopy is considered to be directly related to the transmissivity of the canopy. For example, in Lefsky et al. (1999), it was suggested that canopy fractional cover can be estimated as a function of the ratio of the power reflected from the ground surface divided by the total returned power of the entire waveform. An early attempt at directly quantifying the overhead transmissivity from discrete return laser pulse intensity without any optimization has been presented in Hopkinson and Chasmer (2007). In this study the Beer-Lambert law was applied to intensity transmission losses for individual returns to recreate a intensity power distribution profile throughout the canopy. The ratio of the integrated ground power distribution relative to the total returned intensity was found to have a 1:1 relationship ($R^2 = 0.86$) with DHP overhead gap fraction (Hopkinson and Chasmer, 2007).

10.8.8 Wetland Environments

The properties of a wetland can be classified according to its hydrological interaction with the surrounding landscape, the vegetation species and distribution, and its soil composition (Warner and Rubec, 1997). There have been a small number of attempts to use lidar data to characterize the hydrological interaction and/or the vegetation distribution within various wetland environments. For example, Genc et al. (2004) demonstrated that lidar-derived vegetation height information collected over wetland environment in Florida could be used to classify vegetation communities of tall trees, medium trees, low trees, tall plants, medium plants, low plants, and herb and shrubs. In a study conducted in a forest-covered distributed wetland landscape in central Ontario, Creed et al. (2003) found lidar data to be superior to three other independent DEM sources for the purpose of delineating wetland locations and area coverage. The lidar derived sub canopy wetland coverages were used to drive the estimation of dissolved organic carbon (DOC) export to surrounding lakes and streams. It was found that these sub-canopy wetlands played an important role in modifying the annual variability in DOC export (Creed et al., 2003).

The interaction of terrain-controlled water levels and vegetation properties has been demonstrated in a variety of wetland types from northern Canada (Toyra and Pietroniro, 2005) to coastal regions of the US (Morris et al., 2005; Rosso et al., 2006) and the UK (Blott and Pye, 2004). In the relatively flat Peace Athabasca Delta in northern Canada, the elevational zonation in shrub and graminoid vegetation was studied by comparing a SPOT-4 map of vegetation cover with a lidar DEM (Toyra and Pietroniro, 2005). It was found that the vertical separation of shrubs and graminoids varied from 0.5 m at the south end to 0.73 m at the north end of the delta, suggesting that flood duration and water level are main controls on the vegetation distribution within this environment. An analogous assessment within a tidal salt marsh environment in South Carolina (Morris et al., 2005) cross-referenced a thematic map of Spartina alterniflora habitat and *Juncus roemerianus* marsh with a lidar DEM to study the influence of the tide levels on the frequency distribution of vegetation type.

Two more coastal wetland studies utilized lidar DEM data to monitor (Rosso et al., 2006) and predict (Blott and Pye, 2004) morphological and associated vegetation changes. By comparing two lidar DEMs collected approximately two years apart, Rosso et al. (2006) found that lidar was able to easily discriminate between invasive *Spartina alterniflora* and marsh wetland species, and the *Spartina* vegetation had a marked influence on the patterns of shoreline accretion and erosion. Blott and Pye (2004) were concerned with the reverse situation of the influence of erosion and accretion properties to habitat development. Using a lidar DEM collected along part of the Essex shoreline and local habitat conditions from analogue sites, simulations of tidal inundation behind a breached sea wall were used to predict sediment accumulation and long-term sustainable salt marsh habitat.

10.8.9 Fluvial and Coastal Geomorphology

There is substantial overlap in the literature dealing with lidar research applications within the wetland and geomorphological research communities. This is not surprising given the obvious interaction and feedback between erosion and accretion, and vegetation, particularly within coastal environments. For example, the papers by Lohani and Mason (2001) and Feola et al. (2005) used lidar to investigate tidal zone geomorphology and wetland patterns, and would probably be equally at home in the discussion above. Additionally, French (2003) studied the influence of subtle topographic variations on estuarine tidal flood compartments within marsh environments. However, they are mentioned here as they mostly explore the use of lidar to characterize the coastal morphological landscape rather than the controls of wetland distribution or processes.

Lohani and Mason (2001) presented one of the first examples of a successful semi-automated approach to extract the tidal channel network from airborne lidar data. Their methodology utilized a height thresholding technique to locate channel fragments within several individual sub-sections of the lidar DEM. After extracting the tidal channel network, they were able to compute watershed areas and drainage density, and then compare these observations with data from fluvial river systems. They found that tidal channel networks did not possess the same scaling behavior as terrestrial fluvial drainage networks. A follow up study by Lohani et al. (2007) demonstrated a data fusion technique utilizing lidar and multispectral imagery for the same objective of tidal channel delineation. They found that the data fusion technique provided only a slightly improved channel network map to lidar alone, and stressed the challenges of developing automated channel network extraction routines.

While investigating the morphodynamic evolution of wetlands, lagoons, and estuarine areas, Feola et al. (2005) also studied scaling properties of a tidal drainage network and found similar results to Lohani and Mason (2001). In particular they reported that tidal networks display site specific characteristics and do not conform to the typical fluvial network power laws for probability density functions of watershed area, upstream length to the divide, and unchanneled length to the network. It was concluded that this lack of scalability within tidal networks is due to competing processes operating at similar scales (Feola et al., 2005). Another tidal network scaling study investigated habitat restoration options by extracting channel and intervening island marsh geometries and examining the interrelationships between them (Hood, 2007). It was found that due to an observed disproportionate scaling of the relationship between marsh island area and the length/area properties of tidal channels, the potential channel habitat for juvenile salmon would be maximized by restoring a single large area of marsh as opposed to several smaller areas (Hood, 2007).

Within the terrestrial fluvial context, several studies have utilized lidar data to investigate both contemporary and historical river channel geomorphological processes (Charlton et al., 2003; French, 2003; Challis, 2006; Lane et al., 2003; Magirl et al., 2005, Thoma et al., 2005). Charlton et al. (2003) presented an essentially "proof of concept" study on the use of lidar for the accurate mapping of gravel-bed river environments. They found that while lidar provided excellent geomorphic detail within extended reaches of river channel systems, caution had to be exercised within the UK context when interpreting bed form, channel, and flood plain profiles due to the presence of vegetation and deep water (Charlton et al., 2003). By comparing lidar and photogrammetric DEMs collected over a year apart, Lane et al., (2003) were able to assess erosion and deposition volumes within the gravel bed braided system of the Waimakariri River in New Zealand. After applying an innovative error propagation routine to the DEMs to assess the sensitivity of the change detection analysis, they found that while point measurement accuracy was lower than traditional field cross section surveys, the spatial coverage of the DEMs enabled more accurate and reliable estimates of deposition and erosion volumes (Lane et al., 2003). Similarly, two lidar data sets collected one year apart over a 56 km stretch of the Blue Earth River in Minnesota were compared to quantify river bank erosion to estimate the sediment and phosphorous load within river runoff (Thoma et al., 2005).

The theme of river channel geomorphic change detection using lidar in combination with other data sources is continued by Magirl et al. (2003) where they studied the water surface profile of the Colorado River in the Grand Canyon. USGS traditional leveling survey data from 1923 were obtained and compared with lidar point elevations collected in 2000 and extracted for the same locations by using tie point data from areas that were deemed not to have changed within the 77 year period. It was found that for 80 of the 91 rapids along the river reach examined there was an average aggradation of alluvium within the rapids of 26 cm in height, with an enhanced pool-and-rapid morphology. The increase in height of the water surface profile and increased pool-and-rapid morphology was thought likely due to a more regulated flood regime in the latter half of the period studied (Magirl et al., 2005).

Studies of ancient, as opposed to recent or contemporary, water-related geomorphic features by analyzing subtle spatial variations in lidar DEM data have been presented by Challis (2006), and Kovanen and Slaymaker (2004). Using data collected by the UK Environment Agency over various river floodplains in England, Challis (2006) assessed the viability of lidar DEMs for the classification of peat ground covers into areas of high and low geoarchaeological potential. The criteria were that relict river channels offer high potential for archaeological artifacts, while areas of past river erosion do not. It was found that in mature, middle reach floodplain environments, it was indeed possible to map ancient river channel locations due to the subtle variations of surface expression of desiccating and shrinking peat land surface overlying the channels (Challis, 2006). Kovanen and Slaymaker (2004) used lidar DEM data to investigate a different expression of ancient hydrological processes; namely Pleistocene glacial ice flow patterns and the locations of associated relict shorelines and deltas prior to the isostatic, eustatic, and tectonic shifts that have resulted in the contemporary shoreline position. While the study produced no firm conclusions regarding the processes leading to these features, it did demonstrate the efficacy of lidar as a highly useful source of data for raised post glacial sea level indicators.

10.8.10 Flood and Runoff Modeling

Given the high elevational precision and spatial resolution achievable with lidar DEM data, one of the most obvious applications of the technology in the hydrological sciences is the evaluation or prediction of water levels and floods. Specific applications range from direct measurement of water surface levels (Carter et al., 2001), delineation of the ordinary high water line around lakes (Genc et al., 2005), mapping coastal storm surge risk (Webster et al., 2004; Brown et al., 2007), flood inundation visualization (Neelz et al., 2006), urban flood modeling (Mason et al., 2007), highway and forest road runoff (Hans et al., 2003; Wemple et al., 2003), channel cross sections for hydraulic modeling (Kresch et al., 2002; Jones and Fulford, 2002), land surface imperviousness (Hodgson et al., 2003), and floodplain friction parameterization (Marks and Bates, 2000; Cobby et al., 2003; Davenport et al., 2003; Mason et al., 2003). From an operational prediction perspective, the cost effectiveness of airborne lidar for flood inundation mapping over traditional manually intensive field techniques was demonstrated by Holden (1998). An overview of the use of lidar for flood inundation modeling has been provided in Bates (2004).

10.8.11 Conclusions

From the cross section of materials presented, it is clear that airborne laser scanning data provides more for the hydrological or water resources researcher than elevation data. Lidar data can be used to quantify at a high resolution both the terrain morphology and three dimensional structure of features overlying the land surface that play a critical role in modifying water volumes through melt and evaporation processes, and directly control runoff flow direction. Further, lidar offers the potential to validate hydrological models through water level and soil saturation mapping. There is some justification, therefore, for suggesting that lidar research is opening up a new era in the field of hydrological sciences that more closely marries geomatics technologies to the extraction of key hydrological model input parameters and validation data. Further, while the papers cited in this sub-chapter are drawn from a broad spectrum of academic literature they can all generally be associated with either one of the hydrological or geomatics disciplines. In recognition of this intersection or symbiosis of these two major disciplines, it is postulated that research, teaching and water resources-related operations falling within this sphere of activity could be referred to as the field of "hydrogeomatics."

ACKNOWLEDGMENTS

Laura Chasmer is acknowledged for her assistance in collating the articles cited in this sub-chapter.

A version of this manuscript was originally published in the *Hydroscan Proceedings: Airborne Laser Scanning of Hydrological Features and Resources*, Canadian Water Resources Association, 2008.

REFERENCES

Abdalati, W., W. Krabill, E. Frederick, S. Manizade, C. Martin, J. Sonntag, R. Swift, R. Thomas, J. Yungel, and R. Koerner, 2004. Elevation changes of ice caps in the Canadian Arctic Archipelago, *Journal of Geophysical Research*, Vol. 109.

Arnold, N.S., W.G. Rees, B.J. Devereux, and G.S. Amable, 2006, Evaluating the potential of high-resolution airborne LiDAR data in glaciology, *International Journal of Remote Sensing*, 27:1233–1251.

Baltsavias, E.P., E. Favey, A. Bauder, H. Boesch, and M. Pateraki, 2001. Digital surface modelling by airborne laser scanning and digital photogrammetry for glacier monitoring, *The Photogrammetric Record*, 17(98):243–273.

Baltsavias, E.P. 1999. Airborne laser scanning: Existing systems and firms and other resources, *ISPRS Journal of Photogrammetry and Remote Sensing*, 54:164–198.

Barilotti, A., S. Turco, and G. Alberti, 2006. LAI determination in forestry ecosystems by LiDAR data analysis, *Proceedings of the Workshop on 3D Remote Sensing in Forestry*, 14–15 February, BOKU, Vienna.

Bates, P.D., 2004, Remote sensing and flood inundation modelling, *Hydrological Processes*, 18(13):2593–2597.

Beven, K.J., and M.J. Kirkby, 1979. A physically-based variable contributing area model of basin hydrology, *Hydrological Science Bulletin*, 24(1):43–69.

Blott, S.J., and K. Pye, 2004. Application of lidar digital terrain modelling to predict intertidal habitat development at a managed retreat site: Abbotts Hall, Essex, UK, *Earth Surface Processes and Landforms*, 29:893–903

Bowen, Z.H., and R.G. Waltermire, 2002. Evaluation of light detection and ranging (LIDAR) for measuring river corridor topography, *Journal of the American Water Resources Association*, 38:33–41.

Brown, J.D., T. Spencer, and I. Moeller, 2007. Modelling storm surge flooding of an urban area with particular reference to modelling uncertainties: A case study of Canvey Island, United Kingdom, *Water Resources Research*, 43, W06402.

Bruneau, P., C. Gascuel-Odoux, P. Robin, Ph. Merot, and K. Beven, 1995. The sensitivity to space and time resolution of a hydrological model using digital elevation data, *Hydrological Processes*, 9(1):69–81.

Carter W., R. Shrestha, G. Tuell, D. Bloomquist, and M. Sartori, 2001. Airborne laser swath mapping: Shines new light on earth topography, *EOS Transactions*, 82(46):549–550, 555.

Challis, K., 2006. Airborne laser altimetry in alluviated landscapes, *Archaeological Prospection*,13:103–127.

Charlton, M.E., A.R.G. Large, and I.C. Fuller, 2003. Application of airborne LIDAR in river environments: The River Coquet, Northumberland, UK, *Earth Surface Processes and Landforms*, 28(3):299–306.

Chasmer, L., and C. Hopkinson, 2001, Using airborne laser altimetry and GIS to assess scale induced radiation-loading errors in a glacierized basin, *Proceedings of the 58th Eastern Snow Conference*, 17–19 May, Ottawa, Canada, pp. 195–205.

Chen, J.M., A. Govind, O. Sonnentag, Y. Zhang, A. Barr, and B. Amiro, 2006. Leaf area index measurements at Fluxnet-Canada forest sites, *Agricultural Forest Meteorology*, 140:257–268.

Cobby, D.M., D.C. Mason, and I.J. Davenport, I.J. 2001. Image processing of airborne scanning laser altimetry data for improved river flood modeling, *ISPRS Journal of Photogrammetry and Remote Sensing*, Vol. 56, pp. 121–138.

Cobby, D.M., D.C. Mason, M.S. Horritt, and P.D. Bates, 2003. Two-dimensional hydraulic flood modelling using a finite-element mesh decomposed according to vegetation and topographic features derived from airborne scanning laser altimetry, *Hydrological Processes*, 17(10):1979–2000.

Corsini, A., L. Borgatti, F. Coren, and M. Vellico, 2007. Use of multitemporal airborne lidar surveys to analyse post failure behaviour of earth slides, *Canadian Journal of Remote Sensing*. 33(2):116–120.

Creed, F., S.E. Sanford, F.D. Beall, L.A. Molot, and P.J. Dillon, 2003. Cryptic wetlands: Integrating hidden wetlands into models of dissolved organic carbon export, *Hydrological Processes*, 17(18):3629–3648.

Csatho, B., Y. Ahn, T. Yoon, C.J. van der Veen, S. Vogel, G. Hamilton, D. Morse, B. Smith, and V.B. Spikes, 2005. ICESat measurements reveal complex pattern of elevation changes on Siple Coast ice streams, Antarctica, *Geophysical Research Letters*, Vol. 32, L23S10.

Davenport, I.J., R.B. Bradbury, G.Q.A. Anderson, G.R.F. Hayman, J.R. Krebs, D.C. Mason, J.D. Wilson, and N.J. Veck, 2000. Improving bird population models using airborne remote sensing, *International Journal of Remote Sensing*, 21:2705–2717.

Davenport, K., G. Black, T. Burnell, S. Cross, S. Culloty, B. Ekaratne, M. Furness, M. Mulcahy, and H. Thetmeyer, 2003. *Aquaculture: The Ecological Issues*, Wiley Blackwell Science, Ltd., Hoboken, New Jersey, 96 p.

Deems, J.S., S.R. Fassnacht, and K.J. Elder, 2006. Fractal distribution of snow depth from LiDAR data, *Journal of Hydrometeorology*, 7:285–297.

Echelmeyer, K.A., W.D. Harrison, C.F. Larsen, J. Sapiano, J.E. Mitchell, J. Demallie, B. Rabus, G. Adalgeirsdottir, and L. Sombardier, 1996, Airborne surface profiling of glaciers: A case study in Alaska, *Journal of Glaciology*, 42:538–547.

Fassnacht, S.R., and J.S. Deems, 2006. Measurement sampling and scaling for deep montane snow depth data, *Hydrological Processes*, 20(4):829–838.

Favey, E., A. Wehr, A. Geiger, and H.G. Kahle, 2002. Some examples of European activities in airborne laser techniques and an application in glaciology, *Journal of Geodynamics*, 34:347–355

Favey, E., M. Cerniar, M. Cocard, and M. Geiger, 1999. Sensor attitude determination using GPS antenna array and INS, *Proceedings of ISPRS WG III/1 Workshop: Direct Versus Indirect Methods of Sensor Orientation*, 25–26 November, Barcelona, Spain, pp. 28–38.

Favey, E., M. Pateraki, E.P. Baltsavias, A. Bauder, and H. Bösch, 2000. Surface modelling for alpine glacier monitoring by airborne laser scanning and digital photogrammetry, *ISPRS Archives of Photogrammetry and Remote Sensing*, XIXth ISPRS Congress, Amsterdam, XXXIII(B4):269–277.

Feola, A., E. Belluco, A.D'Alpaos, S. Lanzoni, M. Marani, and A. Rinaldo, 2005. A geomorphic study of lagoonal landforms, *Water Resources Research*, 41, W06019.

French, J.R., 2003. Airborne LiDAR in support of geomorphological and hydraulic modeling, *Earth Surface Processes and Landforms*, 28:321–335.

Gaveau, D., and R.A. Hill, 2003. Quantifying canopy height underestimation by laser pulse penetration in small footprint airborne laser scanning data, *Canadian Journal of Remote Sensing*, 29:650–657.

Geist, T., and J. Stotter, 2002. First results on airborne laser scanning technology as a tool for the quantification of glacier mass balance, *Proceedings of EARSeL-LISSIG-Workshop Observing our Cryosphere from Space*, 11-13 March, Bern.

Genc, L., B. Dewitt, and S. Smith, 2004. Determination of wetland vegetation height with LIDAR, *Turkish Journal of Agriculture and Forestry*, 28:63–71

Genc, L., S.E. Smith, and B.A. Dewitt, 2005. Using satellite imagery and lidar data to corroborate an adjudicated ordinary high water line, *International Journal of Remote Sensing*, 26(17):3683–3693.

Haneberg, W.C., A.L. Creighton, E.W. Medley, and D.A. Jonas, 2005, Use of LiDAR to assess slope hazards at the Lihir gold mine, Papua New Guinea, Landslide Risk Management: *Proceedings of International Conference on Landslide Risk Management* (O. Hungr, R. Fell, R. Couture, and E. Eberhardt, editors), 31 May–3 June, Vancouver, Canada (Joint Technical Committee on Landslides and Engineered Slopes & the Vancouver Geotechnical Society), unpaginated CD-ROM.

Hans, Z., R. Tenges, S. Hallmark, R. Souleyrette, and S. Pattnaik, 2003. Use of LiDAR-based elevation data for highway drainage analysis: A qualitative assessment, *Proceedings of the Mid-Continent Transportation Research Symposium*, 21–23 August, Iowa State University, URL: http://www.ctre.iastate.edu/pubs/midcon2003/ (last date accessed: 16 October 2011).

Hodgson, M.E., and P. Bresnahan, 2004. Accuracy of airborne lidar-derived elevation: Empirical assessment and error budget, *Photogrammetric Engineering & Remote Sensing*, 70(3):331–340.

Hodgson, M.E., J.R. Jensen, G.T. Raber, J.A. Tullis, B. Davis, K. Schuckman, and G. Thompson, 2005. An evaluation of lidar-derived elevation and terrain slope in leaf-off conditions, *Photogrammetric Engineering & Remote Sensing*, 71(12):817–823.

Hodgson, M.E., J.R. Jensen, L. Schmidt, S. Schill, and B. Davis, 2003. An evaluation of lidar- and IFSAR-derived digital elevation models in leaf-on conditions with USGS Level 1 and Level 2 DEMs, *Remote Sensing of Environment*, 84(2):295–308.

Hood, G., 2007. Scaling tidal channel geometry with marsh island area: A tool for habitat restoration, linked to channel formation process, *Water Resources Research*, 43, W03409.

Hopkinson, C. 2007. The influence of flying altitude and beam divergence on canopy penetration and laser pulse return distribution characteristics, *Canadian Journal of Remote Sensing*, 33(4):312–324.

Hopkinson, C. and L.E. Chasmer, 2007. Using discrete laser pulse return intensity to model canopy transmittance, *Photogrammetric Journal of Finland*, 20(2):16–26.

Hopkinson, C., L. Chasmer, K. Lim, P. Treitz, and I. Creed, 2006. Towards a universal LiDAR canopy height indicator, *Canadian Journal of Remote Sensing*, 32(2):139–153.

Hopkinson, C., L.E. Chasmer, G. Zsigovics, I. Creed, M. Sitar, W. Kalbfleisch, and P. Treitz, 2005. Vegetation class dependent errors in LiDAR ground elevation and canopy height estimates in a Boreal wetland environment, *Canadian Journal of Remote Sensing*, 31(2):191–206.

Hopkinson, C., and M.N. Demuth, 2006. Using airborne lidar to assess the influence of glacier downwasting to water resources in the Canadian Rocky Mountains, *Canadian Journal of Remote Sensing*, 32(2):212–222.

Hopkinson, C., M. Demuth, M. Sitar, and L. Chasmer, 2001. Applications of LiDAR mapping in a glacierised mountainous terrain, *Proceedings of the International Geoscience and Remote Sensing Symposium*, 9–14 July, University of New South Wales, Sydney, Australia (IEEE).

Hopkinson, C., M. Sitar, L.E. Chasmer, and P. Treitz, 2004. Mapping snowpack depth beneath forest canopies using airborne lidar, *Photogrammetric Engineering & Remote Sensing*, 70(3):323–330

Jones J.L., and J.M. Fulford, 2002. Near-real time flood modeling and mapping for the internet, *Proceedings of the Second Interagency Hydrologic Modeling Conference*, Las Vegas, Nevada (Subcommittee on Hydrology of the Interagency Advisory Committee on Water Information), unpaginated CD-ROM.

Kennet, M., and T. Eiken, 1997, Airborne measurement of glacier surface elevation by scanning laser altimeter, *Annals of Glaciology*, 24:235–238.

Kovanen, D.J., and O. Slaymaker, 2004. Relict shorelines and ice-flow patterns of the northern Puget Lowland from lidar data and digital terrain modeling, *Geograsika Annaler: Series A, Physical Geography 86A*, pp. 385–400.

Krabill, W.B. R. Thomas, K. Jezek, K. Kuivinen, and S. Manizade, 1995, Greenland ice sheet thickness changes measured by laser altimetry, *Geophysical Research Letters*, 22:2341–2344.

Kresch, D.L., M.C. Mastin, and T.D. Olsen, 2002. Fifty-year flood-inundation maps for Olanchito, Honduras. *US Geological Survey, Tacoma, Washington, Series A, Physical Geography*, 86A:385–400.

Kusakabe, T., H. Tsuzuki, G. Hughes, and T. Sweda, 2000. Extensive forest leaf area survey aiming at detection of vegetation change in subarctic-boreal zone, *Polar Bioscience*, 13:133–146.

Lane, S.N., C.J. Brookes, J. Holden, and M.J. Kirkby, 2004. A network index based version of TOPMODEL for use with high resolution digital topographic data, *Hydrological Processes*, 18:191–201.

Lane, S.N., and J.H. Chandler, 2003. Editorial: The generation of high quality topographic data for hydrology and geomorphology: New data sources, new applications and new problems, *Earth Surface Processes and Landforms*, 28(3):229–230.

Lefsky, M.A., W.B. Cohen, S.A. Acker, G.G. Parker, T.A. Spies, and D. Harding, 1999. Lidar remote sensing of the canopy structure and biophysical properties of Douglas-Fir - Western Hemlock forests, *Remote Sensing of Environment*, 70:339–361.

Lim, K., P. Treitz, M. Wulder, B. St-Onge, and M. Flood, 2003b, Lidar remote sensing of forest structure, *Progress in Physical Geography*, 27:88–106.

Lim, K., P. Treitz, K. Baldwin, I. Morrison, and J. Green, 2003a. Lidar remote sensing of biophysical properties of tolerant northern hardwood forests, *Canadian Journal of Remote Sensing*, 29:658–678.

Lindsey, J., and I. Creed, 2005. Sensitivity of digital landscapes to artifact depressions in remotely-sensed DEMs, *Photogrammetric Engineering & Remote Sensing*, 71(9):1029– 1036.

Lindsey, J.B., I.F. Creed, and F.D. Beall, 2004. Drainage basin morphometrics for depressional Landscapes, *Water Resources Research*, 40, W09307.

Liu, X., J. Peterson, Z. Zhang, and S. Chandra, 2005 Improving soil salinity prediction with high resolution DEM derived from LiDAR data, *Proceedings of the 9th International Symposium on Physical Measurements and Signatures in Remote Sensing*, 17–19 October, Bejing, China (International Society for Photogrammetry and Remote Sensing, Technical Commission VII).

Lohani, B., and D. Mason, 2001. Application of airborne scanning laser altimetry to the study of tidal channel geomorphology, *ISPRS Journal of Photogrammetry and Remote Sensing*, 56:100–120.

Lohani, B., D.C. Mason, T.R. Scott, and B. Sreenivas, 2007. Extraction of tidal channel networks from aerial photographs alone and combined with laser altimetry, *International Journal of Remote Sensing*, 27(1):5–25.

Maclean, G.A., and W.B. Krabill, 1986. Gross-merchantable timber volume estimation using an airborne lidar system, *Canadian Journal of Remote Sensing*, 12:7–18.

Magirl, C.S., R.H. Webb, and P.G. Griffiths, 2005. Changes in the water surface profile of the Colorado River in Grand Canyon, Arizona, between 1923 and 2000, *Water Resources Research*, 41, W05021.

Magnussen, S., and P. Boudewyn, 1998. Derivations of stand heights from airborne laser scanner data with canopy-based quantile estimators, *Canadian Journal of Forest Research*, 28:1016–1031.

Magnussen, S., P. Eggermont, and V.N. LaRiccia, 1999. Recovering tree heights from airborne laser scanner data, *Forest Science*, 45:407–422.

Marks, K., and P. Bates, 2000. Integration of high-resolution topographic data with floodplain flow models, *Hydrological Processes*, 14(11–12):2109–2122.

Mason, D.C., D.M. Cobby, M.S. Horritt, and P.D. Bates, 2003. Floodplain friction parameterization in two-dimensional river flood models using vegetation heights derived from airborne scanning laser altimetry, *Hydrological Processes*, 17:1711–1732.

Mason, D.C., M.S. Horritt, N.M. Hunter, and P. Bates, 2007. Use of fused laser altimetry and digital map data for urban flood modeling, *Hydrological Processes*, 21(11):1436–1447.

Maune, F.D. (editor), 2001. *Digital Elevation Model Technologies and Applications: The DEM Users Manual*, American Society for Photogrammetry and Remote Sensing, Bethesda, Maryland, 539 p.

Means, J.E., S.A. Acker, B.J. Fitt, M. Renslow, L. Emerson, and C.J. Hendrix, 2000. Predicting forest stand characteristics with airborne scanning lidar, *Photogrammetric Engineering & Remote Sensing*, 66(11):1367–1371.

Menenti, M., and J.C. Ritchie, 1994. Estimation of effective aerodynamic roughness of Walnut Gulch watershed with laser altimeter measurements, *Water Resources Research*, 30(5):1329–1337.

Morris, J.T., D. Porter, M. Neet, P.A. Noble, L. Schmidt, L.A. Lapine, and J.R. Jensen. 2005. Integrating lidar elevation data, multi-spectral imagery and neural network modeling for marsh characterization, *International Journal of Remote Sensing*, 26(23): 5221–5234.

Morsdorf, F., B. Kotz, E. Meier, K.I. Itten, and B. Allgower. 2006. Estimation of LAI and fractional cover from small footprint airborne laser scanning data based on gap fraction, *Remote Sensing of Environment*, 104(1):50–61.

Næsset, E., 1997. Determination of mean tree height of forest stands using airborne laser scanner Data, *ISPRS Journal of Photogrammetry and Remote Sensing*, 52:49–56.

Næsset, E., 2002. Predicting forest stand characteristics with airborne scanning laser using a practical two-stage procedure and field data, *Remote Sensing of Environment*, 80: 88–99.

Næsset, E., and T. Økland, 2002. Estimating tree height and tree crown properties using airborne scanning laser in a boreal nature reserve, *Remote Sensing of Environment*, 79:105–115.

Neelz, S., G. Pender, I. Villanueva, M. Wilson, N.G. Wright, P.D. Bates, D. Mason, and C. Whitlow, 2006. Using remotely sensed data to support flood modelling, *Proceedings of the Institution of Civil Engineers, Water Management*, 159(1):35–43.

Parker, G.G., M.A. Lefsky, and D.J. Harding, 2001. Light transmittance in forest canopies determined using airborne laser altimetry and in-canopy quantum measurements, *Remote Sensing of Environment*, 76:298–309.

Pomeroy, J.W., and K. Dion, 1996. Winter radiation extinction and reflection in a boreal pine canopy: Measurements and modeling, *Hydrological Processes*, 10:1591–1608.

Popescu, S.C., R.H. Wynne, and R.F. Nelson, 2002. Estimating plot-level tree heights with lidar: Local filtering with a canopy-height based variable window size, Computers and *Electronics in Agriculture*, 37:71–95.

Quinn, P.F., K.J. Beven, and R. Lamb, 1997. The ln(a/tanb) index: How to calculate it and how to use it within the TOPMODEL framework, *Distributed Hydrological Modelling, Applications of the TOPMODEL Concept* (K.J Beven, editor), Wiley, Chichester, pp. 31–52.

Reutebuch, S., R.J. McGaughey, H.E. Andersen, and W.W. Carson, 2003. Accuracy of a high- resolution lidar terrain model under a conifer forest canopy, *Canadian Journal of Remote Sensing*, 29(5):527–535.

Riaño, D., F. Valladares, S. Condes, and E. Chuvieco, 2004. Estimation of leaf area index and covered ground from airborne laser scanner (Lidar) in two contrasting forests, *Agricultural and Forest Meteorology*, 124(3–4):269–275.

Ritchie, J.C., 1995. Airborne laser altimeter measurements of landscape topography, *Remote Sensing of Environment*, 53(2):91–96.

Ritchie, J.C., M.S. Seyfried, M.J. Chopping, and Y. Pachepsky, 2001. Airborne laser technology for measuring rangeland conditions, *Journal of Rangeland Management*, 54:8-21.

Rosso, R., M.C. Rulli, and G. Vannucchi, 2006. A physically based model for the hydrologic control on shallow landsliding, *Water Resources Research*, 42, W06410.

Schmidt, F., and A. Persson, 2003. Comparison of DEM Data Capture and Topographic Wetness Indices, *Precision Agriculture*, 4(2):179–192.

Stock, J. and W.E. Dietrich, 2003. Valley incision by debris flows: Evidence of a topographic signature, *Water Resources Research*, 39(4):11–25.

St-Onge, B., 2005. Airborne laser scanning altimetry, *Physics in Canada*, 61(5):49–54.

St-Onge, B., P. Treitz, and M. Wulder, 2003. Tree and canopy height estimation with scanning lidar, *Remote Sensing of Forest Environments: Concepts and Case Studies* (S. Franklin and M. Wulder, editors), Kluwer Academic Publishers, pp. 489–509.

Tenenbaum, D.E., L.E. Band, S.T. Kenworthy, and C.L. Tague, 2006. Analysis of soil moisture patterns in forested and suburban catchments in Baltimore, Maryland, using high-resolution photogrammetric and LIDAR digital elevation datasets, *Hydrological Processes*, 20(2):219–240.

Thoma D.P., Gupta S.C., Bauer M.E., Kirchoff C.E., 2005. Airborne laser scanning for riverbank erosion assessment, *Remote Sensing of Environment*, 95:493–501.

Thomas, V., P. Treitz, J.H. McCaughey, and I. Morrison, 2006. Mapping stand-level forest biophysical variables for a mixed wood boreal forest using lidar: An examination of scanning density, *Canadian Journal of Forest Research*, 36(1):34–47.

Todd, K.W., F. Csillag, and P.M. Atkinson, 2003 Three-dimensional mapping of light transmittance and foliage distribution using lidar, *Canadian Journal of Remote Sensing*, 29(5):544–555.

Töyra, J, and A. Pietroniro, 2005. Towards operational monitoring of a northern wetland using geomatics-based techniques, *Remote Sensing of Environment*, 97:174–191.

Töyrä, J., A. Pietroniro, C. Hopkinson, and W. Kalbfleisch, 2003. Assessment of airborne scanning laser altimetry (LiDAR) in a deltaic wetland environment, *Canadian Journal of Remote Sensing*, 29(6):718–729.

Warner, B.G., and C.D.A. Rubec, 1997. *The Canadian Wetland Classification System*, Second edition, University of Waterloo, Waterloo, Ontario, 68 p.

Webster, T.L., D.L. Forbes, S. Dickie, and R. Shreenan, 2004. Using topographic Lidar to map flood risk from storm-surge events for Charlottetown, Prince Edward Island, Canada, *Canadian Journal of Remote Sensing*, 30(1):64–76.

Wehr, A., Lohr, U., 1999. Airborne Laser Scanning - an introduction and overview, *ISPRS Journal of Photogrammetry and Remote Sensing*, 54(2–3):68–82.

Weltz, M.A., J.C. Ritchie, and H.D. Fox, 1994. Comparison of laser and field measurements of vegetation height and canopy cover, *Water Resources Research*, 30(5):1311–1319.

Wemple, B.C., and J.A. Jones, 2003. Runoff production on forest roads in a steep, mountain Catchment, *Water Resources Research*, 39(8):1220.

Witte, C., R. Dowling, D. Weller, R. Denham, and T. Rowland, 2001. Quantifying riparian vegetation and stream bank form through the use of airborne laser scanning and digital video data, *Proceedings of the International Geoscience and Remote Sensing Symposium*, 9–13 July, Sydney, Australia, volume 3, pp. 1545–1548.

10.9 NATURAL HAZARDS

10.9.1 Overview

The Earth is a constantly changing, highly dynamic environment with numerous recurring hazards, occurring at varying spatial scales and rendering significant challenges for existing and future populations and infrastructure. Remotely sensed data is a natural fit to detect, forecast, and monitor these hazards and their impacts. Showalter (2001), in a review of multiple remote-sensing articles related to hazard and disaster research published between 1972 and 1998, determined that the decision making support provided by remote sensing was a key benefit. Lidar systems rapidly capture data at sufficient resolution to enable simultaneous studies ranging from small to large scales, enabling us to see earth processes across the scales at which they occur. Lidar can also be employed to capture each stage of the process, enabling important 4D analyses.

10.9.2 Considerations

Several challenges exist when performing laser scanning for hazard and post-disaster damage assessment:

- Because lidar is new technology, datasets are only available for short term studies, whereas many geologic processes occur over very large time scales.
- Baseline information generally is not available at the same resolution as the acquired data following a disaster; hence, many studies must infer conditions prior to damage analysis. To remedy this, baseline data should be continually collected at higher resolutions to identify at risk areas such that mitigation efforts can be prioritized.
- Limited time and resources are available to collect data immediately following a disaster. While this data can be very helpful to provide valuable information for rescue efforts,

current lags in processing time limit the extent of these analyses. Continued development of near real time change analysis techniques (Olsen et al., In Press, a) will help alleviate this problem. Figure 10.76 shows a developed computer program that georeferences terrestrial scans and performs change analysis, *in-situ*.

- Because hazard prone areas are dynamic, they require repeat monitoring. Given the limited number of lidar acquisition companies and agencies, this can be a difficult task to have sufficient resources to acquire data on a repeat basis. However, advances in technology will continue to enable faster data acquisition and processing.
- Current data dissemination infrastructure limitations lead to significant delays between data acquisition and availability to the public. NOAA Digital Coast, USGS CLICK, and OpenTopography are examples of web-infrastructure that have made these data publicly available, with mixed success.

Despite these limitations (many of which are applicable to any technique), lidar has revolutionized hazard analyses, as will be shown in the following discussion of applications.

Figure 10.76 Screen shot of the Lidar CHange ENgine (Lichen) program for performing rapid, *in-situ* change analysis with TLS data. Red colored points indicate areas of change since the previous survey. Gray points are outside of the extents.

10.9.3 Sample Applications

Due to the evolving nature of laser scan technology, it is not possible to discuss all potential ways that lidar data has been or can be used in hazard analysis and damage assessment applications.

10.9.4 Earth Movements

Considerable work has been undertaken in recent years to document the patterns of earth movements and mechanisms for failure using lidar. Jaboyedoff et al. (2010) provides an overview of several methodologies and case studies to use lidar for landslide, rockfall, and debris flow analyses using airborne and terrestrial lidar. They present four major categories of landslide applications including: detection and characterization, hazard assessment and susceptibility mapping, modeling, and monitoring.

Burns and Madin (2009) demonstrated a methodology to map landslides in northwest Oregon through delineating features from airborne lidar DEMs, ultimately producing landslide hazard maps that could be investigated by local government for planning purposes. Once existing landslides have been mapped, Burns et al. (2012a) have developed a lidar-based analytical method for determining shallow landslide susceptibility. Similarly, Schulz (2005) performed landslide susceptibility estimation from airborne lidar data for the state of Washington.

TLS has been implemented for detailed geological assessments of several landslides, enabling improved understanding of the processes and mechanisms contributing to landslide movement (Jaboyedoff et al. 2010). For example, Collins et al. (2007) used TLS to study failure mechanisms of the coastal Northridge Bluff landslide in California demonstrating significant movement of the slide between 2003 and 2007. Jaboyedoff et al. (2009) studied retrogressive landslides in sensitive clays and rotational landslides occurring along river banks with TLS. Dunning et al. (2010) combined TLS surveying of landslides with numerical modeling. Collins and Sitar (2011) analyzed the stability of steep slopes composed of cemented sand using detailed profiles derived from TLS.

The Johnson Creek Landslide (JCL) creates chronic maintenance problems for the Oregon Department of Transportation (ODOT) where the coastal US Highway 101 crosses the slide mass. The landslide is approximately 360 m wide, 200 m long, and terminates on the beach where its toe is being actively eroded by ocean waves and sub-aerial processes. Priest et al. (2008) performed preliminary laser scan data analysis of cross sections at this site to quantify landslide movements. Olsen et al. (2012) performed additional scanning and used tree displacements to aide in distinguishing landslide movements from erosion. The results confirmed previous studies with traditional measurements that the sections of the slide showing increased erosion also had increased landslide movement.

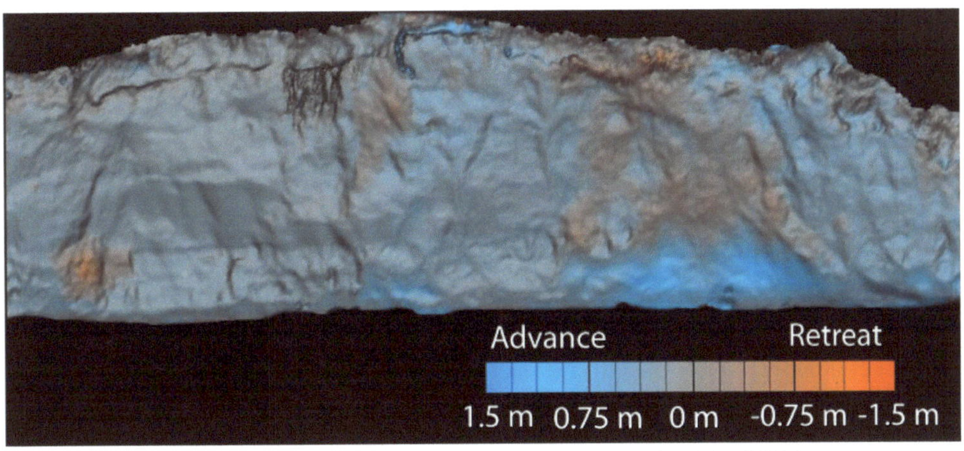

Figure 10.77 Overall change (advance = towards ocean) at the Johnson Creek Landslide between two time-series datasets (December 2010 and August 2011). The cliffs are approximately 20 m tall in this region.

10.9.5 Coastal Hazards

Over half (53 percent) of the US population lives within coastal counties (Crossett et al. 2004), and similar trends are observed in several other countries worldwide. Coastal tourism is an important source of revenue for many coastal states. Unfortunately, the coastal zone faces several hazards including erosion (seacliff, beach, and dune), tsunamis, earthquakes, sea level rise, and flooding. These hazards threaten a substantial amount of development and infrastructure along coastlines and present major challenges for these communities. Table 10.3 compares remote sensing techniques used for coastal analyses.

Table 10.3 Advantages and disadvantages of lidar measurement techniques for coastal erosion studies (modified from Hapke, 2004)

Technique	Advantages	Disadvantages
1. Traditional Ground Surveys	• Very accurate • Easily repeatable	• Poor temporal and spatial coverage • Time consuming/expensive
2. Historical Maps	• Inexpensive • Widely available • Very long temporal coverage (1850s to 1970s) • Good spatial coverage	• Low accuracy • Ambiguous cliff/bluff edge position
3. Aerial Photographs		
a. Un-rectified	• Inexpensive • Widely available • Good temporal coverage (1920s to present) • Good spatial coverage	• Low accuracy • Ambiguous cliff/bluff position in 2D
• Partially rectified	• Widely available • Good temporal coverage (1920's to present) • Good spatial coverage • Improved accuracy over 3a	• Ambiguous cliff/bluff position in 2D • Hardware/software for processing may be expensive
• Fully Rectified	• Widely available • Good temporal coverage (1920s to present) • Good spatial coverage • High accuracy • Cliff/bluff edge digitized in 3D	• Time consuming processing • Required software expensive
4. Lidar		
• Aerial Lidar	• Good spatial coverage • Higher resolution and accuracy than aerial photographs • Rapid Acquisition	• Expensive • Typically no datasets prior to 1990s • Poor coverage of cliff face
• Oblique Helicopter Lidar	• Higher resolution and accuracy than aerial photographs • Captures overhangs • Rapid Acquisition	• Expensive • Typically no datasets prior to the year 2000
• Terrestrial Laser Scanning	• Very high accuracy • Very high resolution • Captures overhangs and caves • Less expensive than 4a and 4b	• Difficult spatial coverage • Typically no datasets prior to the year 2000 • Time consuming/expensive
• Combined Lidar Techniques	• Complete 3D model of coast • Substantial amount of data	• Very expensive • Poor temporal coverage • Surveys must be completed nearly simultaneously

Recent tools have become available to process lidar data for shoreline change assessment. NOAA digital coast (2011) provides several tools to improve processing of lidar data, including a GIS extension. The Digital Shoreline Analysis System (DSAS, Thieler et al., 2009; previous version (DSMS/DSAS, Danforth and Thieler, 1992) utilizes digitized shorelines (from lidar or orthophotography) in a GIS environment to quantify shoreline change. This tool uses regression techniques to calculate the linear change rates at specified locations between multiple digitized shorelines by casting perpendicular transects from the original shoreline. The DSAS tool has been applied successfully to evaluate seacliff top retreat (Hapke and Reid, 2007) and various beach shoreline proxies (Esteves et al., 2006; Morton et al., 2005; Harris et al., 2005; Himmelstoss et al., 2006; Pendleton et al., 2004; Hapke et al., 2010). Olsen et al. (In Press, b) presents the development of a new GIS-based tool, the Topographical Compartment Analysis Tool (TopCAT), for performing a compartmentalized statistical analysis of change along seacliffs, beaches, or rivers. This tool is focused on using DEMs derived from lidar data.

10.9.6 Seacliff Erosion

Seacliffs are constantly threatened through both marine (wave-based) and sub-aerial (rain, groundwater) erosional process. Lidar has been recently employed to understand the geomorphological processes of the seacliffs and overall sediment yields from the coast. Studies have been completed using both airborne lidar and terrestrial-based systems. Once temporal and spatial processes are understood, it is possible to characterize boundary conditions and processes that contribute to cliff erosion.

Aerial lidar is typically performed from a laser scanner mounted on a low flying airplane with a near vertical look angle. Young et al. (2010) compare change detection using DEMs derived from airborne and ground-based lidar to evaluate seacliff erosion. They show that smaller failures can be detected with ground-based lidar from the improved resolution. However, ground-based lidar cannot detect the tops of sea cliffs. Airborne systems are advantageous because data are rapidly acquired as one continuous dataset and a lot of ground can be covered quickly. A limitation of aerial lidar is that vertical cliffs or complex features such as sea caves are not well resolved as a result of the low data density and/or the poor incidence angle; in contrast, terrestrial lidar can capture data from a more orthogonal angle and improve the resolution of imaging of seacliff erosion on the face. However, terrestrial systems cannot efficiently cover the same spatial extent as airborne systems. Terrestrial systems also do not obtain measurements on top of the cliff unless the scanner is setup in those locations. Hence, airborne data provides overall topographic information for a region, but terrestrial systems enable detailed, site-specific or sub-regional studies. Use of lidar mounted on a helicopter or MLS offers a compromise by enabling rapid data acquisition with an improved look angle. Oblique helicopter lidar is currently being investigated by Rosser et al. (2008) for application to seacliff erosion studies. However, complex topography and sea caves still cannot be efficiently captured through this method because the helicopter must be flown above ground away from the cliff. Logistics and costs create substantial difficulty for many organizations to implement oblique helicopter lidar for frequent site and regional surveys.

Terrestrial lidar has been proven to be an effective tool to study seacliff failures and morphology (Collins and Sitar, 2004 and 2005; Rosser et al., 2005; Lim et al., 2005; Young and Ashford, 2007). An ongoing mapping and monitoring program of the seacliffs using TLS was started in 2005 in San Diego County (Young and Ashford, 2007; Olsen et al. 2009), which consisted of continual use of terrestrial lidar to scan and map seacliffs along 17 km of the coastline. Olsen et al. (2009 and 2011) developed a methodology and associated software to rapidly georeference laser scan data in dynamics such as the coastline. Additionally, in conjunction with this effort, a rapid response network (Olsen et al. 2008), to immediately survey the seacliffs following a collapse, was implemented. Substantial change from seacliff failures greater than 30 m^3 can be observed monthly throughout this study area, and the corresponding data must be obtained on a regular basis before becoming overprinted by other events. Young and Ashford (2008) used TLS to evaluate the stability of cantilever cliff sections created by wave notching, showing the usefulness of TLS to provide accurate geometry for stability modeling. Young et al. (2009) compared used airborne lidar data in conjunction with precipitation and wave height data to evaluate processes related to short-term seacliff evolution. Finally, Young et al. (2011) used airborne lidar to perform statistical analyses of a slowly eroding coastal cliff at Point Loma, San Diego, California.

NOAA Digital Coast (2009) also provides repeat airborne lidar survey data for the US coastlines which have been used for a variety of hazard analyses.

Liu et al. (2009) discuss extraction of blufflines from lidar data and ortho-imagery, which can be used for erosion analysis. Young et al. (2009) provide a comparison of common methods for calculating sea cliff retreat and observed that significant differences in values can be obtained whether calculating retreat at the toe or crest of the cliff, or using an average across the cliff face.

From these survey efforts, important information regarding seacliff morphology can be characterized. Young and Ashford (2006a) for example, used lidar to determine that seacliffs provided a much more significant amount of sediment to beaches than was previously thought. Subsequently, Young and Ashford (2006b) evaluated the performance of various types of seawalls and protection structures. They determined that erosional control devices typically reduced erosion rates by approximately 42 percent.

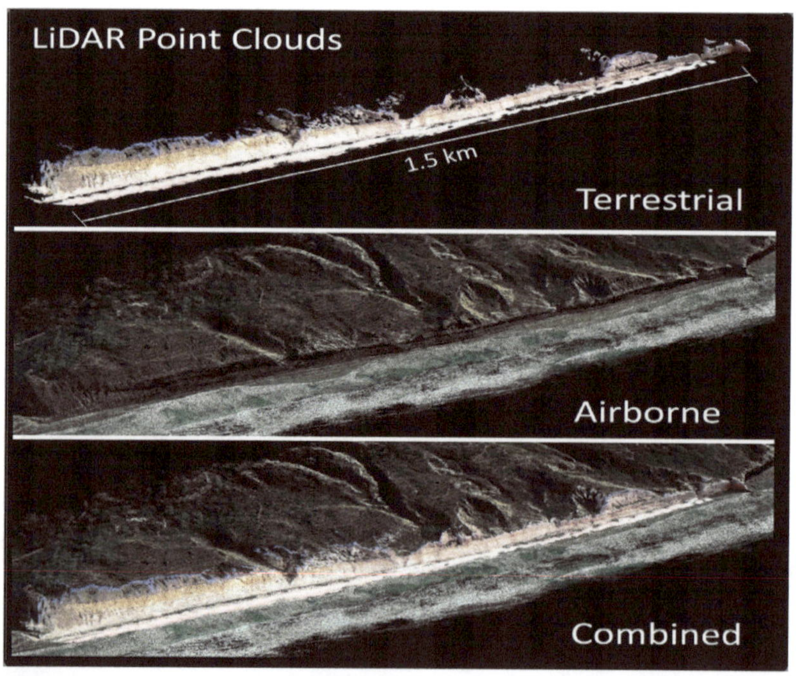

Figure 10.78 Comparison of (a) Terrestrial lidar dataset (February 2008), (b) Aerial lidar dataset (collected in 2002 and available through NOAA Digital Coast with photographic coloring from orthophotographs from USGS, 2008), and (c) combined for Torrey Pines State Reserve (32°55′13.50″N, 117°15′33.70″W)

Olsen (2009) presents analyses at several sites experiencing erosion in San Diego as a result of the previously described TLS rapid response effort. Here-in, we present an example of a site near Flat Rock at Torrey Pines State Reserve, California that has experienced substantial erosion (Figure 10.79). The first TLS documented collapse (23 August 2007) had an original volume of 79 m^3. Note the development of a wave-cut notch in the base of the seacliff. Within 20 days of the collapse, the failed material volume remaining on the beach had dropped to 58 m^3 due to wave-reworking. In early February 2008, a second collapse of 50 m^3 occurred during a major wave event. Finally, in September 2008, the entire upper section seacliff collapsed, unleashing 445 m^3 of material onto the beach.

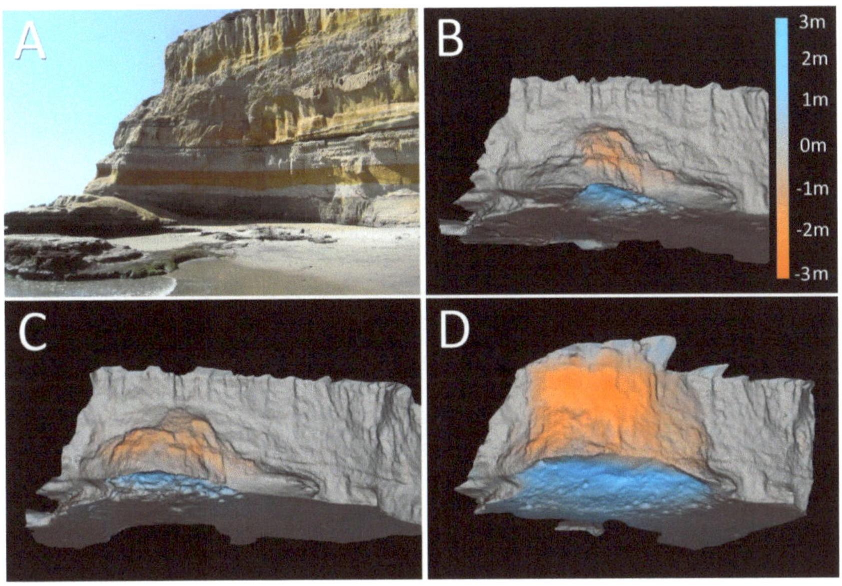

Figure 10.79 Comparison of an eroding site in San Diego, California: (a) before failure (30 June 2006) and after significant failures, (b) 23 August 2007, (c) 17 February 2008, and (d) 23 September 2008. All surface comparisons are relative to a survey completed on 03 November 2006.

As an example of a more energetic coastline, we will show an example of Beverly Beach State Park, Oregon, which is currently being monitored on a repeat basis through 3D laser scanning. Beverly Beach is located about seven miles North of Newport, Oregon along Highway 101. Change analysis reveals the patterns of erosion and accretion along the coastline between May 2010 and January 2011, in some cases, up to 0.5 m. In general, the crest of the seacliffs has retreated from the ocean by nearly 30 centimeters and talus deposits have advanced the base of the cliff by 30 centimeters. The north section shows a more uniform retreat across the cliff face compared to the south section, which shows two dominate erosion trends, one along the top of the cliff face and another above the talus deposits. Continual monitoring will enable more advanced analyses and tracking of geomorphological feature development.

10.9.7 Beach Erosion

Sallenger et al. (2003) evaluated topographical change analysis of beaches using airborne lidar by comparison to GPS surveys. Allan and Hart (2005) present a detailed study and GIS dataset of beach morpho-dynamics derived from analyses of 1997, 1998, and 2002 lidar datasets obtained from NOAA Digital Coast. The study had two objectives: (a) to compare a variety of common gridding techniques, and (b) to develop a detailed assessment of coastal change by analyzing geo-spatial transects (100 m spacing) along the central to northern Oregon coast.

10.9.8 Flood Mapping

Bossak et al. (2005) developed a GIS tool, the Coastal Impact Assessment Tool (CIAT), to predict coastal storm impacts using beach slopes derived from topographic datasets, including lidar datasets. Additionally, the tool utilizes 3D visualization features in GIS for scenario evaluation. The tool calculates estimated water run-ups during storms, which provides useful information for an assessment of likely coastal damage during storm events. The Oregon Department of Geology and Mineral Industries has developed a methodology under FEMA contract to use lidar topographic data to re-delineate flood insurance AE zones, to develop new A zones based on USGS streamstats flow data and lidar-based modeling and mapping, and finally to develop new coastal flooding VE zone maps using new wave climate models, wave transition models, and detailed coastal topographyfrom high-resolution lidar (Allan et al., 2012). These flood zones are then depicted on detailed basemaps developed with layered lidar imagery to provide an easy access to flood hazard information for homeowners and the general public. These tools are available at: http://www.oregongeology.org/FEMACoosCountyHazards/index.html.

10.9.9 Hurricanes

Hurricanes result in significant topographic change to coastal regions. As such, many efforts have been implemented to repeatedly acquire lidar data along coastal regions (NOAA Digital Coast, 2011). Meridith et al. (1999) and Olsen et al. (In Press, b) studied the Cape Hatteras National Seashore as part of an analysis of volumetric change of beach sand on the east coast following Hurricane Bonnie. Seed et al. (2006) studied damages and failure of levees, canals, and other earth structures following Hurricane Katrina. Lidar was a critical part of their investigation to preserve the magnitude and geometry of structural and geotechnical deformations, generate DEMs, obtain critical measurements, and perform stability analyses.

10.9.10 Earthquakes and Tsunamis

10.9.10.1 Post-disaster Reconnaissance

Lidar technology provides a powerful tool to digitally preserve the scene of a disaster so that engineers and scientists can virtually visit the scene to validate models and theories. The powerful 3D data provides a virtual world that scientists and engineers can repeatedly explore, query, and analyze long after restoration and recovery efforts have been completed.

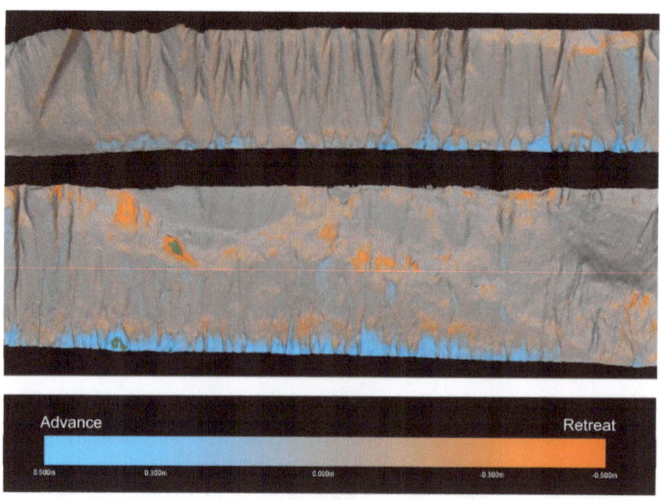

Figure 10.80 Change analysis between terrestrial lidar surveys showing the advance and retreat of the cliff face at Beverly Beach State Park, Oregon between May 2010 and January 2011

Figure 10.81 Screen shots from Coos County, Oregon flood and natural hazard web tool showing lidar-derived flood zones on a lidar basemap.

Kayen et al. (2010) present several applications of lidar to geotechnical earthquake engineering analyses including liquefaction, settlement, and landslide analyses. Olsen et al. (2010) provide examples and workflows for using lidar to perform change detection and damage analysis on full size structural test specimens subjected to seismic loading. Ground-based scan data enables many types of analyses, including deformation maps of partial concrete wall blowouts, scour depths and volumes, lateral and longitudinal displacement in damaged steel frame structures, bridge collapses, and retaining wall failures. Airborne lidar can be used to map surface rupture, flooding zones, landslides, etc. This information is vital to validate numerical models and theories, enabling us to quantify and understand earthquake and tsunami forces, failure mode, and structural response to unprecedented detail. These observations and analyses can then help calibrate proposed building code design methodologies and practices.

The Geo-Engineering Extreme Event Reconnaissance (GEER; http://www.geerassociation.org/) team has been active in deploying ground-based lidar following major events. These and other recent post-disaster reconnaissance efforts have implemented lidar surveys. A partial list is shown below:

- 2002 Denali, Alaska: Kayen et al. (2004) performed lidar, radar, and (Spectral Analysis of Surface Waves) SASW surface wave analyses for a geotechnical investigation.
- 2004 and 2007 Nigata, Japan: Kayen et al. (2006) performed lidar surveys following the earthquake to map surface rupture and liquefaction effects.
- 2009 American Samoa: Olsen and Donahue (2011) demonstrate the potential use of lidar data to analyze post-earthquake and tsunami damage following the 2009 Samoan earthquake and tsunami, including its ability to generate high-resolution topography for analyzing erosion and scour.
- 2010 Haiti: Messinger et al. (2010) discuss collection of lidar data in Haiti following the earthquake and its role in aiding rescue and relief efforts. Cowgill et al. (2010) discuss how these data were used by scientists in an interactive, virtual setting to remotely measure surface rupture and other displacements.
- 2010 Chile: Olsen et al. (Accepted with Revisions) and Olsen et al. (2011) highlight examples of using laser scan data to analyze damaged infrastructure following the 2010 Chilean earthquake, showing that the laser scan data can detect information that is difficult to observe with traditional reconnaissance techniques. Figure 10.82 shows how intensity coloring from TLS data can highlight shear cracks and damage on the building face. From the lidar data, it was determined that 20 percent of the concrete on the south face had spalled from the earthquake. Geotechnical Extreme Events Reconnaissance (GEER) also performed several geotechnical investigations in conjunction with lidar at sites that liquefied and showed site amplification.
- 2011 Tohoku, Japan: GEER (2001) performed extensive lidar scans of several sites damaged during the earthquake and tsunami including sites with lateral spreading, slope failures, settlement, embankment failures, seawall failures, port wharf collapse, scouring, and many other types of damage. Figure 10.83 shows an example of TLS data collected at a hotel damaged by the tsunami.

10.9.10.2 Mt Hood Case Study: Combined Airborne and Terrestrial Scanning to Solve a Geologic Puzzle

Lidar can also be used proactively to identify and discover new fault traces that can be difficult to identify on the ground and through other remote sensing techniques. The use of airborne lidar enabled the Oregon Department of Geology and Mineral Industries (DOGAMI) to detect previously undiscovered fault scarps on Mt. Hood, Oregon. A trench is used at the fault to determine the movement of the soil layers as a result of previous earthquakes.

Triangulated surface models and photo texture mapping (Figure 10.85) were used in the analysis of the fault scarp to unfold its geologic story. DTMs from these scans have also been combined with Gigapan imagery to enable high resolution logging, and have enabled improved interpretation and analysis compared to traditional techniques which use a pre-laid grid of strings across the trench face for measurement reference. Lidar preserves the information digitally to allow for future analyses after the trench is filled back in.

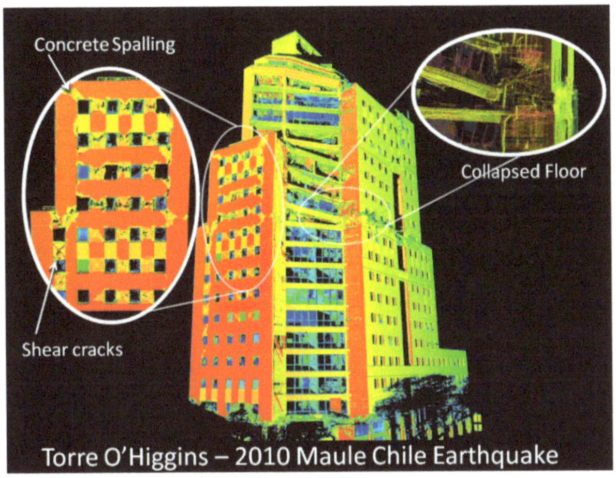

Figure 10.82 3D laser scan showing structural damage assessment from the Torre O Higgen's structure in Consitucion, Chile, (April 2010).

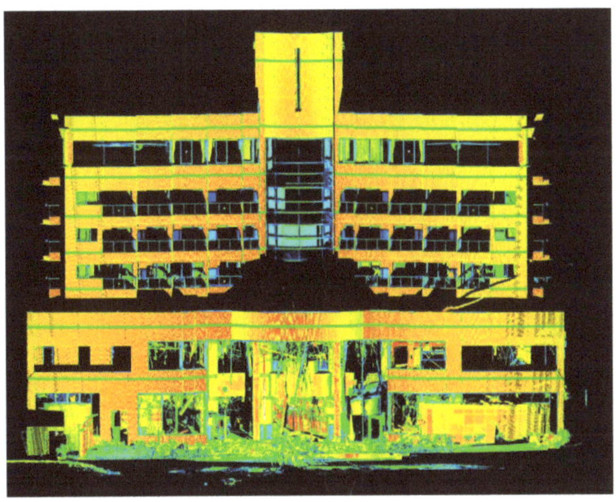

Figure 10.83 Intensity-shaded, 3D laser scan showing structural damage of a hotel in Japan following damage by the 2011 Tohuku Tsunami.

10.9.11 Volcanoes

Mazzarini et al. (2007) used lidar intensity maps to identify and map lava flows from the active Mt. Etna volcano in Italy. They determined that the target-emitter distance and surface roughness (texture) were the two principal parameters that determined the intensity response of lava flows. Mazzarini et al. (2005) evaluated the morphology of lava channels following an eruption of Mt. Etna. The Oregon Department of Geology and Mineral Industries has used high-resolution lidar in several ways in a multi-hazard study of the Mt. Hood volcano in NW Oregon. Lidar topography is used to run simulated lahars down the major drainages on the mountain using the LAHARZ computer model, allowing the development of accurate and detailed inundation zones and inundation depth grids. The hazard zones are coupled with a comprehensive structure inventory extracted from the lidar data, providing accurate exposure information for homes, buildings, and infrastructure. (Burns et al., 2012b).

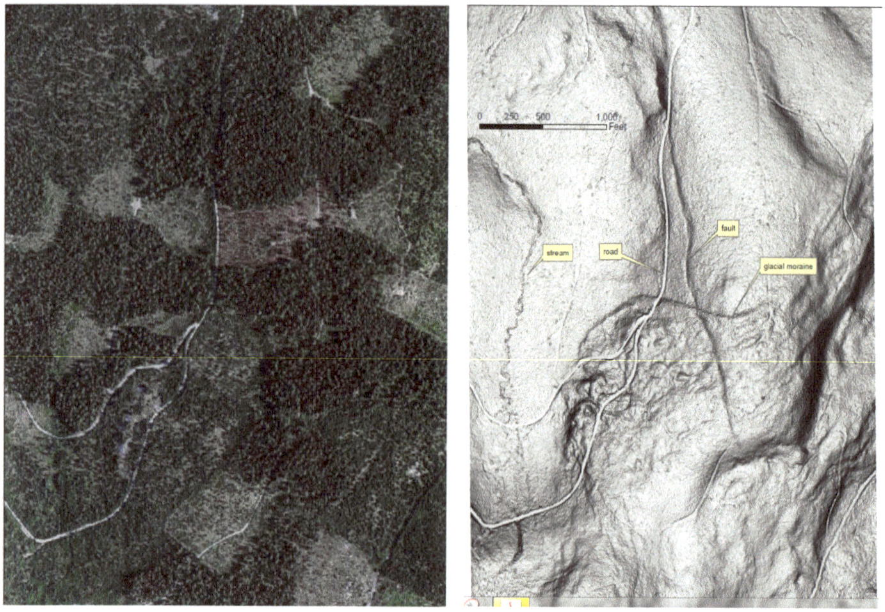

(a) (b)

Figure 10.84 (a) Orthophoto, and (b) lidar slopeshade image of the Blue Ridge Fault scarp north of Mt Hood. The fault was unknown before the collection of high-resolution lidar over the area.

(a)

(b)

Figure 10.85 (a) point cloud acquired at the Blue Ridge Trench on Mt. Hood, Oregon, with (b) the resulting texture-mapped DTM.

Figure 10.86 Perspective view to East of Mt Hood, Oregon, showing modeled inundation zones for lahars of increasing volume. Black rectangles are building footprints extracted from lidar. (Image credit: William Burns and Kaleena Hughes)

ACKNOWLEDGMENTS

The Oregon State University (OSU) Geomatics lab collected much of the scan data used in this section. We thank Leica Geosystems and Maptek I-Site for providing software used for analyzing data. Funding for these case studies was provided by the National Science Foundation (RAPID Grants #1037861, 1138710 and 1138699), California Seagrant (Project #R/OE-39) and the Coastal Environmental Quality Initiative (#04-T-CEQI-06-0046), the Oregon Department of Transportation, and the Oregon Transportation Research Education Consortium (2011-398). Opinions, findings, and conclusions do not necessarily reflect those of sponsoring partners.

REFERENCES

Allan, J.C., P. Ruggiero, and J. Roberts, 2012. *Coastal Flood Insurance Study, Coos County, Oregon*, Oregon Department of Geology and Mineral Industries Special Paper 4X,.

Burns, W.J.B., K.L.B. Hughes, K.V. Olson, J.D. McClaughry, K.A. Mickelson, D.E. Coe, J.T. English, J.T. Roberts, R.R. Lyles-Smith, and I.P. Madin, 2012b. *Open-File Report O-11-16, Multi-Hazard and Risk Study for the Mount Hood Region, Multnomah, Clackamas, and Hood River Counties, Oregon.*

Burns, W.J., and I.A. Madin, 2009. *Protocol for Inventory Mapping of Landslide Deposits from Light Detection and Ranging (LIDAR) Imagery*, DOGAMI Special paper 42.

Burns, W.J., I.A. Madin, and K.A. Mickelson, 2012a, *Protocol for Shallow-Seated Landslide Susceptibility Mapping*, Oregon Department of Geology and Mineral Industries Special Paper 42.

Burns and others, 2012b.

Collins, B., and N. Sitar, 2004. Application of high resolution 3d laser scanning to slope stability Studies, *Proceedings of the 39th Annual Symposium on Engineering Geology and Geotechnical Engineering*, Butte, Montana, pp. 79–92.

Collins, B.D., and N. Sitar, 2005. Monitoring of coastal bluff stability using high resolution 3D laser scanning, *ASCE Geo-Frontiers Special Publication 138: Site Characterization and Modeling, Remote Sensing in Geotechnical Engineering* (E.M. Rathje, editor), ASCE, Austin, Texas, 24-26 January.

Collins, B., R. Kayen, T. Reiss, and N. Sitar, 2007. Terrestrial LIDAR investigation of the December 2003 and January 2007 activations of the Northridge Bluff Landslide, Daly City, California, *USGS Open File Report 2007-1079*, 32 p.

Collins, B., and N. Sitar, 2011. Stability of steep slopes in cemented sands, *Journal of Geotechnical and Geoenvironmental Engineering*, ASCE, 137(1):43–51.

Cowgill, E., T.S. Bernardin, M.E. Oskin, C. Bowles, M.B. Yikilmaz, O. Kreylos, A.J. Elliot, M.S. Bishop, and L.H. Kellog, 2010, Analysis of LIDAR data during rapid scientific response to the January 12, 2010 Haiti earthquake, *2010 Geological Society of America Abstracts with Programs*, 42(5):154.

Dunning, S.A., N.J. Rosser, and C.I. Massey, 2010.The integration of terrestrial laser scanning and numerical modeling in landslide investigations, *Quarterly Journal of Engineering Geology and Hydrogeology*, 43(2):233–247.

Danforth, W.W., and E.R. Thieler, 1992. Digital shoreline analysis system (DSAS) user's guide, version 1.0, US Geological Survey Open-File Report 92-355, 42 p.

Esteves, L.S., J.J. Williams, and S.R. Dillenburg, 2006. Seasonal and interannual influences on the patterns of shoreline changes in Rio Grande do Sul, southern Brazil, *Journal of Coastal Research*, West Palm Beach, Florida, ISSN 0749-0208, 22(5):1076–1093.

Hapke, C., 2004. The measurement and interpretation of coastal cliff and bluff retreat, formation, evolution and stability of coastal cliffs - Status and trends, USGS Professional Paper 1963 (M.A. Hampton and G.B. Griggs, editors), URL: http://pubs.usgs.gov/pp/pp1693/pp1693.pdf (last date accessed: 07 February 2012).

Hapke, C.J., and D. Reid, 2007. National assessment of shoreline change, Part 4: Historical coastal cliff retreat along the California coast, *US Geological Survey Open File Report* 2007-1133.

Hapke, C.J., E.A. Himmelstoss, M. Kratzmann, J.H. List, and E.R. Thieler, 2010, National assessment of shoreline change; historical shoreline change along the New England and Mid-Atlantic coasts: *US Geological Survey Open-File Report* 2010-1118, 57 p., URL: http://pubs.usgs.gov/of/2010/1118/ (last date accessed: 07 February 2012).

Harris, M., J. Brock, A. Nayegandhi, and M. Duffy, 2005. Extracting shorelines from NASA airborne topographic LIDAR-derived digital elevation models, *US Geological Survey Open File Report* 2005-1427.

Himmelstoss, E.A., D.M. Fitzgerald, P.S. Rosen, and J.R. Allen, 2006. Bluff evolution along coastal drumlins: Boston Harbor Islands, Massachusetts, *Journal of Coastal Research*, West Palm Beach (Florida), ISSN 0749-0208, 22(5):1230–1240.

Kayen, R., W. Barnhardt, B. Carkin, B.D. Collins, E.E. Grossman, D. Minasian, and E. Thompson, 2004. Imaging the M7.9 Denali fault earthquake 2002 rupture at the Delta River using LIDAR, RADAR, and SASW surface wave geophysics, *Eos Transactions AGU*, Fall Meeting Supplement, Abstract S11A-0999, 85(47).

Kayen, R., R.T. Pack, J. Bay, S. Sugimoto, and H. Tanaka, 2006. Terrestrial-LIDAR visualization of surface and structural deformations of the 2004 Nigata Ken Chuetsu Japan, Earthquake, *Earthquake Spectra*, EERI, 22(S1):S147–S162.

Kayen, R., J.P. Stewart, and B. Collins, 2010. Recent advances in terrestrial LIDAR applications in geotechnical earthquake engineering, *Proceedings of the 5th International Confer-*

ence on Recent Advances in Geotechnical Earthquake Engineering and Soil Dynamics, San Diego, California.

Lim, M., D.N. Petley, N.J. Rosser, R.J. Allison, and A.J. Long, 2005. Combined digital photogrammetry and time-of-flight laser scanning for monitoring cliff evolution, *The Photogrammetric Record*, 20(110):109–129.

Mazzarini, F., M.T. Pareschi, M. Favalli, I. Isola, S. Tarquini, and E. Boschi, 2007. Lava flows identification and aging by means of LIDAR intensity: Mount Etna case, *Journal of Geophysical Research*, doi: 10.1029/2005JB004166.

Mazzarini, F., M.T. Pareschi, M. Favalli, I. Isola, S. Tarquini, and E. Boschi, 2005. Morphology of basaltic lava channels during the Mt. Etna September 2004 eruption from airborne laser altimeter data, *Geophysical Research Letters*.

Meridith, A., D. Eslinger, and D. Aurin, 1999. An evaluation of hurricane-induced erosion along the North Carolina Coast using airborne LIDAR surveys, *CSC Technical Report NOAA/CSC/99031-PUB*.

Messinger, D.W., J. van Aardt, D. McKeown, M. Casterline, J. Faulring, J. Raqueño, B. Basener, and M. Velez-Reyes, 2010, High-resolution and LIDAR imaging support to the Haiti earthquake relief effort, *Proceedings of SPIE 7812, 78120L*, doi:10.917/12.862090

Morton, R.A., T. Miller, and L. Moore, 2005. Historical shoreline changes along the US Gulf of Mexico: A summary of recent shoreline comparisons and analyses, *Journal of Coastal Research*, West Palm Beach, Florida, ISSN 0749-0208, 21(4):704–709.

Olsen, M.J., E. Johnstone, A.P. Young, T.J. Hsieh, S.A. Ashford, N. Driscoll, and F. Kuester, 2008. Rapid response to seacliff erosion in San Diego County using terrestrial LIDAR, *Proceedings of Solutions to Coastal Disasters Conference*, ASCE, 573.

Olsen, M.J., 2009. *Methodology for Assessing Coastal Change Using Terrestrial Laser Scanning*, PhD. Dissertation, University of California, San Diego.

Olsen, M.J., E. Johnstone, N. Driscoll, S.A. Ashford, and F. Kuester, 2009. Terrestrial laser scanning of extended cliff sections in dynamic environments: A parameter analysis, *Journal of Survey Engineering*, ASCE, 135(4):161–169.

Olsen, M.J., F. Kuester, B. Chang, and T. Hutchinson, 2010. Terrestrial laser scanning based structural damage assessment, *Journal of Computing in Civil Engineering*, 24(3):264–272.

Olsen, M.J., E. Johnstone, F. Kuester, S.A. Ashford, and N. Driscoll, 2011. New automated point-cloud alignment for ground based LIDAR data of long coastal sections, *Journal of Survey Engineering*, ASCE, 137(1):14–25.

Olsen, M.J., S. Piaskowy, S. Yim, L. Burgos, L., and S. Butcher, 2011. LIDAR investigations of the 2010 Maule Chile Earthquake, *Proceedings of the 6th International Structural Engineering and Construction (ISEC-6)*, Modern Methods and Advances in Structural Engineering and Construction, Zurich Switzerland, pp. 343–348.

Olsen, M.J., and J. Donahue, 2011. A wave of new information: LIDAR investigations of the 2009 Samoan tsunami, *Proceedings of Solutions to Coastal Disasters Confonference*, ASCE, pp. 321–330.

Olsen, M.J., S. Butcher, and E.P. Silvia, (In Press). Real-time change and damage detection of landslides and other earth movements threatening public infrastructure, *OTREC Final Report 2011-22 and ODOT Final Report RS 500-500*, 80 p.

Olsen, M.J., A.P. Young, and S.A. Ashford, (In Press). TopCAT - Topographical Compartment Analysis Tools for ArcGIS®, *Computers and Geosciences*.

Olsen, M.J., J.C. Allan, and G.R. Priest, 2012. Johnson Creek landslide movement and erosion quantification through 3D laser scanning, *ASCE Geoinstitute GeoCongress 2012*, Oakland, California.

Olsen M.J., K.F. Cheung, Y. Yamazaki, S.M. Butcher, M. Garlock, S.C. Yim, S. Piaskowy, I. Robertson, L. Burgos, and Y.L. Young, (Accepted with revisions). Damage assessment of the 2010 Chile Earthquake and Tsunami using ground-based LIDAR, *Earthquake Spectra.*

Pendleton, E.A., S.J. Williams, and E.R. Thieler, 2004. Coastal vulnerability assessment of Assateague Island National Seashore (ASIS) to sea-level rise, *U.S. Geological Survey Open-File Report 2004-1020.*

Rosser, N.J., D.N. Petlety, M. Lim, S.A. Dunning, and R.J. Allison, 2005. Terrestrial laser scanning for monitoring the process of hard rock coastal cliff erosion, *Quarterly Journal of Engineering Geology and Hydrology*, 38:363–375.

Rosser, N.J., M. Lim, E. Norman, and D.N. Petley, 2008. Exploring variations in and controls upon cliff, platform and coastline geometry, Geophysical Research Abstracts, European Geosciences Union, *EGU2008-A-10318*, Vol. 10.

Sallenger, A.H., W.B. Krabill, R.N. Swift, J. Brock, J. List, M. Hansen, R.A. Holman, S. Manizade, J. Sontag, A. Meredith, K. Morgan, J.K. Yunkel, E.B. Frederick, and H. Stockdon, 2003. Evaluation of airborne topographic LIDAR for quantifying beach changes, *Journal of Coastal Research*, 19(1):125–133.

Seed, R.B., R.G. Bea, R.I. Abdelmalak, A.G. Athanasopoulos, G.P. Boutwell, J.D. Bray, J.-L. Briaud, C. Cheung, D. Cobos-Roa, J. Cohen-Waeber, B.D. Collins, L. Ehrensing, D. Farber, M. Hanemann, L.F. Harder, K.S. Inkabi, A. M. Kammerer, D. Karadeniz, R.E. Kayen, R.E.S. Moss, J. Nicks, S. Nimmala, J.M. Pestana, J. Porter, K. Rhee, M.F. Riemer, K. Roberts, J.D. Rogers, R. Storesund, A.V. Govindasamy, X. Vera-Grunauer, J.E. Wartman, C.M. Watkins, E. Wenk, Jr., and S.C. Yim, 2006. Investigation of the Performance of the New Orleans Flood Protection Systems in Hurricane Katrina on August 29, 2005, Volume 1, URL: http://www.ce.berkeley.edu/projects/neworleans/report/VOL_1.pdf (last date accessed: 07 February 2012).

Thieler, E.R., E.A. Himmelstoss, J.L. Zichichi, and A. Ergul, 2009, Digital Shoreline Analysis System (DSAS) version 4.0 - An ArcGIS extension for calculating shoreline change, *U.S. Geological Survey Open-File Report 2008-1278*, URL: http://pubs.usgs.gov/of/2008/1278/ (last date accessed: 07 February 2012).

Young, A.P., R.T. Guza, W.C. O'Reilly, R.E. Flick, and R. Gutierrez, 2011. Short-term retreat statistics of a slowly eroding coastal cliff, *Natural Hazards and Earth Systems Sciences*, 11:205–217.

Young, A.P., R.T. Guza, R.E. Flick, W.C. O'Reilly, and R. Gutierrez, 2009. Rain, Waves, & short-term seacliff evolution, *Marine Geology*, 267:1–7.

Young, A.P., R.E. Flick, R. Gutierrez, and R.T. Guza, 2009. Comparison of short-term seacliff retreat measurement methods in Del Mar, California, *Geomorphology*, 112:318–323.

Young, A.P., M.J. Olsen, N. Driscoll, R.E. Flick, R. Gutierrez, R.T. Guza, E. Johnstone, and F. Kuester, 2010. Comparison of airborne and terrestrial LIDAR estimates of seacliff erosion in southern California, *Photogrammetric Engineering & Remote Sensing*, 76(4):421–427.

Young, A.P., and S.A. Ashford, 2008. Instability investigation of cantilevered seacliffs, *Earth Surface Processes and Landforms*, 33:1661–1677.

Young, A.P., and S.A. Ashford, 2007. Quantifying sub-regional seacliff erosion using mobile terrestrial LIDAR, *Shore and Beach*, 75(3):38–43.

Young, A.P., and S.A. Ashford, 2006a. Application of airborne LIDAR for seacliff volumetric change and beach-sediment budget contributions, *Journal of Coastal Research*, 22(2):307-318.

Young, A.P., and S.A. Ashford, 2006b. Performance evaluation of seacliff erosion control Methods, *Shore and Beach*, 74(2):16–24.

AUTHORS

Section 10.1 was written by Hans Karl Heidemann, Physical Scientist at USGS/EROS.

Section 10.2 was written by Jason Stoker, US Geological Survey Physical Scientist and the Project Manager for the Lidar Science Project at the USGS EROS Center.

Section 10.3.1 was written by David Brown of Terra Remote Sensing (USA), Inc.; section 10.3.2 was written by Michael J. Olsen, School of Civil and Construction Engineering, Oregon State University; Ron Singh, Oregon Department of Transportation; Keith Williams, Oregon Department of Transportation; and Abby Chin, Oregon State University.

Section 10.4 was written by Alvan Karlin, Gordon McClung, and Jamison Janke, all from Southwest Florida Water Management District.

Section 10.5 was written by Jie Shan, KyoHyouk Kim, Aparajithan Sampath, and Serkan Ural, all from Purdue University.

Section 10.6 was written by Christopher E. Parrish, Lead Physical Scientist, NOAA, National Geodetic Survey.

Section 10.7 was written by Christopher E. Parrish, Kirk Waters, Jennifer Wozencraft, Christopher L. Macon, John Brock, and C. Wayne Wright, all from NOAA.

Section 10.8 was written by Chris Hopkinson, Centre for Marine & Atmospheric Research, CSIRO; and Alain Pietroniro, Centre for Hydrology, University of Saskatchewan.

Section 10.9 was written by Michael J. Olsen, School of Civil and Construction Engineering, Oregon State University; Ian Madin, Oregon Department of Geology and Mineral Resources; Abby Chin, Oregon State University; and Jeremy Conner, Oregon Department of Geology and Mineral Resources.

APPENDIX A
GEOID HEIGHT MODELS AND VERTICAL DATUMS

DANIEL R. ROMAN

BACKGROUND

Geoid height models take advantage of a simplified relationship between the ellipsoidal and orthometric heights. This appendix will provide a brief overview of the relationships between ellipsoidal, orthometric, and geoid heights followed by a discussion of the applicability of current and future geoid height models as it applies to lidar work.

GPS-derived ellipsoidal coordinates represent an efficient and accurate mechanism for determining your location. The ellipsoidal reference frame provides a first order model of the Earth accurate to about 99.999 percent. However, that little bit of misfit can mean about ±100 meters difference. This is because the ellipsoid neglects most variations in mass due to the continents and other sources, which eventually diminish to zero at satellite orbits. Only the oblateness of the Earth's shape and gravity field significantly affects the satellite orbits and is therefore accounted for in an ellipsoidal reference frame. However, we live at the surface, not in orbit.

All those features that attenuate at satellite elevations remain real and significant at the surface. Orthometric heights seek to describe elevations in that real, physical system along the surface. Mass and local gravity variations are accounted for to better describe variations in geopotential surfaces and heights above it. Such a surface serves better as a Vertical Datum since it better describes how water will flow (i.e., the "down" in downhill). This surface then also serves as the best estimate of global mean sea level, since water normally does not flow below the level of the sea. A model that describes the difference between an ellipsoid and a Vertical Datum would, therefore, be very useful.

GEOID HEIGHTS

As has long been known in physical geodesy, knowledge of those mass and local gravity variations can be used to describe this difference. The ellipsoidal reference frame describes a very smooth model of the Earth's gravity field and heights that relate to it. This is only a limited estimate of the true field described by a Vertical Datum. Observation of related elements of these mass variations (gravity observations) permits development of a geoid height model that describes this relationship that can be simplified to a linear expression that the ellipsoidal height (h) represents the sum of the orthometric height (H) and the geoid height (N):

$$h = H + N \qquad \text{or, equivalently,} \qquad H = h - N.$$

The geoid heights then are the separation between the ellipsoid and geoid. Note that since the ellipsoid is a first order surface fit to the Earth, there will be regions where the geoid is above ellipsoid and regions where it is below. This is because, as mentioned above, the geoid surface is more closely following the variations in the Earth's mass. The resulting geoid heights would be positive or negative, correspondingly, since the geoid height is actually an ellipsoid height measured from the ellipsoid to the geoid. In North America, most geoid heights are negative.

APPLICATION TO LIDAR WORK

The most critical thing to note about geoid height models is that they represent the difference between the zero height surface in a specific *ellipsoidal* datum and the zero height surface in a specific *orthometric* datum. For GEOID09, geoid heights in the conterminous United States (CONUS) and Alaska describe the separation between the ellipsoid of NAD83 (NSRS2007) and the zero height surface of NAVD88. For American Samoa, the transformation is between ASVD02 and NAD83 (PACP00). Similar datums are used for Puerto Rico (PRVD02), the U.S. Virgin Islands (VIVD09, not yet released), Guam (GUVD04), and the Commonwealth of the Northern Marianas Islands (NMVD03). For each of these regions, a locally determined tide gage was selected as the basis for the local Vertical Datum. Since it is impossible to level across the water between these regions, each region has its own Vertical Datum. It should be noted that NAVD88 is not valid in those outlying regions. Several maps have been produced that might indicate that heights on it relate to NAVD88 for some of the outlying territories when this is, in fact, not true.

Caution should always be taken to use only the official Vertical Datum (and the associated geoid height model) for a given region. The name GEOID09 is applied to all such models for the United States and its territories. Hence, the only concern for the user should be to obtain the GEOID09 model designed for their specific region, as the way to convert the ellipsoid heights (in NAD83) that inherently come out of lidar into orthometric heights in the official Vertical Datum of the region. This only becomes a potential problem when in Southeast Alaska area where the CONUS model and Alaska model overlap. Both refer to NAVD 88, but the two regions were developed separately and may differ. Note that no official Vertical Datum has ever been defined by NGS for Hawaii. As such, the GEOID09 model there describes the separation between NAD83 (PACP00) and the gravimetric geoid surface defined by NGS (which is the theoretical definition of a true orthometric zero-height surface anyway).

This leads to a further caution. NGS produces two sets of geoid height models for all regions: USGG2009 and GEOID09. For most applications, you should use only GEOID09 products as they describe the separation between the official ellipsoidal and orthometric (vertical) datums for given locations. USGG2009 models are developed entirely from gravity and terrain data. They are tied into the EGM2008 geopotential model (ref), which is itself based on data from the GRACE gravity field mission (ref). These models use a GRS-80 ellipsoid centered in the ITRF00 reference framework (ref). This is roughly equivalent (+/-5 cm) to WGS-84 (G1150). The USGG2009 models define the separation between that ellipsoid and geoid surface adopted by NGS as the best fitting to global mean sea level. These two surfaces both differ greatly from those referenced in GEOID09 models. The USGG2009 models may provide more realistic assessment of heights for scientific and engineering applications that require high accuracy. However, they do not provide transformations between the official datums of a region. Do not use USGG2009 to transform between NAD83 and NAVD88 (or other official Vertical Datums). Use only GEOID09 for that purpose.

AVAILABLE MODELS

USGG2009 and GEOID09 models are available through the NGS website. These geoid height models and their antecedents can be downloaded from links on the NGS Tool Kit at http://www.ngs.noaa.gov/TOOLS/. A more thorough description of the development of these models may be found on associated webpages located on the NGS Geoid Page at http://www.ngs.noaa.gov/GEOID/. Available models, their extents, and their respective datums are given in the below tables for GEOID09 and USGG2009 further below.

Table 1 Extents and datums for regions covered by GEOID09.

Area	Latitude		Longitude		Ellipsoidal Datum	Vertical Datum
	Min	Max	Min	Max		
Conterminous US (CONUS)	24N	58N	60W	130W	NAD83 (NSRS2007)	NAVD88
Alaska	49N	72N	126W	188W	NAD83 (NSRS2007)	NAVD88
Hawaii	18N	24N	154W	161W	NAD83 (PACP00)	<gravimetric geoid surface>
Guam/ Northern Marianas Is.	11N	18N	143E	146E	NAD83 (MARP00)	GUVD04 (Guam) NMVD03 (CNMI)
American Samoa	17S	11S	186E	192E	NAD83 (PACP00)	ASVD02
Puerto Rico/ Virgin Islands	15N	21N	64W	69W	NAD83 (NSRS2007)	PRVD02 (PR) VIVD09 (USVI)

Table 2 Extents and datums for regions covered by USGG2009.

Area	Latitude		Longitude		Ellipsoidal Datum	Vertical Datum
	Min	Max	Min	Max		
Conterminous US (CONUS)	24N	58N	60W	130W	ITRF00 (WGS-84)	<gravimetric geoid surface>
Alaska	49N	72N	126W	188W	ITRF00 (WGS-84)	<gravimetric geoid surface>
Hawaii	18N	24N	154W	161W	ITRF00 (WGS-84)	<gravimetric geoid surface>
Guam/ Northern Marianas Isl.	11N	18N	143E	146E	ITRF00 (WGS-84)	<gravimetric geoid surface>
American Samoa	17S	11S	186E	192E	ITRF00 (WGS-84)	<gravimetric geoid surface>
Puerto Rico/ Virgin Islands	15N	21N	64W	69W	ITRF00 (WGS-84)	<gravimetric geoid surface>

As you can see from Table 2, USGG2009 provides a common set of models for comparison across all regions. However, it is not tied to official datums.

APPENDIX B
QC PROCESS FOR DATA USERS

US GEOLOGICAL SURVEY
(TAKEN FROM UPCOMING ASPRS GUIDELINES
BEING DEVELOPED IN CONJUNCTION WITH USGS)

6 QC PROCESS FOR DATA USERS

6.1 QC Testing procedures

"Quality control (QC)" is used to denote post-mission procedures for evaluating the quality of the final LiDAR data product (Habib et al., 2010). The user of the data is more concerned with the final product quality, than the system level QA procedures that may vary depending on the type of instrument in use. Thus, as far as the user is concerned, the acquisition system is a black box. The QC procedures, therefore, should be system agnostic and work with only the point clouds delivered by the data provider. The QC procedures suggested in this document adhere to this ideal. However, if more information on the data (including raw measurements) is available, these procedures can be extended to correct the calibration of the instrument.

The QC procedures for LiDAR data are distinguished as "Internal" or "External". The internal QC (IQC) procedures refer to the comparison of the positional accuracy of features extracted from different scans. External QC (EQC) procedures are analogous to the check point analysis of photogrammetric data collection procedures. The IQC and the EQC procedures are described below.

6.1.1 Parameters for accuracy assessment

The accuracy of LiDAR data can be estimated based on certain parameters. In the case of the internal/relative accuracy of the data, as seen in the next section, the relative shifts between two adjacent scans are measured. Therefore, the parameters for measuring the internal/relative accuracy of two overlapping scans should be the relative translation and rotation necessary to make the scans coincident. The parameters are defined in the next section and expressed in Eq. 1.

The parameters for expressing the absolute/external accuracy of the data are the same as those for other photogrammetric products, i.e. the horizontal and the vertical RMSE values, and they have been described in

6.1.3 QC methods for Internal (Relative) Accuracy of LiDAR data (IQC)

The purpose of the intra-swath IQC procedure is to measure and quantify the relative discrepancy between two (adjacent) swaths of LiDAR data. Methods to determine the relative *vertical* accuracy of the LiDAR scans have been used to quantify the accuracy of the data (described later in this section). However, quantifying the relative rotational and translational shifts between LiDAR scans are more difficult, because of the nature of the data. In recent years, there have been encouraging developments in this regard, and a few procedures to quantify the same are described below.

A detailed analysis of the IQC methods can be obtained from Habib et al., 2010. The conceptual basis of the IQC methodologies is that conjugate surface elements in overlapping strips should perfectly match in the absence of random and systematic errors. If consistent discrepancies are detected, then one can infer the presence of biases in the system parameters and/or measurements. Once the biases are removed, one can evaluate the noise level in the data by quantifying the goodness of fit between conjugate surface elements after removing systematic discrepancies.

Two overlapping scans can be (locally) described by the following equation:

$$\begin{pmatrix} X \\ Y \\ Z \end{pmatrix}_{Scan1} = \begin{pmatrix} X \\ Y \\ Z \end{pmatrix}_{T} + R_{\omega\varphi\kappa} \begin{pmatrix} X \\ Y \\ Z \end{pmatrix}_{Scan2} \quad (1)$$

Where $\begin{pmatrix} X \\ Y \\ Z \end{pmatrix}_{Scan1}$ and $\begin{pmatrix} X \\ Y \\ Z \end{pmatrix}_{Scan2}$ refer to the coordinates of the same feature from the two scans, $\begin{pmatrix} X \\ Y \\ Z \end{pmatrix}_{T}$ refers to the translational discrepancy between the two scans and $R_{\omega\varphi\kappa}$ is the rotational matrix derived from the angles ω,φ,κ. The values for the translational vector and the rotational angles should be as small as possible, and are the measures of discrepancy between two scans.

6.3.2.1 IQC using Iterative Closest Point method

The Iterative Closest Point (ICP) method (Besl and McKay, 1992; Chen and Medioni, 1992; Zhang, 1994) can be used to determine the transformation parameters between two point clouds/scan data of the same object. The ICP method, however, assumes that the points may be coincident. Since this is usually not the case, the method can be an approximate indicator of the differences in the scans. The proposed ICP method can be carried out for the entire overlap area between neighboring strips or it can be conducted for an aggregated set of local areas to speed up the process. A user can select a number of overlapping regions that he/she wants to check and those regions could be used to establish the parameters described in Eq. 1.

6.3.2.2 IQC using conjugate features (Planar and Linear)

The process of extracting planar features from point clouds is well studied. With assistance from an operator to determine regions in the point cloud where planar features are present, they can be extracted. IQC using planar and linear features can be more accurate than other methods. This is because of the availability of redundant data (points) over the planar features, which makes their (the planar parameters) estimates accurate (Kim et. al., 2007; Sampath and Shan, 2010). For IQC using planar features, conjugate planes from adjacent/overlapping scans can be extracted, and the location of their centroids and their normal vectors compared. More details on the procedure can be obtained in Habib et.al., 2010.

Linear features (e.g. roof ridges) can be defined as the intersection of two adjacent planar features and accurately extracted (Kim et. al., 2007; Sampath and Shan, 2010). Conjugate linear features extracted from overlapping scans can be compared on their directions (Habib et. al., 2010).

Point features can be extracted from planar features by determining the intersection of any three extracted (non-parallel) planes. While the (three or more) extracted planes may or may not intersect in reality, the coordinates obtained from the overlapping swaths should be comparable.

Another possibility is to compare the coordinates of the intersection of the perpendicular line to the extracted planar features.

The comparisons can be used to determine the parameters described in Eq. 1.

6.3.2.3 IQC using Point to Patch correspondence

The above method of QC using planar and linear features is easily accomplished in urban and semi-urban areas. However, in places where there are not enough of these features, a 'point to patch' correspondence scheme can be used to obtain the parameters in Eq. 1. In the ideal scenario, two overlapping LiDAR scans will match perfectly, i.e. the features formed from one scan will match the features formed from the other scan with regards to both shape and position. To measure the deviations from this ideal, a point to patch correspondence can be used.

To satisfy the objectives of having primitives that can be derived in the majority of ground-cover types with minimal preprocessing of the LiDAR point cloud, one can chose to represent one strip using the original points, while the other strip is represented by triangular patches, which can be derived from a TIN generation procedure. TIN generation from a given set of points is an automated procedure , which is available in several software packages. As an example, Fig. 1 shows the case where the strip denoted by (S2) is represented by a set of points while the other strip denoted by (S1) is represented by a set of triangular patches. Due to the high density of the LiDAR data, as well as the relatively smooth characteristics of terrain and man-made structures, using TIN patches to describe the physical surface is quite acceptable.

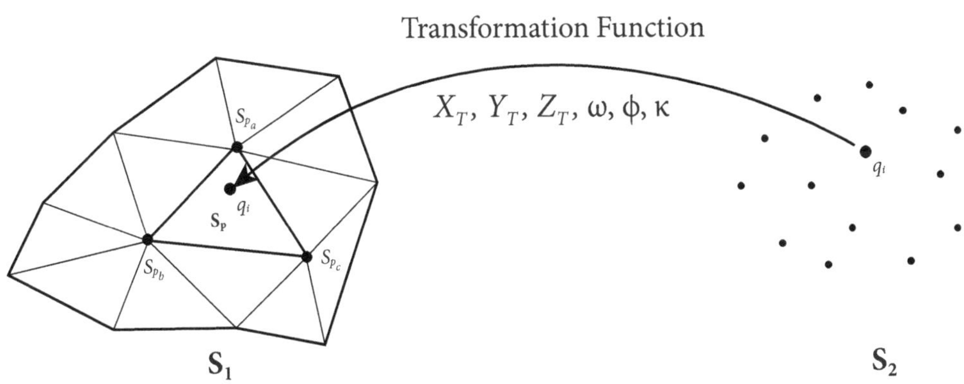

Figure 1 Conceptual basis of the proposed point-to-patch correspondence procedure.

There are some exceptions where the TIN patches would not represent a physical surface. These exceptions are caused by the irregular and sparse distribution of the LiDAR points. In vegetation areas, for example, planar patches through the LiDAR points do not represent the vegetation physical surface. Another instance of such a deviation will take place at building boundaries where the TIN patches are connecting points along the building rooftop and neighboring ground. In such a case, the TIN patches do not represent the building walls since the LiDAR point cloud would not necessarily coincide with the building boundaries and its footprint along the ground. Although one cannot assume that there is a point-to-point correspondence in overlapping strips, one can argue that we have a point-to-patch correspondence as long as the patch represents the physical surface.

This point to patch correspondence can be represented by the following equations:

$$\begin{vmatrix} X_{qi1} & Y_{qi1} & Z_{qi1} & 1 \\ X_{pa} & Y_{pa} & Z_{pa} & 1 \\ X_{pb} & Y_{pb} & Z_{pb} & 1 \\ X_{pc} & Z_{pc} & Z_{pc} & 1 \end{vmatrix} = 0 \tag{2}$$

$$\begin{pmatrix} X \\ Y \\ Z \end{pmatrix}_{qi1} = \begin{pmatrix} X \\ Y \\ Z \end{pmatrix}_{T} + R \begin{pmatrix} X \\ Y \\ Z \end{pmatrix}_{qi} \tag{3}$$

Where $\begin{pmatrix} X \\ Y \\ Z \end{pmatrix}_{qi}$ is the coordinate of Point qi in Scan2 (S2) and $\begin{pmatrix} X \\ Y \\ Z \end{pmatrix}_{qi1}$ is the corresponding transformed coordinate of the point in Scan1's (S1) coordinate system. Eq. 2 represents the constraint that the volume of the tetrahedron formed by the points S_{pa}, S_{pb}, S_{pc} and q_{i1} should be zero. This can be used to obtain the parameters expressed in Eq. 1.

6.3.2.4 IQC for relative height discrepancy

In many cases it is required to only determine the relative height discrepancy between two overlapping scans. It is assumed in these cases that the two scans are very well aligned planimetrically and discrepancy exists only in the vertical direction. In such cases, one of the scans can be treated as the reference, and a TIN model can be generated for the data. The 'Z' values for the two scans can be compared by interpolating the value of 'Z' from the TIN for a given (X, Y, Z) from the second scan and comparing the values. An RMSE can be generated for the entire overlap region by sampling randomly. The guidelines for reporting the RMSE is described in the ASPRS Guidelines on Vertical Accuracy Reporting for Lidar Data.

6.3.3 QC methods for external (absolute) accuracy of the data (EQC)

6.3.3.1 The OSU method

One method of measuring the absolute horizontal and vertical accuracy of LiDAR point clouds, developed at the Ohio State University (OSU), is based on using specially designed targets that can be sensed from the data. The first implementation was developed in collaboration with the Ohio Department of Transportation (ODOT), and, subsequently, other organizations have adopted the technique later. A detailed description and analysis of the method is provided in Csanyi and Toth (2007). The method also uses the intensity data collected with LiDAR.

The design and methodology are explained in Csanyi and Toth (2007). For point density of about 5 pts/sq.m and greater, a 1-m circle radius can already provide sufficient accuracy and further increasing the target size will not lead to significant improvements. A two-concentric-circle design (the inner circle has half the radius of the outer circle) with different coatings can result in significant accuracy improvements in the determined horizontal position since it provides additional geometric constraint in contrast to the one-circle design. The optimal coating pattern is a special white (high intensity return) coating for the inner circle and black (low intensity return) for the outer ring (Fig. 2).

The targets are deployed at or before the data collection, and the center points are surveyed using GPS; typically RTK or VRS measurements are performed at the perimeter and the center position is computed. To automate the processing of the target-based LiDAR data quality check and correction, a software toolbox was developed. The main modules include the initial batch processing, which is followed by an interactive analysis, and then the actual batch correction of the LiDAR data takes place. In the first phase, the program without human intervention selects the target areas from the LiDAR strips, finds the LiDAR points on the targets, and determines the target positions. The extraction of the points falling on the LiDAR targets is accomplished in two steps. First, LiDAR points in the vicinity of the targets are windowed out based on the known (surveyed) target coordinates and the maximum expected errors in the LiDAR data. In the second step, the points not falling on targets are filtered based on vertical elevation differences, and intensity information and subsequently the remaining target candidate points are checked for geometry. A detailed methodology to determine the target positions are described in Csanyi and Toth (2007). They are simply paraphrased here for the sake of completion. Since the targets are typically leveled, the vertical position of the target can be determined by fitting a horizontal plane to the points fallen on the target. A solution also exists for not leveled targets, but it is not discussed here. The horizontal position is determined by determining the center of gravity of the points that falls on the target. This is accomplished by adapting a generalized version of the Hough-transform (Hough, 1959 and 1962, Duda and Hart, 1972)

(a) (b) (c)

Figure 2 OSU/ODOT Targets for measurement of Horizontal and vertical accuracy of LiDAR data

6.3.3.2 The Purdue University/InDOT method

The Purdue University method (Bethel et. al., 2006) was developed in collaboration with the Indiana Department of Transportation (InDOT) and involves the use of targets as shown in Figure 2. The methodology is similar to the one developed by OSU/ODOT described above. It consists of surveying the interior vertex of the L-shaped target. The methodology also requires the use of intensity of LiDAR data. An intensity image of the data is generated, and the targets are determined based least squares based template matching techniques (Fig. 3). A more detailed description of the matching techniques are provided in (Bethel et. al., 2006). Once the template matching process locates the painted target, the horizontal and the vertical accuracy are determined in two steps.

The vertical accuracy is measured by generating a TIN model of the data, and measuring the vertical offset between the matched location and the surveyed values of the same point. Similarly, the horizontal accuracy is measured by comparing the planimetric coordinates of the matched location and the corresponding surveyed values

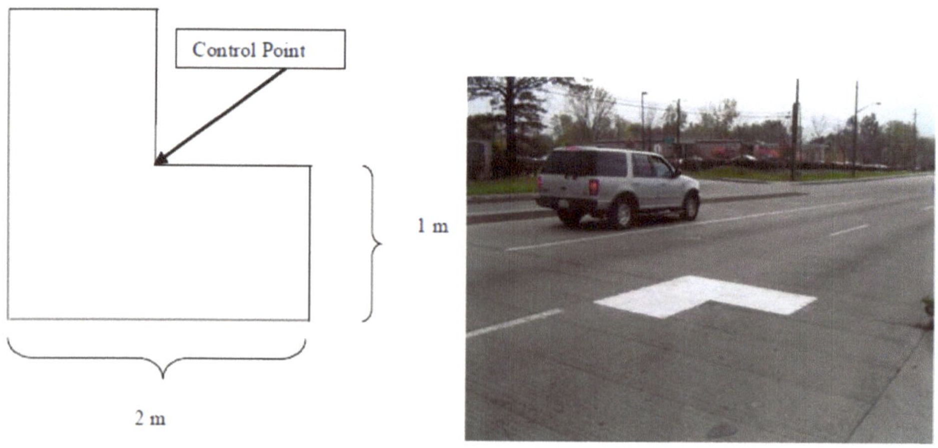

Figure 3 Targets for LiDAR point QC

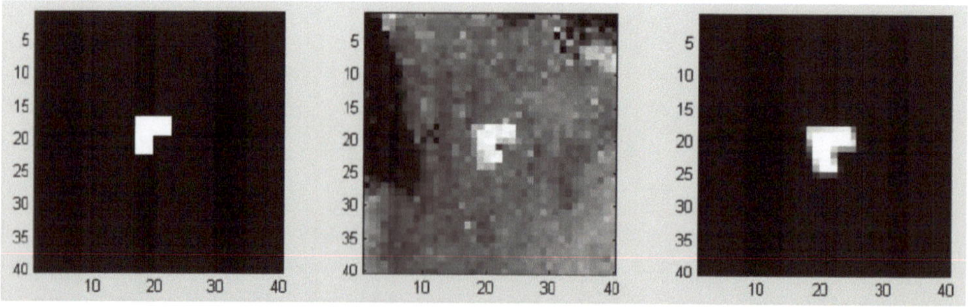

Figure 4 Template matching to determine the inner vertex

6.3.3.3 3D Pyramid shaped targets

Three dimensional targets, whose shape is accurately known, can also be used to measure the external accuracy of LiDAR data. Fig.5 shows the design of a 3D target in the shape of a pyramid (Jason Stoker, USGS, 2011). These targets could be made from plywood or alternatively using Mylar and stretched over a 3D frame. The center of the targets should be accurately surveyed and recorded. In the LiDAR scan, the four planar sides of the target should be extracted using feature extraction algorithms (similar to 6.3.2.2) and the intersection of these planes extracted from the LiDAR data should be compared with the surveyed coordinates of the target.

Alternatively, the planar sides could be measured using a total station instrument, and the surface of the sides could be determined. These surfaces could be compared with the surfaces extracted from the LiDAR data in a manner similar to methods described in 6.3.2.2.

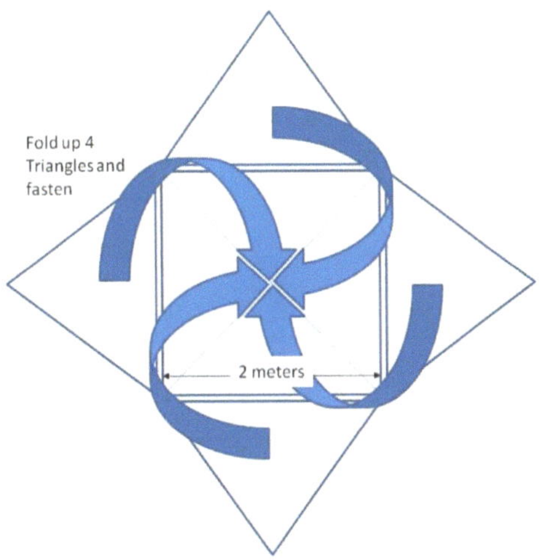

Figure 5 Design of targets in the shape of pyramids

ACKNOWLEDGEMENT

The content for Appendix B will be a sub chapter in the upcoming *ASPRS Guidelines on QA/ QC of Geometric Accuracy of Lidar Data.* These guidelines are being developed under the leadership of USGS EROS with input and review from experts in industry, academia, and government. The content is unedited and unchanged from the original, including section numbers and the capitalization of "LiDAR."

REFERENCES

Bethel, J. S., B. v. Gelder; A. F. Cetin, and A. Sampath. *Corridor Mapping Using Aerial Technique.* Publication FHWA/IN/JTRP-2006/23. Joint Transportation Research Program, Indiana Department of Transportation and Purdue University, West Lafayette, Indiana, 2006. doi: 10.5703/1288284313396

Bethel, J. S. and N. D. McKay, 1992. "A method of registration of 3-D shapes," IEEE Trans. Pattern Anal. Mach. Intell., vol. 14, no. 2, pp. 239–256, Feb. 1992.

Chen, Y. and G. Medioni, "Object modelling by registration of multiple range images," Image Vis. Comput., vol. 10, no. 3, pp. 145–155, Apr. 1992.

Csanyi, N. and Toth, C., 2007. Improvement of lidar data accuracy using lidar-specific ground targets. Photogrammetric Engineering & RemoteSensing 73(4), 385–396

Habib, A.; Kersting, A.P.; Ki In Bang; Dong-Cheon Lee; , "Alternative Methodologies for the Internal Quality Control of Parallel LiDAR Strips," Geoscience and Remote Sensing, IEEE Transactions on , vol.48, no.1, pp.221-236, Jan. 2010, doi: 10.1109/TGRS.2009.2026424

Kim , C., A. Habib; and P.Mrstik, New approach for planar patch segmentationusing airborne laser data, in Proc. ASPRS, Annu. Conf., Tampa, FL,2007. CD-ROM.

Lee, J., K. Yu, Y. Kim; and A. Habib, "Adjustment of discrepancies between LiDAR data strips using linear features," IEEE Geosci. RemoteSens. Lett., vol. 4, no. 3, pp. 475–479, Jul. 2007.

Sampath, A., Jie Shan; , "Segmentation and Reconstruction of Polyhedral Building Roofs From Aerial Lidar Point Clouds," Geoscience and Remote Sensing, IEEE Transactions on , vol.48, no.3, pp.1554-1567, March 2010, doi: 10.1109/TGRS.2009.2030180

Shin, S. W., A. Habib; M. Ghanma; C. Kim; and E. Kim, "Algorithms for multi-sensor and multi-primitive photogrammetric triangulation," ETRI J., vol. 29, no. 4, pp. 411–420, Aug. 2007.

Stoker, J. 2011. Design of targets for measurement of accuracy of LiDAR point cloud. Internal USGS communication.

Zhang, Z., "Iterative point matching for registration of free-form curves and surfaces," Int. J. Comput. Vis., vol. 13, no. 2, pp. 119–152, Oct. 1994

APPENDIX C
LIDAR METADATA

LIDAR METADATA EXAMPLE


```
<?xml version="1.0" encoding="UTF-8"?>
<!--DOCTYPE metadata SYSTEM "fgdc-std-001-1998.dtd"-->
<metadata>
    <idinfo>
    <citation>
      <citeinfo>
        <origin>We Map 4U, Inc.</origin>
        <pubdate>20101208</pubdate>
        <title>Lidar data for Phelps and Dent Counties, MO
        MO_Phelps-Dent-CO_2010</title>
        <geoform>Lidar point cloud</geoform>
      </citeinfo>
    </citation>
    <descript>
      <abstract>Geographic Extent: This dataset is lidar point cloud data
      which encompasses a 1000 meter buffer around Phelps and Dent Counties
      in Missouri, approximatly 829 square miles.

      Dataset Description: This dataset consists of 457 lidar point cloud
      .LAS swath files. Each .LAS file contains lidar point information which
      has been calibrated, controlled, and classified. Each file represents a
      separate swath of lidar. Collected swath files that were larger than
      2GB were initially written in multiple sub-swath files, each less than
      2GB.

      Ground Conditions: water at normal levels; no unusual innundation; no
      snow; leaf off
      </abstract>

      <lidar>
        <ldrinfo>
        <ldrspec>USGS NGP Base Specification v1</ldrspec>
          <ldrsens>Optech Gemini Airborne Laser Terrain Mappers (ALTM)
          </ldrsens>
          <ldrmaxnr>4</ldrmaxnr>
          <ldrnps>0.7071</ldrnps>
          <ldrdens>2</ldrdens>
          <ldrfltht>3000</ldrfltht>
          <ldrfltsp>115</ldrfltsp>
          <ldrscana>26</ldrscana>
          <ldrscanr>40</ldrscanr>
          <ldrpulsr>120</ldrpulsr>
          <ldrpulsd>10</ldrpulsd>
          <ldrpulsw>3</ldrpulsw>
          <ldrwavel>1064</ldrwavel>
          <ldrmpia>0</ldrmpia>
          <ldrbmdiv>4.5</ldrbmdiv>
          <ldrswatw>1200</ldrswatw>
          <ldrswato>15</ldrswato>
        <ldrcrs> NAD_1983_UTM_Zone_15N_Meters</ldrcrs>
        <ldrgeoid>National Geodetic Survey (NGS) Geoid09</ldrgeoid>
        </ldrinfo>
        <ldraccur>
          <ldrchacc>0.5</ldrchacc>
```

<rawfva>0.11</rawfva>
<clsfva>0.09</clsfva>
<clscva>0.188</clscva>
<clssvas>
<svalctyp>Forest</svalctyp>
<svavalue>0.38</svavalue>
</clssvas>
<clssvas>

<laswheld>Withheld (ignore) points were identifed in these files using the standard LAS Withheld bit
</laswheld>
<lasolap>Swath "overage" points were identified in these files by adding 16 to the standard classification values.
</lasolap>
<lasintrz>11-bit</lasintrz>
<lasclass>
<clascode>0</clascode>
<clasitem>Never Processed</clasitem>
</lasclass>
<lasclass>
<clascode>1</clascode>
<clasitem>Undetermined/Unclassified</clasitem>
</lasclass>
<lasclass>
<clascode>2</clascode>
<clasitem>Bare-earth</clasitem>
</lasclass>
<lasclass>
<clascode>4</clascode>
<clasitem>All vegetation</clasitem>
</lasclass>
<lasclass>
<clascode>6</clascode>
<clasitem>All structures except bridges</clasitem>
</lasclass>
<lasclass>
<clascode>7</clascode>
<clasitem>All noise</clasitem>
</lasclass>
<lasclass>
<clascode>8</clascode>
<clasitem>Bare-earth Key Points</clasitem>
</lasclass>
<lasclass>
<clascode>9</clascode>
<clasitem>Water</clasitem>
</lasclass>
<lasclass>
<clascode>10</clascode>
<clasitem>Bridges</clasitem>
</lasclass>
</lasinfo>
</lidar>

<purpose>The purpose of these LiDAR data was to produce high accuracy 3D hydro-flattened Digital Elevation Model (DEM) with a 1.0 foot cell size. The data will be used by FEMA for floodplain mapping.
These raw lidar point cloud data were used to create classified lidar LAS files, intensity images, 3D breaklines, hydro-flattened DEMs as necessary.
</purpose>
<supplinf>
USGS Contract No. G10PC01234
CONTRACTOR: We Map4U, Inc.
SUBCONTRACTOR: Aerial Scanning Services, LLC

```
    Lidar data was aquired and calibrated by Aerial Scanning Services. All
    follow-on processing was completed by the prime contractor.
    </supplinf>
  </descript>
  <timeperd>
    <timeinfo>
      <rngdates>
        <begdate>20100216</begdate>
        <enddate>20100218</enddate>
      </rngdates>
    </timeinfo>
    <current>ground condition</current>
  </timeperd>
  <status>
    <progress>Partial: Lot 2 of 5</progress>
    <update>None planned</update>
  </status>

  <spdom>
    <bounding>
      <westbc>-91.750000</westbc>
      <eastbc>-91.250000</eastbc>
      <northbc>38.000000</northbc>
      <southbc>37.250000</southbc>
    </bounding>
    <lboundng>
      <leftbc>584800.00</leftbc>
      <rightbc>664800.00</rightbc>
      <topbc>4225400.00</topbc>
      <bottombc>4141400.00</bottombc>
    </lboundng>
  </spdom>

  <keywords>
    <theme>
      <themekt>None</themekt>
      <themekey>Elevation data</themekey>
      <themekey>Lidar</themekey>
      <themekey>Hydrology</themekey>
    </theme>
    <place>
      <placekt>None</placekt>
      <placekey>Missouri</placekey>
      <placekey>Phelps County</placekey>
      <placekey>Dent County</placekey>
      <placekey>Mark Twain National Forest</placekey>
    </place>
    </keywords>
  <accconst>No restrictions apply to this data.</accconst>
  <useconst>None. However, users should be aware that temporal changes
  may have occurred since this data set was collected and that some parts
  of this data may no longer represent actual surface conditions. Users
  should not use this data for critical applications without a full
  awareness of its limitations. Acknowledgement of the U.S. Geological
  Survey would be appreciated for products derived from these data.
    </useconst>
  <ptcontac>
    <cntinfo>
      <cntorgp>
        <cntorg>We Map 4U, Data Aquisition Department</cntorg>
        <cntper>Jane Smith</cntper>
      </cntorgp>
      <cntaddr>
        <addrtype>mailing and physical</addrtype>
        <address>123 Main St.</address>
        <city>Anytown</city>
```

```
          <state>MO</state>
          <postal>61234</postal>
          <country>USA</country>
        </cntaddr>
        <cntvoice>555-555-1234</cntvoice>
        <cnttdd>555-555-1122</cnttdd>
        <cntfax>555-5550-1235</cntfax>
        <cntemail>jsmith@wemap4u.com</cntemail>
        <hours>Monday through Friday 8:00 AM to 4:00 PM(Central Time)
          </hours>
        <cntinst>If unable to reach the contact by telephone, please send an
        email. You should get a response within 24 hours.
          </cntinst>
      </cntinfo>
    </ptcontac>
    <native>Optech DASHMap 4.2200; ALS Post Processor 2.70 Build#15; GeoCue
    Version 6.1.21.4; Windows XP Operating System
    \\server\directory path\*.LAS
    17 GB
    </native>
  </idinfo>

  <dataqual>
    <logic>Data covers the entire area specified for this project.</logic>
    <complete>These raw LAS data files include all data points collected. No
    points have been removed or excluded. A visual qualitative assessment
    was performed to areas or missing data. The raw point cloud is of good
    quality and data passes Fundamental ertical Accuracy specifications.
    </complete>
  <posacc>
    <vertacc>
    <vertaccr>The specifications require that only Fundamental Vertical
    Accuracy (FVA) be computed for raw lidarpoint cloud swath files. The
    vertical accuracy was tested with 25 independent survey located in open
    terrain. These check points were not used in the calibration or post
    processing of the lidar point cloud data. The survey checkpoints were
    distributed throughout the project area. Specifications for this project
    require that the FVA be 25 cm or better AccuracyZ at 95% confidence
    level.
    </vertaccr>
    <qvertpa>
      <vertaccv>0.19 meters Accuracyz at 95% Confidence Interval</vertaccv>
      <vertacce>The FVA was tested using 25 independent survey located
      in open terrain terrain.  The survey checkpoints were distributed
      throughout the project area. The 25 independant check points were
      surveyed using xxxx technique.  Elevations from the unclassified lidar
      surface were measured for the x,y location of each check point.
      Elevations interpolated from the lidar surface were then compared to
      the elevation values of the surveyed control. The RMSE was computed to
      be 0.097cm.
        Accuracyz has been tested to meet 9.7 cm Fundamental Vertical
      Accuracy at 95% confidence level using RMSE(z) x 1.9600 as defined by
      the National Standards for Spatial Data Accuracy (NSSDA); assessed
      and reported using National Digital Elevation Program (NDEP)/ASPRS
      Guidelines.
        </vertacce>
      </qvertpa>
    </vertacc>
  </posacc>

  <lineage>
    <srcinfo>
      <srccite>
        <citeinfo>
          <origin>Jiffy Survey, Inc</origin>
          <pubdate>20100115</pubdate>
```

```
    <title>Ground Control for Phelps and Dent County, MO lidar
    project</title>
    <geoform>vector digital data and tabular data</geoform>
    <pubinfo>
      <pubplace>Jiffy Survey, Inc.</pubplace>
      <publish>Jiffy Survey, Inc., GPS department</publish>
    </pubinfo>
    <othercit>None</othercit>
    <onlink>ftp://JiffySurveyftp.com/data/outgoing/Task1/ </onlink>
  </citeinfo>
</srccite>
<srcscale>50</srcscale>
<typesrc>CD-ROM</typesrc>
<srctime>
  <timeinfo>
    <sngdate>
      <caldate> 201001003</caldate>
      </sngdate>
    </timeinfo>
    <srccurr>ground condition</srccurr>
  </srctime>
  <srccitea>Phelps_Co_lidar_gnd_ctrl</srccitea>
  <srccontr>This data source was used (along with the airborne GPS/IMU
  data) to georeference the lidar point cloud data.
  </srccontr>
</srcinfo>
<srcinfo>
  <srccite>
    <citeinfo>
      <origin>USDA</origin>
      <pubdate> 20090606</pubdate>
      <title>NAIP Imagery for Phelps and Dent County, MO lidar
      project</title>
      <geoform>raster orthoimagery</geoform>
      <pubinfo>
        <pubplace>USGS-EROS</pubplace>
        <publish>USGS-EROS</publish>
      </pubinfo>
      <othercit>None</othercit>
      <onlink></onlink>
    </citeinfo>
  </srccite>
  <srcscale>50</srcscale>
  <typesrc>online</typesrc>
  <srctime>
    <timeinfo>
      <sngdate>
        <caldate>20090101</caldate>
      </sngdate>
    </timeinfo>
    <srccurr>ground condition</srccurr>
  </srctime>
  <srccitea>Phelps-Dent_Co_NAIP_Imagery</srccitea>
  <srccontr>This data source was used (along with the lidar intensity
  imagery) to classify the lidar point cloud data.
  </srccontr>
</srcinfo>
<srcinfo>
  <srccite>
    <citeinfo>
      <origin>We Map 4U, Inc.</origin>
      <pubdate>20101208</pubdate>
      <title>Lidar Intensity Imagery for Phelps and Dent County, MO
      </title>
      <geoform>raster orthoimagery</geoform>
      <pubinfo>
```

```
        <pubplace>USGS-EROS</pubplace>
        <publish>USGS-EROS</publish>
      </pubinfo>
      <othercit>None</othercit>
      <onlink></onlink>
    </citeinfo>
  </srccite>
  <srcscale>50</srcscale>
  <typesrc>online</typesrc>
  <srctime>
    <timeinfo>
  <rngdates>
    <begdate>20100216</begdate>
    <enddate>20100218</enddate>
  </rngdates>
    </timeinfo>
    <srccurr>ground condition</srccurr>
  </srctime>
  <srccitea>Phelps-Dent_Co_Lidar_Intensity_Imagery</srccitea>
  <srccontr>This data source was used (along with NAIP imagery) to
  classify the lidar point cloud data.
  </srccontr>
</srcinfo>

<procstep>
  <procdesc>Lidar Pre-Processing: Airborne GPS and IMU data were
  merged to develop a Single Best Estimate (SBET) of the lidar
  system trajectory for each lift. Lidar ranging data were initially
  calibrated using previous best parameters for this instrument and
  aircraft. Relative calibration was evaluated using advanced plane-
  matching analysis and parameter corrections derived. This was
  repeated iteratively until residual errors between overlapping
  swaths, across all project lifts, was reduced to 2cm or less. Data
  were then block adjusted to match surveyed calibration control. Raw
  data FVA was checked using independently surveyed checkpoints. Swath
  overage points were identified and tagged within each swath file.
  The results of the final calibration, FVA and horizontal accuracy
  assessments, and the "raw" swaths were forwarded to the client to
  obtain a Notice To Proceed on classification and derivative product
  generation.
  </procdesc>
  <srcused>Phelps_Co_lidar_gnd_ctrl</srcused>
  <procdate>20100131</procdate>
  <proccont>
    <cntinfo>
      <cntorgp>
        <cntorg>We Map 4U, Data Aquisition Department</cntorg>
        <cntper>Manny Puntas</cntper>
    </cntorgp>
      <cntaddr>
      <addrtype>mailing and physical</addrtype>
      <address>123 Main St.</address>
      <city>Anytown</city>
      <state>MO</state>
      <postal>61234</postal>
      <country>USA</country>
    </cntaddr>
    <cntvoice>555-555-556</cntvoice>
    <cntfax>555-5550-1236</cntfax>
    <cntemail>mpuntas@wemap4u.com</cntemail>
    <hours>Monday through Friday 8:00 AM to 4:00 PM (Central Time)
      </hours>
    <cntinst> If unable to reach the contact by telephone, please send
    an email. You should get a response within 24 hours.
    </cntinst>
  </cntinfo>
```

```
      </proccont>
    </procstep>
    <procstep>
      <procdesc>Lidar Post-Processing: The calibrated and controlled
      lidar swaths were processed using automatic point classification
      routines in proprietary software. These routines operate against the
      entire collection (all swaths, all lifts), eliminating character
      differences between files. Data were then distributed as virtual tiles
      to experienced lidar analysts for localized automatic classification,
      manual editing, and peer-based QC checks. Supervisory QC monitoring
      of work in progress and completed editing ensured consistency of
      classification character and adherence to project requirements across
      the entire project area.
      All classification tags are stored in the original swath files. After
      completion of classification and final QC approval, the FVA, SVAs, and
      CVA for the project are calculated. Sample areas for each landcover
      type present in the project area were extracted and forwarded
      to the client, along with the results of the accuracy tests. Upon
      acceptance, the complete classified lidar swath files were delivered to
      the client.
      </procdesc>
      <srcused>Phelps-Dent_Co_NAIP_Imagery</srcused>
      <srcused>Phelps-Dent_Co_Lidar_Intensity_Imagery</srcused>
      <procdate>20100530</procdate>
      <proccont>
        <cntinfo>
          <cntorgp>
            <cntorg>We Map 4U, Data Aquisition Department</cntorg>
            <cntper>Manny Puntas</cntper>
        </cntorgp>
          <cntaddr>
          <addrtype>mailing and physical</addrtype>
          <address>123 Main St.</address>
          <city>Anytown</city>
          <state>MO</state>
          <postal>61234</postal>
          <country>USA</country>
        </cntaddr>
        <cntvoice>555-555-556</cntvoice>
        <cntfax>555-5550-1236</cntfax>
        <cntemail>mpuntas@wemap4u.com</cntemail>
        <hours>Monday through Friday 8:00 AM to 4:00 PM (Central Time)
        </hours>
        <cntinst> If unable to reach the contact by telephone, please send
        an email. You should get a response within 24 hours.
        </cntinst>
       </cntinfo>
     </proccont>
   </procstep>
 </lineage>
 </dataqual>

 <spdoinfo>
   <direct>Vector</direct>
   <ptvctinf>
     <sdtsterm>
       <sdtstype>Point</sdtstype>
       <ptvctcnt>764,567,423</ptvctcnt>
     </sdtsterm>
   </ptvctinf>
 </spdoinfo>

 <spref>
   <horizsys>
     <planar>
       <gridsys>
```

```
        <gridsysn>Universal Transverse Mercator</gridsysn>
        <utm>
          <utmzone>15</utmzone>
            <transmer>
              <sfctrmer>0.9996</sfctrmer>
              <longcm>-117.000000</longcm>
              <latprjo>0.0</latprjo>
              <feast>500000</feast>
              <fnorth>0.0</fnorth>
            </transmer>
        </utm>
      </gridsys>
        <planci>
          <plance>coordinate pair</plance>
          <coordrep>
            <absres>0.01</absres>
            <ordres>0.01</ordres>
          </coordrep>
          <plandu>meters</plandu>
        </planci>
    </planar>
    <geodetic>
      <horizdn>North American Datum of 1983</horizdn>
      <ellips>Geodetic Reference System 80</ellips>
        <semiaxis>6378137</semiaxis>
        <denflat>298.257222101</denflat>
    </geodetic>
  </horizsys>
  <vertdef>
    <altsys>
      <altdatum>North American Vertical Datum of 1988</altdatum>
      <altres> 0.01</altres>
      <altunits>meters</altunits>
      <altenc>Explicit elevation coordinate included with horizontal
      coordinates</altenc>
    </altsys>
  </vertdef>
</spref>

<distinfo>
  <distrib>
    <cntinfo>
      <cntperp>
        <cntper>Jim Brooks, GISP</cntper>
        <cntorg>Phelps-Dent Council of Government (PDCOG), GIS and Data
        Division</cntorg>
      </cntperp>
      <cntpos>Director</cntpos>
      <cntaddr>
        <addrtype>mailing and physical address</addrtype>
        <address>PDCOG, GIS Division</address>
        <address>123 ABD Street</address>
        <address> Suite 456</address>
        <city>Sometown</city>
        <state>MO</state>
        <postal>99999</postal>
        <country>USA</country>
      </cntaddr>
      <cntvoice>555-555-9999</cntvoice>
      <cntemail>jim.brooks@PDCOG.org</cntemail>
    </cntinfo>
  </distrib>
  <resdesc>The PDCOG distributes data directly to program partners. Public
  access to the data are available from the USGS as listed below.
  </resdesc>
```

```
  <distliab>In no event shall the creators, custodians, or distributors
  of these data be liable for any damages arising out of its use, or from
  the inability of the customer to use these data for their intended
  application.
</distliab>
</distinfo>

<metainfo>
  <metd> 20101206</metd>
  <metrd> 20101207</metrd>
  <metc>
    <cntinfo>
      <cntorgp>
        <cntorg>We Map 4U, Data Aquisition Department</cntorg>
        <cntper>John Smith</cntper>
      </cntorgp>
      <cntaddr>
        <addrtype>mailing and physical</addrtype>
        <address>123 Main St.</address>
        <city>Anytown</city>
        <state>MO</state>
        <postal>61234</postal>
        <country>USA</country>
      </cntaddr>
      <cntvoice>555-555-1234</cntvoice>
      <cnttdd>555-555-1122</cnttdd>
      <cntfax>555-5550-1235</cntfax>
      <cntemail>jsmith@wemap4u.com</cntemail>
      <hours>Monday through Friday 8:00 AM to 4:00 PM (Central Time)</hours>
      <cntinst> If unable to reach the contact by telephone, please send an
      email. You should get a response within 24 hours.
      </cntinst>
    </cntinfo>
  </metc>
  <metstdn>FGDC Content Standard for Digital Geospatial Metadata</metstdn>
  <metstdv>FGDC-STD-001-1998</metstdv>
  <metac>None</metac>
  <metuc>None</metuc>
  <metsi>
    <metscs>None</metscs>
    <metsc>Unclassified</metsc>
    <metshd>None</metshd>
  </metsi>
  <metextns>
    <onlink>None</onlink>
    <metprof>None</metprof>
    </metextns>
  </metainfo>
</metadata>
```


LIDAR METADATA TEMPLATE


```
<?xml version="1.0" encoding="UTF-8"?>
<!--DOCTYPE metadata SYSTEM "fgdc-std-001-1998.dtd"-->
<metadata>
  <idinfo>
    <citation>
      <citeinfo>
        <origin> EXAMPLE: We Map 4U, Inc.
        <!--REQUIRED Element: Originator
        Name of the contractor that developed the data set.
        Domain: "Unknown" free text-->
        </origin>
        <pubdate> 20101208
```

```
        <!--REQUIRED Element: Publication Date
        Date that the dataset was RELEASED. The field MUST be formatted
        YYYYMMDD
        Domain: "Unknown" "Unpublished Material" YYYYMMDD free text-->
        </pubdate>
        <title> EXAMPLE: Lidar data for Phelps and Dent Counties, MO
                MO_Phelps-Dent-CO_2010
          <!--REQUIRED Element: Title
          The name by which the data set is known.
          IF A PROJECT ID IN THE FOLLOWING FORMAT HAS BEEN ISSUED FOR THIS
          PROJECT, INCLUDE IT in the title element [State_description_
          aquisition-date]
        Domain: free text-->
        </title>
        <geoform> EXAMPLE: Lidar point cloud
          <!--REQUIRED Element: Geospatial Data Presentation Form
          The mode in which the geospatial data are represented.
          Domain:  free text-->
        </geoform>
      </citeinfo>
    </citation>
    <descript>
      <abstract> EXAMPLE: Geographic Extent: This dataset is lidar point
      cloud data which encompasses a 1000 meter buffer around Phelps and
      Dent Counties in Missouri, approximatly 829 square miles.

      Dataset Description: This dataset consists of 457 lidar point cloud
      .LAS swath files. Each .LAS file contains lidar point information which
      has been calibrated, controlled, and classified. Each file represents a
      separate swath of lidar. Collected swath files that were larger than 2GB
      were initially written in multiple sub-swath files, each less than 2GB.

      Ground Conditions: water at normal levels; no unusual innundation; no
      snow; leaf off
        <!--REQUIRED Element: Abstract
        A brief narrative summary of the dataset.

        It is desireable for the Abstract to include a consolidated summary
        other elements that are included elsewhere in this metadata file, for
        ease of use.

        Domain: free text-->
      </abstract>

      <lidar>
      <!-- REQUIRED Section: for Project, Swath, and Classified LAS metadata
      files -->
        <ldrinfo>
          <!--REQUIRED Group:  This group of tags contains metadata about
          the sesnor and collection conditions.-->
          <ldrspec>EXAMPLE: USGS-NGP Base Lidar Specification v1.0
          <!--REQUIRED Element: the lidar specification applicable to the
          point cloud --></ldrspec>
          <ldrsens>EXAMPLE: Optech Gemini Airborne Laser Terrain Mappers
          (ALTM)
          <!--REQUIRED Element: the lidar sensor make and model -->
          </ldrsens>
        <ldrmaxnr>EXAMPLE: 4
          <!--REQUIRED Element: the maximum number of returns per pulse
          --></ldrmaxnr>
        <ldrnps>EXAMPLE: 1.2
          <!--REQUIRED Element: the Nominal Pulse Spacing, in Meters -->
          </ldrnps>
        <ldrdens>EXAMPLE: 2
          <!--REQUIRED Element: the Nominal Pulse Spacing, in Points Per
          Square Meter --></ldrdens>
```

```
<ldrfltht>EXAMPLE: 3000
  <!--REQUIRED Element: the nominal flight height for the collection,
  in Meters --></ldrfltht>
<ldrfltsp>EXAMPLE: 115
  <!--REQUIRED Element: the nominal flight speed for the collection,
  in Knots --></ldrfltsp>
<ldrscana>EXAMPLE: 26
  <!--REQUIRED Element: the sensor scan angle, total, in Degrees
  --></ldrscana>
<ldrscanr>EXAMPLE: 40
  <!--REQUIRED Element: the scan frequency of the scanner, in Hertz
  --></ldrscanr>
<ldrpulsr>EXAMPLE: 120
  <!--REQUIRED Element: the pulse rate of the scanner, in Kilohertz
  --></ldrpulsr>
  <ldrpulsd>EXAMPLE: 10
  <!--REQUIRED Element: the pulse duration of the scanner, in
  Naonseconds --></ldrpulsd>
  <ldrpulsw>EXAMPLE: 3
  <!--REQUIRED Element: the pulse width of the scanner, in Meters
  --></ldrpulsw>
  <ldrwavel>EXAMPLE: 1064
  <!--REQUIRED Element: the central wavelength of the sensor laser,
  in Nanometers --></ldrwavel>
<ldrmpia>EXAMPLE: 0
  <!--REQUIRED Element: Whether the sensor was operated with
  Multiple Pulses In The Air, 0=No; 1=Y --></ldrmpia>
<ldrbmdiv>EXAMPLE: 0.3
  <!--REQUIRED Element: the beam divergence, in Milliradians --></
  ldrbmdiv>
<ldrswatw>EXAMPLE: 1200
  <!--REQUIRED Element: the nominal swath width on the ground, in
  Meters --></ldrswatw>
<ldrswato>EXAMPLE: 15
  <!--REQUIRED Element: the nominal swath overlap, as a Percentage
  --></ldrswato>
<ldrcrs> EXAMPLE: NAD_1983_UTM_Zone_15N_Meters
<!--REQUIRED Element: This is the Horizontal Coordinate Reference
System for the data.
Details for the desired format are below.
Go to the website:  http://spatialreference.org/ref/
Search for your CRS, then select the name that reflects your CRS,
Select ESRI WKT
Use the name that is within the first set of quotation marks.
This name should consist of datum, coordinate system, and sometimes
units all separated by underscores instead of spaces.
When units are not already present, the must be appended then the
units to the end of the string:
_Meters  _Feet_US   _Feet_Intl   -->
</ldrcrs>
<ldrgeoid>EXAMPLE: National Geodetic Survey (NGS) Geoid09
<!--REQUIRED Element: Geoid used for vertical reference. -->
</ldrgeoid>
</ldrinfo>

 <ldraccur>
  <!-- REQUIRED Group: This group of tags contains information
  on point cloud accuracy. Not all tags within this group are
  mandatory.
  The FVA of the raw point cloud is required. CVA and SVA values
  for the classified point cloud are optional, but are required to
  be reported if they are available. ALL Values are reported in
  METERS-->
  <ldrchacc>EXAMPLE: 0.5
  <!--REQUIRED Element: the required vertical accuracy for the point
  cloud data --></ldrchacc>
```

```
    <rawfva>EXAMPLE: 0.11
    <!--REQUIRED Element: the calculated fundamental vertical accuracy
    of the raw point cloud data --></rawfva>
    <clsfva>EXAMPLE: 0.09
    <!--OPTIONAL Element: the calculated fundamental vertical accuracy
    of the classified point cloud data (required if available) -->
    </clsfva>
    <clscva>EXAMPLE: 0.188
    <!--OPTIONAL Element: the calculated consolidated vertical
    accuracy of the classified point cloud data (required if available)
    --></clscva>
    <clssvas>
      <!--OPTIONAL Group for Raw swath points, REQUIRED Group for
      Classified LAS Tiles:
      Each land cover besides open ground will have a Supplemental
      Vertical Accuracy.
      The two tags under <clssvas> document the land cover type and
      the SVA value (Forest, in this case)
      This clssvas group will be repeated for each landcover class -->
      <svalctyp>EXAMPLE: Forest</svalctyp>
      <svavalue>EXAMPLE: 0.38 </svavalue>
    </clssvas>
    <clssvas>
      <svalctyp>EXAMPLE: Agriculture</svalctyp>
      <svavalue>EXAMPLE: 0.21 </svavalue>
    </clssvas>
    <clssvas>
      <svalctyp>EXAMPLE:Pasture/Meadow</svalctyp>
      <svavalue>EXAMPLE: 0.15 </svavalue>
    </clssvas>
    <clssvas>
      <svalctyp>EXAMPLE: Scrub/Shrub</svalctyp>
      <svavalue>EXAMPLE: 0.11 </svavalue>
    </clssvas>
    <clssvas>
      <svalctyp>EXAMPLE: Urban</svalctyp>
      <svavalue>EXAMPLE: 0.13 </svavalue>
    </clssvas>
  </ldraccur>

  <lasinfo>
    <!-- REQUIRED Group: This group of tags contains information on
    the LAS version and classification values for the point cloud.-->
  <lasver>EXAMPLE: 1.3
  <!-- REQUIRED Element: The version of the LAS Standard applicable to
  this data set.--></lasver>
  <lasprf>EXAMPLE: 6
    <!-- REQUIRED Element: The point record format used for the point
    cloud.--></lasprf>
  <laswheld>EXAMPLE: Withheld (ignore) points were identifed in these
  files using the standard LAS Withheld bit
  <!-- REQUIRED Element: Describe how withheld points are
  identified.--></laswheld>
  <lasolap>EXAMPLE: Swath "overage" points were identified in these
  files by adding 16 to the standard classification values.
  <!-- REQUIRED Element: This describes how overage points are
  identified.--></lasolap>
  <lasintr>EXAMPLE: 11
  <!-- REQUIRED Element: This specifies the radiometric resolution of
  intensity values, in Bits.--></lasintr>
<lasclass>
  <!-- REQUIRED Section if LAS data is classified: All lasclass
  sections provide a code value and a description for that code.-->
      <clascode>EXAMPLE: 0</clascode>
      <clasitem>EXAMPLE: Never Processed</clasitem>
  </lasclass>
```

```
    <lasclass>
        <clascode>EXAMPLE: 1</clascode>
        <clasitem>EXAMPLE: Undetermined/Unclassified</clasitem>
    </lasclass>
    <lasclass>
        <clascode>EXAMPLE: 2</clascode>
        <clasitem>EXAMPLE: Bare-earth</clasitem>
    </lasclass>
    <lasclass>
        <clascode>EXAMPLE: 4</clascode>
        <clasitem>EXAMPLE: All vegetation</clasitem>
    </lasclass>
    <lasclass>
        <clascode>EXAMPLE: 6</clascode>
        <clasitem>EXAMPLE: All structures except bridges</clasitem>
    </lasclass>
    <lasclass>
        <clascode>EXAMPLE: 7</clascode>
        <clasitem>EXAMPLE: All noise</clasitem>
    </lasclass>
    <lasclass>
        <clascode>EXAMPLE: 8</clascode>
        <clasitem>EXAMPLE: Bare-earth Key Points</clasitem>
    </lasclass>
    <lasclass>
        <clascode>EXAMPLE: 9</clascode>
        <clasitem>EXAMPLE: Water</clasitem>
    </lasclass>
    <lasclass>
        <clascode>EXAMPLE: 10</clascode>
        <clasitem>EXAMPLE: Bridges</clasitem>
    </lasclass>
    </lasinfo>
  </lidar>

  <purpose>The purpose of these LiDAR data was to produce high accuracy
  3D hydro-flattened Digital Elevation Model (DEM) with a 1.0 foot cell
  size. The data will be used by Fema for flood plane mapping.
  These raw lidar point cloud data were used to create classified lidar
  LAS files, intensity images, 3D breaklines, hydro-flattened DEMs as
  necessary.
    <!--REQUIRED Element: Purpose
    Why was the data set was created? For what applications?
    What other products this dataset will be used to create: tiled
    classified LAS, DEM, etc, both required deliverables, or interim
    products necessary tocomplete the project.  What scales is it
    appropriate or inappropriate for use?
    Domain: free text-->
  </purpose>
  <supplinf>
    USGS Contract No. G10PC01234
    CONTRACTOR: We Map4U, Inc.
    SUBCONTRACTOR:  Aerial Scanning Services, LLC
    Lidar data was aquired and calibrated by Aerial Scanning Services.
    All follow-on processing was completed by the prime contractor.
    <!--OPTIONAL Element: Supplemental Information
    Enter other descriptive information about the data set.
    Desirable information includes any deviations from project
    specifications and reasons. It may alos include any other information
    that the contractor finds necessary or useful, such as contract
    number, summary of LiDAR technology. Remove this tag or clear the
    contents of this tag if none.
    Domain: free text-->
  </supplinf>
</descript>
<timeperd>
```

```
    <timeinfo>
    <!--REQUIRED Group: Time info:will be either:
                            single date,

                            OR multiple dates,
                            OR a range of dates.
                            Examples are provided for all three formats.
                            Delete the ones that do not apply.-->
      <sngdate>
        <!--Begin the example of Single Date-->
          <caldate> 20100216
            <!--REQUIRED Element: Calendar Date
            This is the date of the LiDAR collection, if the collection
            was completed in one day. &#13
          The field MUST be formatted YYYYMMDD-->
          </caldate>
      </sngdate>
      <mdattim>
        <!-- Begin example of a multiple dates -->
        <sngdate>
          <caldate> 20100216
            <!--REQUIRED Element: Calendar Date
            This is the first date of the LiDAR collection, when multiple
            collection dates are specified.
            The field MUST be formatted YYYYMMDD-->
            </caldate>
          </sngdate>
        <sngdate>
          <caldate> 20100218
            <!--REQUIRED Element: Calendar Date
            This is the second date of the LiDAR collection, when multiple
            collection dates are specified.
            The field MUST be formatted YYYYMMDD
            REPEAT the sngdate and caldate tags for each collection
            date-->
          </caldate>
        </sngdate>
      </mdattim>
      <rngdates>
        <!-- Begin example of a date range -->
          <begdate> 20100216
            <!--REQUIRED Element: Beginning Date
            This is the beginning date of LiDAR collection.
            The field MUST be formatted YYYYMMDD-->
          </begdate>
          <enddate> 20100218
            <!--REQUIRED Element: Ending Date
            This is the ending date of LiDAR collection.
            The field MUST be formatted YYYYMMDD-->
          </enddate>
        </rngdates>
      </timeinfo>
    <current> EXAMPLE: ground condition
      <!--REQUIRED Element: Currrentness Reference
      Enter the basis on which the time period of content information is
      determined.
      Domain: "ground condition" "publication date" free text-->
    </current>
  </timeperd>
  <status>
    <progress> EXAMPLE: Partial: Lot 2 of 5
      <!--REQUIRED ELEMENT: Progress
      Enter the state of the data set.
      Domain: "Complete" "Partial: Lot x of n"-->
    </progress>
    <update> EXAMPLE: None planned
```

```
    <!--REQUIRED ELEMENT: Maintenance and Update Frequency
    Enter the repeat cycle for the project.
    Domain: "Annually" "Unknown" "None planned" free text-->
  </update>
</status>
<spdom>
  <bounding>
    <westbc> -91.750000
      <!--REQUIRED Element: West Bounding Coordinate
      This value is the Western-most limit of coverage of the dataset
      expressed as longitude.
      This value will be negative in the United States.
      This value MUST be expressed in Decimal Degrees.
      Domain: -180.0 <= West Bounding Coordinate < 180.0-->
    </westbc>
    <eastbc> -91.25000
      <!--REQUIRED Element: East Bounding Coordinate
      This value is the Eastern-most limit of coverage of the dataset
      expressed as longitude.
      This value will be negative in the United States.
      This value MUST be expressed in Decimal Degrees.
      Domain: -180.0 <= East Bounding Coordinate <= 180.0-->
    </eastbc>
    <northbc> 38.00000
      <!--REQUIRED Element: North Bounding Coordinate
      This value is the Northern-most limit of coverage of the dataset
      expressed as latitude.
      This value will be positive in the United States.
      This value MUST be expressed in Decimal Degrees.
      Domain: -90.0 <= North Bounding Coordinate <= 90.0-->
    </northbc>
    <southbc> 37.250000
      <!--REQUIRED Element: South Bounding Coordinate
      This value is the Southern-most limit of coverage of the dataset
      expressed as latitude.
      This value will be positive in the United States.
      This value MUST be expressed in Decimal Degrees.
      Domain: -90.0 <= South Bounding Coordinate <= 90.0-->
    </southbc>
  </bounding>
    <lboundng>
      <leftbc> 584800
        <!--REQUIRED Element: This value is the Western-most coordinate
        of limit of coverage of the dataset expressed in the Coordinate
        Reference System in which the data is delivered.  -->
      </leftbc>
      <rightbc> 664800
        <!--REQUIRED Element: This value is the Eastern-most coordinate
        of limit of coverage of the dataset expressed in the Coordinate
        Reference System in which the data is delivered.  -->
      </rightbc>
      <topbc> 4225400
        <!--REQUIRED Element: This value is the Northern-most coordinate
        of limit of coverage of the dataset expressed in the Coordinate
        Reference System in which the data is delivered.  -->
      </topbc>
      <bottombc> 4141400
        <!--REQUIRED Element: This value is the Southern-most coordinate
        of limit of coverage of the dataset expressed in the Coordinate
        Reference System in which the data is delivered.  -->
      </bottombc>
    </lboundng>
  </spdom>
  <keywords>
    <theme>
      <themekt> EXAMPLE: None
```

```
        <!--REQUIRED Element: Theme Keyword Thesaurus
        A formally registered thesaurus or a similar authoritative
        source of theme keywords.
          Domain: "None" free text-->
      </themekt>
      <themekey> EXAMPLE: Elevation data
        <!--REQUIRED Element: Theme Keyword: Elevation data
        (required)-->
      </themekey>
      <themekey> EXAMPLE: Lidar
        <!--REQUIRED Element: Theme Keyword: Lidar (required)-->
      </themekey>
      <themekey> EXAMPLE: Hydrology
        <!--Enter any additional applicable theme keywords, use only ONE
        keyword for each themekey tag. Repeat the themekey tag as many
        times as necessary.
        Domain: free text-->
      </themekey>
    </theme>
    <place>
      <placekt> EXAMPLE: None
        <!--REQUIRED Element: Place Keyword Thesaurus
        Reference to a formally registered thesaurus or a similar
        authoritative source of place keywords.
          Domain: "None" "Geographic Names Information System" free
          text-->
      </placekt>
      <placekey> EXAMPLE: Missouri
        <!--REQUIRED Element:  Place Keyword
        For multi-state projects, make a separate entry for each state.
        List only one state for each placekey tag. -->
      </placekey>
      <placekey> EXAMPLE: Phelps County
        <!--REQUIRED Element:  Place Keyword
        For multi-county projects, make a separate entry for each
        county.
        List only one county for each placekey tag. -->
      </placekey>
      <placekey> EXAMPLE: Dent County
      </placekey>
      <placekey> EXAMPLE: Mark Twain National Forest
        <!--Enter any additional applicable place keywords, for example
        cities or landmarks.
        Use only one keyword for each placekey tag.
        Repeat the placekey tag as many times as necessary.

        Domain: free text-->
      </placekey>
    </place>
  </keywords>
<accconst> EXAMPLE: No restrictions apply to this data.
  <!--REQUIRED Element: Access Constraints.
  Enter restrictions and legal prerequisites for accessing the data set.
  These include any access constraints applied to assure the protection
  of privacy or intellectual property, and any special restrictions or
  limitations on obtaining the data set.
  Domain: "None" free text-->
</accconst>
<useconst> EXAMPLE: None. However, users should be aware that temporal
changes may have occurred since this data set was collected and
that some parts of this data may no longer represent actual surface
conditions.  Users should not use this data for critical applications
without a full awareness of it's limitations.  Acknowledgement of the
U.S. Geological Survey would be appreciated for products derived from
these data.
```

```
<!--REQUIRED Element: Enter restrictions and legal prerequisites for using
the data set after access is granted. These include any use constraints
applied to assure the protection of privacy or intellectual property, and
any special restrictions or limitations on using the data set.
  Domain: "None" free text-->
</useconst>
<ptcontac>
  <cntinfo>
    <cntorgp>
      <cntorg> EXAMPLE: We Map 4U, Data Aquisition Department
        <!--REQUIRED Element: Contact Organization:
        Name of the organization that created the data and is
        knowledgeable about the data.
        Domain: free text-->
      </cntorg>
      <cntper> EXAMPLE: Jane Smith
        <!--REQUIRED Element: Contact Person
        The name of the individual who is knowledgeable about the data.
        Domain: free text-->
      </cntper>
    </cntorgp>
    <cntaddr>
      <addrtype> EXAMPLE: mailing and physical
        <!--REQUIRED Element:  Address Type
        The type of address that follows.
        Only required for "mailing" or "mailing and physical".  If the
        contractor has a different mailing and physical address, the
        physical address does not need to be included.  This section may
        be repeated if you would like to provide a separate physical
        address.
          Domain: "mailing" "physical" "mailing and physical", free
          text-->
      </addrtype>
      <address> EXAMPLE: 123 Main St.
        <!--REQUIRED Element: Address
        The address of the contractor.
        For multiple line addresses the address tag may be repeated as
        many times as needed.
        Domain: free text-->
      </address>
      <city> EXAMPLE: Anytown
        <!--REQUIRED Element: City
        The city of the address.
        Domain: free text-->
      </city>
      <state> EXAMPLE: MO
        <!--REQUIRED Element: State
        The state or province of the address.
        Domain: free text-->
      </state>
      <postal> EXAMPLE: 61234
        <!--REQUIRED Element: Postal Code
        Enter the ZIP or other postal code of the address.
          Domain: free text-->
      </postal>
      <country> EXAMPLE: USA
        <!--OPTIONAL Element: Country
        The country of the address.
          Domain: free text-->
      </country>
    </cntaddr>
    <cntvoice> EXAMPLE: 555-555-1234
      <!--REQUIRED Element: Contact Voice Telephone
      The telephone number by which individuals can speak to the
      organization or individual responsible for the data.
        Domain: free text-->
```

```
      </cntvoice>
      <cnttdd> EXAMPLE: 555-555-1122
        <!--OPTIONAL Element: Contact TDD/TTY Telephone
        The telephone number by which hearing-impaired individuals can
        contact the organization or individual.
          Domain: free text-->
      </cnttdd>
      <cntfax> EXAMPLE: 555-5550-1235
      <!--OPTIONAL Element: Contact Fax
        The telephone number of a facsimile machine of the organization or
        individual.
          Domain: free text-->
      </cntfax>
      <cntemail> EXAMPLE: jsmith@wemap4u.com
        <!--OPTIONAL Element: Contact E-mail Address
        The email address of the organization or individual.
        Domain: free text-->
      </cntemail>
      <hours> EXAMPLE: Monday through Friday 8:00 AM to 4:00 PM (Central
      Time)
        <!--OPTIONAL Element: Hours of Service
        The time period when individuals can speak to the organization or
        individual.
          Domain: free text-->
      </hours>
      <cntinst> EXAMPLE: If unable to reach the contact by telephone,
      please send an email.  You should get a response within 24 hours.
        <!--OPTIONAL Element: Contact Instructions
        Supplemental instructions on how or when to contact the individual
        or organization.
          Domain: free text-->
      </cntinst>
    </cntinfo>
  </ptcontac>
  <native> EXAMPLE: Optech DASHMap 4.2200; ALS Post Processor 2.70
  Build#15; GeoCue Version 6.1.21.4; Windows XP Operating System
  \\server\directory path\*.LAS
     17 GB
    <!--REQUIRED: Native Dat Set Environment
    Description of the data set in the producer's processing
    environment, including items such as the name of the software
    (including version), the computer operating system, file name
    (including host-, path-, and filenames), and the data set size.

      Domain: free text-->
  </native>
</idinfo>

<dataqual>
  <logic> EXAMPLE:  Data covers the entire area specified for this project.
    <!--REQUIRED Element: Logical Consistency Report
    Describe the fidelity of relationships in the data structure of the
    lidar data: tests of valid values or topological tests. Identify
    software used and the date of the tests.
    Domain: free text-->
  </logic>
  <complete> EXAMPLE: These raw LAS data files include all data points
  collected.
  No points have been removed or excluded.
  A visual qualitative assessment was performed to ensure data
  completeness.
  There are no void areas or missing data. The raw point cloud is of good
  quality and data passes Fundamental Vertical Accuracy specifications.
    <!--REQUIRED Element: Completeness Report
    Document the inclusion or omissions of features for the dataset.
    Minimum width or area thresholds.
```

```
      Selection criteria or other rules used to derive the dataset.
        Domain: free text-->
    </complete>
    <posacc>
      <vertacc>
      <vertaccr> EXAMPLE: The specifications require that only Fundamental
      Vertical Accuracy (FVA) be computed for raw lidarpoint cloud swath
      files The vertical accuracy was tested with 25 independent survey
      located in open terrain.  These check points were not used in the
      calibration or post processing of the lidar point cloud data. The
      survey checkpoints were distributed throughout the project area.
      Specifications for this project require that the FVA be 25 cm or better
      Accuracyz at 95% confidence level.
        <!--REQUIRED Element: Vertical Positional Accuracy Report
        An explanation of the accuracy of the vertical coordinate
        measurementsand a description of the tests used.
          Domain: free text-->
      </vertaccr>
      <qvertpa>
        <vertaccv> EXAMPLE: 0.19 meters Accuracyz at 95% Confidence Interval
          <!--REQUIRED Element: Vertical Positional Accuracy Value
          Vertical accuracy expressed in (ground) meters.
          Clearly state wether this value is RMSEz or Accuracyz
            Domain: free text-->
        </vertaccv>
        <vertacce> The FVA was tested using 25 independent survey located in
        open terrain  terrain.  The survey checkpoints were distributed
        throughout the project area.
        The 25 independant check points were surveyed using  xxxx technique.
        Elevations from the unclassified lidar surface were measured for the
        x,y location of each check point.
        Elevations interpolated from the lidar surface were then compared to
        the elevation values of the surveyed control.  The RMSE was computed
        to be 0.097cm.
        Accuracyz has been tested to meet 9.7 cm Fundamental Vertical
        Accuracy at 95% confidence level using RMSE(z) x 1.9600 as defined by
        the National Standards for Spatial Data Accuracy (NSSDA); assessed
        and reported using National Digital Elevation Program (NDEP)/ASRPS
        Guidelines.
        <!--REQUIRED Element: Vertical Positional Accuracy Explanation
          Identification of the test that yielded the Vertical Positional
          Accuracy Value.
            Domain: free text-->
        </vertacce>
      </qvertpa>
    </vertacc>
  </posacc>
  <lineage>
    <srcinfo>
      <!-- The srcinfo section of the metadata MUST be repeated for each
      data source that contributed making this unclassified LAS swath
      dataset. This includes, but is not limited to, 1) ground control used
      for calibrating the lidar data, 2) the actual lidar aquistion data,
      and 3) independent ground control used to assess the accuracy of the
      lidar point cloud. -->
        <srccite>
          <citeinfo>
            <origin> EXAMPLE: Jiffy Survey, Inc
            <!--REQUIRED Element:  Originator
            This the name of an organization or individual that developed
            the data set. If the creation of this data source was created by
            a subcontractor, the subcontractors name and contact information
            should be entered as the source for that contributing dat set.
              Domain: "Unknown" free text -->
            </origin>
            <pubdate> 20100115
```

```
          <!--REQUIRED element: Date of Publication
          Enter the date when the data set is published or otherwise
          made available for release.
          The format of this date must be YYYMMDD.
            Domain: "Unknown" "Unpublished material" free date-->
        </pubdate>
        <title> EXAMPLE: Ground Control for Phelps and Dent County, MO
        lidar project
          <!--REQUIRED Element: Title
          The name by which the first contributing dataset is known.
            Domain: free text-->
        </title>
        <geoform> EXAMPLE: vector digital data and tabular data
          <!--OPTIONAL Element: Enter the mode in which the geospatial
          data are represented.
            Domain:(the listed domain is partially from pp. 88-91 in
            Anglo-American Committee on Cataloguing of Cartographic
            Materials, 1982, Cartographic materials: A manual of
            interpretation for AACR2: Chicago, American Library
            Association):
              "atlas" "audio" "diagram" "document" "globe" "map" "model"
              "multimedia presentation" "profile" "raster digital data"
              "remote-sensing image" "section" "spreadsheet" "tabular
              digital data" "vector digital data" "video" "view" free
              text-->
        </geoform>
        <pubinfo>
          <pubplace> EXAMPLE: Jiffy Survey, Inc.
            <!--REQUIRED Element: Publication Place
            The name of the city (and state or province, and country, if
            needed to identify the city) the originator of the data set.
              Domain: free text-->
          </pubplace>
          <publish> EXAMPLE: Jiffy Survey, Inc., GPS department
            <!--Enter the name of the individual or organization that
            published the data set.
              Domain: free text-->
          </publish>
        </pubinfo>
        <othercit> EXAMPLE: None.
          <!--OPTIONAL Element: Other Citation Details
          Other information required to complete the citation.
            Domain: free text-->
        </othercit>
        <onlink> EXAMPLE:  ftp://JiffySurveyftp.com/data/outgoing/Task1/
          <!--OPTIONAL Element: Online Linkage
          IF APPLICABLE: The URL of an online computer resource that
          contains the data set.
            Domain: free text-->
        </onlink>
      </citeinfo>
    </srccite>
    <srcscale> Example: 50
      <!--OPTIONAL Element: Source Scale Denominator
      IF APPLICABLE: The denominator of the representative fraction on
      a map (for example, on a 1:24,000-scale map, the Source Scale
      Denominator is 24000).
        Domain: Source Scale Denominator > 1-->
    </srcscale>
    <typesrc> EXAMPLE: CD-ROM
      <!--REQUIRED Element:  Type of Source Media
      The medium of the first source data set.
        Domain: "paper" "stable-base material" "microfiche"
        "microfilm" "audiocassette" "chart" "filmstrip" "transparency"
        "videocassette" "videodisc" "videotape" "physical model"
        "computer program" "disc" "cartridge tape" "magnetic tape"
```

```
        "online" "CD-ROM" "electronic bulletin board" "electronic mail
        system" free text-->
    </typesrc>
    <srctime>
      <timeinfo>
        <sngdate>
          <caldate> 201001003
            <!--REQUIRED Element: Calendar Date
            This is the date of the first source dataset was created.
            The field MUST be formatted YYYYMMDD-->
          </caldate>
        </sngdate>
      </timeinfo>
      <srccurr> EXAMPLE: ground condition
        <!--REQUIRED Elemenet: Source Currentness Reference
        The basis on which the source time period of content information
        of the source data set is determined.
          Domain: "ground condition" "publication date" free text-->
      </srccurr>
    </srctime>
    <srccitea> EXAMPLE: Phelps_Co_lidar_gnd_ctrl
      <!--REQUIRED Element: Source Citation Abbreviation
      Enter short-form alias for the source citation. Each source MUST
      HAVE A UNIQUE ID.
      This ID will be used to reference this source data in the Process
      Step sections below.
        Domain: free text-->
    </srccitea>
    <srccontr> EXAMPLE:  This data source was used (along with the
    airborne GPS/IMU data to georeferencing of the lidar point cloud data.
      <!--REQUIRED Element: Source Contribution
      Brief statement identifying the information contributed.
        Domain: free text-->
    </srccontr>
</srcinfo>

<srcinfo>
  <srccite>
    <citeinfo>
      <origin>USDA</origin>
      <pubdate> 20090606</pubdate>
      <title>NAIP Imagery for Phelps and Dent County, MO lidar project
      </title>
      <geoform>raster orthoimagery</geoform>
      <pubinfo>
        <pubplace>USGS-EROS</pubplace>
        <publish>USGS-EROS</publish>
      </pubinfo>
      <othercit>None</othercit>
      <onlink></onlink>
    </citeinfo>
  </srccite>
  <srcscale>50</srcscale>
  <typesrc>online</typesrc>
  <srctime>
    <timeinfo>
      <sngdate>
        <caldate>20090101</caldate>
      </sngdate>
    </timeinfo>
      <srccurr>ground condition</srccurr>
  </srctime>
    <srccitea>Phelps-Dent_Co_NAIP_Imagery</srccitea>
    <srccontr>This data source was used (along with the lidar intensity
    imagery) to classify the lidar point cloud data.
    </srccontr>
```

```
</srcinfo>

<srcinfo>
  <srccite>
    <citeinfo>
      <origin>We Map 4U, Inc.</origin>
      <pubdate>20101208</pubdate>
      <title>Lidar Intensity Imagery for Phelps and Dent County, MO
      </title>
      <geoform>raster orthoimagery</geoform>
      <pubinfo>
        <pubplace>USGS-EROS</pubplace>
        <publish>USGS-EROS</publish>
      </pubinfo>
      <othercit>None</othercit>
      <onlink></onlink>
    </citeinfo>
  </srccite>
  <srcscale>50</srcscale>
  <typesrc>online</typesrc>
  <srctime>
    <timeinfo>
  <rngdates>
    <begdate>20100216</begdate>
    <enddate>20100218</enddate>
  </rngdates>
    </timeinfo>
    <srccurr>ground condition</srccurr>
  </srctime>
  <srccitea>Phelps-Dent_Co_Lidar_Intensity_Imagery</srccitea>
  <srccontr>This data source was used (along with NAIP imagery) to
  classify the lidar point cloud data.
  </srccontr>
</srcinfo>
<procstep>
  <procdesc> EXAMPLE: Lidar Pre-Processing: Airborne GPS and IMU data
  were merged to develop a Single Best Estimate (SBET) of the lidar
  system trajectory for each lift.
  Lidar ranging data were initially calibrated using previous best
  parameters for this instrument and aircraft. Relative calibration
  was evaluated using advanced plane-matching analysis and parameter
  corrections derived. This was repeated iteratively until residual
  errors between overlapping swaths, across all project lifts, was
  reduced to 2cm or less. Data were then block adjusted to match
  surveyed calibration control. Raw data FVA was checked using
  independently surveyed checkpoints. Swath overage points were
  identified and tagged within each swath file. The results of the final
  calibration, FVA and horizontal accuracy assessments, and the "raw"
  swaths were forwarded to the client to obtain a Notice To Proceed on
  classification and derivative  product generation.
  <!--Enter an explanation of the event and related parameters or
  tolerances.
  Domain: free text-->
  </procdesc>
  <srcused> EXAMPLE: Phelps_Co_lidar_gnd_ctrl
    <!--Enter the Source Citation Abbreviation of a data set used in the
    processing step.
      Domain: Source Citation Abbreviations from the Source Information
      entries for the data set.-->
  </srcused>
  <procdate> 20100131
    <!--Enter the date when the event was completed.
      Domain: "Unknown" "Not complete" free date-->
  </procdate>
```

```
    <srcprod> EXAMPLE: LiDAR datasets with USGS classifications
    <!--Enter the Source Citation Abbreviation of an intermediate data set
    that (1) is significant in the opinion of the data producer, (2) is
    generated in the processing step, and (3) is used in later processing
    steps.
      Domain: Source Citation Abbreviations from the Source Information
      entries for the data set.-->
    </srcprod>
    <proccont>
      <cntinfo>
        <cntorgp>
          <cntorg> EXAMPLE: We Map 4U, Data Aquisition Department
            <!--Enter the name of the organization to which the contact
            type applies.
              Domain: free text-->
          </cntorg>
          <cntper> EXAMPLE: Manny Puntas
            <!--Enter the name of the individual to which the contact type
            applies.
              Domain: free text-->
          </cntper>
        </cntorgp>

        <cntaddr>
          <addrtype>mailing and physical</addrtype>
          <address>123 Main St.</address>
          <city>Anytown</city>
          <state>MO</state>
          <postal>61234</postal>
          <country>USA</country>
        </cntaddr>
        <cntvoice>555-555-556</cntvoice>
        <cntfax>555-5550-1236</cntfax>
        <cntemail>mpuntas@wemap4u.com</cntemail>
        <hours>Monday through Friday 8:00 AM to 4:00 PM (Central Time)</
        hours>
        <cntinst> If unable to reach the contact by telephone, please send
        an email. You should get a response within 24 hours.
        </cntinst>
      </cntinfo>
    </proccont>
  </procstep>
  <procstep>
    <procdesc>Lidar Post-Processing: The calibrated and controlled
    lidar swaths were processed using automatic point classification
    routines in proprietary software. These routines operate against the
    entire collection (all swaths, all lifts), eliminating character
    differences between files. Data were then distributed as virtual tiles
    to experienced lidar analysts for localized automatic classification,
    manual editing, and peer-based QC checks. Supervisory QC monitoring
    of work in progress and completed editing ensured consistency of
    classification character and adherence to project requirements across
    the entire project area. All classification tags are stored in the
    original swath files. After completion of classification and final QC
    approval, the FVA, SVAs, and CVA for the project are calculated.
    Sample areas for each landcovertype present in the project area were
    extracted and forwarded to the client, along with the results of the
    accuracy tests. Upon acceptance, the complete classified lidar swath
    files were delivered to the client.
    </procdesc>
    <srcused>Phelps-Dent_Co_NAIP_Imagery</srcused>
    <srcused>Phelps-Dent_Co_Lidar_Intensity_Imagery</srcused>
    <procdate>20100530</procdate>
    <proccont>
      <cntinfo>
        <cntorgp>
```

```
            <cntorg>We Map 4U, Data Aquisition Department</cntorg>
            <cntper>Manny Puntas</cntper>
        </cntorgp>
          <cntaddr>
          <addrtype>mailing and physical</addrtype>
            <address>123 Main St.</address>
            <city>Anytown</city>
            <state>MO</state>
            <postal>61234</postal>
            <country>USA</country>
        </cntaddr>
        <cntvoice>555-555-556</cntvoice>
        <cntfax>555-5550-1236</cntfax>
        <cntemail>mpuntas@wemap4u.com</cntemail>
        <hours>Monday through Friday 8:00 AM to 4:00 PM (Central Time)</
        hours>
        <cntinst> If unable to reach the contact by telephone, please send
        an email. You should get a response within 24 hours.
        </cntinst>
      </cntinfo>
      </proccont>
    </procstep>

  </lineage>
  <!--Enter area of a data set obstructed by clouds, expressed as a
  percentage of the spatial extent.
  Domain: 0 <= Cloud Cover <= 100 "Unknown"-->
  </dataqual>

  <spdoinfo>
    <direct> EXAMPLE: Vector
    <!--REQUIRED Element: Enter the system of objects used to represent
    space in the data set.
      Domain: "Point" "Vector" "Raster" -->
    </direct>
    <ptvctinf>
      <sdtsterm>
        <sdtstype> EXAMPLE: Point
          <!--REQUIRED Element: SDTS Point and Vector Object Type
          Enter  name of point and vector spatial objects used to l
          ocate zero-, one-, and two-dimensional spatial locations
          in the data set.

Domain: (The domain is from "Spatial Data Concepts," which is Chapter 2 of
Part 1 in Department of Commerce, 1992, Spatial Data Transfer Standard
(SDTS) (Federal Information Processing Standard 173): Washington,
Department of Commerce, National Institute of Standards and Technology):
"Point" "Entity point" "Label point" "Area point" "Node, planar graph"
"Node, network" "String" "Link" "Complete chain" "Area chain"
"Network chain, planar graph" "Network chain, nonplanar graph"
"Circular arc, three point center" "Elliptical arc" "Uniform B-spline"
"Piecewise Bezier" "Ring with mixed composition"
"Ring composed of strings" "Ring composed of chains"
"Ring composed of arcs" "G-polygon" "GT-polygon composed of rings"
"GT-polygon composed of chains"
"Universe polygon composed of rings"
"Universe polygon composed of chains"
"Void polygon composed of rings" "Void polygon composed of chains"-->
        </sdtstype>
        <ptvctcnt> EXAMPLE: 764,567,423
          <!--OPTIONAL Element: Point and Vector Count
          Enter the total number of the point or vector object type
          occurring in the data set.

Domain: Point and Vector Object Count > 0-->
        </ptvctcnt>
```

```
      </sdtsterm>
    </ptvctinf>
  </spdoinfo>

  <spref>
    <horizsys>
      <planar>
        <gridsys>
        <!--REQUIRED Section: The section should be filled out with the
        relevant parameters for the coordinate reference system for the
        data. Usually it will be UTM or a State Plane Zone. Delete the
        irrelevant section below.-->
          <gridsysn> EXAMPLE: Universal Transverse Mercator
            <!--Enter name of the grid coordinate system.

Domain: "Universal Transverse Mercator" "Universal Polar Stereographic"
"State Plane Coordinate System 1927" "State Plane Coordinate System 1983"
"ARC Coordinate System"
"other grid system" -->
          </gridsysn>
          <utm>
            <utmzone> EXAMPLE: 15
              <!--Enter the  identifier for the UTM zone.
Type: integer

Domain:
1 <= UTM Zone Number <= 60 for the northern hemisphere;
-60 <= UTM Zone Number <= -1 for the southern hemisphere-->
            </utmzone>
            <transmer>
              <sfctrmer>0.9996
                <!--Enter a multiplier for reducing a distance obtained from
                a map by computation or scaling to the actual distance along
                the central meridian.

Domain: Scale Factor at Central Meridian > 0.0-->
              </sfctrmer>
              <longcm>-117.000000
                <!--Enter the line of longitude at the center of a map
                projection generally used as the basis for constructing the
                projection.
Type: real
Domain: -180.0 <= Longitude of Central Meridian < 180.0-->
              </longcm>
              <latprjo>0.0
                <!--Enter latitude chosen as the origin of rectangular
                coordinates for a map projection.

Domain: -90.0 <= Latitude of Projection Origin <= 90.0-->
              </latprjo>
              <feast>500000
                <!--Enter the value added to all "x" values in the
                rectangular coordinates for a map projection. This value
                frequently is assigned to eliminate negative numbers.
                Expressed in the unit of measure identified in Planar
                Coordinate Units.

Domain: free real-->
              </feast>
              <fnorth>0.0
                <!--Enter the value added to all "y" values in the
                rectangular coordinates for a map projection. This value
                frequently is assigned to eliminate negative numbers.
                Expressed in the unit of measure identified in Planar
                Coordinate Units.
```

```
Domain: free real-->
            </fnorth>
          </transmer>
        </utm>

        <spcs>
          <spcszone>
            <!--Enter identifier for the SPCS zone.

Domain: Four-digit numeric codes for the State Plane Coordinate Systems
based on the North American Datum of 1927 are found in Department of
Commerce, 1986, Representation of geographic point locations for information
interchange (Federal Information Processing Standard
70-1): Washington: Department of Commerce, National Institute of Standards
and Technology.
Codes for the State Plane Coordinate Systems based on the North American
Datum of 1983 are found in Department of Commerce, 1989 (January),
State Plane Coordinate System of 1983 (National Oceanic and Atmospheric
Administration Manual NOS NGS 5): Silver Spring, Maryland, National Oceanic
and Atmospheric Administration, National Ocean Service, Coast and Geodetic
Survey.-->
          </spcszone>
          <lambertc>
            <stdparll>
              <!--Enter line of constant latitude at which the surface of
              the Earth and the plane or developable surface intersect.

Domain: -90.0 <= Standard Parallel <= 90.0-->
            </stdparll>
            <longcm>
              <!--Enter the line of longitude at the center of a map
              projection generally used as the basis for constructing the
              projection.

Domain: -180.0 <= Longitude of Central Meridian < 180.0-->
            </longcm>
            <latprjo>
              <!--Enter latitude chosen as the origin of rectangular
              coordinates for a map projection.

Domain: -90.0 <= Latitude of Projection Origin <= 90.0-->
            </latprjo>
            <feast>
              <!--Enter the value added to all "x" values in the
              rectangular coordinates for a map projection. This value
              frequently is assigned to eliminate negative numbers.
              Expressed in the unit of measure identified in Planar
              Coordinate Units.

Domain: free real-->
            </feast>
            <fnorth>
              <!--Enter the value added to all "y" values in the
              rectangular coordinates for a map projection. This value
              frequently is assigned to eliminate negative numbers.
              Expressed in the unit of measure identified in Planar
              Coordinate Units.

Domain: free real-->
            </fnorth>
          </lambertc>
          <transmer>
            <sfctrmer>
              <!--Enter a multiplier for reducing a distance obtained from
              a map by computation or scaling to the actual distance along
              the central meridian.
```

```
Domain: Scale Factor at Central Meridian > 0.0-->
                </sfctrmer>
                <longcm>
                  <!--Enter the line of longitude at the center of a map
                  projection generally used as the basis for constructing the
                  projection.
Type: real
Domain: -180.0 <= Longitude of Central Meridian < 180.0-->
                </longcm>
                <latprjo>
                  <!--Enter latitude chosen as the origin of rectangular
                  coordinates for a map projection.

Domain: -90.0  <= Latitude of Projection Origin <= 90.0-->
                </latprjo>
                <feast>
                  <!--Enter the value added to all "x" values in the
                  rectangular coordinates for a map projection. This value
                  frequently is assigned to eliminate negative numbers.
                  Expressed in the unit of measure identified in Planar
                  Coordinate Units.

Domain: free real-->
                </feast>
                <fnorth>
                  <!--Enter the value added to all "y" values in the
                  rectangular coordinates for a map projection. This value
                  frequently is assigned to eliminate negative numbers.
                  Expressed in the unit of measure identified in Planar
                  Coordinate Units.

Domain: free real-->
                </fnorth>
              </transmer>
              <obqmerc>
                <sfctrlin>
                  <!--Enter a multiplier for reducing a distance obtained from
                  a map by computation or scaling to the actual distance along
                  the center line.

Domain: Scale Factor at Center Line > 0.0-->
                </sfctrlin>
                <obqlazim>
                  <azimangl>
                    <!--Enter angle measured clockwise from north, and
                    expressed in degrees.

Domain: 0.0  <= Azimuthal Angle < 360.0-->
                  </azimangl>
                  <azimptl>
                    <!--Enter longitude of the map projection origin.

Domain: -180.0 <= Azimuth Measure Point Longitude < 180.0-->
                  </azimptl>
                </obqlazim>
                <obqlpt>
                  <obqllat>
                    <!--Enter latitude of a point defining the oblique line.

Domain: -90.0  <= Oblique Line Latitude <= 90.0-->
                  </obqllat>
                  <obqllong>
                    <!--Enter longitude of a point defining the oblique line.

Domain: -180.0 <= Oblique Line Longitude < 180.0-->
                  </obqllong>
```

```
              </obqlpt>
              <latprjo>
                <!--Enter latitude chosen as the origin of rectangular
                coordinates for a map projection.

Domain: -90.0 <= Latitude of Projection Origin <= 90.0-->
              </latprjo>
              <feast>
                <!--Enter the value added to all "x" values in the
                rectangular coordinates for a map projection. This value
                frequently is assigned to eliminate negative numbers.
                Expressed in the unit of measure identified in Planar
                Coordinate Units.

Domain: free real-->
              </feast>
              <fnorth>
                <!--Enter the value added to all "y" values in the
                rectangular coordinates for a map projection. This value
                frequently is assigned to eliminate negative numbers.
                Expressed in the unit of measure identified in Planar
                Coordinate Units.

Domain: free real-->
              </fnorth>
            </obqmerc>
            <polycon>
              <longcm>
                <!--Enter the line of longitude at the center of a map
                projection generally used as the basis for constructing the
                projection.

Domain: -180.0 <= Longitude of Central Meridian < 180.0-->
              </longcm>
              <latprjo>
                <!--Enter latitude chosen as the origin of rectangular
                coordinates for a map projection.

Domain: -90.0 <= Latitude of Projection Origin <= 90.0-->
              </latprjo>
              <feast>
                <!--Enter  the value added to all "x" values in the
                rectangular coordinates for a map projection. This value
                frequently is assigned to eliminate negative numbers.
                Expressed in the unit of measure identified in Planar
                Coordinate Units.

Domain: free real-->
              </feast>
              <fnorth>
                <!--Enter the value added to all "y" values in the
                rectangular coordinates for a map projection. This value
                frequently is assigned to eliminate negative numbers.
                Expressed in the unit of measure identified in Planar
                Coordinate Units.

Domain: free real-->
              </fnorth>
            </polycon>
          </spcs>
        </gridsys>
        <planci>
          <plance>coordinate pair</plance>
          <coordrep>
            <absres>0.01</absres>
            <ordres>0.01</ordres>
```

```
          </coordrep>
          <plandu>meters</plandu>
        </planci>
      </planar>
    <geodetic>
        <horizdn>EXAMPLE: North American Datum of 1983
          <!--REQUIRED Element: Enter the identification given to the
          reference system used for defining the coordinates of points.

Domain: "North American Datum of 1927" "North American Datum of 1983" free
text-->
        </horizdn>
        <ellips>EXAMPLE: Geodetic Reference System 80
          <!--REQUIRED Element: Enter identification given to established
          representations of the Earth's shape.

Domain: "Clarke 1866" "Geodetic Reference System 80" free text-->
        </ellips>
        <semiaxis>6378137
          <!--REQUIRED Element: Enter radius of the equatorial axis of the
          ellipsoid.

Domain: Semi-major Axis > 0.0-->
        </semiaxis>
        <denflat>298.257222101
          <!--REQUIRED Element: Enter the denominator of the ratio of the
          difference between the equatorial and polar radii of the ellipsoid
          when the numerator is set to 1.

Domain: Denominator of Flattening > 0.0-->
        </denflat>
      </geodetic>
    </horizsys>
    <vertdef>
      <altsys>
        <altdatum> EXAMPLE: North American Vertical Datum of 1988
        <!--REQUIRED Element: Vertical Datum: The surface of reference from
        which vertical distances are measured.

Domain: "National Geodetic Vertical Datum of 1929" "North American Vertical
Datum of 1988" free text-->
        </altdatum>
        <altres> 0.01
          <!-- REQUIRED Element: Vertical Resolution: The minimum distance
          possible between two adjacent elevation (altitude) values,
          expressed in Altitude Distance Units of measure.

Domain: Altitude Resolution > 0.0-->
        </altres>
        <altunits>  EXAMPLE: meters
          <!--REQUIRED Element: Units in which altitudes are recorded.

Domain: "meters" "feet" free text-->
        </altunits>
        <altenc> EXAMPLE: Explicit elevation coordinate included with
        horizontal coordinates
          <!--REQUIRED Element: Altitude Encoding Method: The means used to
          encode the elevations.

Domain: "Explicit elevation coordinate included with horizontal coordinates"
"Implicit coordinate" "Attribute values"-->
        </altenc>
      </altsys>
    </vertdef>
  </spref>
```

```
<eainfo>
<!--SECTION: Entity and Attribute Information
    THIS SECTION IS NOT REQUIRED FOR LIDAR LAS DELIVERABLES.  It is only
    required for deliverable data classified as a Feature Class. -->
</eainfo>

<distinfo>
<!--SECTION: Distribution Information: Information about the distributor
of and options for obtaining the data set.
THIS SECTION SHOULD ONLY BE POPULATED IF SOME ORGANIZATION OTHER THAN USGS
HAS DISTRIBUTION RIGHTS TO THE DATA. -->
  <distrib>
    <cntinfo>
      <cntorgp>
        <cntorg> Leave blank unless an organization outside of USGS has
        distribution rights to the data.
        </cntorg>
        <cntper> Leave blank unless an organization outside of USGS has
        distribution rights to the data.
        </cntper>
      </cntorgp>
      <cntaddr>
        <addrtype> Leave blank unless an organization outside of USGS has
        distribution rights to the data.
        </addrtype>
        <address> Leave blank unless an organization outside of USGS has
        distribution rights to the data.
        </address>
        <city> Leave blank unless an organization outside of USGS has
        distribution rights to the data.
        </city>
        <state> Leave blank unless an organization outside of USGS has
        distribution rights to the data.
        </state>
        <postal> Leave blank unless an organization outside of USGS has
        distribution rights to the data.
        </postal>
        <country> Leave blank unless an organization outside of USGS has
        distribution rights to the data.
        </country>
      </cntaddr>
      <cntvoice> Leave blank unless an organization outside of USGS has
      distribution rights to the data.
      </cntvoice>
      <cntemail> Leave blank unless an organization outside of USGS has
      distribution rights to the data.
        </cntemail>
      </cntinfo>
    </distrib>
    <resdesc> Leave blank unless an organization outside of USGS has
    distribution rights to the data.
    </resdesc>
    <distliab> Leave blank unless an organization outside of USGS has
    distribution rights to the data.
    </distliab>
</distinfo>

<metainfo>
<!-- REQUIRED SECTION: Metadata Reference Information: Information on the
currentness of the metadata information, and the party responsible for the
metadata. -->
  <metd> 20101206
    <!--REQUIRED Element: Metadata Date: The date that the metadata were
                          created or last updated.
                          Must be in the format YYYYMMDD.-->
  </metd>
```

```
    <metrd> 20101207
      <!--OPTIONAL Element: Metadata Review Date: The date of the latest
                            review of the metadata entry.
                            Must be in the format YYYYMMDD.

Domain: Metadata Review Date later than Metadata Date-->
    </metrd>
    <metc>
      <cntinfo>
        <cntorgp>
          <cntorg> EXAMPLE: We Map 4U, Data Aquisition Department
            <!--REQUIRED Element: Contact Organization: The name of the
                                  organization which is responsible for
                                  creating the metadata.

Domain: free text-->
          </cntorg>
          <cntper> EXAMPLE: John Smith
            <!--REQUIRED Element: Contact Person: The name of the individual
                                  who is the contact person concerning the
                                  metadata.

Domain: free text-->
          </cntper>
        </cntorgp>
        <cntaddr>
          <addrtype> EXAMPLE: mailing and physical
            <!--REQUIRED Element:  Address Type: The type of address that
            follows.
            Only required for "mailing" or "mailing and physical".  If the
            contractor has a different mailing and physical address, the
            physical address does not need to be included.

Domain: "mailing" "physical" "mailing and physical", free text-->
          </addrtype>
          <address> EXAMPLE: 123 Main St.
            <!--REQUIRED Element: Address: The address of the contractor
            responsible for the metadata. For multiple line addresses the
            address tag may be repeated as many times as needed.

Domain: free text-->
          </address>
          <city> EXAMPLE: Anytown
            <!--REQUIRED Element: City: The city of the address.

Domain: free text-->
          </city>
          <state> EXAMPLE: MO
            <!--REQUIRED Element: State: The state or province of the
            address.

Domain: free text-->
          </state>
          <postal> EXAMPLE: 61234
            <!--REQUIRED Element: Postal Code: Enter the ZIP or other postal
            code of the address.

Domain: free text-->
          </postal>
          <country> EXAMPLE: USA
            <!--OPTIONAL Element: Country: The country of the address.

Domain: free text-->
          </country>
        </cntaddr>
        <cntvoice> EXAMPLE: 555-555-1234
```

```
          <!--REQUIRED Element: Contact Voice Telephone: The telephone
          number by which individuals can speak to the organization or
          individual responsible for the metadata.

Domain: free text-->
        </cntvoice>
        <cnttdd> EXAMPLE: 555-555-1122
          <!--OPTIONAL Element: Contact TDD/TTY Telephone: The telephone
          number by which hearing-impaired individuals can contact the
          organization or individual.

Domain: free text-->
        </cnttdd>
        <cntfax> EXAMPLE: 555-5550-1235
          <!--OPTIONAL Element: Contact Fax: The telephone number of a
          facsimile machine of the organization or individual.

Domain: free text-->
        </cntfax>
        <cntemail> EXAMPLE: jsmith@wemap4u.com
          <!--OPTIONAL Element: Contact E-mail Address: The email address of
          the organization or individual.

Domain: free text-->
        </cntemail>
        <hours> EXAMPLE: Monday through Friday 8:00 AM to 4:00 PM (Central
        Time)
          <!--OPTIONAL Element: Hours of Service: The time period when
          individuals can speak to the organization or individual.

Domain: free text-->
        </hours>
        <cntinst> EXAMPLE: If unable to reach the contact by telephone,
        please send an email.  You should get a response within 24 hours.
          <!--OPTIONAL Element: Contact Instructions: Supplemental
          instructions on how or when to contact the individual or
          organization.

Domain: free text-->
        </cntinst>
      </cntinfo>
    </metc>
    <metstdn> EXAMPLE: FGDC Content Standard for Digital Geospatial Metadata
      <!--REQUIRED Element: Metadata Standard: Enter the name of the
      metadata standard used to document the data set.

Domain: "FGDC Content Standard for Digital Geospatial Metadata" free text-->
    </metstdn>
    <metstdv> EXAMPLE: FGDC-STD-001-1998
      <!--REQUIRED Element: Metadata Standard Veriosn: Enter identification
      of the version of the metadata standard used to document the data set.

Domain: free text-->
    </metstdv>
    <metac> EXAMPLE: None.
      <!--OPTIONAL Element: Metadata Access Constraints: Restrictions and
      legal prerequisites for accessing the metadata. These include any
      access constraints applied to assure the protection of privacy or
      intellectual property, and any special restrictions or limitations on
      obtaining the metadata.

Domain: free text-->
    </metac>
    <metuc> EXAMPLE: None.
      <!--OPTIONAL Element: Metadata Use Constraints: Restrictions and legal
      prerequisites for using the metadata after access is granted. These
```

```
      include any metadata use constraints applied to assure the protection
      of privacy or intellectual property, and any special restrictions or
      limitations on using the metadata.

Domain: free text-->
    </metuc>
    <metsi>
      <metscs> EXAMPLE: None.
        <!--REQUIRED IF APPLICABLE: Metadata Security Classification System:
        Name of the classification system for the metadata.

Domain: free text-->
      </metscs>
      <metsc> EXAMPLE: Unclassified
        <!--REQUIRED IF APPLICABLE: Metadata Security Classification: Name of
        the handling restrictions on the metadata.

"Top secret" "Secret" "Confidential" "Restricted" "Unclassified" "Sensitive"
free text-->
      </metsc>
      <metshd> EXAMPLE: NONE
        <!--REQUIRED IF APPLICABLE: Metadata Security Handling Description:
        Additional information about the restrictions on handling the
        metadata.

Domain: free text-->
      </metshd>
    </metsi>
    <metextns> <!-- Metadata Extensions Group: REQUIRED IF APPLICABLE. A
    reference to extended elements to the standard which may be defined by a
    metadata producer or a user community. Extended elements are elements
    outside the Standard, but needed by the metadata producer. If extended
    elements are created, they must follow the guidelines in Appendix D,
    Guidelines for Creating Extended Elements to the Content Standard for
    Digital Geospatial Metadata.-->
    <!--This section may be repeated as necessary-->
    <onlink> EXAMPLE: None
            <!--REQUIRED IF APPLICABLE: Online Linkage: URL for the resource
            that contains the metadata extension information for the data
            set. -->
    </onlink>
    <metprof> EXAMPLE: None
              <!--REQUIRED IF APPLICABLE: Profile Name: Name of a document
              that describes the application of the Standard
              to a specific user community.-->
    </metprof>
    </metextns>
    </metainfo>
</metadata>

END
```

REFERENCES AND RESOURCES

http://www.fgdc.gov/metadata/events/metadata-summit/index_html
http://www.fgdc.gov/metadata/csdgm/
http://www.fgdc.gov/metadata/geospatial-metadata-standards
http://www.fgdc.gov/metadata/fgdc-iso-activities
http://www.iso.org/iso/catalogue_detail.htm?csnumber=26020

APPENDIX D
ACRONYMS

1D	One Dimensional
2D	Two Dimensional
3D	Three Dimensional
4D	Four Dimensional
AC	Advisory Circulars
AGC	Automatic Gain Control
AGL	Above Ground Level
AINS	Aided INS
AIP	Airport Improvement Program
ALP	Airport Layout Plans
ALS	Airborne Laser Scanning
ALTM	Airborne Laser Terrain Mapper
AM	Amplitude Modulation
AML	ARC Macro Language
ANUDEM	Austrailian National University's DEM
AOBD	Acousto-Optic Beam Deflector
AOI	Angle Of Incidence
AOI	Area of Interest
AOL	Airborne Oceanographic Lidar
APD	Avalanche Photodiode
ARD	Autonomous Rendezvous and Docking
ARRA	American Reinvestment and Recovery Act
ARV	Autonomous Robitic Vision
ASC	Advanced Scientific Concepts
ASPRS	American Society For Photogrametry and Remote Sensing
AT	Aerotriangulation
ATM	Airborne Topographic Mapper
ATR	Automatic Target Recognition
BATC	Ball Aerospace and Technologies Corp
BED	Hanscom Airfield
CA	Clear Access
CAD	Computer Aided Drafting
CADD	Computer Aided Drafting and Design
CALTRANS	California Department of Transportation

CCRS	Canada Centre for Remote Sensing
CDMA	Code Division Multiple Access
CFD	Constant-Fraction Discriminator
CFIT	Controlled Flight Into Terrain
CFRP	Carbon Fiber Ruggedized Plate
CHARTS	Compact Hydrographic Airborne Rapid Total Survey
CHM	Canopy Height Model
CIAT	Coastal Impact Assessment Tool
CLICK	Center for Lidar Information, Coordination, and Knowledge
CONUS	Conterminous United States
CORS	Continuously Operating Reference System
COTS	Commercial Off-The-Shelf
CPU	Central Processiong Unit
CRS	Coordinate Reference System
CSDGM	Content Standards for Digital Geospatial Metadata
CVA	Consolidated Vertical Accuracy
CW	Continuous Wave
CZMIL	Coastal Zone Mapping and Imaging Lidar
DAS	Data Acquisition System
DBH	Diameter at Breast Height
DCM	Direction Cosine Matrix
DEM	Digital Elevation Model
DESDynI	Deformation, Ecosystem Structure, and Dynamics of Ice
DG	Direct Georeferencing
DGNSS	Differential GNSS
DGPS	Differential GPS
DMI	Distance Measurement Indicator
DOC	Dissolved Organic Carbon
DOGAMI	Oregon Department of Geology and Mineral Industries
DOP	Dilution of Precision
DOQQ	Digital Orthophoto Quarter-Quad
DOT	Department of Transportation
DSAS	Digital Shoreline Analysis System
DSM	Digital Surface Model
DSMS	Digital Shoreline Manangement System
DTED	Digital Terrain Elevation Data
DTM	Digital Terrain Model
DTO	Development Test Objective
DVD	Digital Video Disk

EAARL	Experimental Advanced Airborne Research Lidar
ECEF	Earth Centered Earth Fixed
EDME	Electronic Distance Measuring Equipment
EDNA	Elevation Derivatives for National Applications
EEZ	Exclusive Economic Zone
EGI	Embedded GNSS-Inertial
EGM	Earth Gravitational Model
EGNOS	European Geostationary Navigation Overlay System
EPA	Environmental Protection Agency
EPSG	European Petroleum Survey Group
EQC	External Quality Control
ESFL	Electronically Steerable Flash Lidar
EVLR	Extended Variable Length Record
FAA	Federal Aviation Administration
FDC	Frequency-to-Distance Conversion
FDMA	Frequency Domain Multiple Access
FEMA	Federal Emergency Management Agency
FERC	Federal Energy Reliability Commission
FFPA	Flash-Focal-Plane Array
FGDC	Federal Geographic Data Committee
FIPAR	Fraction of Intercepted Photosynthetically Active Radiation
FIRM	Flood Insurance Rate Map
FLTA	Forward Looking Terrain Avoidance
FM	Frequency Modulation
FMS	Flight Management System
FOG	Fber-Optic Gyros
FOR	Field Of Regard
FOV	Field Of View
FPA	Focal Plane Array
FPGA	Field-Programmable Gate-Arrays
FPS	Frames Per Second
FVA	Fundamental Vertical Accuracy
FW	Full Waveform
FWHM	Full Width at Half Maximum
GB	Gigabyte
GDOP	Geometric Dilution of Precision
GEER	Geo-Engineering Extreme Event Reconnaissance
GIS	Geographic Information System
GLONASS	Globalnaya navigatsionnaya sputnikovaya sistema

GmAPD	Geiger Mode Avalanche Photodiode
GNSS	Global Navigation Satellite System
GPS	Global Positioning System
GPWS	Ground Proximity Warning Systems
GRACE	Gravity Recovery And Climate Experiment
GUID	Globally Unique Identifier
GWIS	Geographic Watershed Information System
H&H	Hydrologic and Hydraulic
HARN	High Accuracy Reference Network
HLDA	Hazard Detection and Avoidance
HLS	Helicopter Lidar System
IAG	International Association of Geodesy
ICAO	International Civil Aviation Organization
ICP	Iterative Closest Point
ICP	Iterative Closest Point
ICP	Iterative Closest Point
ICPR	Interconnected Channel and Pond Routing
ICPR	Interconnected Channel and Pond Routing
IDW	Inverse Distance Weighted
IERS	International Earth Rotation and Reference Services
IFSAR	Interferometric Synthetic Aperature Radar
IGS	International GNSS Service
IIP	Instrument Incubator Program
IMU	Inertial Measurement Unit
InDOT	Indiana Department of Transportation
InGaAs	Indium-Gallium-Arsenide
INS	Inertial Navigation Systems
IQC	Internal Quality Control
IR	Infrared
IRI	International Roughness Index
ISDT	Integrated Scanner, Detector, and Telescope
ISO	International Organization for Standardization
ISP	Inferred Stereo Pair
ISR	Intelligence, Surveillance, and Reconnaissance
ISS	International Space Station
JALBTCX	Joint Airborne Lidar Bathymetry Technical Center of Expertise
JCL	Johnson Creek Landslide
JPL	Jet Propulsion Lab
JPO	Joint Program Office

KAR	Kinematic Ambiguity Resolution
LADAR	LAser radar
LAHARZ	lahar-inundation hazard zone
LAI	Leaf Area Index
LAPF	Laser Altimeter Processing Facility
LAS	Lidar Archive Standard
LA-WASM	Large-Aperture Wide-Angle Scanning Mirror
LDART	Lidar Data Retrieval Tool
LIF	Laser Induced Fluorescence
LINS	Laser Inertial Measurement system
LOLC	Length Of Live Crown
LRF	Laser Rangefinder
LVIS	Laser Vegetation Imaging Sensor
LWG	LAS Working Group
LWIR	Long-Wave Infrared
MASER	Microwave Amplification by Stimulated Emission of Radiation
MDL	Minimum Description Length
MHW	Mean High Water
MLLW	Mean Lower Low Water
MLP	Multiple Laser Profiler
MLS	Mobile Lidar System
MPIA	Multiple Pulses in Air
MSE	Mechanically Stabilized Earth
MWIR	Mid-Wave Infrared
NAD	North American Datum
NASA	National Aeronautics and Space Administration
NAVD	North American Vertical Datum
NC	North Carolina
NCMP	National Coastal Mapping Program
NCS	Navigation computer system
Nd:YAG	Neodymium-Doped Yttrium Aluminum Garnet
NDEP	National Digital Elevation Program
NED	National Elevation Dataset
NERC	National Energy Regulatory Corporation
NGP	National Geospatial Program
NGS	National Geodetic Survey
NHC	National Hurricane Center
NIR	Near Infrared
NIST	National Institute of Standards and Technology

NMAS	National Map Accuracy Standard
NPS	Nominal Point Spacing; Nominal Post Spacing
NPS	Navigation Processor
NPS	Nominal Pulse Spacing
NSDI	National Spatial Data Infrastructure
NSRS	National Spatial Reference System
NSRS	National Spatial Reference System
NSSDA	National Standard for Spatial Data Accuracy
NW	North West
OCONUS	Outside the Conterminous United States
ODOT	Oregon Department of Transportation
OIS	Obstruction Identification Surfaces
OMCMFNC	Organic Mine Countermeasures Future Naval Capabilities
OPUS	Online Positioning User Service
OSTA	Optical Storage Technology Association
OSU	Oregon State University
OSU	The Ohio State University
PDOP	Position Dilution of Position
PDRF	Pont Data Record Format
POS	Positioning Orientation System
POS LV	Position and Orientation System for Land Vehicles
POV	Point of View
PPK	Post Processed Kinematic
PPP	Precise Point Positioning
PPS	Pulses Per Second
PPSM	Points Per Square Meter
PRF	Pulse Repetition Frequency
PRN	Pseudo-Random Number
PROXOPS	Proximity Operations
PRR	Pulse Repetition Rate
QA	Quality Assurance
QC	Quality Control
RAMS	Real-time Aerial Mapping System
RANSAC	RANdom SAmple Consensus
RASCAL	RAster SCanning Airborne Laser Altimeter
RF	Radio Frequency
RFP	Request For Proposal
RGB	Red-Green-Blue
RINEX	Receiver Independent Exchange Format

RJMCMC	Reversible Jump Markov Chain Monte Carlo
RLG	Ring Laser Gyro
RMN	Stafford Regional Airport
RMSE	Root Mean Square Error
RO	Relative Orientation
ROAR	Rapid Overt Airborne Reconnaissance
ROIC	Read-Out Integrated Circuit
ROW	Right-Of-Way
RTK	Real-Time Kinematic
RVS	Raytheon Vision Systems
SA	Selective Availability
SAR	Synthetic Aperture Radar
SBAS	Satellite Based Augmentation System
SFQ	Suffolk Executive Airport
SLA	Shuttle Laser Altimeter
SLICER	Scanning Lidar Imager of Canopies by Echo Recovery
SLOSH	Overland Surges from Hurricanes
SNR	Signal-to-Noise Ratio
SPACEX	Space Exploration Corporation
STC	Standard Technical Conversion
StDEV	Standard Deviation
STORRM	Sensor Test for Orion RelNav Risk Mitigation
SVA	Supplemental Vertical Accuracy
SVS	Synthetic Vision Systems
SWAP	Size, Weight And Power
SWAT	Soil and Water Management Tool
SWFWMD	Southwest Florida Water Management District
SWIR	Short-Wave Infrared
SWMM	Storm Water Management Model
SZ	Surf Zone
TAWS	Terrain Awareness and Warning Systems
TDOP	Time Dilution of Position
TEC	Total Electron Count
TIM	Time Interval Meter
TIN	Triangulated Irregular Network
TLS	Terrestrial Lidar System
TMFL	Topographic Mapping Flash Lidar
TOF	Time-Of-Flight
TopCAT	Topographical Compartment Analysis Tool

TRB	Transportation Research Board
TRN	Terrain-Relative Navigation
UAS	Unmanned Aerial Ssytem
UAV	Unmanned Aerial Vehicle
UGV	Unmanned ground vehicles
UPS	Uninterruptible Power Supply
URI	Uniform Resource Indicator,
USACE	US Army Corps of Engineers
USGG	US Gravimetric Geoid
USGS	United States Geological Surbey
USNO	United States Naval Observatory
UTC	Coordinated Universal Time
UTM	Universal Transverse Mercator
UUID	Universally Unique Identifer
UUV	Unmanned underwater vehicles
UV	Ultraviolet
VIS	Visible Light
VLR	Variable Length Record
VLWIR	Very-long Wave Infrared
VNS	Vision Navigation System
VO	Vertical Objects
VRS	Virtual Reference Station
VRS	Virtual Reference Station
VSW	Very Shallow Water
VTUAV	Vertical Takeoff Unmanned Airborne Vehicle
WAAS	Wide Area Augmentation System
WDM	Waveform Digitizer Module
WGS	World Geodetic System
WKT	Well Known Text
WSEL	Water Surface Elevations
XML	eXtensible Markup Language

INDEX

E

G

I

J

M

U

www.ingramcontent.com/pod-product-compliance
Lightning Source LLC
LaVergne TN
LVHW071239210926
848486LV00007B/306
* 9 7 9 8 8 9 6 9 1 4 3 5 8 *